Thermo-Mechanical Processing of Metallic Materials

PERGAMON MATERIALS SERIES

Series Editor: Robert W. Cahn FRS
Department of Materials Science and Metallurgy, University of Cambridge, Cambridge, UK

Vol. 1 **CALPHAD** by N. Saunders and A. P. Miodownik
Vol. 2 **Non-Equilibrium Processing of Materials** edited by C. Suryanarayana
Vol. 3 **Wettability at High Temperatures** by N. Eustathopoulos, M. G. Nicholas and B. Drevet
Vol. 4 **Structural Biological Materials** edited by M. Elices
Vol. 5 **The Coming of Materials Science** by R. W. Cahn
Vol. 6 **Multinuclear Solid-State NMR of Inorganic Materials** by K. J. D. MacKenzie and M. E. Smith
Vol. 7 **Underneath the Bragg Peaks: Structural Analysis of Complex Materials** by T. Egami and S. J. L. Billinge
Vol. 8 **Thermally Activated Mechanisms in Crystal Plasticity** by D. Caillard and J. L. Martin
Vol. 9 **The Local Chemical Analysis of Materials** by J. W. Martin
Vol. 10 **Metastable Solids from Undercooled Melts** by D. M. Herlach, P. Galenko and D. Holland-Moritz

Thermo-Mechanical Processing of Metallic Materials

Bert Verlinden
Department of Metallurgy and Materials Engineering, Katholieke Universiteit Leuven, Heverlee, Belgium

Julian Driver
Materials Centre, Ecole des Mines de Saint-Etienne, Saint Etienne Cedex, France

Indradev Samajdar
Department of Metallurgical Engineering and Materials Science, Indian Institute of Technology Bombay, Powai, Mumbai, India

Roger D. Doherty
Drexel University, Philadelphia, PA, USA

Edited by
Robert W. Cahn

ELSEVIER

Amsterdam • Boston • Heidelberg • London • New York • Oxford
Paris • San Diego • San Francisco • Singapore • Sydney • Tokyo
Pergamon is an imprint of Elsevier

Pergamon is an imprint of Elsevier
Linacre House, Jordan Hill, Oxford OX2 8DP, UK
Radarweg 29, PO Box 211, 1000 AE Amsterdam, The Netherlands

First edition 2007

British Library Cataloguing in Publication Data
A catalogue record for this book is available from the British Library

Library of Congress Cataloging-in-Publication Data
A catalog record for this book is availabe from the Library of Congress

ISBN: 978-0-08-044497-0

For information on all Pergamon publications
visit our website at books.elsevier.com

Transferred to digital print, 2007
Printed and bound by CPI Antony Rowe, Eastbourne

Dedication

This book is dedicated to the late Robert W. Cahn (1924–2007), who encouraged the writing of this volume and edited it with his usual clear insights into both Materials Science and good writing. One of the co-authors (RDD) was mentored by Robert over many years, offering him invaluable help, insights and lively discussion on many topics.

Robert W. Cahn made many important contributions to Materials Science, including the successful hypothesis, made while he was still a graduate student, that recrystallization nuclei grew from, what were then called, polygonized cells, and so had orientations produced by the prior deformation. That idea has been at the heart of recrystallization research for the last 60 years.

Contents

Preface

The present volume is the eleventh in the Pergamon Materials Series, but only the first with avowedly technological subject matter. The thermo-mechanical processing of metals and alloys is right at the heart of materials engineering, and at the same time it is firmly linked with the underlying science of plastic deformation, recrystallization and texture formation.

Accordingly, the book includes a broad spectrum of subject matter from basic science to production engineering . . . in fact, Materials Science and Engineering (MSE).

It is barely more than a century since Walter Rosenhain uncovered the basic mechanism of plastic deformation of crystalline metals; another 20 years had to elapse before X-ray diffraction had become available to study the formation of textures, and another 30 years after that before electron microscopy was mature enough to reveal fine detail of the microstructural processes involved in recrystallization. All these came together, together with modern developments in mechanical engineering, to create the methods used in the shaping of containers and other components made of aluminium alloys, steels, titanium and zirconium, the description of which represents the culmination of this intriguing book.

I have pleasure in commending to our readership this exemplification of MSE at its best.

Robert W. Cahn (Series Editor)
Cambridge

List of Symbols

This list contains the most important symbols used in the text.
Occasionally some letters or symbols are used for other purposes; this is specifically stated in the text.

b	Burgers vector
B	Magnetic induction
C	Concentration
d	Subgrain diameter
D	Diameter (grain, roll, . . .)
D_{gb}, D_b	Grain boundary and bulk diffusion coefficients
E	Young's modulus
E'	Plane strain Young's modulus
f	Frequency
F	Load
G	Shear modulus
H	Magnetic field
J	Energy
k	Boltzmann constant
$K, K_1, K_2, \ldots, K_n$	Constants defined in the text
l	Length
m	Strain rate sensitivity coefficient
M	Mass (also mobility)
n	Strain hardening coefficient
P	Driving pressure
p	Pressure, hydrostatic pressure
Q	Activation energy
R	Gas constant
R_{ext}	Extrusion ratio
$R(\alpha)$	Resistance against thinning in direction α
$\bar{R}$	Normal anisotropy
RA	Reduction in area
r	Radius of precipitate
s	Thickness
t	Time
v	Velocity
V	Volume
w	Width

W	Work
Z	Zener–Hollomon parameter
α	Drawing angle in wire drawing, angle to tensile direction
δ	Boundary thickness
ΔR	Planar anisotropy
γ_{gb}	Grain boundary energy
γ_{SFE}	Stacking fault energy
ε	Strain in general
$\bar{\varepsilon}$	Von Mises equivalent strain
ε_c	Critical strain
ε_{fr}	Fracture strain
ε_u	Uniform deformation
ε_t	Thickness strain
ε_{total}	Total deformation
ε^o or $\dot{\varepsilon}$	Strain rate
μ	Friction coefficient
ν	Poisson's ratio
θ	Misorientation across a boundary
$\bar{\theta}$	Work hardening
ρ	Dislocation density
$\bar{\sigma}$	Von Mises equivalent stress
σ	Stress in general
σ_f	Flow stress during forming
σ_y	Initial flow stress (also YS)
σ_F	Drawing stress in wire drawing
σ_{fr}	Fracture stress
σ_s	Saturation stress
τ	Shear stress
τ_c	Critical shear stress
Ω	Atomic volume

Abbreviations

bcc	Body-centred cube
CDRX	Continuous dynamic recrystallization
CSL	Coincidence site lattice
DRV	Dynamic recovery
DSC	Differential scanning calorimetry
DRX	Dynamic recrystallization
EBSD	Electron backscatter diffraction
fcc	Face-centred cube
FLD	Forming limit diagram
HAGB	High-angle grain boundary
hcp	Hexagonal closed packed
HIP	Hot isostatic pressing
JMAK	Johnson–Mehl–Avrami–Kolmogorov kinetic model
LDR	Limiting drawing ratio
ND, RD, TD	Normal, rolling, transverse direction
ODF	Orientation distribution function
PSN	Particle-stimulated nucleation
SEM	Scanning electron microscope
SFE	Stacking fault energy
SRX	Static recrystallization
TEM	Transmission electron microscope
UTS	Ultimate tensile strength
YS	Yield strength

PART I: SCIENCE

Chapter 1
General Introduction

Chapter 1
General Introduction

Most metal-forming operations have two direct consequences. On a 'macroscopic scale' the desired shape change is obtained and on a 'micro scale' the microstructure of the material is changed. This microstructure is related to the mechanical and physical properties of the metal and is heavily dependent on external parameters such as temperature, strain, strain rate, deformation mode, lubricants, etc.

The blacksmith in ancient times was mainly concerned with the macroscopic shape change, but modern metal industry requires thermo-mechanical processes that lead to a product with optimal properties, not only for dimensional precision and appearance, but also with respect to the mechanical and physical properties (Figure 1.1).

Thermo-mechanical processing (TMP) describes the set of heating and shaping operations by which relatively simple, basic materials are converted into high-quality components. The term is principally applied to the processing of metallic alloys but extensions to ceramics, polymers and many of their combinations are increasingly being developed. It is as old as mankind and as new as the latest microelectronic device. One of the earliest examples is of course the forging of bronze (copper–tin alloys) for both tools and personal decoration in the Bronze Age (~3000 BC); as-cast alloys were subjected to a sequence of heating and plastic deformation operations to shape the component and simultaneously improve the material properties, usually strength and tenacity. The basic idea has changed

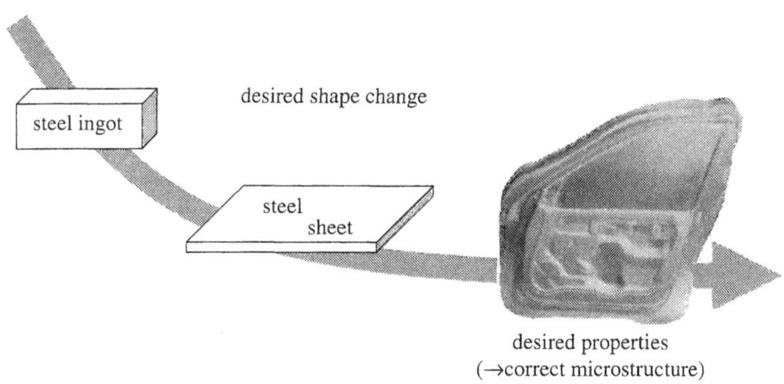

Figure 1.1. Metal forming: a combination of a desired shape change with a controlled microstructure.

little over the centuries but the industrial machines and levels of scientific understanding have undergone a dramatic transformation, essentially in the last 50 years. In particular, the role of microstructure evolution during shaping is now widely accepted as critical for understanding and relating the process/material/property combinations.

The subject is usually treated in two separate parts: mechanical processing (as a branch of mechanics) and physical metallurgy (as part of materials science). One of the aims of the present book is to bring the two disciplines closer together to develop a more unified approach that could facilitate new advances in the field. Although TMP is a dynamic subject of modern research, books on the subject are rare, if non-existent. The aim of the present text is to give an overview of the main manufacturing processes, but also to provide some understanding of the complicated relations between a forming process and the microstructure of a metal on one hand, and between this microstructure and the mechanical properties on the other.

In Part I, *Science*, the microstructural science of the subject is treated. It is assumed that most readers have already acquired some typical undergraduate-level knowledge of material science. Those readers can probably skip Chapters 2 and 3, which provide an elementary survey and suggestions, for further reading about microstructures and properties of materials and about plasticity. The other chapters of Part I treat, at a more advanced level, important issues such as work hardening, softening mechanisms (recrystallization), texture and residual stresses. Deformation mechanisms like creep and superplasticity are introduced and the influence of phase transformations on TMP is highlighted. Since more and more predictive models concerning TMP operations are being developed, the last chapter will focus on this rapidly expanding field.

In Part II, *Technology*, the most important metal-forming operations are described. The paragraphs on primary forming processes such as rolling, extrusion, forging and wire drawing are mainly focussed on the 'macro scale' (shaping a metallic part) although here also some connections with microstructural control will be made. In the chapter about 'sheet metal forming', some forming operations like deep drawing, bending, stretching and incremental forming will be explained. A key point in this chapter is the relation between plastic anisotropy and formability. Some advanced forming operations like hydroforming, hipping and superplastic forming will end this extensive chapter on forming operations. An overview of the major macroscopic problems that may arise during metal forming, is provided in Chapter 12 and as far as possible, the microstructural origins of those macroscopic problems are discussed. Finally, Chapter 13 will illustrate the possibilities of measuring material properties and studying material behaviour during TMP-type deformation on a labscale.

The third part of this book, *Case studies*, is mainly devoted to the 'micro scale'. In a number of examples the TMP of some products is analysed and discussed. These case studies show how properties can be successfully optimized by a careful control of the microstructure during processing. Examples from the aluminium industry include the processing of beverage cans, capacitor foil, extrusion of metal matrix composites and rolling of thick plates for aerospace applications. The TMP of steel products is illustrated with a case study on steel for car bodies, a short description of the development of dual phase and TRIP steels, the concept of controlled rolling of HSLA steel, a discussion on the production of electrical steel and the patenting of steel wires. Two examples of successful processing of hexagonal metals are provided with a discussion about Zr for the nuclear industry and Ti for aero-space applications. Finally, some new developments are discussed: the production of sub-micron materials by severe plastic deformation and an example of 'grain boundary engineering' to improve the local corrosion resistance of stainless steel.

This textbook is written at the level of an advanced Master course. It is our hope that it can serve as a course textbook for students taking a course on TMP. On the other hand, this book is also intended for engineers in industry. It will offer them a coherent overview of the field and provide them references to the most important literature.

Chapter 2

Microstructure and Properties

Chapter 2

Microstructure and Properties

2.1. INTRODUCTION

The response or interactions of a metal to an external stimulus or energy can be generalized as 'a property' of this metal. To a designer of devices or components, the 'property' is an important design parameter. 'Property' can be classified as intrinsic and as microstructure sensitive. The former is expected to depend on chemistry, atomic bonding and electronic configuration, while microstructure-sensitive properties would depend on various aspects of the microstructure. A successful design automatically involves two steps: (i) selection of the metal or alloy with requisite intrinsic properties and (ii) optimized processing to optimize microstructure-sensitive properties. (ii) is often the basis for thermo-mechanical processing (TMP). This chapter is initiated with different aspects of the microstructure. 'Properties' of metals are then summarized. Finally, the 'microstructure sensitivities' of the properties are highlighted.

2.2. MICROSTRUCTURE

The microstructure, a term popular with all metallurgists, cannot be explained simply as 'structures visible microscopically'. The so-called microstructures can vary significantly in scale and also in resolution. As shown in Figure 2.1, the size of the grains may range from nanometres to millimetres – the field of observation and the visibility of details[1] would depend on the microscopic technique used. The microstructural features, in a crystalline metal, can be generalized (Hornbogen [1984]) in two categories – phase/grain and defects. Table 2.1 summarizes structures typical for these two categories. The size, shape, distribution and orientation of the microstructural features determine the structure-sensitive 'properties'. Some of the microstructural features, mentioned in Table 2.1, are discussed in detail in subsequent chapters.[2] The present section tries to bring out two aspects of microstructure: the solidification structure and the interfaces.

[1] This depends on the resolution, but also on the source of contrast.

[2] Aspects of phase transformations and related microstructures are covered in Chapter 7, while microstructural changes during deformation and recrystallization, including changes in dislocation substructure, are covered respectively in Chapters 4 and 5.

Table 2.1. Summarizing microstructural features.

Microstructural feature	Details
Phase–grain structure	
Phase	Uniform composition, structure and properties. Phase structure can be altered through phase transformation (Chapter 7)
Grain	Same phase (or crystal structure) – but different orientations of the crystals. Grain structure can be altered through phase transformation (Chapter 7), plastic deformation (Chapter 4), recrystallization and grain growth (Chapter 5)
Defect structure	
Composition	Segregation or difference in composition. Depending on the scale, this can be termed as macro or micro
Point	Vacancies, interstitial and substitutional atoms
Line	Dislocations. For microstructure property and TMP, of interest is dislocation sub-structure (Chapter 4)
Planar	Stacking faults, solid–vapour interface, grain and phase boundaries
Volume	Voids (includes gas bubbles and cavities) and inclusions

The features are generalized into two categories: phase/grain structure and defect structure. These are also divided into sub-categories.

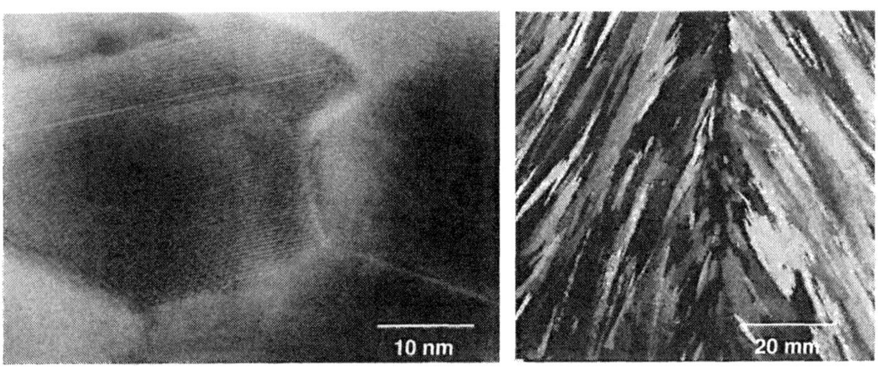

(a) (b)

Figure 2.1. Microstructures at different scales and resolution. (a) High-resolution TEM image of nanocrystalline grains with stacking faults and anti-phase boundaries in crystallized Zr-based metallic glass. Courtesy of G.K. Dey, BARC, India. (b) Solidified grains after electro-slag refining of Fe – 4.5 wt% Si alloy. In (b), the field of observation is large – but the resolution is poor.

2.2.1 Solidification

The starting structure for TMP is often a solidification structure. Knowledge of the cast structure, including an understanding of the solidification basics, is essential and is covered in this section. Further details on solidification can be obtained in a range of text books (Flemings [1974], Porter and Easterling [1986], Kurz and Fisher [1989], Martin *et al.* [1997]).

2.2.1.1 Solidification basics – nucleation and growth. Solidification is a process of nucleation and growth. The crystalline nuclei may form in the amorphous liquid by thermally activated 'random fluctuations' (an example of crystalline nuclei in bulk metallic glass is shown in Figure 2.2a). The stability of such nuclei is dependent on the energy balance between the volume driving force and the surface energy term. The driving force per unit volume for solidification ($\Delta G = G^{S} - G^{L} < 0$ at $T < T_e$), see Figure 2.2b, can be approximated as

$$\Delta G = \frac{L \Delta T}{T_e} \tag{2.1}$$

where L is the latent heat of fusion, T_e the melting temperature and $\Delta T(T_e - T_1)$ the undercooling, T_1 being the actual solidification temperature. The latent heat of solidification in typical metallic systems is of the order of −10 kJ/mol. Assuming

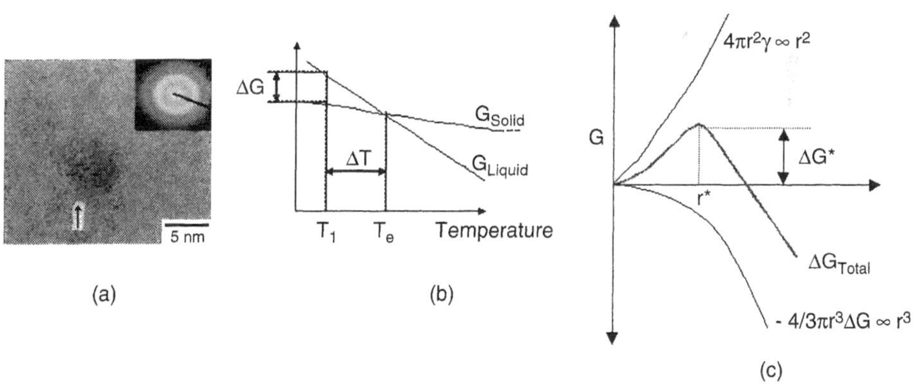

Figure 2.2. (a) Supercooled Zr-based bulk metallic glass showing crystalline nuclei – lattice fringes marked by arrow head. Courtesy of G.K. Dey, BARC, India. (b) Free energy (G) vs. temperature, for solid and liquid. T_e and T_1 are respectively the melting and the actual solidification temperatures, while ΔG and ΔT are respectively the driving force for solidification and the thermal undercooling. (c) G vs. r, r being the radius of the solidifying nuclei. The energy balance is given in Eq. (2.2).

a spherical nuclei of radius r, an energy balance for the surface energy created and the volume free energy change gives

$$\Delta G_{\text{total}} = (4/3\pi r^3)\Delta G + (4\pi r^2)\gamma \tag{2.2}$$

ΔG is *negative* at $T < T_e$. As shown graphically in Figure 2.2c, the critical radius r^* emerges out as 'limiting' size. Nuclei larger than r^* will be stable, as an increase in size would reduce the energy, and hence the nuclei are expected to grow. To reach r^*, an activation energy barrier ΔG^* has to be crossed. Differentiating Eq. (2.2), r^* and ΔG^* can be obtained as

$$r^*_{\text{hom}} = \frac{2\gamma}{\Delta G} \tag{2.3}$$

$$\Delta G^*_{\text{hom}} = \frac{16\pi\gamma^3}{3(\Delta G)^2} \tag{2.4}$$

These are valid for homogeneous nucleation, where any atom in the melt is a potential site for nucleation. If the nucleation occurs on the mould wall, ΔG^* can be significantly reduced by heterogeneous nucleation. As shown in Figure 2.3a,

$$\gamma_{\text{ML}} = \gamma_{\text{SM}} + \gamma_{\text{SL}} \cos\theta \tag{2.5}$$

It can also be shown that $\Delta G^{\text{het}}_{\text{total}}$ for heterogeneous nucleation (Porter and Easterling [1986]) is

$$\Delta G^{\text{het}}_{\text{total}} = \{-(4/3\pi r^3)\Delta G + (4\pi r^2)\gamma_{\text{SL}}\}S(\theta) \tag{2.6}$$

where $S(\theta)$, the shape factor, is

$$S(\theta) = \tfrac{1}{4}(2 + \cos\theta)(1 - \cos\theta)^2 \le 1 \tag{2.7}$$

r^* and ΔG^* for heterogeneous nucleation are

$$r^*_{\text{het}} = \frac{2\gamma}{\Delta G} \tag{2.8}$$

$$\Delta G^*_{\text{het}} = \frac{16\pi\gamma^3}{3(\Delta G)^2} S(\theta) \tag{2.9}$$

Note that for $\theta = 45°$, $S(\theta)$ is only 0.06! The nucleation rate (N), for both homogeneous and heterogeneous nucleation, is proportional to exp $(-\Delta G^*/\text{kT})$.

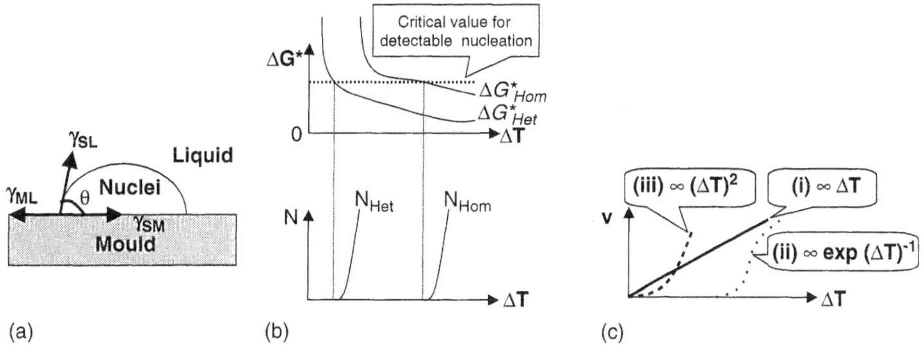

Figure 2.3. (a) Heterogeneous nucleation on a mould wall. The subscripts for γ represent respective interfacial energies between ML (mould–liquid), SM (solid–mould) and SL (solid–liquid), θ being the contact angle. (b) ΔG^* and nucleation rates for heterogeneous (Het) and homogeneous (Hom) nucleation. (c) Growth after nucleation. Growth velocity as a function of ΔT for (i) continuous growth, (ii) surface nucleation and (iii) spiral growth; (i) is valid for rough interfaces (e.g. metals), while (ii) and (iii) are applicable for smooth interfaces (e.g. intermetallics).

Hence, a direct effect of lower ΔG^* for heterogeneous nucleation is much faster N_{het} or the dominance of heterogeneous nucleation at smaller undercoolings ΔT (see Figure 2.3b). The nucleation is succeeded by growth. As shown in Figure 2.3c, the growth velocity varies as a function of ΔT. The exact relationship depends on the nature of the solid–liquid interface. Metals typically have rough interfaces,[3] while intermetallics (or material with high entropy of mixing) have smooth interfaces. In the latter case, surface nucleation or ledge creation and spiral growth are typically the means to improve the accommodation of atoms being received from the liquid.

Two other important aspects of growth, in metals and alloys, during solidification are dendritic growth and solute partitioning (see Figure 2.4). Dendritic growth is a consequence of the instability of the solid–liquid interface by the development of thermal dendrites, mostly by constitutional undercooling. Growth of the primary dendrite arm then takes place along specific crystallographic directions, e.g. along the direction of thermal gradient (Kurz and Fisher [1989]). 'Solute partitioning' can be explained easily from a binary phase diagram (see Figure 2.4b). As shown in this figure, solidifying regions will reject solute into the surrounding liquid and hence the liquid or regions solidifying at later stages will be enriched in solute. The other important aspect of solidification is that of shrinkage – originating from the volume difference between solid and liquid. Almost all metals shrink

[3] It is also important to point out that the lean alloys may grow as twin-columnar structure – twin boundaries at the solid–liquid interface providing a permanent source of steps for easy growth.

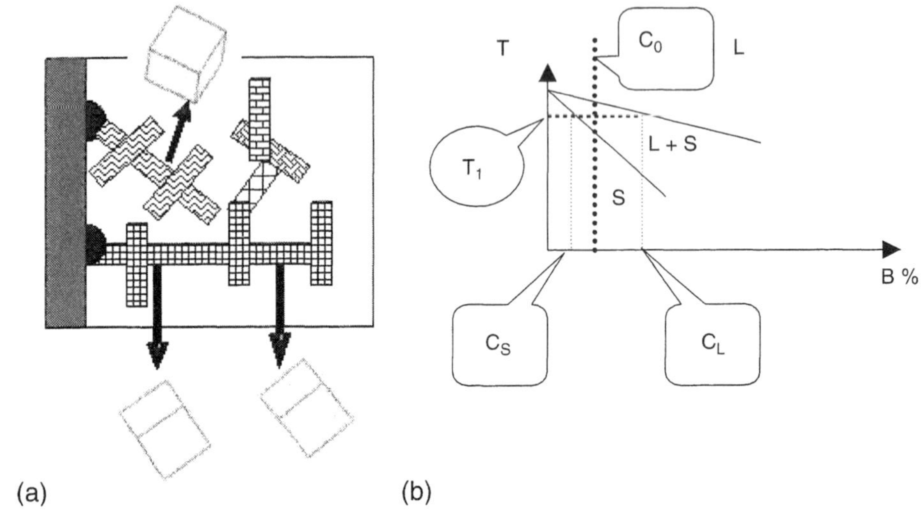

Figure 2.4. (a) Nuclei from mould wall growing dendritically. The orientation may differ, but not significantly, inside a single dendrite arm – while orientations may be completely different between different dendrites. (b) The gap between the solidus and liquidus lines in a phase diagram would ensure 'solute partitioning'. C_0 and C_S, respectively, being the composition of the original melt and the solid – C_L is the solute-enriched liquid in contact/equilibrium with the solid.

by approximately 3% by volume during solidification. However, for a few metals, e.g. gallium, bismuth and antimony (metals with lower coordination), a contraction during melting takes place.

2.2.1.2 Solidification structure. The solidification structure depends on the alloy chemistry and on the solidification process. The chemistry, especially the nature of the phase transformation (eutectic, off eutectic, peritectic, etc.), can dictate the solidification structure (Flemings [1974], Porter and Easterling [1986], Martin *et al.* [1997]). The solidification process, specifically the thermal gradient, solidification rate and fluid flow conditions (Boettinger *et al.* [2000]), also have very strong effects. For example, the structure of an electro-slag-refined ingot (Figure 2.1b) is completely different from the structure of a conventional laboratory ingot (see Figure 2.5). The electro-slag-refined ingot has a stronger thermal gradient (approximately in the direction of the columnar grains) and fluid flow conditions. As shown in Figure 2.5, the important aspects of the solidification structure are grain structure, segregation, inclusions and gas porosity.

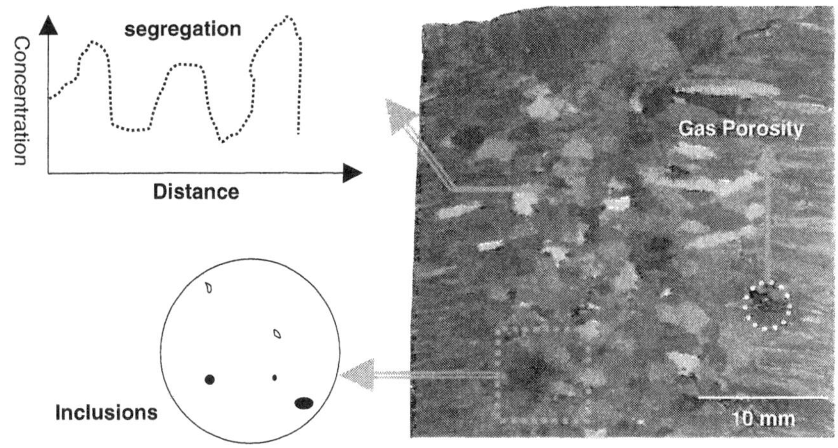

Figure 2.5. Macrostructure of a laboratory ingot, cast in a permanent mould, of Fe–4.5 wt% Si. Equiaxed grains near the centre and columnar grains close to the mould walls are visible. Also visible are pockets of gas porosities and coarse inclusions. Segregations of alloying elements (shown schematically) are also typical.

2.2.1.2.1 Grain structure. Immediately adjacent to the mould wall, fine equiaxed grains are observed – the so-called chill zone.[4] Large undercooling ensures high heterogeneous nucleation rate and in turn is responsible for the fine equiaxed morphology. A large undercooling of the melt (or low pouring temperature) and turbulent flow may lead to a fully equiaxed ingot – the so-called big-bang nucleation (Flemings [1974], Porter and Easterling [1986]), where crystals swept from the mould wall fill the melt. This is, however, more of an exception. Usually, columnar grains or dendrites grow in the direction opposite to that of heat flow – in the case of ingot castings, towards the centre of the ingot (Figure 2.5). The centre of the ingot often contains equiaxed grains, originating from melted-off dendrite side arms.

The grains nucleating on the mould wall are reported to be randomly oriented (Takatani *et al.* [2000]). Subsequent growth of these grains, in the form of the columnar structure, is expected to bring growth selection mechanism(s), where preferentially oriented grains grow, truncating the growth of non-preferred orientations (Henry *et al.* [1998]). The selection mechanism is expected to be a function of temperature gradient and growth velocity (Versnyder and Shank [1970], Borisov *et al.* [1991], Henry *et al.* [1998]). For example, high growth velocities (large temperature gradients) are expected to promote dendrites growing in preferential crystallographic

[4] This is not clearly visible in Figure 2.5.

direction, whereas at lower velocities dendrites are reported to grow without any preferred orientation (Borisov *et al.* [1991]).

2.2.1.2.2 Segregation. Variations of composition over distances comparable to the dendrite arm spacing (DAS) happen as a direct effect of 'solute partitioning' (see Figure 2.4b). This is the basis of the so-called micro-segregation. Macro-segregation, on the other hand, can be formed by combination of factors, ranging from shrinkage to convection currents and density differences (Flemings [1974], Kurz and Fisher [1989]).

2.2.1.2.3 Inclusions. Non-metallic or ceramic inclusions are often a by-product of a casting process originating from furnace/casting refractories and/or through reactions in the melt. The nature and the severity of such inclusions are usually identified by comparing unetched microstructures with standard charts.[5] It is, however, important to point out that modern liquid metal processing routes ensure much better inclusion control than was feasible a few decades ago.

2.2.1.2.4 Gas porosity. The solubilities of gasses in the liquid metal are orders of magnitude more than in the solid castings. The dissolved gasses in the melt can leave gas porosities and also inclusions, the latter in the form of reaction products. Various vacuum and inert gas melting and degassing techniques are primarily applied to control the content of harmful gasses in the melt.

The solidification structures and the solidification technology have strong implications on the science and technology of TMP. Two examples are given below:

- *Homogenization*: A homogenization or soaking treatment is often applied to the ingot to reduce[6] the extent of micro-segregation, a signature of the solidification structure.
- *Breaking the cast structure*: One of the first steps of TMP, for example a roughing operation during hot rolling, is mainly intended to break up or refine the as-cast structure. Effective use of alternative technologies, such as electro-magnetic stirring, may reduce the scope of such operations.

Similarly, new developments in solidification technology can have strong implications for TMP. The biggest leap in steel TMP in the recent past, has been in the form of thin slab casting, a solidification technology. A similar, if not bigger,

[5] An example of such a chart for steel: J.K.Chart of ASTM E-45 Standard.
[6] It is impossible to eliminate segregation fully – homogenization can, however, reduce the extent of compositional fluctuations.

revolution in TMP technology is often speculated for the future – the development of thin strip casting.

2.2.2 Interfaces

As discussed in the earlier section, the solid–liquid interface plays a major role in the nucleation and growth of solidification. Other types of interface also strongly influence microstructural changes during thermal and thermo-mechanical treatments. The types of interfaces in a metallic microstructure may range from the free surface or the solid–vapour interface to phase and grain boundaries and these are discussed separately in this section.

2.2.2.1 Solid–vapour interface. The energy of a solid–vapour interface (γ_{SV}) is due to 'broken bonds'. γ_{SV} per surface atom can be approximated (Porter and Easterling [1986]) as

$$\gamma_{SV} = \frac{n_n L_S}{z N_a} \tag{2.10}$$

where z represents the coordination number (or the number of nearest neighbours to each atom), N_a the Avogadro's number, L_S the heat of sublimation (summation of the latent heats of melting and vaporization) and n_n the number of broken bonds per surface atom.

L_S and correspondingly γ_{SV}, see Figure 2.6, does strongly depend on the melting temperature. Z or the coordination number, on the other hand, is expected to depend on the crystal structure, while n should depend on the crystal orientation. For example, the nature of the (hkl) plane at the interface will determine the number of missing neighbours per surface atom. As shown in Figure 2.7a, a non-closed packed plane has more broken bonds than a closed packed plane. Corresponding the energy of the former is also more (see Figure 2.7b). The γ_{SV} can be plotted in a polar plot, see Figure 2.6c, where the radius from origin represents the γ_{SV} for respective (hkl)s, both closed packed and non-closed packed planes.[7] This is the so-called Wulff plot and can be used to determine the equilibrium shape. The equilibrium shape is given by the Wulff planes with low γ_{SV} – the inner envelop or dotted lines in Figure 2.6c. The three-dimensional (3D) equilibrium shape would be represented by $\Sigma A_i \gamma_i$ = minimum. For more details on Wulff plots, the reader may refer (Christian [1975], Murr [1991], Martin *et al.* [1997]).

[7] Closed packed planes represent the cusps in the plot.

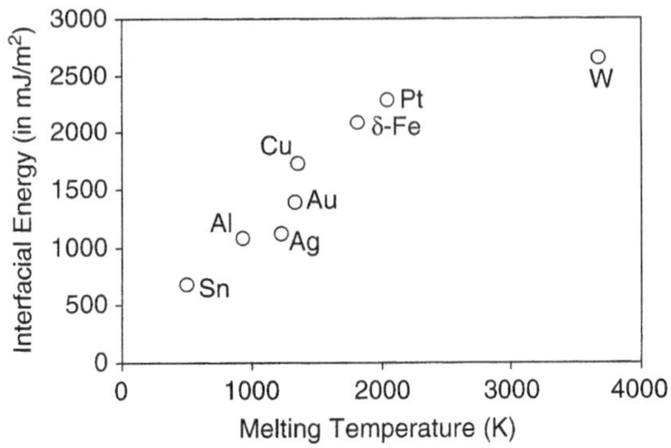

Figure 2.6. γ_{SV} as a function of melting temperature. After Jones [1971].

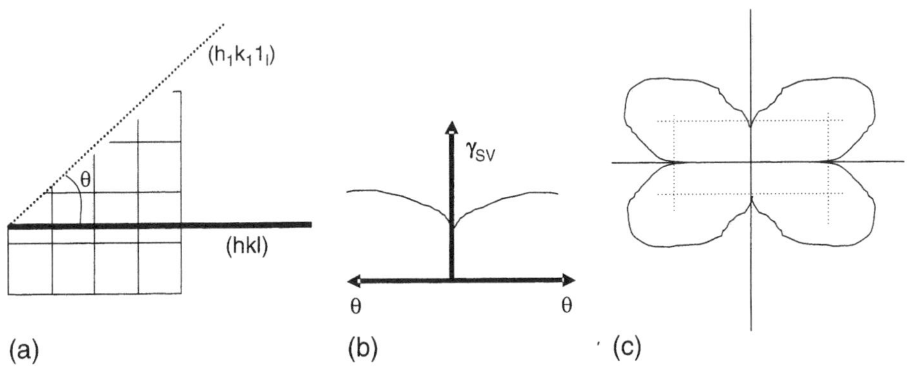

Figure 2.7. Schematics of (a) two-dimensional (2D) configuration of a non-closed packed plane. The non-closed packed plane $(h_1k_1l_1)$, at an angle θ with closed packed plane (hkl). $(h_1k_1l_1)$ has more broken bonds – $f(\theta) = (1 + \tan\theta)/a$, where a is interatomic spacing. (b) γ_{SV} as a function of θ. (c) A possible section through the interfacial energy plot of a cubic crystal. The solid lines represent γ_{SV} at different orientations or θ, while the dotted lines show the equilibrium shape or shape with lowest γ_{SV}.

2.2.2.2 Grain boundary. The boundary between two crystals or grains is referred as a grain boundary. The energy of grain boundaries can differ between metals, see Figure 2.8, and also between boundaries of the same system (Murr [1991], Randle [1996]). For example, the grain boundary energy, among the so-called high angle boundaries, in type 304 austenitic stainless steel may vary from 20 to 835 mJ/m^2 (Murr [1991]). The difference in grain boundary energy can have strong implications for processing (e.g. superplastic deformation (Chapter 6), phase transformation (Chapter 7), recrystallization and grain growth (Chapter 5)) and for properties

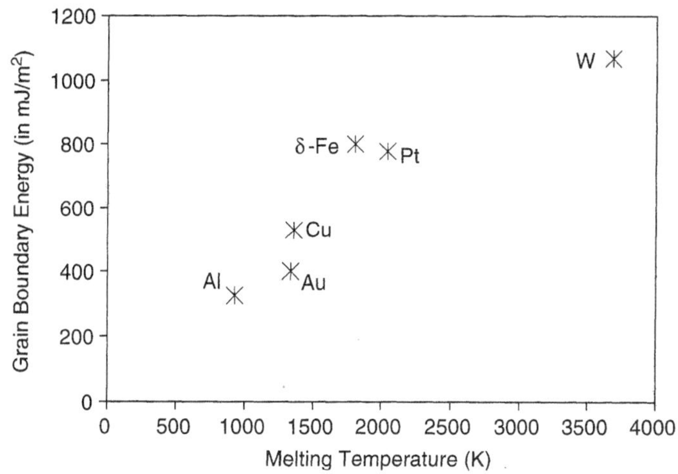

Figure 2.8. Grain boundary energy as a function of melting temperature. After Hondros [1969]. Metals of higher melting point or bond strength, in general, have higher grain boundary energies – grain boundary energy $\approx 1/3\,\gamma_{SV}$ (Murr [1991]).

(e.g. localized corrosion in austenitic stainless steel (Section 17.2)). This section summarizes the present understanding on grain boundary nature.

As shown in Figure 2.9, the first index of distinguishing grain boundaries is the so-called tilt or twist nature (Randle [1996]). This not only marks a difference in relative orientations (as indicated by the two-dimensional (2D) schematics – Figure 2.9a, b), but also in the basic nature of the respective boundaries. For example, a low angle (low θ – as in Figure 2.9c) tilt and twist boundary correspond respectively to an array of parallel edge dislocations (as shown in Figure 2.9a) and a cross-grid of two sets of screw dislocations (Porter and Easterling [1986]). The tilt and twist nature of grain boundaries are rarely talked about while discussing experimental data on grain boundary nature. The reason is apparent. The data from microtexture measurements can be translated easily into rotation axis $<uvw>$ and the minimum angle of rotation (misorientation angle θ), but the data does not reveal the tilt and twist nature of the boundaries.[8] It is, however, important to note that two otherwise identical (i.e. with similar $\theta<uvw>$) boundaries can have completely different energies based on their tilt or twist nature (Randle [1996]).

Normally, the first step in identifying grain boundary nature from experimental data is based on θ, the misorientation angle. As shown in Figure 2.10a, a misorientation of 10–15° is usually taken as the borderline between two types of

[8] Even lattice imaging, as in Figure 2.1a, is incapable of bringing out twist grain boundaries.

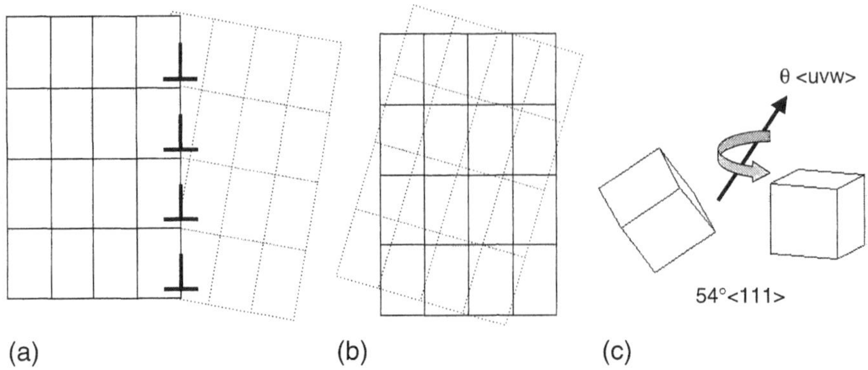

Figure 2.9. Two-dimensional schematics of (a) tilt and (b) twist boundary – distinction based on relative orientations of the crystals. (c) Introducing axis–angle (<*uvw*> and θ) notation for a grain boundary. The axis of rotation <*uvw*> is indicated with arrow and the minimum angle of rotation (the so-called orientation distance or misorientation angle) is given as θ.

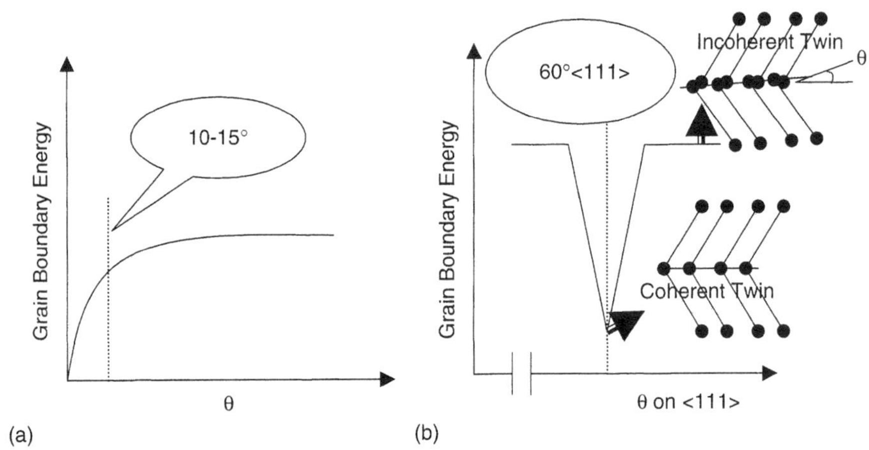

Figure 2.10. (a) Schematic of grain boundary energy as a function of θ. θ of 10–15° marks the demarcation between low and high angle boundaries. (b) In the high angle boundary, there are specific orientations (θ<*uvw*> – for example 60° <111> fcc twin boundary) with low energy. These are the special boundaries. Schematics of coherent (energy cusp) and incoherent twin boundaries are also given.

boundaries – low and high angle, the former representing low energy. The high angle boundaries, however, do not necessarily relate to high energy, as there are specific orientations (θ<*uvw*> combinations – see Figure 2.10b) or types of high angle boundaries with low energy. Such boundaries are called special boundaries, while high angle and high-energy boundaries are termed as random boundaries.

This classification of the grain boundaries, in terms of their relative energy, is summarized in Figure 2.11.

It is important to appreciate the physics of such a difference (see Figures 2.10 and 2.11) in energy. The low misorientation angles, in a tilt boundary, can be accommodated by an array of parallel edge dislocations. A larger misorientation angle would bring more dislocations and a corresponding rise in energy. The rate of rise, however, diminishes at higher energy or dislocation content – as the dislocation strain fields cancel each other. Beyond a misorientation angle or dislocation spacing, overlapping of dislocation cores would make grain boundary energy nearly independent of θ. This is the transition from low to high angle boundary. Certain orientations of the high angle boundaries can also represent low energies – the so-called special boundaries. These are the orientations where the two crystals fit with relatively low distortion of the interatomic bonding (see the coherent twin boundary of Figure 2.10b). Slight difference in θ will bring in mismatch or distortion and hence an increase in the energy, as shown for the incoherent twin boundary in Figure 2.10b.

Figure 2.11 relates energy and grain boundary nature. This generalization leaves question on the identity of the so-called special boundaries. The CSL or coincident site lattice model happens to be the only tool available to explain or identify the low energy special boundaries. The theory of CSL (Brandon [1966], Warrington and Boon [1975], Smith and Pond [1976], Bollman [1982], Randle [1996]) considers coincidence of the lattice points between the two crystals or grains. 'The inverse of the common lattice points' defines the CSL number Σ. For example, if all the lattice points are common to both grains (1 out of 1) then it is $\Sigma 1$, if 1 out of 3 is common then it is $\Sigma 3$, $\Sigma 5$ (see Figure 2.12) represents

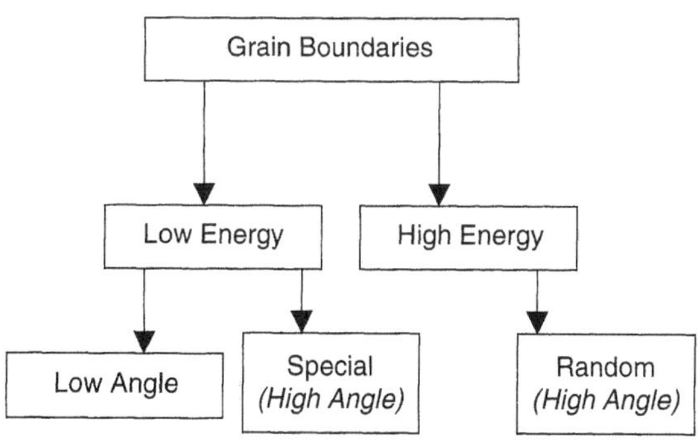

Figure 2.11. Summarizing different types of grain boundaries. The classification is made in terms of relative energy of the boundaries.

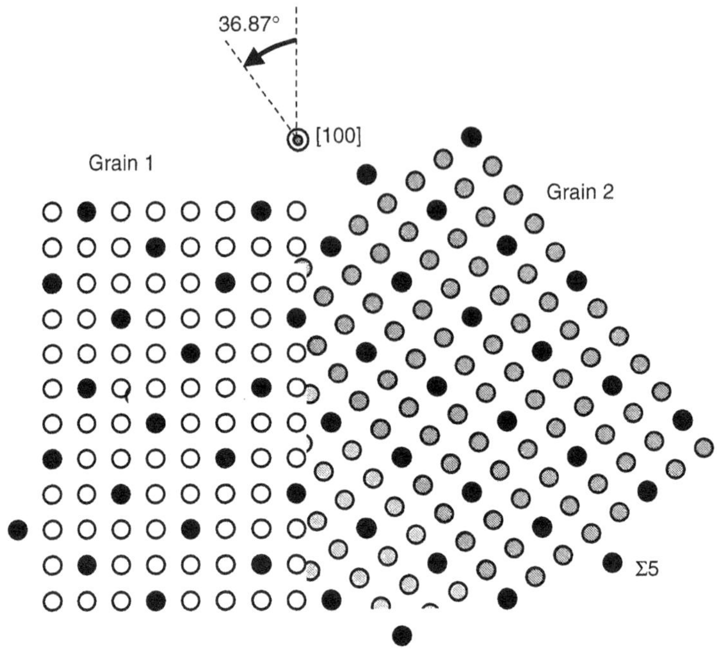

Figure 2.12. Schematic of a Σ5 boundary, 37°<100>, in cubic system. One out of every 5 lattice points between the two grains or crystals is common.

1 out of 5 common lattice points, and so on. Beyond a particular number, often considered as Σ29 or Σ33 in cubic system, the CSL notation has no meaning and the boundary nature or energy would approximate that of a random boundary. Rarely a boundary is an exact CSL. The tolerable deviation from exact CSL, $\Delta\theta$, is defined as

$$\Delta\theta = k\Sigma^{-n} \tag{2.11}$$

where the constants k and n are defined by some criteria – for example, Brandon's (Brandon [1966]) criterion considers k and n respectively as 15° and 0.5. The CSL theory, however, has several drawbacks:

- It effectively considers boundaries with 2D misfits, while real boundaries have 3D misfits. This makes a CSL classification incomplete – Σ3 twin boundaries have completely different energies based on their tilt or twist nature, though the CSL notation in both cases remains the same (Randle [1996]).

- Estimation of CSL from microtexture data is far too difficult in non-cubic systems (Bollman [1982]). For example, in a hexagonal system, CSL can be formed only if $(c/a)^2$ is some rational number (Bollman [1982]) and analytical methods to identify CSL nature in lower symmetry crystal structures are difficult (though more rigorous numerical methods can be suitably adopted – see Mani *et al.* [2006]) to implement in a computer algorithm.
- In cubic system the energy of a CSL boundary can be approximated as a fraction $(1-1/\sqrt{\Sigma})$ of the random boundary energy (Aleshin *et al.* [1978]). No such straight correlation between CSL number and grain boundary energy is possible for lower symmetry crystal structures (Bollman [1982]).

In spite of these limitations, CSL remains the only theory explaining the special boundary nature and often linking it with both processing and properties.

2.2.2.3 Phase boundary. Like the grain boundaries, phase boundaries are also classified based on their energy. As shown in Figure 2.13, the range and the relative origin of interphase energy grossly classify three different types of phase boundaries – coherent, semi-coherent and non-coherent. Figure 2.14 shows the HREM images of coherent and non-coherent interfaces. The origin of energy for a coherent interface is γ_{chemical} or chemistry/bonding difference across the interface and γ_{misfit} or lattice parameter difference between the two phases. The γ_{misfit} can be accommodated, partly or in full, by interfacial dislocations or misfit dislocations, see Figure 2.13, and incorporation of such dislocations (and correspondingly introducing $\gamma_{\text{dislocation}}$ as a source of interfacial energy) would transform a coherent interface into semi-coherent. More γ_{misfit} would require more misfit dislocations – dislocations with reduced spacing. If the spacing of misfit dislocations is lower than 4 interplanar spacings, then the dislocation cores will overlap – converting the interface into a non-coherent type. The energy of three types of interfaces are generalized:

$$\Delta G_{\text{coherent}} = A\gamma_{\text{chemical}} + V\gamma_{\text{misfit}} \tag{2.12}$$

$$\Delta G_{\text{semi-coherent}} = A\gamma_{\text{chemical}} + V\gamma_{\text{misfit}} + A\gamma_{\text{dislocation}} \tag{2.13}$$

$$\Delta G_{\text{non-coherent}} = A\gamma_{\text{chemical}} + A\gamma_{\text{distortion}} \tag{2.14}$$

where A and V represent the interfacial area and the volume of the 2nd phase, respectively. As shown in Figure 2.15a, the coherent interface would convert or transform into semi-coherent if the size exceeds r_{critical}.[9] Existence of r_{critical} is purely a consequence of the energy balance, Eqs. (2.12 and 2.13) being equivalent, but

[9] A similar r_{critical} would also exist between semi-coherent and non-coherent phases or interfaces.

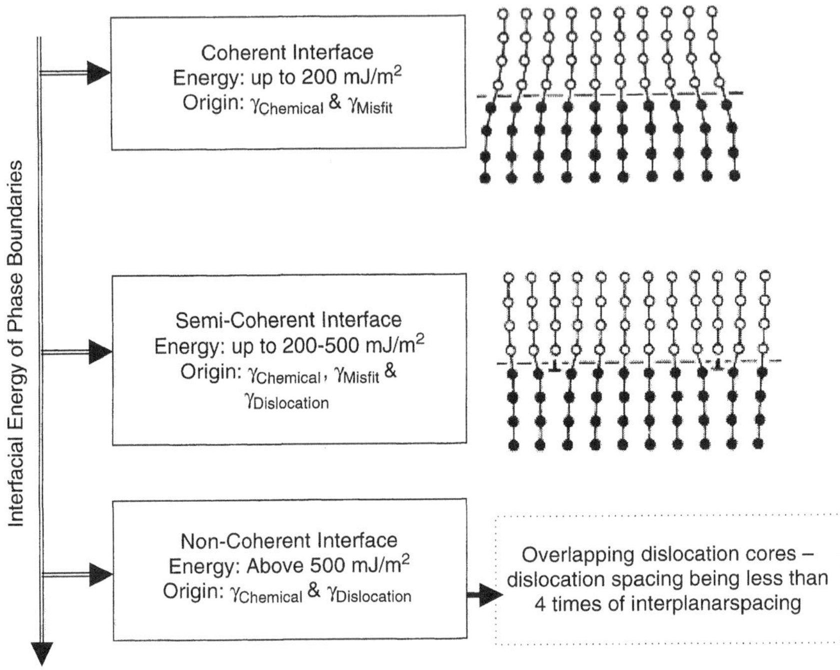

Figure 2.13. Summarizing the phase boundaries in terms of their energy. The range of energy (approximate) and the origin of energy are marked.

actual conversion of the interfaces also would depend on the availability of matrix dislocations. The other important aspect of coherent and semi-coherent phases is the orientation relationship. As shown in Figure 2.14b, significant misfit strains can be avoided if the two phases, matrix and the 2nd phase, are with identical crystal structure and lattice parameter. As this is never the case, the matrix and coherent or semi-coherent 2nd phases, has to follow a particular fitting or orientation relationship – for example $(hkl)_{\text{matrix}}//(hkl)_{\text{2nd phase}}$ and $<uvw>_{\text{matrix}}//<uvw>_{\text{2nd phase}}$.

The concepts of the phase boundary energy will be used in establishing metastable phase transformations in Chapter 7. In the same chapter, the concepts of glissile interface will be introduced to address the martensitic transformation.

2.3. PROPERTIES

For most practicing engineers, the 'structure', and all the intricacies of structure, are only important in so far as they are relevant to material 'properties'. It should be recalled that all the significant properties of metallic systems have strong

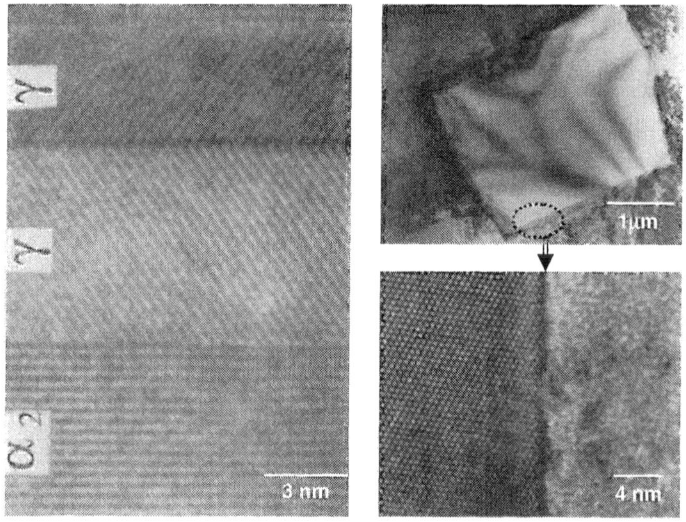

(a) (b)

Figure 2.14. HREM images of (a) coherent and (b) non-coherent interfaces. Courtesy of G.K. Dey, BRC, India. (a) Coherent interface between α_2 and γ phases, in TiAl. $\Sigma3$ twin boundary in the γ region is also visble. (b) Incoherent interface between Ti_2Ni and matrix (NiTi) phases. In (b), matrix lattice fringes are not visble.

(a) (b) (c)

Figure 2.15. (a) Schematic of the conversion from coherent into semi-coherent interface as the size exceeds $r_{critical}$. Coherent interface with (a) two phases of identical crystal structure and lattice parameter and (b) two phases of different crystal structure and/or lattice parameter. (b) would require an orientation relationship between the phases.

intrinsic components – selection of the alloy chemistry being vital (Shackelford and Alexander [2001]). Almost all these properties, however, also have structure sensitive components so that controlling the microstructure is of major practical interest (Smallman and Bishop [1999]).

Table 2.2 collates the typical properties (Shackelford and Alexander [2001]), which are of relevance in metallic systems, and highlights their relative structure sensitivities. A more extended discussion is included in the following sub-sections.

2.3.1 Physical properties

Density is an important index for alloy design. An initial guideline for 'minimum weight design' (Ashby [1989]) would depend on strength vs. density. Density is typically an intrinsic property but it can also depend on the structure. For example, porosity and/or inclusion content can affect the density – by default, as in the case of solidification structures, or by design, as in the case of 'porous' metals, metal foams and metal matrix composites.

The diffusivity has strong intrinsic components – bonding or melting temperature is important for both interstitial and substitutional diffusion, while size of the diffusing species is important for interstitial diffusion (Porter and Easterling [1986]). Diffusivity also depends on defect structure[10] and, in case of non-cubic systems, on crystallographic orientations.

2.3.2 Chemical properties

Of particular relevance to metals is the corrosion resistance. This is normally classified as bulk and local. Though corrosion (bulk/local or any electro-chemical attack), in general, depends on the structural indices,[11] typically the bulk corrosion resistance is considered as an intrinsic property. The local corrosion resistance (e.g. intergranular corrosion, etc.), on the other hand, is often structure dependent and can be optimized by appropriate microstructural engineering. This topic is covered in detail in Section 17.2. This behaviour is affected by both alloy chemistry and also by grain/phase boundaries and defects (Smallman and Bishop [1999]).

2.3.3 Mechanical properties

Elastic modulus and anelasticity (or internal friction) are the typical elastic properties of a metal. The elastic modulus is a 6×6 matrix and is dependent on

[10] Free surface, grain/phase boundaries, dislocations and vacancy clusters. These are the so-called diffusion short-cuts or high diffusivity paths. The diffusivity also can depend on lattice strain. For example, diffusivity of carbon in steel increases with carbon content – an effect attributed to the straining of the lattice (Porter and Easterling [1986]).

[11] Microstructure and stresses.

Table 2.2. The relevant properties in metals and alloys and their relative structure sensitivity.

Property	Brief description	Structure sensitivity
Physical		
Density	Mass per unit volume; ↑with atomic number	Depends primarily on chemistry. The porosity and 2nd phase content can also have an effect
Diffusivity	Quantity diffusing per unit area per unit time	Depends on defect structure, and also on crystallographic orientation for non-cubic systems
Chemical		
Bulk and local	Resistance to bulk and local corrosion	Bulk corrosion is primarily considered as an intrinsic property, while local corrosion is strongly microstructure sensitive
Mechanical		
Elastic and plastic	Response to elastic and plastic deformation	In addition to the intrinsic nature, these are strongly affected by structure
Electrical		
Resistivity	Resistance over unit length per unit area	Strongly depends on alloy chemistry. Also depends on the presence of defects. Affected by the presence of fully coherent phases
Superconductivity	Zero resistivity at low temperature	Superconducting nature depends on atomic mass, while critical current density depends on structure
Magnetic		
Saturation magnetization, permeability, core loss, etc.	Various characteristics of the *B–H* loop	Other than chemistry, magnetic properties also have strong structure sensitivity
Thermal		
Thermal expansion	Increase in crystal dimensions through increase in amplitude of atomic vibrations	Strongly depends on structure. Basis of dilatometric measurements for phase ID
Specific heat	Mostly vibrational motion of ions	Primarily intrinsic, but also depends on structure

The properties are broadly classified as physical, chemical, mechanical, electrical, magnetic and thermal. It is to be noted that all the properties have strong intrinsic character (dependent on chemistry, bonding, etc.), but they also do have different levels of structure sensitivity.

crystallographic orientation (Mura [1987]). The relaxation of part of the elastic strain with time is termed anelasticity (Smallman and Bishop [1999]). In cyclic loading, the direct result of anelasticity is a drop in the amplitude of vibration – a dissipation of energy by internal friction, usually termed as damping. This behaviour is affected by both alloy chemistry and also by grain/phase boundaries and defects (Smallman and Bishop [1999]). Plastic deformation depends on slip and twinning – both being structure dependent. An entire range of mechanical properties,[12] which depend on dislocation activity and plastic deformation, are structure sensitive. The role of structure on plastic deformation is illustrated in many chapters of this book.

2.3.4 Electrical properties

Resistivity or resistance offered to the motion of conduction electrons, is primarily an intrinsic property, which is also affected by irregularities in lattice structure. The irregularities typically represent defects. In metallic systems, the presence of solute atoms has strong effects on resistivity and controlling the alloy chemistry is often the first and the most important step – examples range from OFHC copper (where low oxygen content is important) to electrical steel (where alloying additions are tailored to increase resistivity – Section 15.4). Formation of a 2nd phase, or depletion of solute in the solid solution, typically decreases the resistivity. This rule has one exception – as shown in Figure 2.16a, the presence of coherent precipitates increases resistivity.

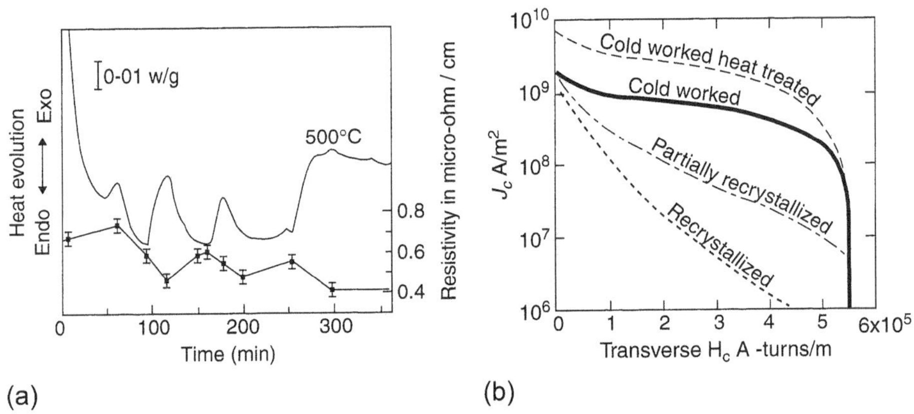

Figure 2.16. (a) DSC (differential scanning calorimetry) and resistivity plots showing phase transformations in 316L austenitic stainless steel (Wasnik *et al.* [2003]). DSC peaks and resistivity troughs correspond to each other and represent different phases, exception being the first phase involving coherent precipitation. The coherent precipitates are associated with increased resistivity. (b) Effects of structure on J_C vs. H plot in Nb – 25% Zr superconductor. Courtesy of Rose *et al.* [1966].

[12] Ranging from fatigue, creep, fracture behaviour, tensile and impact properties to hardness and formability.

Below 20K, some metals have near-zero resistivity and are called superconductors. In spite of the hype of high temperature oxide superconductors, metallic superconductors remain the dominant class of engineering materials for superconducting applications. The applications often involve introduction of a strong magnetic field resulting in a high current density. The superconducting properties are T_C, H_C and J_C, respectively representing the critical temperature below which the zero resistivity is appreciated and the critical magnetic field and current density. Superconductivity has both intrinsic and structure-sensitive components. For example, T_C is proportional to (atomic mass)$^{-1/2}$, while J_C can strongly be affected by structure (see Figure 2.16b).

2.3.5 Magnetic properties

A magnetism expert is often interested in coercivity (H_c) and remanence (B_r) values related to a $B-H$ loop (see Figure 2.17). Typically, H_c is structure sensitive, while B_r is more an intrinsic property. The applications of metals as magnetic material, however, require three specific properties – saturation magnetization, core loss and permeability. All these have strong structure-sensitive components. These properties and their relative structure sensitivities will be covered in detail in Section 15.4.

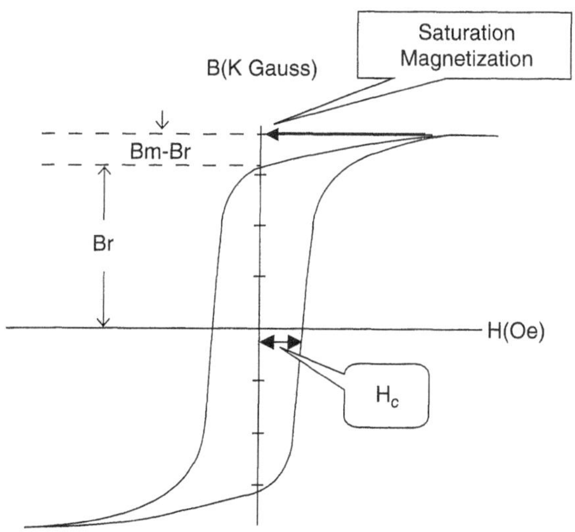

Figure 2.17. Schematic of a $B-H$ loop, where H and B represent respectively the strength of the imposed and induced magnetic field. The important magnetic properties, especially those relevant to metals, are saturation magnetization, permeability (slope of the $B-H$) and core loss (area under $B-H$). These are typically estimated at a particular field strength (e.g. 1.5 or 1.7 T).

2.3.6 Thermal properties

Specific heat originates mostly from the vibrational motion of ions, with only a small contribution possible from free electrons. It can be used to monitor phase changes. For example, the slope of a DSC (differential scanning calorimetry – where heat evolution, ΔH, is recorded as a function of temperature and time) plot (as in Figure 2.15a) represents $\mathrm{d}H/\mathrm{d}T$ or C_P, the specific heat at constant pressure, and can be used to monitor phase transformations. Thermal expansion has, however, even more relevance to structure sensitivity. The linear coefficient of thermal expansion is defined as $(1/L)(\mathrm{d}L/\mathrm{d}T)$, where L is the original length and T the absolute temperature. To a physicist, this represents a temperature-dependent increase in the amplitude of atomic vibrations, while a metallurgist often uses this as a property to identify phases and phase transformations (e.g. dilatometry). Typically, the coefficient of thermal expansion is anisotropic, but positive. A negative coefficient is, however, reported in nuclear fuels along certain crystallographic directions (Smallman and Bishop [1999]).

LITERATURE

Flemings M.C., "Solidification Processing", McGraw-Hill, New York, USA (1974).
Hornbogen E., Acta Metall., vol. 32 (1984), 615.
Kurz W. and Fisher D.J., "Fundamentals of Solidification", 3rd edition, Trans. Tech., Switzerland (1989).
Martin J.W., Doherty R.D. and Cantor B., "Stability of Microstructures in Metallic System", 2nd Edition, Cambridge University Press, UK (1997).
Porter D.A. and Easterling K.E., "Phase Transformations in Metals and Alloys", Van Nostrand Reinhold, UK (1986).
Smallman R.E. and Bishop R.J., "Modern Physical Metallurgy and Materials Engineering", 6th edition, Butterworth-Heinemann, UK (1999).

Chapter 3
Plasticity

Chapter 3

Plasticity

3.1. INTRODUCTION

Plastic flow is one of the fundamental characteristics of metals that has led to their historical and current technological importance. The ability to form metals to particular shapes by plastic deformation has been in use since the Bronze Age and is still a critical feature of many productive operations in the automobile, aerospace, packaging and general engineering industries. Of course, plastic deformation is used extensively in the primary production industries, such as hot rolling steel ingots down to plate, sheet, etc. before the final forming operations. These are the industries in which thermo-mechanical processing (TMP) has made the greatest impact; the 'mechanical' part of TMP essentially refers to the mechanics of plastic deformation.

There are a number of important features of metal plasticity that are used everyday in major applications, and therefore often taken for granted, but need emphasizing. The first concerns the relatively high stresses required to initiate plastic deformation at room temperature in most metallic alloys; this leads to their extensive use as structural materials. The second is their capacity to work harden during plastic straining; this gives rise to their tenacity or capacity to resist fracture. Finally, when heated, plastic flow becomes relatively easy so the material can be shaped into components or semi-finished products by operations such as rolling or forging involving very large strains. In other words, the critical relations between stress and strain during plastic flow vary enormously with temperature (and as will be seen below with other, microstructural, parameters), so the material can be conveniently shaped and then put into service. The fundamental relations between flow stresses and strain are called constitutive laws and depend closely on the microscopic mechanisms of plastic flow in crystalline materials.

This chapter briefly recalls the basic descriptions, definitions and elementary constitutive laws used to describe plastic deformation. It then goes on to analyse some particular features of plastic deformation that are important for large strains, such as macroscopic hardening, anisotropy and failure.

3.2. FUNDAMENTALS

3.2.1 Flow stresses and strains

During progressive loading in a simple tensile test, a metal first deforms elastically, Figure 3.1a, and then at the yield stress (denoted σ_y) begins to deform plastically, such that on unloading to zero force, the metal retains a permanent shape change which determines the amount of plastic deformation. The important quantities are the stress σ, usually taken as the force per unit area, and the strain, for example, the relative elongation in a tensile test. For plastic strains greater than about 10%, the true strain ε is defined by $\ln(l/l_0)$, which has the important property of additivity for successive strains. The schematic tensile stress–strain curves of Figure 3.1 illustrate different but typical behaviours: (a) room temperature deformation for two materials with different hardening rates during plastic deformation and (b) high-temperature deformation of a given alloy at two straining rates. In Figure 3.1a, the slope of the curve in the plastic regime is used to define the work hardening coefficient n:

$$n = \frac{d(\ln \sigma)}{d(\ln \varepsilon)} \tag{3.1}$$

The value of the work hardening coefficient, n is of major importance for forming operations since it controls the amount of uniform plastic strain the material can undergo during a tensile test before strain localization, or necking, sets in leading to failure. If the onset of necking occurs at maximum load, it is easily shown that, to first order, the amount of uniform plastic strain then equals the work hardening coefficient, n; this relation is known as the Considère criterion. This then limits the amount of uniform plastic strain that can be applied in a room temperature tensile test to maximum values of about 40%.

At room temperature, or more exactly at temperatures below about $0.4T_m$ (homologous temperature, T_m being the melting temperature in K), the stress–strain curve is virtually independent of the straining rate, at least for conventional rates.

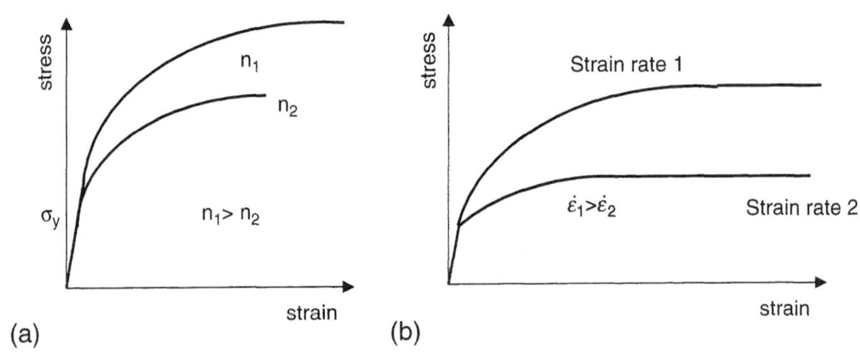

Figure 3.1. Schematic work hardening curves, (a) 2 alloys at $T_m < 0.4$ and a constant strain rate, (b) 1 alloy at $T > 0.4T_m$ and two different strain rates.

However, at higher temperatures, the straining rate ($\dot{\varepsilon}$) has a significant influence on the plastic flow stresses. As illustrated in Figure 3.1b, higher strain rates lead to higher flow stresses. The material strain rate sensitivity, m is defined as follows:

$$m = \frac{\mathrm{d}\ln\sigma}{\mathrm{d}\ln\dot{\varepsilon}} \tag{3.2}$$

where m takes values from about 0.05 at moderate temperatures to 0.5 or higher in some specific materials at very high temperatures (see Chapter 6).

3.2.2 Generalized stresses and strains

Of course, during a shaping or forming operation, plastic flow is no longer limited to one direction, as in a tensile test, so the stresses and strains are defined in a more general, multiaxial, way by tensor quantities. In the following, we shall use the conventional description of stress as the Cauchy stress tensor (σ), where σ_{ij} is the ith component of the force acting on unit area of a plane of normal $\mathbf{x}_j$.

There are several definitions of strain; here for simplicity, the small strain tensor (ε) is adopted, defined from the relative incremental displacement components $\mathbf{du}_i$ along the $\mathbf{x}_j$ direction:

$$\varepsilon_{ij} = \frac{1}{2}\left(\frac{\partial u_i}{\partial x_j} + \frac{\partial u_j}{\partial x_i}\right) \tag{3.3}$$

In most forming operations the strain rate components are the important quantities, defined, in a similar manner to the above, from the gradients of the velocity components, $\mathbf{v}_i$:

$$\dot{\varepsilon}_{ij} = \frac{1}{2}\left(\frac{\partial v_i}{\partial x_j} + \frac{\partial v_j}{\partial x_i}\right)$$

The final (large) strain is obtained by integrating over the small strain increments.

Both the above stress and strain rate quantities are symmetrical, second rank, tensors, which obey the standard tensor transformation rules. For further descriptions of stress/strain tensor analysis, the reader is referred to Spencer [1980] and Wagoner and Chenot [1997].

When the stresses and the strains refer only to the plastic part, i.e. after subtraction of the relatively small elastic strains, there is an additional simplification since one can derive reduced stress variables, **s** whose trace is null (the sum of the diagonal elements equals zero). This is because to a good approximation the hydrostatic pressure $p = -(\sigma_{11} + \sigma_{22} + \sigma_{33}/3)$ has no, or negligible, influence on plastic flow stresses so that the latter can be represented by the *deviatoric stress* **s**, whose components are given by

$$s_{ij} = \sigma_{ij} - \sigma_p \cdot \delta_{ij} \tag{3.4}$$

where δ_{ij} is the Kronecker delta (=1 or 0 for $i \neq j$).

Finally, since plastic strain occurs at constant volume, then its trace is also null so both plastic strain and deviatoric stress have only five independent components; they can often be represented by five component vectors.

It is now becoming standard to use a generalized description of strain increments (and strain rate), i.e. the von Mises effective or equivalent strain defined as

$$d\bar{\varepsilon} = \sqrt{\frac{2}{3} d\varepsilon_{ij} d\varepsilon_{ij}} \quad \text{or strain rate} \quad \dot{\bar{\varepsilon}} = \sqrt{\frac{2}{3} \dot{\varepsilon}_{ij} \dot{\varepsilon}_{ij}} \tag{3.5}$$

adopting the repeated indices summation convention.

As an example, to a rough approximation for macroscopic sheet, rolling deformation is plane strain compression for which the transverse strain ($\dot{\varepsilon}_{22}$) is zero and the averaged shear components are taken as zero so that, in terms of strain rate components,

$$\dot{\bar{\varepsilon}} = \sqrt{\frac{2}{3}\left(\dot{\varepsilon}_{11}^2 + \dot{\varepsilon}_{22}^2 + \dot{\varepsilon}_{33}^2 + 2\dot{\varepsilon}_{12}^2 + 2\dot{\varepsilon}_{13}^2 + 2\dot{\varepsilon}_{23}^2\right)} = \frac{2}{\sqrt{3}} \dot{\varepsilon}_{11} \tag{3.6}$$

where $\dot{\varepsilon}_{11}$ is the strain rate along the rolling direction.

Similarly a general stress, described by the Cauchy stress tensor (σ), is often reduced to a scalar equivalent stress or von Mises stress $\bar{\sigma}$, defined in a similar way to the equivalent strain (Eq. (3.5)). During plastic strain, this equivalent flow stress is identical to the flow stress of a material as measured during a standard, uniaxial, tensile test. Therefore, the tensile test directly gives $\bar{\sigma} = f(\bar{\varepsilon})$ curves. In the following, unless stated otherwise, stress and strain are taken as the von Mises equivalent values.

The total effective strain for any complex shaping operation is obtained by evaluating a path or line integral, which using the principal strains is written as

$$\Delta\bar{\varepsilon} = \int_{\bar{\varepsilon}(1)}^{\bar{\varepsilon}(2)} d\bar{\varepsilon} = \left(\frac{2}{3}\right)^{1/2} \int_{\text{strainpath}} \left(d\varepsilon 1^2 + d\varepsilon 2^2 + d\varepsilon 3^2\right)^{1/2} \tag{3.7}$$

It is useful to have an idea of the order of magnitude of these strains and strain rates as practiced in typical shaping and forming operations. Table 3.1 indicates

Table 3.1. Some typical strains, strain rates and strain paths involved in shaping Al alloys.

Mechanical process or test	Deformations	Strain rates (s^{-1})	Deformation modes	Practical examples
Traction	10^{-3}–0.4	10^{-4}–1	Extension	
Torsion	10^{-2}–50	10^{-2}–50	Shear (variable through thickness)	Cutting and trimming
Creep	10^{-3}–10^{-1}	10^{-8}–10^{-4}	Extension, imposed σ	Age forming
Hot rolling	0.2–0.8 per pass	1–20	Plane strain compression + shear near surface	Hot band down to 2 mm
Cold rolling	0.5–0.8 per pass	10^2–10^3	Plane strain compression	Sheet and foil down to 5 μm
Hot pressing	10^{-1}–2	10^{-2}–3	Biaxial expansion	Structural components
Extrusion	10	10^{-1}–10	Extension + surface shear	Profiles
Wire drawing	0.35 per pass	10	Extension	Wire
Stamping	10^{-3}–1	10^{-1}–1	Expansion	Car bodies, cans

some typical average values for aluminium processing (given that in many deformation processes the deformations and their rates are very inhomogeneous, so that the local values can be quite different from the average values given in the table). Note also that for many multipass operations such as rolling and wire drawing, typically involving up to 20 passes, the final average strain can be very large, of the order of 3–6 compared with a maximum value of 0.4 in a tensile test.

3.2.3 Yield criteria

In the case of the simple tensile test, there is a direct relation between the scalar quantities of flow stress and strain (see, e.g. Section 3.3). In the case of multiaxial plastic flow, these relations are described in terms of their tensor components. In theory, this implies the five independent components of stress as a function of the five components of the plastic strain rate. This is often simplified to measurable two-dimensional (2D) stress and strain rate components, as in biaxial stressing of thin sheet, where the normal stress component is zero.

In the 2D case, one can plot the yield stresses for a series of biaxial loadings (varying, e.g. σ_{11} and σ_{22}) on a stress diagram. Figure 3.2 illustrates the typical form of a 2D yield surface of an isotropic material. The form of this particular yield surface, known as the von Mises surface, is an ellipse with its major axis along the line given by

$$\sigma_{11} = \sigma_{22}$$

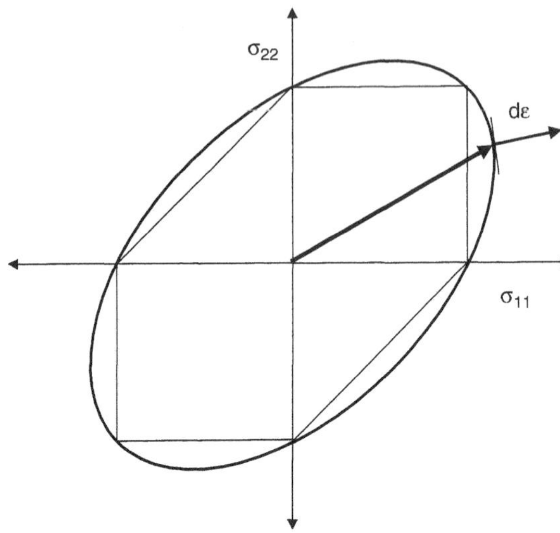

Figure 3.2. Schematic plane stress-yield surfaces for an isotropic material; von Mises ellipse and Tresca maximum shear stress criteria.

Another commonly used yield surface is based on the Tresca, maximum shear stress, criterion:

$$\tau_{max} = 1/2(\sigma_1 - \sigma_2) = \sigma_f \text{ (the flow stress)}$$

where σ_1 and σ_2 are respectively, the largest and the smallest principal stress components. This Tresca yield surface is made of facets (straight lines in 2D), which are circumscribed by the von Mises surface. The latter has the advantage, compared with the Tresca surface, of being continuously differentiable.

For any stress state within the yield surface, deformation is purely elastic but when the stress attains this surface, as shown in Figure 3.2 by the first arrow from the origin, plastic deformation occurs with an incremental amount **dε**. The stress vector (σ_{11}, σ_{22}) corresponds to different loading conditions such as

Uniaxial: σ_{11} or $\sigma_{22}=0$

Equal biaxial expansion: $\sigma_{11}=\sigma_{22}$

The properties of yield surfaces are described in several textbooks on the mechanics of plastic flow (Wagoner and Chenot [1997], Calladine [2000]). The important properties of interest here are the shape of the yields surface and the requirements of strain vector normality. The shape of the yield surface is described by a yield function of the general form

$$f(\sigma_{ij}) = f(\sigma_{11}, \sigma_{22}, \sigma_{33}, \sigma_{23}, \sigma_{13}, \sigma_{12}) = k$$

which for isotropic materials can be reduced to the von Mises yield function written in principal stresses as

$$f(\sigma_i) = c = (\sigma_1 - \sigma_2)^2 + (\sigma_1 - \sigma_3)^2 + (\sigma_2 - \sigma_3)^2 \tag{3.8}$$

with $c = 2\sigma_f^2$. In a general coordinate system this becomes

$$c = (\sigma_{11} - \sigma_{22})^2 + (\sigma_{11} - \sigma_{33})^2 + (\sigma_{22} - \sigma_{33})^2 + 6\sigma_{12}^2 + 6\sigma_{13}^2 + 6\sigma_{23}^2 \tag{3.9}$$

For the plane stress state used above in Figure 3.2, this is obviously an ellipse with the major axis along $\sigma_1 = \sigma_2$.

The strain vector normality rule states that the *incremental strain vector*, **dε** *is perpendicular to the yield surface* at the point corresponding to the stress state on the surface giving rise to plastic flow. In Figure 3.2, this is represented by the second arrow **dε** out from the yield surface using the same (1, 2) coordinate system as for the stresses. The condition of strain vector normality applied to the von Mises yield function gives some useful stress conditions for different strain paths. For example, if the flow stress is σ_f:

$$\text{Plane strain: } (\varepsilon_2 = 0, \varepsilon_3 = -\varepsilon_1) \quad \sigma_1 = \frac{2}{\sqrt{3}}\sigma_f; \quad \sigma_2 = \frac{1}{\sqrt{3}}\sigma_f$$

$$\text{Drawing: } (\varepsilon_3 = 0, \varepsilon_2 = -\varepsilon_1) \quad \sigma_1 = \frac{1}{\sqrt{3}}\sigma_f; \quad \sigma_2 = \frac{1}{\sqrt{3}}\sigma_f$$

However, for the case of anisotropic materials, typical of heavily deformed metals and alloys, then the yield function can be non-elliptical although in principle always convex. The corresponding plastic strain vector respects the normality rule and can vary substantially with stress state; this is discussed further in the section on anisotropy.

The plastic work per unit volume involved during straining is the product of the equivalent stress and the equivalent strain, i.e.

$$\bar{\sigma}\, d\bar{\varepsilon} = \sigma_{ij}\, d\varepsilon_{ij} \tag{3.10}$$

This means that the plastic work done in a unidirectional test is equal to the plastic work done in a general state for the same values of equivalent stress and strain.

3.3. STRESS–STRAIN RELATIONS

In this section, we shall only be concerned with the macroscopic stress–strain relations measured in a unidirectional deformation test (tensile, compression or torsion test).

Stress–strain curves are sometimes represented by the simple Ludwik power law

$$\sigma = \sigma_y + k_1(\bar{\varepsilon})^n \tag{3.11}$$

where k_1 is the flow stress increase at $\bar{\varepsilon}=1$, and n the work hardening coefficient which for most metals takes values between 0.05 and 0.5.

If the initial yield stress is low compared with the work hardening, the above relation can be reduced to the Hollomon law:

$$\sigma = k_2(\bar{\varepsilon})^{n'} \tag{3.12}$$

These power laws apply reasonably well to the parabolic part of the stress–strain curve, typically for strains ≤0.5. These are the strains that can be attained in a normal tensile test. However, at higher strains, as occurs in many shaping operations, the hardening rate decreases so that the flow stress tends to vary nearly linearly with the strain. Two or more pairs of (k, n) values are then required to describe the stress–strain curve over a wide range of deformations. Alternatively, and preferably for large strains, one can use the fact that the flow stress tends to saturate at a value denoted σ_s. In this case, the constitutive law can be approximately described by an exponential relation initially proposed by Voce [1948]:

$$\sigma = \sigma_s - (\sigma_s - \sigma_y)\exp(-\alpha\bar{\varepsilon}) \tag{3.13}$$

where α is a dimensionless constant characteristic of the work hardening behaviour.

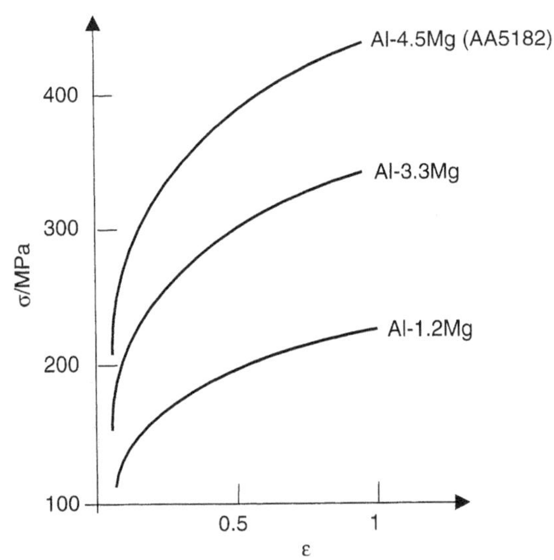

Figure 3.3. Stress–strain curves of Al–Mg alloys. From Lloyd and Kenny [1982]. With kind permission of Springer Science and Business Media.

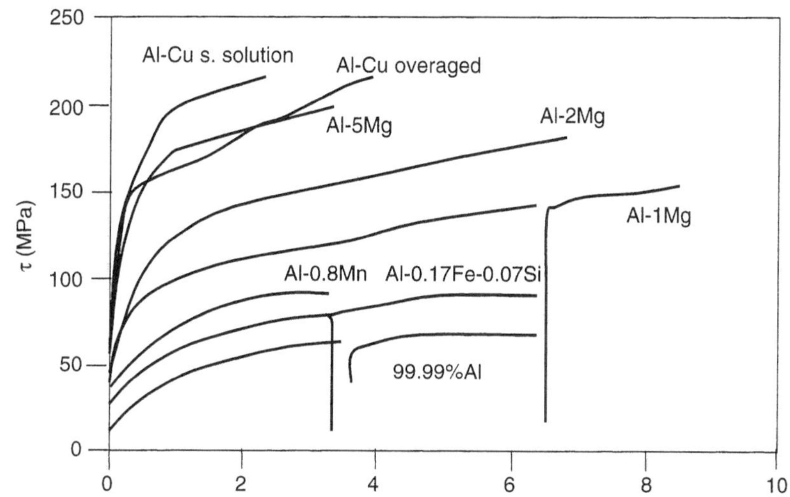

Figure 3.4. Large strain shear stress–strain curves of several Al alloys as measured in room temperature torsion tests, adapted from Rollett and Kocks [1994].

In fact, as can be seen in Figure 3.4, the saturation stress of many alloys deformed at room temperature is not defined with any great accuracy. Other variants of the Voce law have therefore been proposed one of which, initially due to Hockett and Sherby [1975], is

$$\sigma = \sigma_s - (\sigma_s - \sigma_y)\exp(-\alpha\bar{\varepsilon}^p) \tag{3.14}$$

where for aluminium alloys, the exponent p takes values ~0.5, see Lloyd and Kenny [1982].

Deriving the Voce law gives the current work hardening rate $\bar{\theta}$ as a linearly decreasing function of the flow stress

$$\bar{\theta} = \frac{d\sigma}{d\varepsilon} = \bar{\theta}_0\left(1 - \frac{\sigma_f}{\sigma_s}\right) \tag{3.15}$$

where $\bar{\theta}_0$ is the initial work hardening rate.

The experimental values of this law are identified by plotting the work hardening rate $\bar{\theta}$ as a function of the flow stress σ_f. Both parameters are often normalized by the shear modulus, i.e. $\theta/G = f(\sigma_f/G)$. Figure 3.5 illustrates this type of analysis for Cu at different temperatures ($<0.4T_m$). This plot, often termed the Kocks–Mecking plot, illustrates some of the complexity of work hardening behaviour of many metals and alloys. After an initial, roughly constant work hardening rate at low stresses and strains (not shown in Figure 3.5), $\bar{\theta}$ decreases linearly with stress, in accordance with

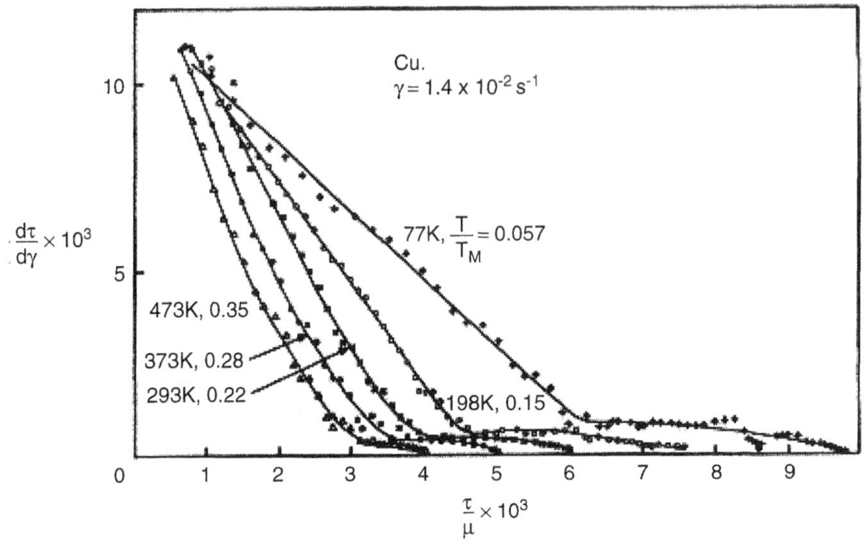

Figure 3.5. Normalized work hardening rates of Cu plotted as a function of the reduced flow stress showing stages III and IV at temperatures between 77 and 473 K (here the shear modulus is denoted μ), from Alberdi [1984].

the above Voce law; at a given flow stress, its value decreases with temperature. Also it is apparent that at high stresses (and therefore large strains), the work hardening rate does not go completely to zero; there is a roughly constant residual $\bar{\theta}_{IV} \approx 10^{-4}G$, independent of temperature for which an additional term is often introduced into what is known as the generalized Voce work hardening law:

$$\bar{\theta} = \bar{\theta}_{IV} + \bar{\theta}_{III}\left(1 - \frac{\sigma_f}{\sigma_s}\right) \tag{3.16}$$

The different stages of work hardening are conventionally classified as stages II (constant $\bar{\theta}$), III (linearly decreasing $\bar{\theta}$) and IV (low, constant $\bar{\theta}$); their microscopic origins are described in Section 4.3. Obviously, work hardening decreases with increasing temperature and these strong temperature effects are developed in more detail in Chapters 4–6.

3.4. PLASTIC ANISOTROPY

Virtually, all large strain shaping and forming operations lead to some form of material plastic anisotropy such that the mechanical, physical and sometimes chemical properties vary with testing direction. This is basically due to the formation of directional microstructures (e.g. particle or grain alignment), and in particular, preferential

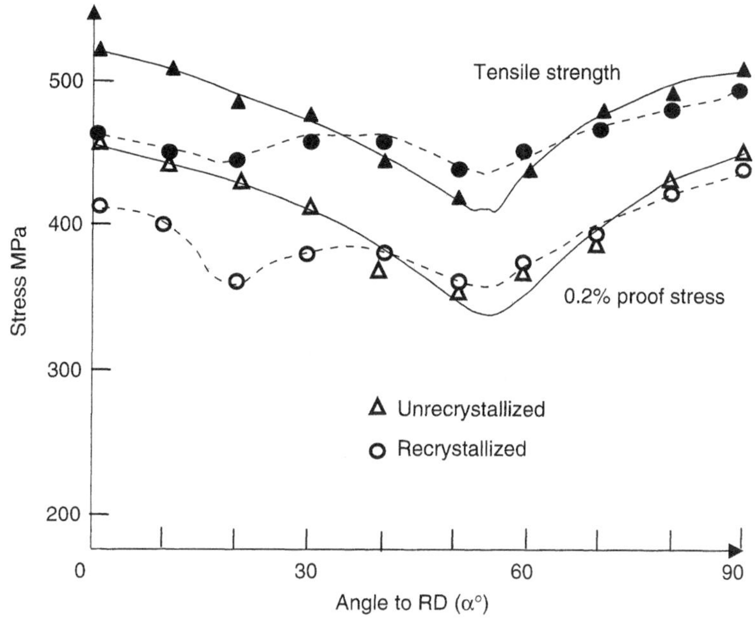

Figure 3.6. Anisotropic yield strength of a high-strength aluminium alloy sheet, from Bowen [1990]. Permission obtained from Maney.

grain orientations or crystallographic textures during, for example, rolling deformation and/or eventually subsequent recrystallization (Chapter 8). An example is the variation, with direction in the sheet plane, of the yield strength of an Al–Li alloy as illustrated in Figure 3.6.

A second case concerns the plastic anisotropy during a forming operation such as deep drawing. A circular blank stamped out of a sheet is drawn down by punching through a die into a cup shape, but the resulting 'cup' develops irregular wall heights commonly denoted 'ears' (see Figure 3.7). The ears are situated at certain angles to the sheet directions according to the plastic anisotropy, or the texture, of the sheet (see Chapter 11).

Plastic anisotropy of sheet material is measured experimentally by the contraction ratios of flat tensile samples, taken at angles α to the rolling direction, and elongated by tensile testing. The ratio of the width (plastic) strain ε_w to the through-thickness strain ε_t is denoted the Lankford coefficient or the R value.

$$R(\alpha) = \frac{\varepsilon_w}{\varepsilon_t} \quad \text{or} \quad r = \frac{\dot{\varepsilon}_w}{\dot{\varepsilon}_t} \tag{3.17}$$

Figure 3.7. Ear formation during cupping of a strongly textured aluminium alloy sheet. Courtesy of J. Hirsch, Hydro Aluminium.

In theory, R varies from 0 (no widening) to ∞ (no thinning) and equals 1 for isotropic materials. Note that some authors use the contraction ratio $q = -\dot{\varepsilon}_w / \dot{\varepsilon}_t$ measured in terms of the strain rates along the sample width and length, i.e. $q = r/1 + r$ and varies from 0 to 1. To characterize the overall resistance to thinning of a sheet, an average of the R values at three α angles is usually determined[1]:

$$\bar{R} = \frac{R_0 + R_{90} + 2R_{45}}{4} \tag{3.18}$$

and to measure the amplitude of the in-plane anisotropy (related to the ear heights) the difference value is used

$$\Delta R = \frac{R_0 + R_{90} - 2R_{45}}{2} \tag{3.19}$$

An $\bar{R}$ value greater than 1 means that during elongation the sheet contracts more than it thins and often indicates good formability. In some interstitial free (IF) steels (Chapter 15.1), the $\bar{R}$ value can attain 2.5. Average $\bar{R}$ values greater than 1 are not

[1] This is justified because most cups have four ears with maxima/minima at 0, 45 and 90°. In some exceptional cases, cups with six ears can be obtained and then another definition of $\bar{R}$ should be adopted, e.g. measuring $R(\alpha)$ every 15°.

common in most aluminium alloys. Only in materials that have been produced to include pronounced shear textures have $\bar{R}$ values of about unity been obtained.

Plastic anisotropy can be understood at a macroscopic level by the particular form of the yield surface and the resulting plastic strain vector for different stress states. Figure 3.8 illustrates schematically the influence of a slight change of the yield surface, compared with the isotropic one of Figure 3.2 (in this case a small rotation). The yield stress along the σ_{22} axis is now higher than that along σ_{11} corresponding to the type of yield stress anisotropy of Figure 3.6. Furthermore, the strain increment vector $\mathbf{d\varepsilon}$ is now parallel to σ_{11} so that ε_{22} is expected to be zero although the stress component along σ_{22} is non-zero (and by constant volume, the strain along X_1 equals the strain along X_3, the sheet thickness direction).

The actual form of the anisotropic yield surface for different materials and degrees of anisotropy has been the subject of much research. It obviously depends on the type of symmetry of the property anisotropy. Many rolled metals possess orthotropic symmetry (along the three orthogonal directions RD, TD and ND), and so the von Mises yield criterion has been modified by Hill [1948] to predict yielding along these axes using three coefficients (*F*, *G*, *H*):

$$\sigma_y = \left[\frac{3/2}{F+G+H}\right]^{1/2} \left[F(\sigma_2-\sigma_3)^2 + G(\sigma_3-\sigma_1)^2 + H(\sigma_1-\sigma_2)^2\right]^{1/2} \quad (3.20)$$

which reduces to the von Mises criterion for $F=G=H$. This criterion is only strictly valid when the principal stress axes are aligned with the material anisotropy

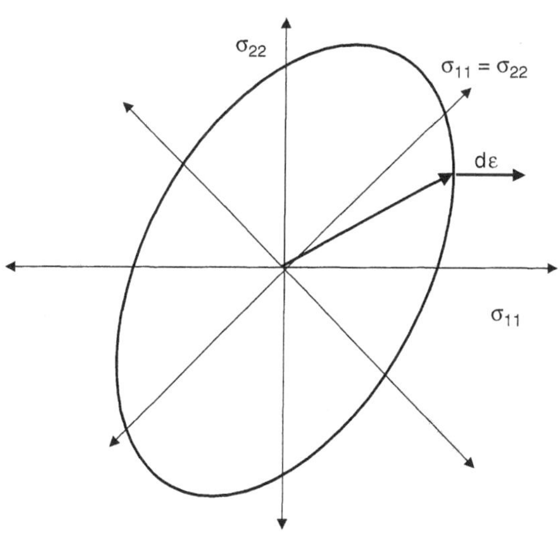

Figure 3.8. Schematic 2D anisotropic yield surface.

directions. Despite many limitations, this criterion has been extensively used for numerical, finite element, simulations because of its simplicity.

More recent developments in anisotropy theory include more general formulations, e.g. the Barlat *et al.* [1991] criterion. Alternatively, but often at the risk of large computing requirements, the crystal plasticity analysis developed in Chapter 10 can be applied to predict yield surfaces from first principles. This method can also allow for the change in orientations of the grains during plastic deformation. As described in more detail in Chapters 8 and 10, most grains change crystal orientation during plastic deformation by crystal slip and/or twinning mechanisms; at large strains (typically more than unity) their orientations tend towards one or more stable deformation orientations known as 'crystallographic texture components'. It is due to the presence of these texture components, which possess anisotropic properties, that the plastically deformed material becomes anisotropic.

3.5. FRACTURE

During the shaping and forming operations typical of many thermo-mechanical processes, rupture of the material is a constant threat; so, the conditions leading to rupture have been the subject of a large number of studies. The treatment here is an elementary introduction to the problem and the reader can find more detailed analyses in the books by Semiatin and Jonas [1984], Dodd and Bai [1987] and Thomason [1990].

The schematic stress–strain curves of Figure 3.9 illustrate the three major, typical, fracture behaviours of materials during tensile loading. Figure 3.9a represents purely elastic stressing before brittle failure, as occurs, for example, in ceramics and some brittle metallic alloys at low temperatures. This brittle failure mode will not be examined here in any detail since we are primarily concerned with the more ductile failure characteristic of deformation processing methods. Figure 3.9b illustrates the typical behaviour of a metal when deformed at room temperature as in a forming operation; the material undergoes a certain amount of plastic strain before failure. Figure 3.9c describes typical ductile stress–strain behaviour during high-temperature straining; plasticity can take place over larger strains with a gradual stress decrease after the maximum load, as failure occurs progressively in the material.

The ductility that can be obtained before fracture is usually given by the sample cross-section at failure A_f relative to the initial value A_0, i.e. ductility = $\ln(A_0/A_f)$.

The respective influences of work hardening and strain rate sensitivity on ductility can be understood on the basis of the modified Considère model due to Hart [1967]. During tensile straining, one analyses the behaviour of a small, local, variation in the specimen cross-section, ΔA to determine if this local change in area

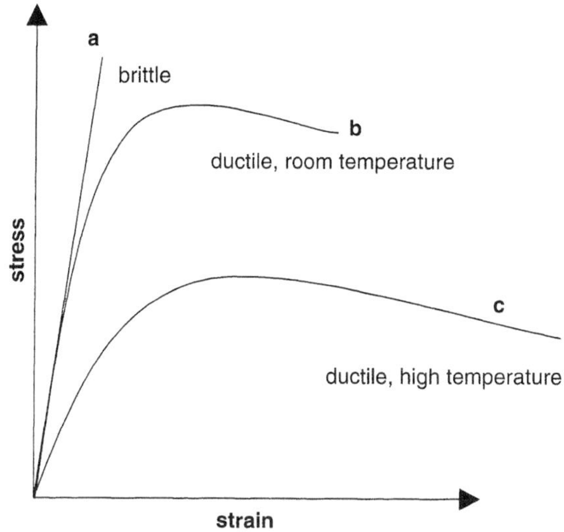

Figure 3.9. Schematic stress–strain curves for common failure modes (a) brittle, (b) ductile at room temperature and (c) high temperature.

will be amplified (leading to failure) or reduced (leading to continued plastic strain). We therefore need the rate of change of the local area compared with the surrounding 'average' area. The load transmitted through the specimen is the same everywhere so

$$P = \sigma A = (\sigma + \Delta\sigma)(A + \Delta A) \quad \text{and therefore} \quad A\Delta\sigma + \sigma\Delta A + \Delta\sigma\Delta A = 0$$

As a differential equation and ignoring second-order terms this gives

$$\frac{\mathrm{d}\sigma}{\sigma} = -\frac{\mathrm{d}A}{A}(= \mathrm{d}\varepsilon)$$

where, as indicated, the second term corresponds to the difference in strain increment ($\mathrm{d}\varepsilon$) between the local and average sections. The variation in stress $\mathrm{d}\sigma$ is due to strain hardening and strain rate effects:

$$\begin{aligned} \mathrm{d}\sigma &= \left(\frac{\partial\sigma}{\partial\varepsilon}\right)_{\dot{\varepsilon}} \mathrm{d}\varepsilon + \left(\frac{\partial\sigma}{\partial\dot{\varepsilon}}\right)_{\varepsilon} \mathrm{d}\dot{\varepsilon} \\ \text{but } \dot{\varepsilon} &= -\frac{\dot{A}}{A} \quad \text{and} \quad \mathrm{d}\dot{\varepsilon} = -\frac{\mathrm{d}\dot{A}}{A} + \frac{\dot{A}}{A^2}\mathrm{d}A \end{aligned} \tag{3.21}$$

Substituting for $\mathrm{d}\varepsilon$ and $\mathrm{d}\dot{\varepsilon}$ gives a relation for $\mathrm{d}\sigma$ in terms of A, $\mathrm{d}A$, etc. and the partial derivates for stress. The latter can be written $1/\sigma\left(\partial\sigma/\partial\varepsilon\right)_{\dot{\varepsilon}}=\left(\partial\ln\sigma/\partial\varepsilon\right)_{\dot{\varepsilon}}$, etc. to obtain the following expression for the relative rate of change of section:

$$\frac{\mathrm{d}\dot{A}}{\mathrm{d}A}=\frac{1-\left(\partial\ln\sigma/\partial\varepsilon\right)_{\dot{\varepsilon}}-\left(\partial\ln\sigma/\partial\dot{\varepsilon}\right)_{\varepsilon}\dot{\varepsilon}}{\left(\partial\ln\sigma/\partial\ln\dot{\varepsilon}\right)_{\varepsilon}}$$

Assuming constant values for the work hardening and strain rate sensitivity coefficients, n and m as defined in Section 3.2, this can be written

$$\frac{\mathrm{d}\dot{A}}{\mathrm{d}A}=\frac{1-\left(n/\varepsilon\right)-m}{m} \tag{3.22}$$

The critical condition between local necking and plastic stability is $\mathrm{d}\dot{A}=0$ so that the uniform strain to this condition is then simply:

$$\varepsilon_{\mathrm{u}}=\frac{n}{1-m} \tag{3.23}$$

High work hardening and strain rate sensitivity coefficients clearly stabilize plastic flow against incipient flaws. Obviously for rate-insensitive materials, as frequently occurs at room temperature, then $\varepsilon_{\mathrm{u}}=n$ which is the original Considère criterion.

3.5.1 Failure mechanisms

The range of possible fracture mechanisms for most metals and alloys is shown schematically in Figure 3.10. At low temperatures, they extend from brittle cleavage through plastic growth of voids to macroscopic necking or shearing. At higher temperatures, cleavage is replaced by intergranular creep fracture and void growth becomes controlled by power-law creep. Some typical fracture surfaces associated with these types of failure are depicted in Figure 3.11. As noted above, we shall ignore brittle failure for hot working operations and concentrate on the more typical ductile fracture modes due to void nucleation and growth. It should also be recalled that complete necking down to a near point may occur in some high-purity metals, but most industrial alloys undergo some form of void growth. The exception concerns the higher temperature regimes of many alloys ($>0.7T_{\mathrm{m}}$), Figure 3.10, when concurrent, dynamic, recovery or recrystallization (Chapter 5) can soften the material sufficiently to avoid void growth and permit this type of necking, termed rupture.

The respective influences of stress and temperature on the failure mechanisms can be represented in a fracture mechanism map, Ashby *et al.* [1979], plotting out domains of fracture mechanism on a map with the normalized stress (σ/E) along the ordinate and homologous temperature on the abscissa. Figure 3.12

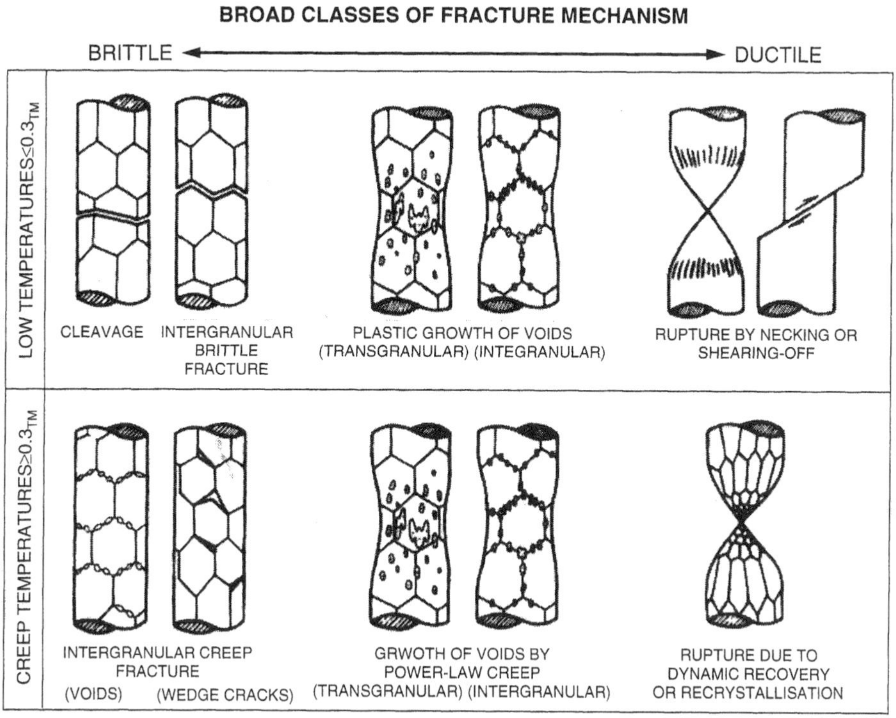

Figure 3.10. Classes of fracture mechanisms, after Ashby *et al.* [1979]. Permission obtained from Elsevier.

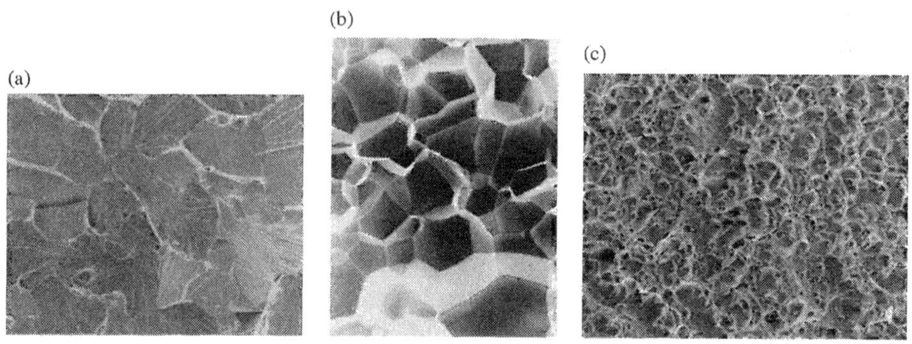

Figure 3.11. Typical SEM (scanning electron microscope) fracture surfaces: (a) brittle cleavage of a martensitic 0.38%C steel (b) brittle intergranular fracture of W at room temperature and (c) ductile rupture fracture surface showing cusps formed by void growth around particles (quenched and tempered 0.38%C steel).

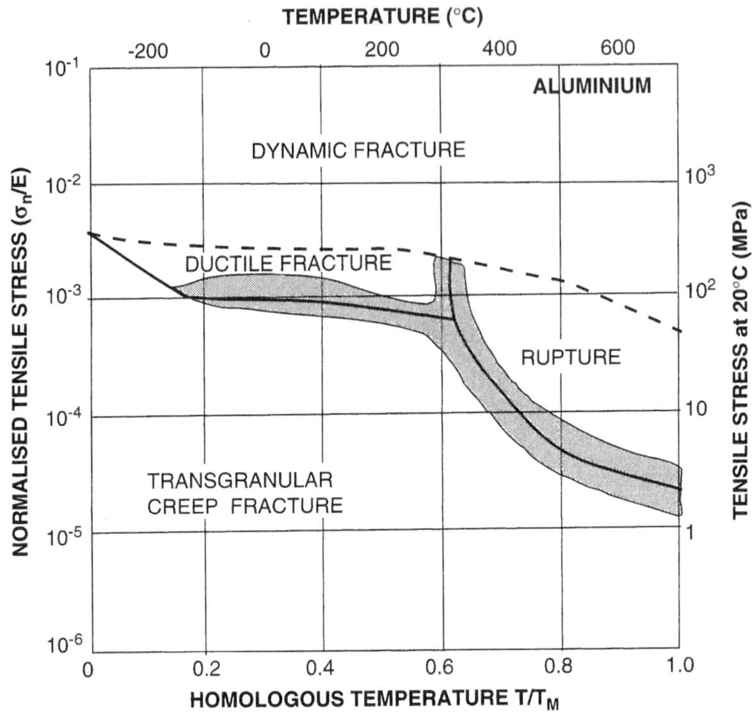

Figure 3.12. Failure mechanism map for Al, from Ashby *et al.* [1979]. Permission obtained from Elsevier.

gives an example for commercial purity aluminium containing five domains: dynamic fracture, ductile fracture, transgranular and intergranular fracture, and rupture. The very high-stress dynamic failure mode corresponds to exceptional conditions for rupture without preceding plastic flow. For most straining conditions, the plastic flow stress is of the order of 2×10^{-4}–2×10^{-3} E (or 10–100 MPa for Al) so ductile fracture, transgranular creep fracture and rupture predominate. This is typical of many face-centred cube (fcc) metals and alloys such as Ni, Cu and Ag. On this type of map the time to failure, t_f, can also be represented by contour lines of constant t_f. An alternative method is to plot the normalized stress as a function of time to failure and give contours of constant temperature. This second type of map is illustrated in Figure 3.13, for the case of an austenitic stainless steel. This immediately shows that for typical deformation processing of metals, involving times between fractions of seconds and at most a few minutes, then the major mechanisms are ductile fracture (by void growth) or rupture at high temperatures.

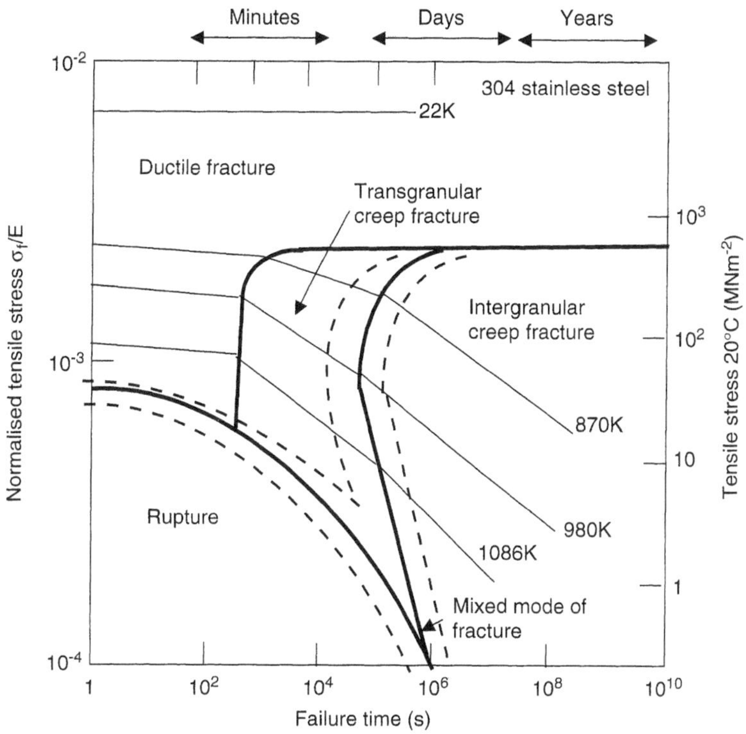

Figure 3.13. Time to failure map for 304 stainless steel, from Fields *et al.* [1980]. With kind permission of Springer Science and Business Media.

In the case of iron, the fracture mechanism map is somewhat complicated by the $\alpha/\gamma/\delta$ allotropic transformations. As for most body-centred cube (bcc) metals, it contains a significant cleavage domain at temperatures $\leq 0.3T_m$.

Figure 3.14 shows a schematic of the ductile fracture mechanism for void growth after nucleation at second-phase particles. It comprises microscopic crack initiation, usually at the particle–matrix interface, then growth by plastic deformation and finally localized necking between the voids to rupture. It is the plastic strain and the local necking which give rise to the characteristic fine cusps on the fracture surface (Figure 3.11c), each of which initiates at a second-phase particle. At higher temperatures, void growth is controlled by creep or viscoplastic flow leading to transgranular creep fracture.

Clearly, ductility will be sensitive to both the particle content, i.e. size and volume fractions and the temperature regimes for plastic flow. It should also be pointed out that the ductility depends on the degree of triaxiality, i.e. the ratio of average stress to the von Mises equivalent flow stress (or $-p/\overline{\sigma}$). This can be illustrated by the

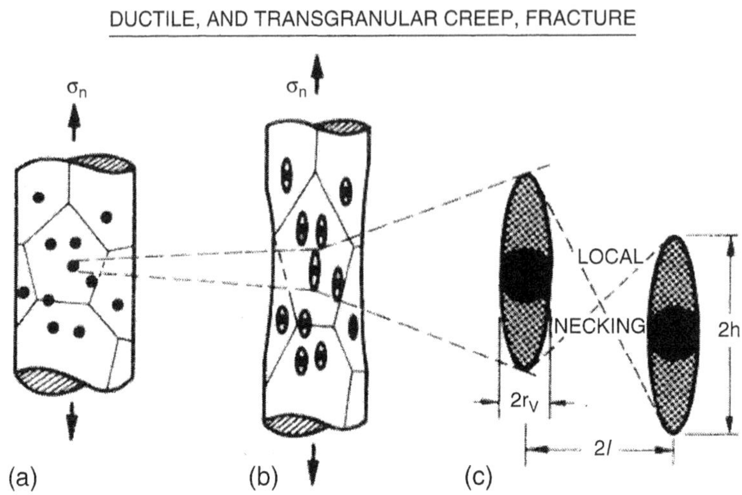

Figure 3.14. Void growth mechanisms, after Ashby *et al.* [1979]. Permission obtained from Elsevier.

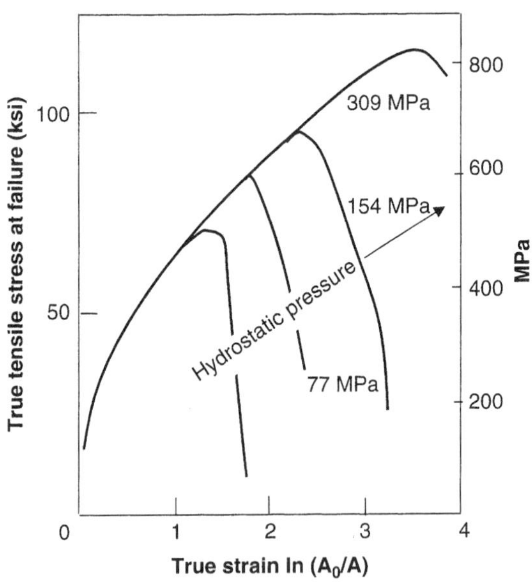

Figure 3.15. Influence of superimposed hydrostatic stress on tensile ductility of Cu, adapted from Pugh [1964].

results of Figure 3.15 pertaining to copper deformed in tension under different compressive pressures; the ductility can be increased by a factor of about 3 under very high, positive, pressures although the work hardening behaviour is unaffected. This ductility improvement under high pressures is, of course, one of the major advantages of shaping by rolling or extrusion. It should be noted that negative pressures (hydrostatic tension), which frequently occur in many final forming operations, strongly decrease the ductility (see, e.g. the forming limit diagrams of Chapter 11).

Nucleation of voids occurs at the particle/matrix interface and using an energy balance approach (similar to the Griffith model of brittle fracture) Gurland and Plateau [1963] derived a critical matrix stress for void initiation as

$$\sigma = \frac{1}{\alpha}\sqrt{\frac{E \cdot \gamma}{d_{\mathrm{p}}}} \tag{3.24}$$

where E is Young's modulus, γ the specific surface energy, α a stress concentration factor and d_{p} the particle diameter. It follows that small particles require higher matrix stresses for void nucleation. In practice, particles much smaller than 1 μm do not readily initiate voids in metallic systems, but particles greater than about 1 μm can do so easily.

Growth of these voids has been the subject of much research and elementary formulae are not available. The difficulty resides in correctly describing the plastic deformation around and between particles and groups of particles and the onset of localized necking between them. The reader is referred to the works of Rice and Tracy [1969], Gurson [1977], Tvergaard [1982] and Klocker and Tvergaard [2000].

Apart from the triaxiality effects described above, the important features are the volume fractions of second-phase particles and the temperature. The influence of the volume fractions is depicted in Figure 3.16 showing the large decrease in ductility of various copper alloys as the particle fraction, or inclusion content, increases to about 5% and above.

The temperature controls the mechanisms of plastic flow between the voids after their nucleation at particles. At high temperatures, where viscoplastic processes dominate, one can distinguish between flow controlled by matrix creep (see Chapter 4) and diffusional flow along grain boundaries. The distance within which diffusional flow (rather than matrix creep) dominates is given by

$$\Lambda = \left(\frac{D_{\mathrm{b}} \cdot \delta_{\mathrm{b}} \cdot \Omega \sigma_{\infty}}{kT \dot{\varepsilon}_{\infty}} \right)^{1/3} \tag{3.25}$$

where D_{b} is the grain boundary diffusivity, δ_{b} the boundary thickness, Ω the atomic volume and k the Boltzman constant. At high values of this parameter, diffusional

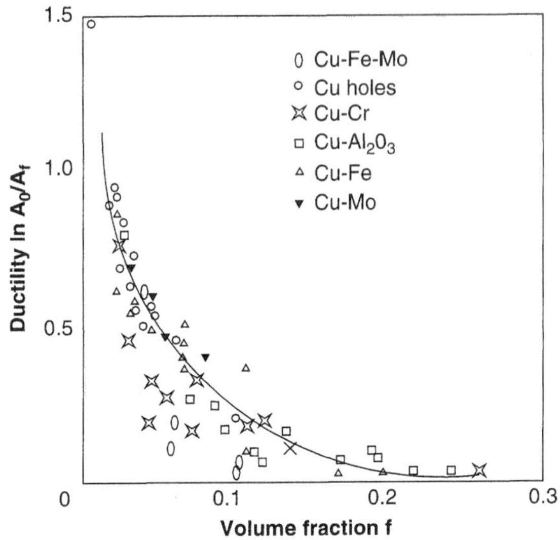

Figure 3.16. Influence of particle volume fraction on ductility of copper, adapted from Edelson and Baldwin [1962].

flow tends to accelerate void growth rates compared with those expected of matrix creep. This parameter is clearly temperature dependant and indicates the transition temperature for diffusion-controlled creep (Needleman and Rice [1980]).

LITERATURE

Calladine C.R., "Plasticity for Engineers – Theory and Applications", Horwood, Chichester, England (2000).

Dodd B. and Bai Y., "Ductile Fracture and Ductility – With Applications to Metalworking", Academic Press (1987).

Hill R., "The Mathematical Theory of Plasticity", Oxford University Press, London (1950).

Johnson W. and Mellor P.B., "Plasticity for Mechanical Engineers", Van Nostrand, Princeton, NJ (1962).

Semiatin S.L. and Jonas J.J., "Formability and Workability of Metals; Plastic Instability and Flow Localization", ASM, Ohio (1984).

Spencer A.J.M., "Continuum Mechanics", Longman (1980).

Thomason P.F., "Ductile Fracture of Metals", Pergamon Press (1990).

Wagoner R.H. and Chenot, J.-L. "Fundamentals of Metal Forming", Wiley (1997).

Chapter 4

Work Hardening

Chapter 4
Work Hardening

4.1. INTRODUCTION

Work hardening is one of the characteristic properties of most metallic alloys; it is also probably the most useful since it is the fundamental cause of their tenacity, or their capacity, in the presence of internal flaws or geometrical defects, to resist loading. Thus using the very simple example outlined in Chapter 3, according to the Considère criterion, the maximum amount of uniform plastic deformation in tensile straining is given by the work hardening coefficient. A high coefficient therefore facilitates complex-forming operations without premature failure. Work hardening of course strongly influences the mechanical energy expended to shape a material by plastic deformation, for example during rolling, etc. Work hardening also controls the amount of energy stored in the material as a consequence of plastic deformation. It therefore strongly governs the behaviour of the metal during subsequent softening by annealing (Chapter 5). Finally, many high-volume components such as beverage cans are directly used in the work hardened state, so the hardening capacity and the stability of the work hardened state are important practical issues.

Modern theories of work hardening started with Taylor [1934] and the concept of crystal dislocations and have continued ever since. There is no universally accepted theory of work hardening but much empirical knowledge and many models which share a certain number of common features; these are the points that are developed here.

This chapter starts with a description of work hardening at relatively low temperatures where thermally activated processes do not play a major role, typically for homologous temperatures below $0.4T_m$. It then continues with an analysis of plastic flow at high temperatures where material viscosity effects are important. In both the cases, the relation to dislocation behaviour will be emphasized.

4.2. 'LOW' TEMPERATURE

4.2.1 Basic microscopic mechanisms

Plastic flow of crystalline materials takes place by the movement of dislocations along crystal planes under the influence of an applied stress. Work hardening is a consequence of the fact that the stress required for dislocation movement usually increases during plastic flow as the dislocations become increasingly hindered by microstructural obstacles. In order of increasing size these obstacles are solute atoms, dislocations, precipitates and grain boundaries. The most important variation in obstacle density is usually due to the dislocations themselves. Their behaviour in crystals under stress is therefore of paramount importance for understanding and modelling work hardening.

Dislocation theory – geometry, mechanics, crystallography, kinetics, etc. – has been the subject of many treatises (see, e.g. Cottrell [1953], Hirth and Lothe [1968], Nabarro [1967], Kocks *et al.* [1975]), and the reader is advised to consult them for greater detail. Many useful experimental results on work hardening to large strains can be found in the review by Hecker and Stout [1982]. The following only summarizes the essential features necessary for understanding the fundamentals of thermo-mechanical processes.

Dislocations slip in crystallographic directions on planes of easy glide under the influence of a critical applied shear stress τ_c. The emergence of each dislocation creates a displacement along the slip direction of quantity b (the magnitude of the Burgers vector). The amount of shear strain due to the emergence of length l of dislocation in a small volume element is $d\gamma = \rho b l$, where ρ is the dislocation density (in units of [m/m^3] or [m^{-2}]). In terms of the shear strain rate this is written $\dot{\gamma} = \rho b v$, where v is the average speed of the mobile dislocations. According to the von Mises law, plastic deformation of a crystal requires the operation of as many slip systems as imposed strain rate components ($\leqslant 5$), so usually several systems (combinations of slip planes and directions) operate, and interact, in each grain. The shear stress for plastic flow then evolves with strain since the interaction of a moving dislocation with other dislocations usually constitutes a barrier to further movement. This requires an increase in flow stress for the dislocation to continue moving either by circumventing the barrier, or more often by creating a new segment of dislocation which is then activated in the vicinity. The flow stress therefore increases with dislocation density and the relation between them is usually written

$$\tau_c = \tau_0 + \alpha G b \sqrt{\rho} \qquad (4.1)$$

where τ_0 is the intrinsic glide stress of the material at near-zero ρ and α a material constant (0.2–0.5). α is generally considered to vary with the crystal structure and the type of dislocation substructure which is formed, typically about 0.2–0.3 for fcc and 0.4 for bcc metals.

This widely accepted square-root law is a consequence of the fact that the dislocation glide stress is inversely proportional to the spacing of the obstacles, l, i.e.

$$\tau_c = \tau_0 + \frac{\alpha G b}{l} \qquad (4.2)$$

For a homogeneous dislocation distribution, the dislocation spacing is given by the reciprocal of the square root of the dislocation density ($l=1/\sqrt{\rho}$), hence the above law.

For aluminium at 20°C the shear modulus $G=26$ GPa, $b=0.286$ nm and $\alpha \approx 0.3$ so that $\Delta\tau = \tau - \tau_0 \approx 2.23\sqrt{\rho}$ Pa (ρ in m^{-2}). For dislocation densities typical of a softened, recrystallized, metal $\rho \approx 10^{10}/\mathrm{m}^2$ and so $\Delta\tau \approx 0.2$ MPa, which is negligible. For $\rho \approx 10^{15}/\mathrm{m}^2$, as in heavily cold-worked alloys, $\Delta\tau \approx 70$ MPa, which is obviously significant. Note that the applied macroscopic tensile stress would be about 3 times τ_c (cf. Chapter 10), i.e. $\Delta\sigma \approx 210$ MPa, which is of the order of the difference in flow stress between the annealed and work hardened states of many work hardening aluminium alloys.

This elementary, order of magnitude analysis of work hardening can be developed further to characterize the evolution of the dislocation density, and hence flow stress, with strain, alloy content, microstructure and, at high temperatures, with T and $\dot{\varepsilon}$.

The alloy content, when fairly high, often intervenes through the influence of alloying elements on the stacking fault energy (SFE). The latter parameter controls the extent to which dislocations are dissociated into stacking faults in the slip plane (see, e.g. Hirth and Lothe [1968]). Table 4.1 gives some typical SFE values for common metals and alloys. Low SFE values, typically less than 50 mJ/m^2, lead to widely dissociated dislocations whose movements are therefore confined to the slip planes so that localized cross-slip out of them is extremely difficult. At low strains they then tend to pile-up in the slip planes creating 'planar' dislocation structures, associated with strong interactions between dislocations on different slip systems and high local stresses (Figure 4.1a). Common examples are austenitic stainless steels, and some Cu–Al or Cu–Zn alloys.

At higher strains these low SFE alloys often undergo extensive microtwinning (or eventually localized martensitic transformations) and shear banding, i.e. the formation of very pronounced planar faults in the crystal structure and/or local orientations. Allowing for this dependency on SFE, the evolution of the flow stress and dislocation structure with strain can be represented schematically by four stages (Gil Sevillano *et al.* [1980]) as follows.

4.2.1.1 The work hardening stages. *Plastic yielding and stage 1*: During Stage I, which is most easily seen in single crystals oriented for single slip, the dislocations are usually confined to their slip planes and do not interact with each other so that the work hardening rate is very low. However, as the crystal rotates by plastic deformation (see Chapter 10) it tends to reorient towards double slip orientations, which then favour the stronger dislocation interactions of Stage II.

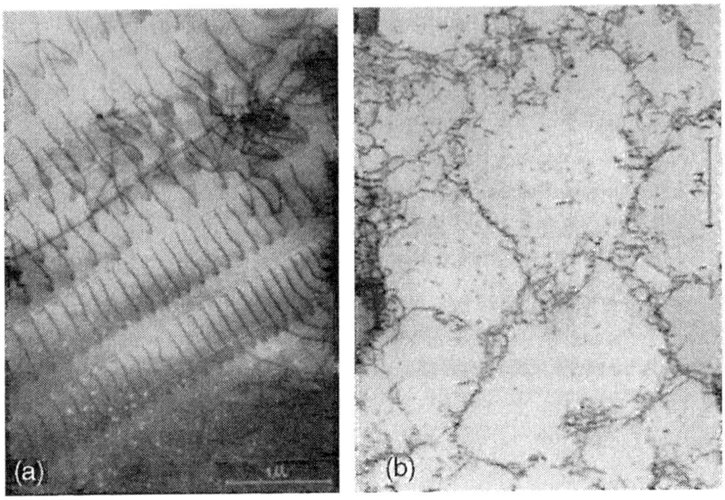

Figure 4.1. (a) Dislocation pile-ups in a lightly deformed austenitic stainless steel from Dupouy and Perrier [1962] by courtesy of Journal de Microscopie, (b) dislocation tangles in lightly deformed Al.

Table 4.1. Stacking fault energies of metals, from Murr [1975].

Metal	γ_{SFE} (mJ/m^2)	Metal	γ_{SFE} (mJ/m^2)
Aluminium	166	Zinc	140
Copper	78	Magnesium	125
Silver	22	Zirconium	240
Gold	45	91Cu:9Si	5
Nickel	128	70Cu:30Zn	20
Cobalt (fcc)	15	304 Austenitic stainless steel	21

In polycrystals, which begin to deform plastically, Stage I is negligible since the movement of the first few dislocations is restricted by the grain boundaries at which they often pile-up. The resistance of the grain boundaries to dislocation movement gives rise to the well-known Hall–Petch grain size (D) hardening relation:

$$\sigma(D) = \sigma_\infty + k_1 D^{-1/2} \tag{4.3}$$

where σ_∞ is the flow stress of the material at a very large grain size. This relation is very important for hardening many polycrystalline alloys particularly when the grain size can be reduced to 20 μm or less. Typical values of the k_1 coefficient (in MNm$^{-3/2}$) vary from about 0.7 to 1 for carbon steels, through 0.4−0.2 for the hcp metals and down to 0.1−0.07 for fcc Cu and Al, respectively. The high value of k_1 for carbon-containing steels constitutes the physical basis for the development of high-strength low alloy (HSLA) steels with fine grain sizes (see Section 15.3 of Chapter 15). Note that the presence of carbon in solution significantly increases k_1 compared with the pure metal case which is 0.2 for pure iron (Armstrong [1970]).

Stage II: The dislocation interactions on different slip systems give rise to a rapid multiplication of the dislocations and thereby a high, and roughly constant, work hardening rate $(d\tau/d\gamma) \sim 30-50\times10^{-4}$ G. In low SFE metals, Taylor networks composed of 3D arrays of dislocation multipoles are formed. In the higher SFE metals dislocation tangles develop and often adopt a cellular pattern.

This Stage II extends up to strains of $0.05-0.2$ (or even more at low temperatures and/or low SFEs) and to first order is independent of temperature.

Stage III: Subsequently, and up to strains of order unity, the flow curve becomes parabolic as the work hardening rate decreases progressively down to values almost an order of magnitude lower than Stage II. In this stage, the dislocation multiplication processes are counterbalanced by local dislocation annihilations (dynamic recovery due to localized cross-slip, climb and/or annihilation of segments of opposite sign). These recovery mechanisms and therefore the work hardening rates are strongly temperature dependent as shown in Figure 4.2. The microstructure evolves towards a well-defined cell substructure composed of dislocation cell walls which delimit cell interiors of low dislocation density. The cell walls are initially complex tangles of dislocations but which tend to collapse into thinner, neater, boundaries during Stage III. The cells have dimensions which decrease during deformation, typically from a few microns to some tenths of microns. Simultaneously, the misorientation between adjacent cells increases from about 1 to 3 or 4° (and often higher for heterogeneous deformations).

Stage IV: At higher strains $\geqslant 1$ typical of many rolling and extrusion processes, many grains break up into bands of different orientations, separated by transition zones and grain boundaries. Figure 4.3 shows the onset of this process in cold-rolled iron where the TEM micrograph reveals a cell structure which is starting to break up by the formation of microbands more or less inclined to the rolling direction. At very high strains this develops into a lamellar structure composed of misoriented microbands parallel to RD (see, e.g. Hughes and Hansen [1993]). Note that the grain boundary area per unit volume increases significantly by large plastic deformations. All these features are characteristic of fibrous microstructures. As noted above, the work hardening rate in Stage IV is relatively low, but stays at a near-constant value ($\sim 1\times10^{-4}$ G) over large strains so that at very high strains the flow stress increase can be considerable.

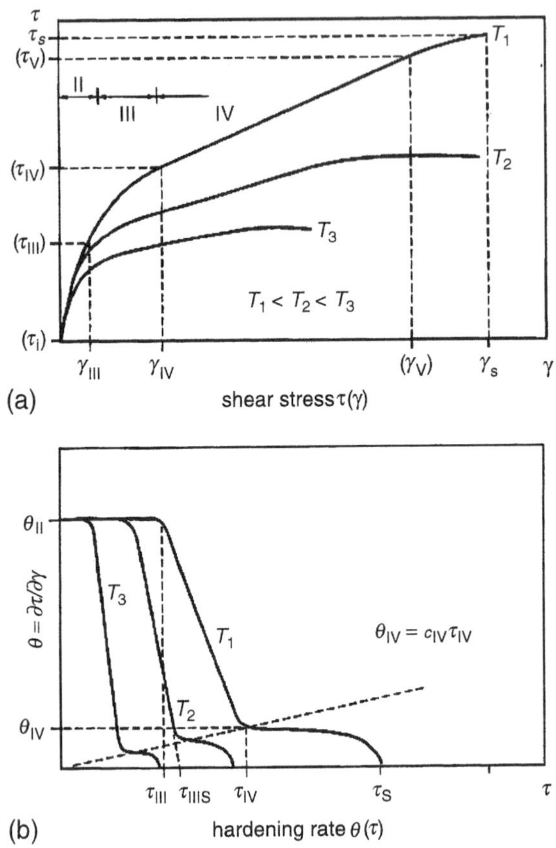

Figure 4.2. Schematic stress–strain $\tau(\gamma)$ and work hardening $(\mathrm{d}\tau/\mathrm{d}\gamma)(\tau)$ plots at three temperatures $T < 0.5T_m$, from Nes [1998]. Permission obtained from Elsevier.

The crystal defects, particularly the dislocations, generated by plastic deformation possess high elastic energies that are stored in the deformed material essentially in the stress fields around the dislocations. The energy per unit length of dislocation is

$$E_1 = \frac{Gb}{K_2} \ln\left(\frac{R}{R_0}\right) \tag{4.4}$$

with $K_2 = 4\pi$ for screws or $4\pi(1-\nu)$ for edge-type dislocations. R is the outer cut-off radius of the dislocation stress field ($\approx$ the distance between dislocations or $\rho^{-1/2}$) and R_0 is the core radius ($\approx 2b$). For typical densities of 10^{-14} to 10^{-15}/m^2,

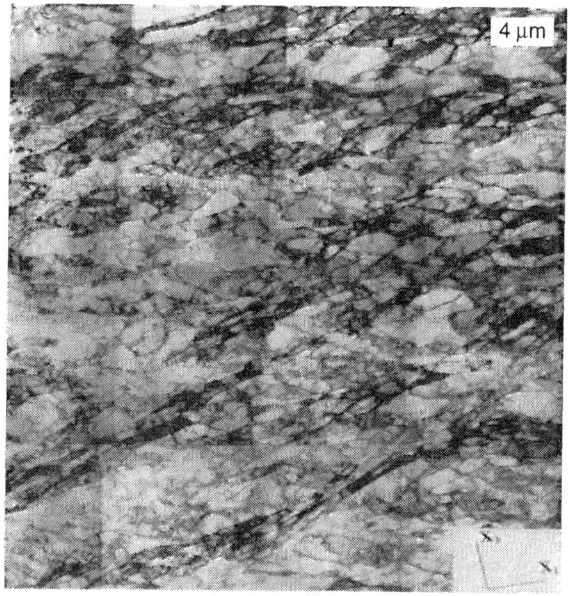

Figure 4.3. TEM micrograph of high purity Fe cold rolled 80% showing cells and microbands (D.A. Hughes and J. H. Driver, unpublished).

$E_l \approx 0.5Gb^2$ and the stored energy/unit volume $E_v \approx 0.5\rho Gb^2$. This value varies from 50 kJ/m^3 for lightly deformed aluminium to 2×10^4 kJ/m^3 for heavily deformed steel (or in molar quantities respectively, 0.5 and 200 J/mol).

4.2.1.2 Inhomogeneity of plastic deformation. Beyond the low-to-medium strain regimes plastic deformation is rarely homogeneous and tends to localize in regions of the material (often as bands), in a manner which depends strongly on its SFE, the deformation temperature and the strain path. In this short survey of the problem, we shall separate out deformation heterogeneities into two major classes: deformation banding and shear banding. Deformation bands could be described as regions within grains which deform relatively homogeneously but with slip characteristics that are different from those of adjacent regions. Shear bands are usually regions of strong localized shears, both within grains and over many grains, often extending to macroscopic dimensions. Both lead to what is known as grain subdivision but by somewhat different mechanisms.

4.2.1.2.1 Deformation banding. Over the length scale of a grain, plastic deformation is influenced by the variations in stress states that result from the interactions of dislocations with grain boundaries and other obstacles. As outlined in Chapter 10,

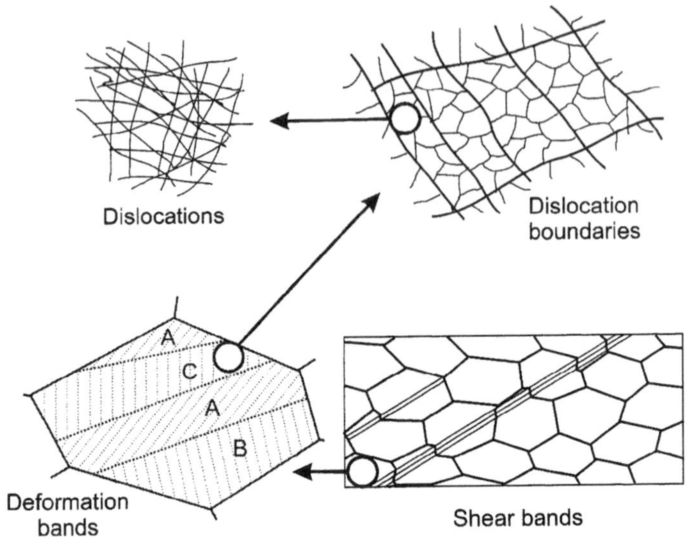

Figure 4.4. Schematic of different types of deformation heterogeneities. From Humphreys and Hatherly [2004]. Permission obtained from Elsevier.

plastically deforming grains possess different 'hardnesses' and therefore stress states, so that any one grain deforms in a quite heterogeneous environment composed of the adjacent, deforming grains. For given conditions of strain path and temperature, etc., its resulting deformation substructure is dependent on the latter as described above, but also, to a certain extent, on the grain orientation and the requirements of local strain compatibility. The local strain tensors can vary somewhat within a grain to accommodate these different properties. The slip distribution will then be heterogeneous over the grain (Figure 4.4).

As deformation proceeds this slip heterogeneity at the grain scale can be amplified, particularly during cold deformation, by the misorientations that tend to develop between the deformation bands. The grain can then be considered to be separated into domains denoted cell blocks (using the Riso nomenclature) or deformation bands. There is a tendency to restrict the terminology 'deformation bands' to the coarse bands seen in some large grains associated with divergent lattice rotations and which often start from high symmetry orientations. The adjacent bands rotate away from each other in opposite senses and are separated by micron-sized zones of high orientation gradients termed transition bands. The cell blocks, on the other hand, are separated by walls which develop misorientations at lower rates and are variously termed dislocation boundaries, geometrically necessary boundaries (GNBs) or rotation walls. More detailed descriptions of these features in relation to deformation banding are given by Kuhlmann-Wilsdorf [1999a],

Table 4.2. Summary of main deformation substructure features.

Substructure type	Cells	Cell blocks	Deformation bands
Spacing (μm)	0.5–5	2–20	10–100
Interface type	Dislocation walls	GNBs, rotation walls	Transition bands
Misorientations (°)	0.5–5	2–20	5–50

Hansen [2001] and Humphreys and Hatherly [2004]. At the risk of extreme oversimplification, Table 4.2 summarizes the essential characteristics of these deformation structures.

Individual dislocation cells and their walls are essentially transient features which are continuously created and destroyed by random dislocation interactions during the deformation (sometimes denoted incidental dislocation boundaries – IDBs). The cell block walls and transition bands are permanent features of the deformation substructure and evolve with plastic deformation. Apart from their gradual increase of misorientation they tend to become aligned with the macroscopic straining direction. During plane strain compression or rolling they may start at $25-40°$ to the rolling plane but progressively align themselves with this plane at deformations >1. At higher deformations they form long lamellar boundaries parallel to the rolling plane with misorientations typical of high angle boundaries (the well-known fibrous structure).

In contrast to shear banding, the formation and development of deformation bands is a phenomenon of plastic deformation which can explain certain aspects of the resulting microstructures but does not, *per se*, constitute a form of damage to the material.

4.2.1.2.2 Shear bands. Bands of localized shear can develop in strongly deformed metals over a wide range of scales from the macroscopic (cm) to the microscopic (micron). The extended Considère analysis given in Chapter 3 derives the general conditions for plastic instability or strain localization. When the material hardening rates decrease during plastic flow to values which respect this condition, plastic deformation becomes localized into what are usually shear bands, initially fairly wide, diffuse bands but which rapidly concentrate the deformation into intense localized shear zones leading to failure. This is classical macroscopic shear banding during tensile straining.

However, shear bands can also initiate at the grain scale and then develop to cross several grains before eventually becoming macroscopic bands. Alloys hardened by fine scale precipitation, irradiation, fine twinning or extensive dislocation accumulation are particularly prone to this type of shear banding. It occurs widely

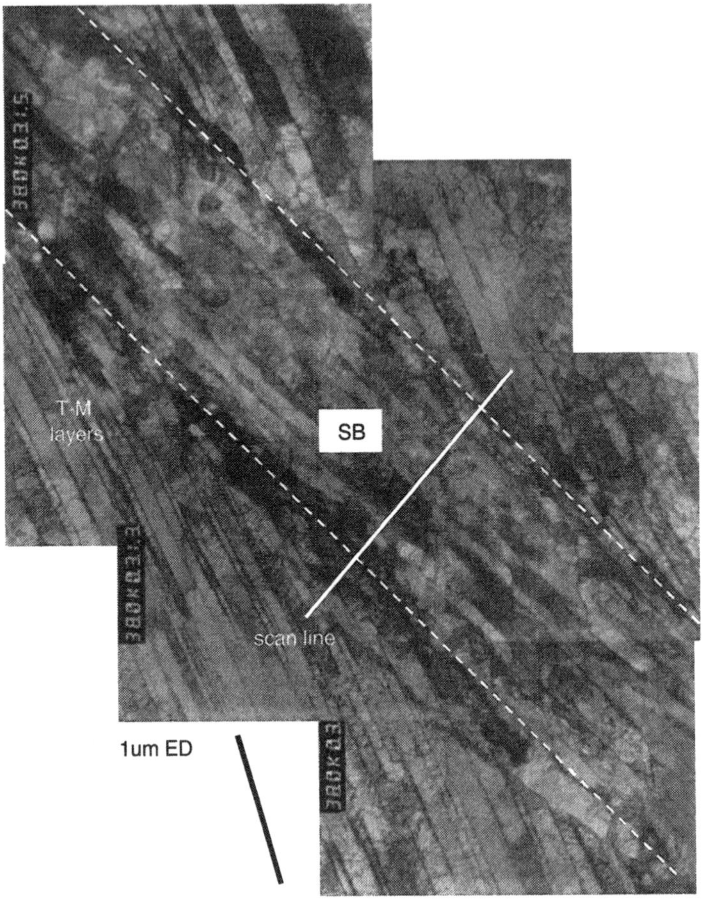

Figure 4.5. Ag crystal after plane strain compression to a strain of 1. The TEM micrograph shows a shear band (SB) superimposed on the lamellar deformation structure composed of alternating twin and matrix layers. Courtesy of H. Paul, Krakow.

during cold rolling of low SFE alloys and/or previously work hardened metals where the bands tend to be aligned about 35° to the rolling direction.

An example of a shear band in a deformed Ag single crystal (SFE = 22 mJ/m^2) is given in Figure 4.5. Here the initial plane strain compression has transformed the crystal into a structure of fine twin lamellae or alternating twin-matrix lamellae which is very sensitive to shear-induced plastic instability. Fine shear bands then develop readily and cluster into macroscopically visible bands. The fine bands in these low SFE materials are often called 'Brass-type' shear bands since they were first characterized in cold-rolled 70/30 Brass which twins heavily. Other materials which form them are austenitic stainless steels and Cu–Al alloys. The bands of width 1–3 μm accommodate high plastic strains of order 1–5 or more (Duggan *et al.* [1978]). Once they have propagated over the sample thickness they effectively become incipient cracks.

Shear bands in moderate SFE metals such as Cu, Al and Fe are also developed at rolling strains of about 1. In this case they tend to be restricted to one or a few grains without developing over the sample scale. These 'Copper-type' shear bands are mostly limited to conditions of cold rolling and are not often observed during hot deformation where dynamic recovery effects inhibit their formation (Duckham *et al.* [2001]).

4.2.2 Influence of alloying elements

The strong influence of alloying elements on the work hardening behaviour is related to the metallurgical state of the material and depends particularly on whether the elements are in solid solution or as a dispersion of 2nd phase particles.

4.2.2.1 Solute atom effects. In solid solution, many elements such as Mg and Cu in Al or C in Fe are attracted to the dislocation cores and significantly reduce their mobility thereby increasing the stress required for plastic flow. The solute hardening is expressed in terms of the increment of critical resolved shear stress $\Delta\tau_s$ compared with the pure metal flow stress. The effect is ascribed to two major types of dislocation–solute interactions, respectively atom size and shear modulus. The size effect is usually the strongest and produces an attractive force between the dislocation core and the solute atoms as a consequence of the elastic distortion of the core which can accommodate solute atoms of sizes different from those of the solvent matrix. For example, edge dislocations are characterized by expanded zones below the extra half plane (and compressed above) which will attract larger atoms (or smaller above). Rigorously, this is not the case of undissociated screw dislocations but is true of dissociated screws, which behave like two edge dipoles, so the same reasoning can be generally applied. The size interaction energy between a dislocation and an immobile solute atom can be estimated with the assumption of spherical distortions which is valid for substitutional solute atoms, Table 4.3 gives some typical values together with those due to modulus variations. The latter becomes significant when the solvent and solute atom sizes are similar. In this case, the variation of the shear modulus of the matrix with solute concentration gives rise to an additional interaction energy as a result of the dependence of flow stress on shear modulus.

Table 4.3. Some examples of dislocation–solute interaction energies.

Dislocation type	Solute interaction	Cu–Ag/eV	Cu–Sb/eV	Al–Si/eV
Edge	Size	0.37	0.85	0.09
	Modulus	0.04	0.06	0.14
Screw	Size	0.02	0.04	0.10
	Modulus	0.11	0.25	0.03

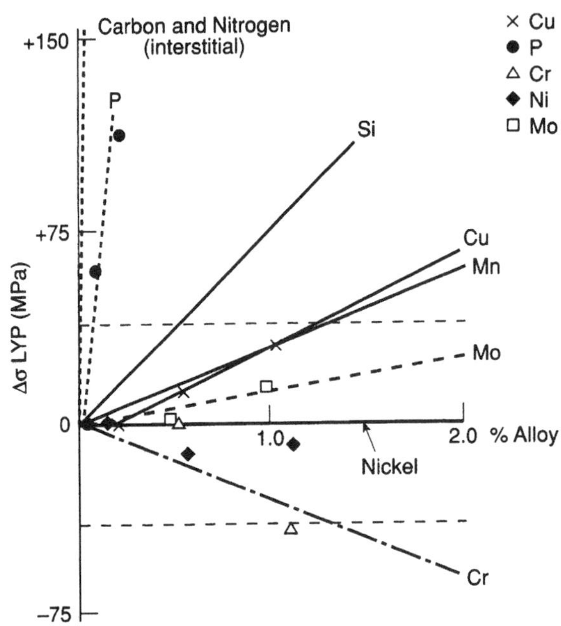

Figure 4.6. Solute hardening in ferritic iron from Gladman [1997]. Permission obtained from Maney.

A particular form of the size effect is due to the tetragonal distortion of the lattice by interstitial atoms in cubic structures, e.g. carbon in bcc iron. Here the carbon atoms occupy the octahedral interstitial sites and induce elastic displacements which are larger along the directions of closest distance to the Fe atoms than along the other two orthogonal directions. The resulting interaction energy can be very large (see Figure 4.6).

The overall effect of solute hardening has been estimated for these different situations with a variety of dislocation – solute atom interaction models. Most of them predict a (low temperature $T < 0.5T_m$) hardening which varies with the square root of the solute concentration, e.g. for spherical distortions Fleischer [1963] gives

$$\Delta\tau_s = \left(\frac{G\varepsilon_s^{\ 3/2}}{700}\right)\sqrt{C} \tag{4.5}$$

where the parameter ε_s, of order unity, is a function of the fractional change of lattice parameter and shear modulus with concentration C (expressed as atomic fraction). Solute atoms creating tetragonal distortions give strengthening effects about an order of magnitude greater than predicted by Eq. (4.5).

After initial yielding and during subsequent plastic deformation, the solute atoms tend to confine the dislocations to slip planes, reduce their capacity to recover dynamically by local climb and cross-slip, inhibit the formation of 'clean' cell structures and thereby increase the dislocation density for a given deformation. In general, Stage II work hardening is prolonged and Stages III and IV retarded, so the alloy work hardens extensively (see the example of Al–Mg, Figure 3.3). As shown by Doherty and Baumann [1993a] and Doherty and McBride [1993b], the solute-enhanced strain hardening in Al alloys essentially increases the k parameter in the Hollomon-type power law $\sigma = k\varepsilon^{n'}$ without affecting the work hardening exponent; the strain hardening becomes more sensitive to the amount of plastic strain.

At moderate temperatures, which allow local diffusion processes, the solute atoms can begin to move and can therefore 'catch up' with slowly moving dislocations so that very strong interactions are possible (typically between RT and ~300°C). The reduced dislocation mobility by solute segregation tends to favour heterogeneous deformation by Lüders bands and Portevin LeChatelier(PLC) serrations in the stress–strain curves (Figure 4.7).

4.2.2.2 Second phase particles. Alloying elements in the form of 2nd phase particles almost always harden the material by requiring the dislocations to expend additional energy either by cutting the particle (fine particles of radius $r_p \leqslant 10$ nm) or by looping around them (r_p typically $\geqslant 20$ nm). The effect of these processes on the yield strength is well documented since they control, in part, the final properties (Figure 4.8).

As is well known, dislocation cutting of fine particles requires a critical shear stress which increases with the particle size. The exact relation between $\Delta\tau_p$ and r_p depends upon the details of the cutting mechanism. There are in fact several basic mechanisms for hardening crystals by a dispersion of fine particles through which dislocations can pass. Excellent descriptions of these mechanisms can be found in the books by Martin [1980] and Courtney [2000]. All of them give rise

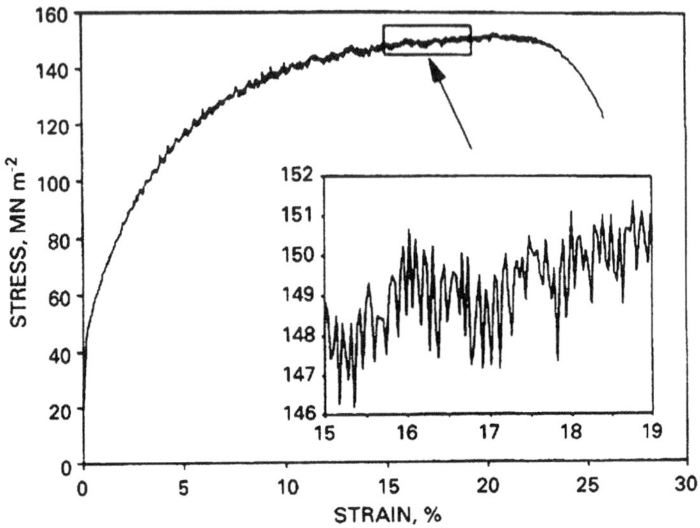

Figure 4.7. PLC serrations in Al–Mg from Robinson [1994]. Permission obtained from Maney.

to variations in dislocation energies as they interact with and then cut through the particles. The parameters which are considered to influence these processes are the particle coherency, its shear modulus, the particle–matrix interface energy and the degree of order. Each of these parameters give rise to a specific hardening effect but which is dependent upon the particle size (r_p) and volume fraction f in the following form:

$$\Delta\tau_p = k_3(G\varepsilon_p^{3/2})\sqrt{(fr/b)} \tag{4.6}$$

where ε_p is a strengthening factor function of the particular hardening mechanism. This expression has a similar form to that of solute hardening but the proportionality constants k_3 and ε_p are significantly greater in the case of particle hardening.

Clearly, during an ageing treatment to precipitate out 2nd phase particles from supersaturated solid solutions, the material hardens as f and r increase until the equilibrium fraction f is attained. The particle size then continues to increase by coarsening and then so does the interparticle spacing $L = r\sqrt{(\pi/f)}$ for spheres. However, at a critical interparticle spacing, the dislocation may begin to loop around the particles at a critical stress given by the Orowan criterion:

$$\Delta\tau_p = \frac{Gb}{L-2r} \tag{4.7}$$

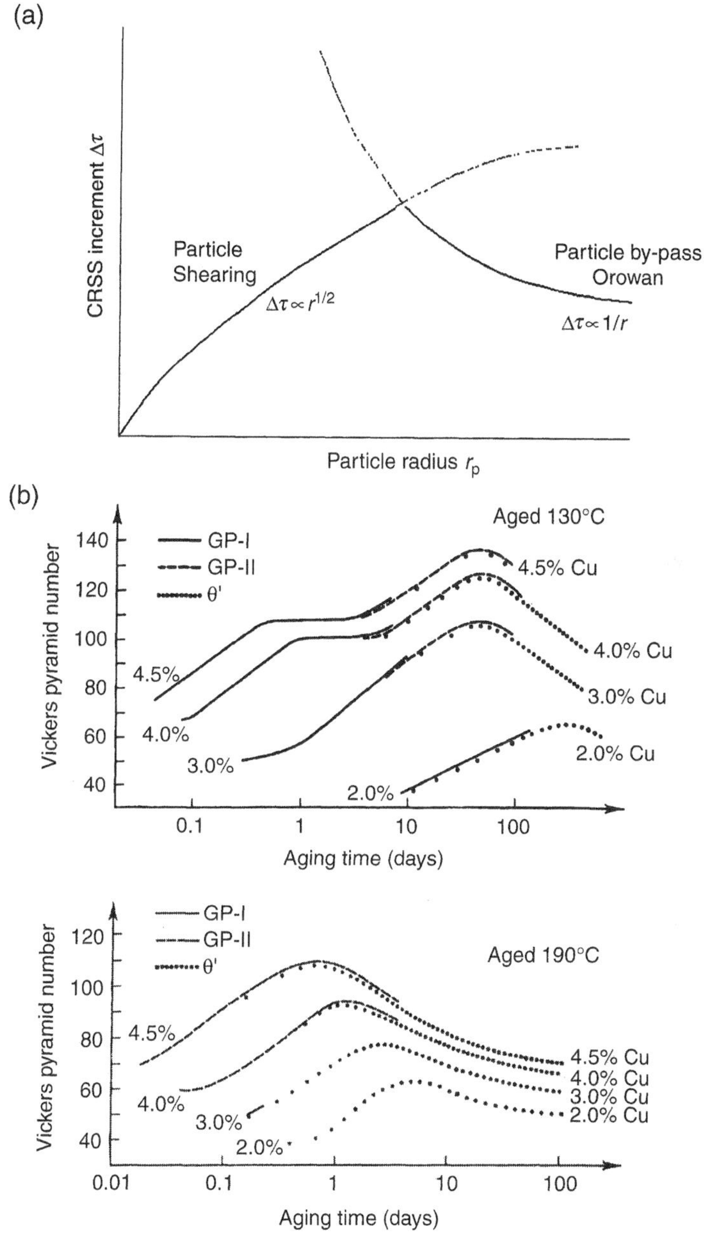

Figure 4.8. (a) Schematic variation of flow stress with particle size, (b) example of hardness evolution during ageing of precipitation hardened Al–Cu from Silcock [1953]. Permission obtained from Maney.

The amount of hardening will then decrease as the interparticle spacing increases on overageing. The combination of both the cutting and looping mechanisms means that the particle strengthening goes through a maximum – the peak hardness – as a function of particle radius (or of time during an ageing treatment) as shown in Figure 4.8.

The effect of 2nd phase particles on work hardening, i.e. after yield, depends critically on the particle size but in a rather different way. The work hardening rate tends to be rather low for very fine particles (including the peak hardness distributions) since the dislocations, once they have cut through one particle, can cut through entire fields of particles at roughly the same stress. This leads to the formation of shear bands in which plastic deformation is heavily concentrated, e.g. certain Al–Cu alloys. On the other hand, if the particles have dimensions of $\geqslant 1$ μm the dislocations loop around them and on continued straining build up high local dislocation densities near the particles and therefore relatively high work hardening rates. In fact, a considerable fraction of the dislocation population is required to accommodate the difference in plastic strain between the hard particles and the soft matrix; they are termed GNDs (Ashby [1970]). The accumulation of dislocations in these particle deformation zones creates high local stresses if the particles are sufficiently large to withstand them ($\geqslant 1$ μm). This process creates a strongly dislocated substructure with a dimension characteristic of the interparticle distance and is particularly pronounced if the particles are non-spherical, e.g. discs (Al–Cu) or fibres (perlitic steels).

4.2.3 Microscopic hardening laws

The macroscopic hardening laws of Chapter 3 are essentially empirical descriptions of some parts of the flow curves; they are incapable of describing a complete through-process material behaviour during rolling schedules. It is now recognized that microscopic-based models of work hardening using internal variables such as the dislocation density are required for accurate modelling of plastic flow. The particular microscopic laws are often complex so the relatively simple analysis given below in terms of the total dislocation density can be considered as an illustration of the principles.

The microscopic work hardening rate $(\mathrm{d}\tau_c/\mathrm{d}\gamma)$ can be written in terms of the variation of the dislocation density as

$$\frac{\mathrm{d}\tau}{\mathrm{d}\gamma} = \frac{\mathrm{d}\tau}{\mathrm{d}\rho}\frac{\mathrm{d}\rho}{\mathrm{d}\gamma} \qquad (4.8)$$

Using the standard $\tau_c(\rho)$ relation, Eq. (4.1), the variation of the critical shear stress with dislocation density is

$$\frac{d\tau}{d\rho} = \frac{\alpha G b}{2\sqrt{\rho}} \tag{4.9}$$

According to the analysis of Kocks [1976] and Mecking and Kocks [1981], the rate of creation of dislocations ($d\rho^+$) during a small strain increment $d\gamma$ is inversely proportional to the mean dislocation slip distance λ so that $\left(d\rho^+/d\gamma\right) = \left(1/b\lambda\right)$ and the mean slip distance $\lambda = \left(C_1/\sqrt{\rho}\right)$, where C_1 is the average number of obstacles that the dislocation meets before stopping.

$$\frac{d\rho^+}{d\gamma} = \frac{\sqrt{\rho}}{C_1 b} \tag{4.10}$$

It can be noted that this relation between the rate of creation of dislocations and $\sqrt{\rho}$ is not accepted by all authors, some of whom prefer a constant rate of dislocation creation or some other function (Table 4.4).

During Stage II hardening, where dislocation annihilation is small and can be neglected, the above model gives a constant hardening rate:

$$\frac{d\tau^{II}}{d\gamma} = \frac{\alpha G b}{2\sqrt{\rho}}\frac{\sqrt{\rho}}{C_1 b} = \frac{\alpha G}{2C_1} \tag{4.11}$$

Table 4.4. Some one-parameter work hardening laws.

Authors	Evolution law	Macroscopic hardening law	Flow stress
Kocks [1976]	$\frac{d\rho}{d\varepsilon} = h\sqrt{\rho} - r\sigma$	$\theta = \frac{d\sigma}{d\varepsilon} = \theta_0\left(1 - \frac{\sigma}{\sigma_S}\right)$	$\sigma = \sigma_S - (\sigma_S - \sigma_e)\exp(-\alpha\varepsilon)$
Estrin and Mecking [1984], Laasraoui and Jonas [1991]	$\frac{d\rho}{d\varepsilon} = h - r\rho$	$\theta = \frac{d\sigma}{d\varepsilon} = \left(\frac{A}{\sigma} - B\sigma\right)$	$\sigma = \sqrt{\sigma_S^2 - (\sigma_S^2 - \sigma_e^2)\exp(-\alpha\varepsilon)}$
Roberts [1982]	$\frac{d\rho}{d\varepsilon} = h - r\sqrt{\rho}$	$\theta = \frac{d\sigma}{d\varepsilon} = P\left(\frac{\sigma_S}{\sigma} - 1\right)$	$\sigma_S \ln\left(\frac{\sigma_S}{\sigma_S - \sigma}\right) - \sigma = P\varepsilon$
Lagneborg *et al.* [1993], Stüwe [1965]	$\frac{d\rho}{d\varepsilon} = h - r\rho^2$		$\sigma = \sigma_S\sqrt{\tanh\left(\alpha(\varepsilon + \varepsilon_0)\right)}$

As an example, if the experimental value of this rate is 5×10^{-3} G, C_1 is about 50 μm^{-1} and assuming $\rho\approx10^{14}$/m^2, the average slip length would be about 5 μm.

During Stage III the hardening rate decreases continuously as some of the dislocations are annihilated by dynamic recovery at a rate written $(\mathrm{d}\rho^-/\mathrm{d}\gamma)$. The exact mechanisms of annihilation (cross-slip, climb) are the subject of current research but one can write that the rate of annihilation is proportional to the current density ρ and a probability of elimination $P(T)$, which is strongly temperature dependent:

$$\frac{\mathrm{d}\rho^{\mathrm{III}}}{\mathrm{d}\gamma}=\frac{\mathrm{d}\rho^+}{\mathrm{d}\gamma}-\frac{\mathrm{d}\rho^-}{\mathrm{d}\gamma}=\frac{\sqrt{\rho}}{bC_1}-C_2P(T)\rho \tag{4.12}$$

Consequently, Stage III work hardening rate becomes linear in τ_c:

$$\frac{\mathrm{d}\tau_c^{\ \mathrm{III}}}{\mathrm{d}\gamma}=\frac{\alpha G}{2C_1}-\frac{C_2P(T)}{2}\tau_c \tag{4.13}$$

This law can be expressed in the following form for the macroscopic stress:

$$\frac{\mathrm{d}\sigma^{\mathrm{III}}}{\mathrm{d}\varepsilon}=\theta_0\left(1-\frac{\sigma}{\sigma_s}\right)$$

which is the hardening rate in the Voce law (Eq. (3.15)).

At large strains and at low temperatures dislocation annihilation is insufficient to completely balance the rate of creation. This results in the low, but non-zero, work hardening rate of Stage IV; there is no general agreement on the basic physical causes of this stage.

The detailed descriptions of work hardening are controversial, and the reader is referred to some recent extended treatises on the subject by Nes [1998], Kuhlmann-Wilsdorf [1999b] and Kocks and Mecking [2003].

A more general method of describing work hardening is to write down the constitutive law as a function of the internal variables S_i: $\sigma=f\left(T,\dot{\varepsilon},S_1\cdots S_n\right)$, together with a set of evolution laws for each of the internal variables S_i: $\left(\partial S_i/\partial\varepsilon\right)=g_i(T,\dot{\varepsilon},S_{j\neq i})$. The analysis described above assumes only 1 internal variable, the total dislocation density. Other evolution laws can be used and one is not necessarily limited to a single internal variable. For example, several recent two- or three-parameter models take account of the different dislocation distributions (cell walls, cell interiors, etc.). The added complexity usually enables one to better define transition behaviour (varying T, $\dot{\varepsilon}$, etc.), see for example Mughrabi [1987] and Nes [1998] for two- and three-parameter models, respectively.

To give some idea of these laws, Table 4.4 summarizes the essential features of a set of commonly used one-parameter models.

4.3. HOT DEFORMATION

4.3.1 Flow stresses

At homologous temperatures above $0.4T_m$, plastic deformation is strongly influenced by thermally activated processes so that the flow stress becomes temperature and strain rate dependent (viscoplastic). The processes involved are mostly controlled by local atomic diffusion and give rise to strong dynamic recovery of the dislocation substructure as the dislocations are continuously annihilated during deformation by climb and cross-slip mechanisms. This strongly reduces the flow stresses which then tend to saturate at values function of T, $\dot{\varepsilon}$ and alloy content. It is generally recognized that the T and $\dot{\varepsilon}$ terms can be regrouped into a temperature-compensated strain rate known as the Zener–Hollomon parameter $Z = \dot{\varepsilon}\exp(Q/RT)$, where R is the gas constant and Q an apparent activation energy for plastic flow. Q takes values in the range 150–200 kJ/mol for aluminium alloys, 280 kJ/mol for ferritic steels, 300 kJ/mol for austenite and 400–500 kJ/mol for austenitic stainless steels (McQueen [2002]). When atoms are added to the alloy, the resistance to dislocation movement will increase and so does the value of Q (Ryan and McQueen [1987], Verlinden and Voith [1994]).

For the austenitic stainless steels, and also for many other low-or-medium SFE fcc metals, the rate of dynamic recovery is limited even at elevated temperatures (the standard mechanism of annihilation by cross-slip is hampered by the inability of dissociated dislocations to avoid local obstacles by cross-slipping out of their slip plane). The flow stress increases significantly and the dislocation densities attain values where recrystallization can occur during hot plastic deformation. The resulting dynamic recrystallization is described in Chapter 6.

At low Z values (typically $\ln Z \leq 26$ for Al), plastic flow is purely viscous after an initial transition strain of about 0.1; at a given imposed strain rate the material deforms at constant flow stress. This also corresponds to the creep regime where constant imposed stresses give rise to a continuous shape change, i.e. a roughly constant strain rate.

At higher Z ($26 \leq \ln Z \leq 50$) the flow stress also depends on the applied strain, i.e. $\sigma = f(T, \dot{\varepsilon}, \varepsilon)$. Note that in typical aluminium hot rolling schedules ln Z varies from about 22 in the first passes to about 33 near the end. Steel hot rolling schedules are similar (e.g. $24 < \ln Z < 37$ for carbon steels rolled in the austenite regime).

In the low Z regime many studies have shown that the saturation flow stress σ_s can be related to the Zener–Hollomon parameter by a power law:

$$\sigma_s = AZ^m \tag{4.14}$$

where m is the strain rate sensitivity $(\partial \ln \dot{\varepsilon}/\partial \ln \dot{\varepsilon})$, typically 0.1–0.3.

Blum [1991] has shown that the saturation flow stresses of different materials can be correlated by normalizing with respect to the self-diffusion coefficient D and the shear modulus G (Figure 4.9):

$$\frac{\sigma_s}{G} = \left(\frac{\dot{\varepsilon} k_B T}{DGb}\right)^m \tag{4.15}$$

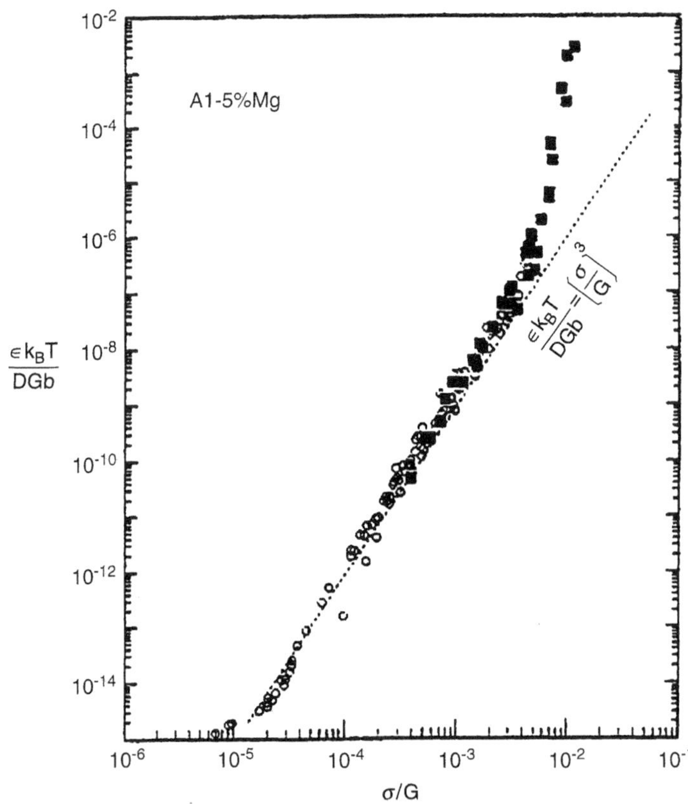

Figure 4.9. A plot of Eq. (4.15) for Al–Mg from McQueen *et al.* [1994]. Permission obtained from TMS.

This relation is then general to a wide range of materials for both creep and hot working in the low Z regime.

In the higher Z regime, typical of most hot rolling schedules, the relations $\sigma(T,\varepsilon,\dot{\varepsilon})$ are more complex and often represented by a hyperbolic sine law due to Sellars and Tegart [1966]:

$$Z = A_2(\sinh \omega\sigma)^{n'} \quad \text{or} \quad \dot{\varepsilon} = A_2(\sinh\omega\sigma)^{n'} \exp\left(-\frac{Q}{RT}\right) \quad \text{or} \quad \sigma = \text{arcsinh}[(Z/A)^{1/n'}]/\omega \tag{4.16}$$

where A_2, ω and n' are parameters to be evaluated experimentally. In a recent review, McQueen [2002] gives ω values of about 0.052 (MPa^{-1}) for aluminium alloys, 0.014 for ferritic and C steels and 0.012 for austenitic stainless steels.

At low Z values the sinh law is similar to the above power law. At high Z the sinh law can be reasonably well approximated by an exponential function:

$$Z \cong A_3 \exp[\beta\sigma(\varepsilon)] \tag{4.17}$$

where $\sigma(\varepsilon)$ is the flow stress at a given value of the applied strain and $\beta \approx \omega n'$.

To describe an entire work hardening curve in this regime there are two possible approaches:

(i) A purely empirical approach, using a variant of the Voce law where the stresses are functions of Z, e.g. the law used recently by Shi *et al.* [1997a]:

$$\sigma = \sigma_e + (\sigma_s - \sigma_e)\left[1 - \exp\left(\frac{-\varepsilon}{\varepsilon_r}\right)\right]^{m'} \tag{4.18}$$

ε_r is a transition deformation and the exponent $m' \sim 0.5$ for aluminium. An example of this type of representation for the stress–strain curves of an Al–1%Mn alloy at different T and $\dot{\varepsilon}$ is given in Figure 4.10.

(ii) A modified Kocks–Mecking approach which allows for the effect of different microscopic variables on the rates of accumulation and annihilation of the dislocation populations (usually more than one). The reader is referred to Mughrabi [1987], Nes [1998] and the reviews of Gil Sevillano [1993] and Argon [1996].

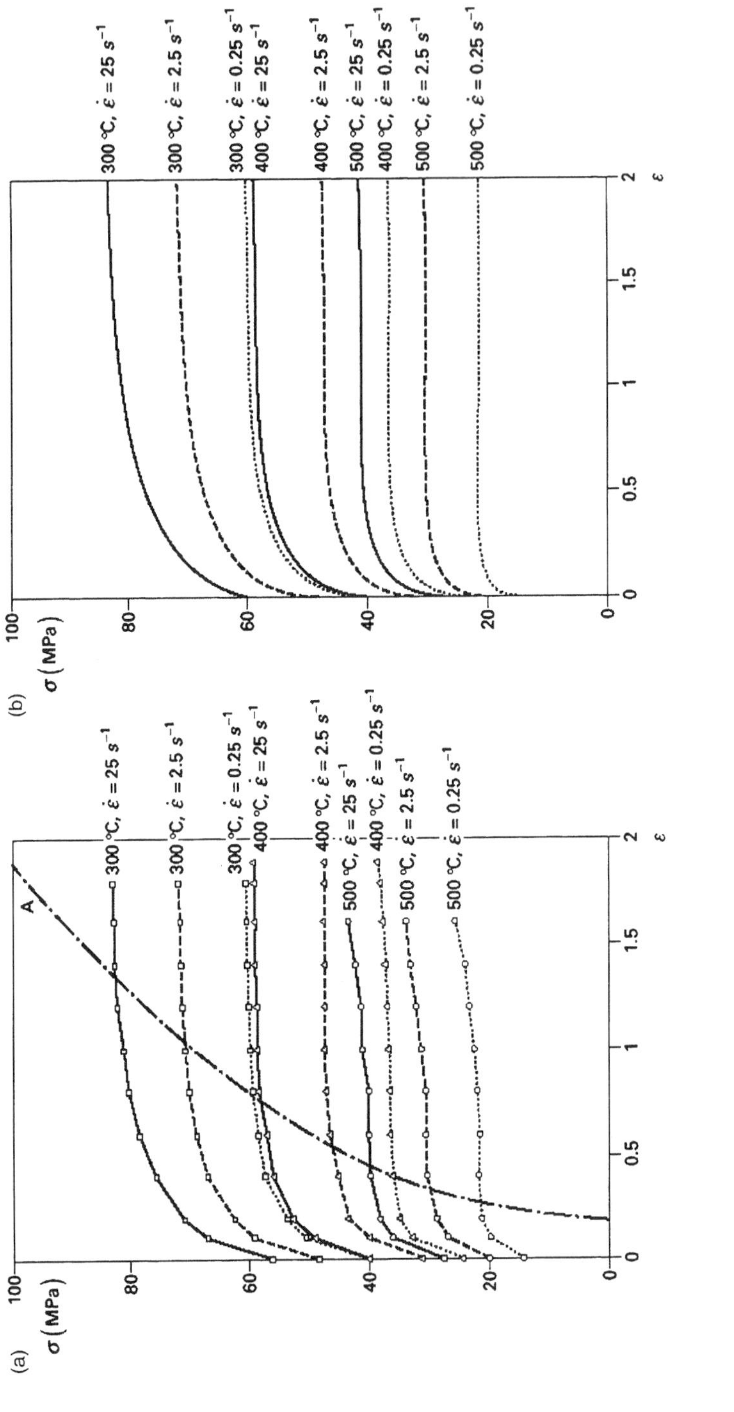

Figure 4.10. Example of high-temperature stress–strain curves of Al–1% Mn from Shi *et al.* [1997a], (a) measured, (b) plotted from equation (4.18). Permission obtained from Maney.

4.3.2 Hot deformation microstructures

The hot deformation microstructures of most metals and alloys are essentially composed of dislocation subgrains within each of which there is a relatively low density of free dislocations. Subgrains are made up of relatively clean dislocation walls (i.e. mostly, but not entirely, of edge dislocations of the same sign which accommodate the misorientation across them). Dislocations can be rapidly annihilated by the interactions of segments of opposite sign through cross-slip and local climb, etc. These processes of dynamic recovery limit the accumulation of free dislocations, so the flow stress saturates rapidly at a level controlled by the subgrain size (δ) and the dislocation contents of the subgrain walls and interiors. These microstructural parameters depend essentially on the T, $\dot{\varepsilon}$ and solute atom–dislocation interactions. Figure 4.11 illustrates subgrains developed in hot plane strain compressed single crystal samples of Al–1%Mn at different temperatures and strain rates.

During plastic flow the subgrain boundaries (unlike the grain boundaries) are continually annihilated by growth processes and replaced by new ones via dynamic polygonization of the free dislocations, so that their average size d and equiaxed shape remain constant over very large strains. The ability of dislocation subgrains to rearrange under stress to maintain a constant deformation structure (at constant Z) is a critical feature of the dynamic recovery behaviour of high SFE metals during hot deformation. This is the fundamental cause of the relatively simple relations between flow stress, d and Z. For example, to first order, the subgrain size varies inversely with log Z as first demonstrated by McQueen *et al.* [1967].

A frequently used relation between subgrain size and flow stress, confirmed several times, is of the form

$$\sigma = \sigma_0 + \frac{k_4 Gb}{d^c} \tag{4.19}$$

where the constant $k_4 \sim 28$ for aluminium and the exponent c is between 0.75 and 1.5 according to the alloy and deformation conditions. To a reasonable approximation and for many metallic materials the exponent c can often be taken as 1 and the constant k_4 as 23 (Raj and Pharr [1986]). This then implies that the flow stress varies with log Z as in Eq. (4.17) for high Z.

A more detailed version of the above is due to Nes [1998] and allows for the influence of free dislocations within the cells ρ_i, i.e.

$$\sigma = \sigma_0 + \alpha_1 Gb\sqrt{\rho_i} + \alpha_2 \frac{Gb}{d} \tag{4.20}$$

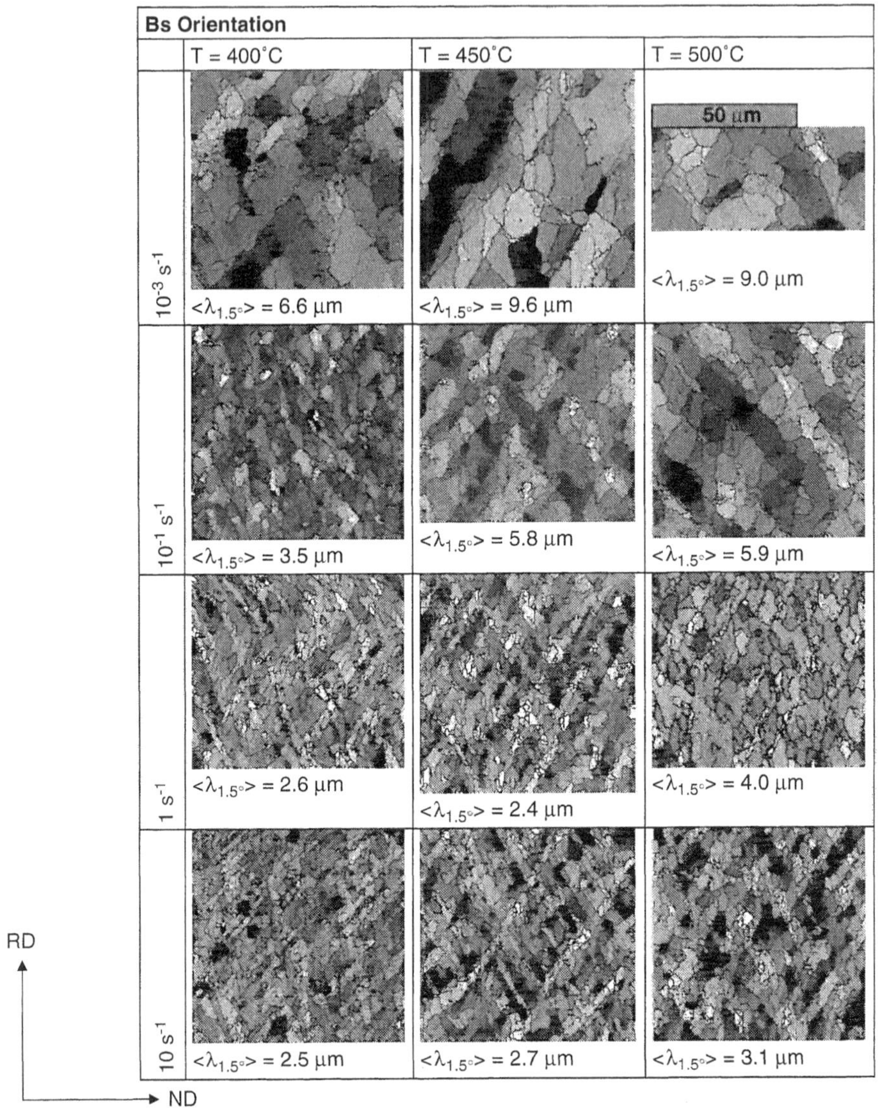

Figure 4.11. Subgrain structures in hot plane strain compressed Al–1%Mn crystals over a range of temperatures and strain rates from Glez and Driver [2003]. The subgrain size is determined using a cutoff misorientation angle λ of 1.5°. Permission obtained from Elsevier.

In many cases the dislocation density within the subgrains, ρ_i, and the average subgrain size, d, have been found to be linked through the relation $\sqrt{\rho_i} = (C/d)$, where C is an alloy-dependent constant of the order of 2.

The average misorientation of the subgrain boundaries in high SFE materials tends to saturate at strains about unity (see, e.g. McQueen and Kassner [2004]) but may continue to increase for deformation modes involving complex strain paths. The subgrain behaviour of the low SFE alloys is often heterogeneous; in regions near the grain boundaries, where localized strain associated with multiple slip occurs, well-defined subgrains develop more readily than in the grain centres. Consequently, a structure of subgrains surrounding the original grain boundaries and encircling more uniform dislocation populations is frequently observed.

The subgrains in contact with the grain boundaries create serrations along the boundary by interfacial tension effects. This scalloping leads to a higher density of high angle boundaries and eventually, during very large plastic strains typical of many hot working processes, favours dynamic geometric recrystallization (Chapter 5). This occurs when the serration amplitudes (or the subgrain size) are about half the grain thickness as a consequence of the very large shape change of the original grains.

An important mechanism controlling the subgrain structure evolution and hence flow stresses during hot deformation of most commercial alloys is the interaction between moving dislocations and solute atoms, or solute drag. Second phase particles do not have the same strong effect as at room temperature since dislocations can rapidly climb around them (and often, because of their higher solubility, the particle volume fraction is reduced). At high temperatures solute atoms diffuse with the dislocations, either completely at low stresses or partially at high stress and thereby control the rate of dynamic recovery of the dislocation substructure. This has been mostly studied in creep but the case of hot working has been treated recently by Nes [1998]. In many cases, during deformation at high-temperature solute hardening is the major hardening mechanism.

LITERATURE

Krauss G. (ed.), "Deformation Processing and Structure", ASM, St. Louis (1984).

Chapter 5
Softening Mechanisms

Chapter 5
Softening Mechanisms

5.1. INTRODUCTION

The deformation structures developed during thermo-mechanical processing (TMP) are intrinsically unstable so that on annealing, after deformation processing, substructure evolution (excluding classical phase transformations) often occurs by thermally activated processes, leading to a reduction of stored energy. These processes usually induce a significant softening of the plastically deformed material as shown in Figure 5.1.

At low temperatures (in commercial alloys, typically between 0.4–0.5 of the melting temperatures, T_m), recovery dominates and often leads to a slow, logarithmic, decrease of the hardness. At high temperatures (more than $0.7T_m$), recrystallization can occur very rapidly, without much prior recovery. At intermediate temperatures, both mechanisms can contribute significantly to softening.[1]

The associated microstructural changes are defined based on the driving force and the mechanism(s) involved. As shown schematically in Table 5.1, there are two possible driving forces – (i) the stored energy of deformation and (ii) the surface energy or the grain boundary energy. During plastic deformation of crystalline materials, part of the plastic work, typically 1–10%, is stored as microstructural energy, mostly as increased dislocation density; the rest is dissipated as heat. The softening processes are usually separated into recovery, recrystallization and eventually grain growth. The latter two necessarily imply the movement of high-angle boundaries, but recovery involves a set of micromechanisms for the motion and annihilation of point defects and dislocations. It is the combination of driving force and mechanism that differentiates these three 'softening' processes:

- recovery (stored energy and movement of dislocations, either individually or collectively as low-angle boundaries),
- recrystallization (stored energy and movement of high-angle boundaries) and
- grain growth (surface energy and movement of high-angle boundaries).

Given its importance for TMP, the subject has been of major interest for over a 100 years. Table 5.2 provides an overview of some important 'recorded' events of its

[1] In very pure metals, recrystallization can occur at much lower temperatures. Pure Cu, for example, may recrystallize at room temperature, as is often observed in microelectronic devices.

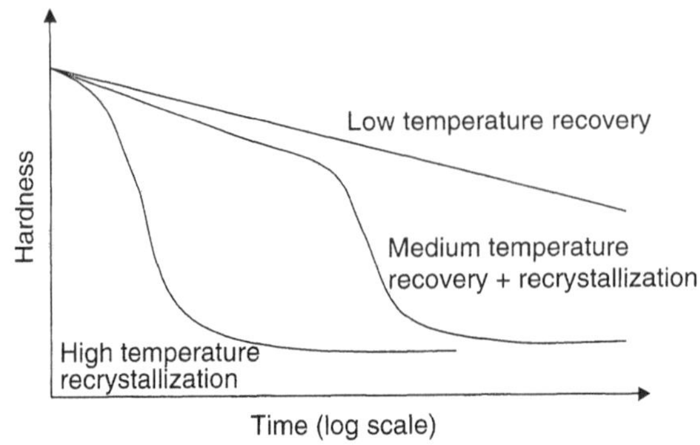

Figure 5.1. Schematic softening kinetics at three temperatures for different combinations of recovery and recrystallization.

Table 5.1. Defining recovery, recrystallization and grain growth, based on the combination of driving force and mechanism.

Mechanism (kinetics)	Driving force (energies)			
	Point defects	Dislocations	Subgrain walls	Grain boundaries
Point defect diffusion	Low temperature			
Dislocation climb		**Recovery**		
Sub-boundary migration and coalascence			High temperature	
Grain boundary migration		**Recrystallization**		**Grain growth**

early history. The field has been well covered by several books and review articles (see, e.g. the literature list). The purpose of this chapter is to provide a 'minimum' knowledge base to students and researchers coming from diverse backgrounds.

5.2. RECOVERY

Softening by recovery can occur if a non-equilibrium concentration of lattice defects is 'reduced', usually by annealing at an appropriate temperature. The defects can be both, point and line defects; but the latter, i.e. dislocations, are more relevant to TMP. Point defects are usually annealed out at relatively low temperatures ($<0.3T_m$),

Table 5.2. A few important 'recorded' events of early history, which shaped the present understanding of softening mechanisms. A more extensive list or description can be found in Beck [1963], Hu *et al.* [1990] and Humphreys and Hatherly [1995].

Recorded events of early history	Details
Evidence of structural changes during deformation and subsequent annealing by acoustic anisotropy	Savart [1829]: A simple tuning fork was used for hearing tones, after casting, deformation and annealing of Pb, Sn, Zn and Cu
Change in grain structure during deformation and after annealing by visual inspection	Percy [1864] and Kalischer [1881]: The change in grain structure during annealing was understood as 'crystallization' from the amorphous state
Identification of deformed structure of 'elongated' grains and creation of new grains during annealing – the first recorded use of the term 'recrystallization'	Sorby [1887]: Sorby, using metallographic technique developed by himself, observed 'elongated' grains in a hammered Fe bar and recognized the deformed structure as a 'state of unstable equilibrium', instability resulting in subsequent 'recrystallization'
'Amorphization' of cold-worked material was discounted	Ewing and Rosenhain [1900]: Argued for the 'continuity' of crystalline structure during cold deformation – slip and twinning being the mechanisms of plastic deformation
Separation of the driving force for recrystallization and grain growth	Carpenter and Elam [1920]: Carpenter and Elam established that grain growth is through boundary migration and not through coalescence
Identification of the mechanism for grain boundary movement	Altherthum [1922]: Altherthum distinguished recrystallization and grain growth as 'cold-worked recrystallization' and 'surface tension recrystallization', respectively

i.e. at temperatures below most TMP processes and can usually be ignored except for very low-temperature deformations. Recovery by dislocation annihilation often involves a combination of several micromechanisms (see Table 5.1). The ratio of recovery to recrystallization, however, depends on several factors – strain, annealing temperature and material. The effect of temperature is illustrated schematically in Figure 5.1. Recovery dominates the low-temperature regime, while recrystallization usually occurs rapidly at higher temperatures. Even at higher temperatures, however, there is always some recovery before recrystallization, recovery kinetics being faster than that of low temperatures.

For single phase materials, the stacking fault energy (SFE) also has a strong influence on the amount of recovery; in deformed high-SFE metals, such as Al and body-centred cube (bcc) iron, dislocation cross slip and local annihilations are sufficiently easy to favour significant amounts of recovery. A classical example is the recovery of cold-drawn Al beverage cans during curing of the varnish at 150–200°C (Section 14.1). On the other hand, the low-SFE metals and alloys, e.g. rolled austenitic stainless steels and α-brass, do not undergo much recovery before recrystallization. This is also generally true for alloys with high solute contents, which reduce the dislocation mobility.

Though recovery is an important issue, it is often difficult to quantify. It can occur immediately after deformation, and also dynamically during deformation. Furthermore, recovery does *not* affect the optical microstructure or the crystallographic texture. Recovery affects properties such as hardness and structural features like, dislocation density, subgrain size and misorientation – but the resolution of such property changes is often poor and statistically quantifying the changes in the affected structural features is difficult.

5.2.1 Recovery mechanisms

The various mechanisms of recovery include (in order of increasing difficulty):

- point defect (vacancies and interstitials) annihilation by diffusion to sinks such as dislocations,
- mutual dislocation annihilation (closely spaced dislocations of opposite sign, or dipoles, which require small amounts of dislocation climb and/or cross slip),
- organization of free, random dislocations into dislocation walls or sub-boundaries (polygonization) (Cahn [1949], Kuhlmann-Wilsdorf [1989]) and
- coalescence of sub-boundary walls during subgrain growth.

The latter three mechanisms are illustrated schematically in Figure 5.2. Figure 5.3 illustrates an example of stored energy reduction in rolled high-purity iron by differential scanning calorimetry (DSC) (calorimetric) measurements at a constant heating rate. The diffuse recovery reaction extends over a wide temperature range before the sharper recrystallization peak.

Recovery mechanisms often operate simultaneously, so that there is no clear demarcation between them. For example, it is impossible to distinguish the collapse of diffuse dislocation cells into well-defined subgrain boundaries and subgrain growth.[2] This is one reason why recovery is difficult to treat analytically. Also, the mechanisms of the later stage of recovery, e.g. the formation of well-defined subgrains, are often the first stages of recrystallization nucleation which

[2] Owing to the heterogeneity of dislocation structures in deformed polycrystals, different recovery mechanisms may occur at different rates between different grains or between different parts of the same grain.

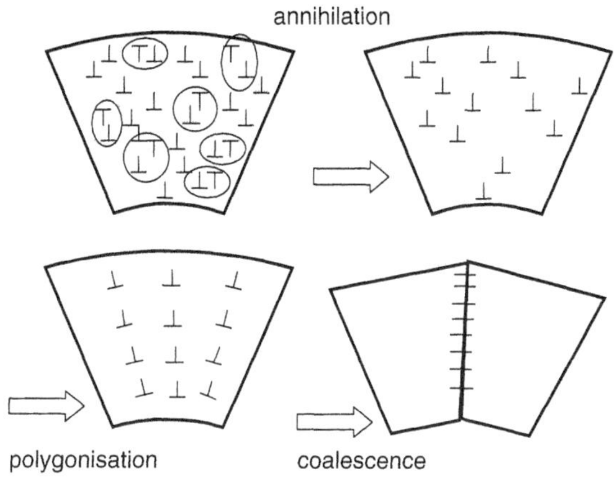

Figure 5.2. Schematic of successive dislocation annihilation mechanisms; cross-section of a bent crystal containing both free (edge) dislocations and dislocations, which accommodate the orientation gradient. During annealing, some dislocations anneal out by climb of opposite sign segments (encircled pairs) then the remainder rearrange into subgrain boundaries.

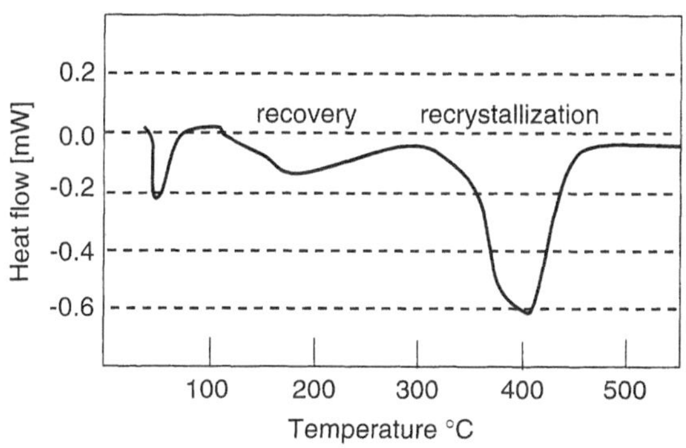

Figure 5.3. DSC plots at 20 K/min of ultrahigh purity Fe cold rolled 80% (Scholz *et al.* [1999]). Negative heat flows correspond to exothermic reactions, which are estimated at ≈4 J/mol for the recovery region (100–300°C) and 15 J/mol for recrystallization (320–450°C).

can lead to rapid recrystallization, stopping any further recovery. Finally, these fundamental mechanisms are also very sensitive to a wide variety of material parameters and processing conditions, such as deformation, temperature, etc.

As noted above, solute atoms play a major role in reducing defect mobility and therefore recovery kinetics. For example, deformed ultrahigh purity metals can recover at about $0.2T_m$ and start recrystallizing at $0.3T_m$ (see Figure 5.3). Solute atoms in commercial alloys push recovery and recrystallization temperatures to

0.4 and $0.6T_m$, respectively. This aspect has been analysed in detail by Nes [1995, 1997], and some parts of this analysis are described in Chapter 10 on modelling. Second phase particles, particularly as very fine distributions, will also inhibit recovery by pinning dislocations.

5.2.2 Recovery kinetics

Obtaining a generalized physical relationship for overall recovery kinetics is difficult. As the process involves several concurrent/consecutive mechanisms (Table 5.1), formulations of all the factors affecting recovery (Figure 5.2) and the heterogeneity of recovery are difficult and a comprehensive formulation does not exist.[3] Two empirical relationships are, however, often used to express the overall kinetics of recovery and denoted types 1 and 2 (Humphreys and Hatherly [1995]).

$$\text{Type 1: } \frac{dX}{dt} = \frac{-k_1}{t} \quad X = k_1' - k_1 \ln t \tag{5.1}$$

$$\text{Type 2: } \frac{dX}{dt} = -k_2 X^m \quad X^{(1-m)} - X_0^{(1-m)} = (m-1)k_2 t \tag{5.2}$$

where X is the fraction recovered (or change in recovery-dependent parameters – mechanical properties, resistivity or enthalpy), t is the annealing time and k_1, k_2 and m are the constants (k_1 and k_2 often scale with the activation energy of self-diffusion). Both relationships have been reported during recovery of single and polycrystalline metallic materials, but type 2 tends to be applicable for the special cases of single crystals with uniform substructures (and single mechanisms). Type 1 empirical kinetics are more typical of the complexities of deformed industrial alloys, where several recovery mechanisms can operate (Humphreys and Hatherly [1995]).

A more physically based kinetics analysis, first proposed by Kuhlmann *et al.* [1949] and later developed by Nes [1995], considers that the activation energy for the process increases during recovery as the internal stress, or the driving force, decreases. The easier, low energy, processes operate first and the higher activation energy mechanisms later, up to the value expected of atomic diffusion. If the activation energy for the process is written $U_F(X)$ and that for diffusion U_d then

$$\frac{dX}{dt} = -c \exp - \left\{ \frac{U_d - U_F(X)}{kT} \right\} \tag{5.3}$$

[3] The effect of recovery on microstructural developments is also difficult to translate into structure-dependent properties.

If $U_F(X)$ is a linear function, this gives a logarithmic time dependency for recovery of the form

$$X = 1 - \frac{kT}{A}\ln(1 + Bt) \tag{5.4}$$

where A and B are constants. This is similar to the type 1 kinetics of Eq. (5.1). The application of this type of analysis to the case of recovery controlled by solute atom diffusion is developed in more detail in Chapter 10.

5.2.3 Structural changes during recovery

Apart from the initial and rapid loss of point defects, the main structural changes taking place during recovery can be categorized as:

- Rearrangement of dislocations into cellular structures (for high-SFE materials and most hot-deformed metals, this occurs simultaneously with deformation).
- Elimination of free dislocations within the cells.
- Collapse of the complicated dislocation cell walls into neat subgrain boundaries – mostly by annihilation of excess or redundant dislocations and rearrangement of the others into low-energy configurations (Figure 5.4).
- *Subgrain growth.* During continued annealing, an increase in subgrain size is theoretically expected since it leads to a reduction of internal energy. The experimentally observed kinetic relationships are usually written

$$d^n - d_0^n = k_3 t \tag{5.5}$$

where d and d_0 are the final and initial subgrain size, n, and k_3 are constants. While k_3 is strongly temperature-dependent (through the activation energy), values of n between 2 and 7 have been reported (Huang and Humphreys [2000]). There is a trend for the lower n values to be found in high-purity metals and at high temperatures, where sub-boundary mobility is the highest. Also, as

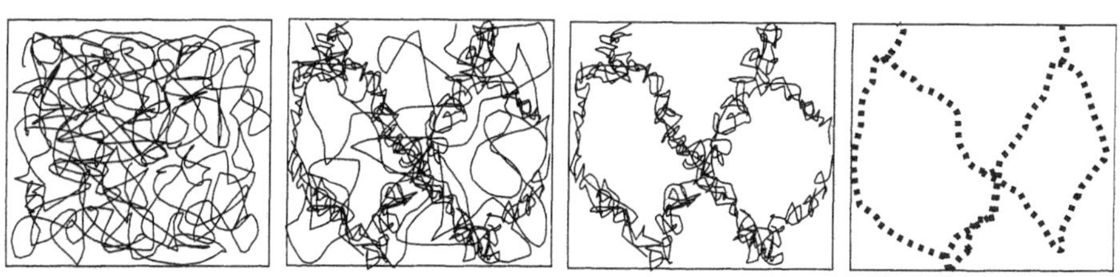

Figure 5.4. Schematics showing dislocation structure changes during recovery – random dislocation tangles through cell structures to subgrains.

discussed by Humphreys and Hatherly [1995], the n exponent may vary due to changes in the subgrain misorientation. Both average misorientation and the misorientation spread are reported to drop slightly and then remain stagnant during recovery, though there is also evidence of misorientation increase with recovery (Furu [1992]). A possible explanation is the role of orientation gradients, which tend to increase the average misorientation.

It should be pointed out that due to experimental difficulties in determing subgrain sizes with good accuracy – and therefore, their exact evolution during recovery before the onset of recrystallization – there have not been many detailed, quantitative studies of subgrain growth.

Another important aspect of recovery, with direct implication to recrystallization, is the issue of orientation dependence. This is also the least understood. Differences in stored energies, or in dislocation substructure, are often related to differences in Taylor factor (Kallend and Huang [1984], Samajdar and Doherty [1995], Rajmohan *et al.* [1997]) and also to differences in the so-called 'textural softening' (Cicalè *et al.* [2002]). Both are, however, related more to deformation than to recovery. The classical example is 'recovery' in rolled Cube $\{001\}\langle 100\rangle$ oriented grains of fcc metals. In this situation, the Burgers vectors of the active edge dislocations are perpendicular to each other, a particular case which avoids strong dislocation interactions and so facilitates rapid Cube recovery (Ridha and Hutchinsen [1982]) (see Figure 5.5).

The quantitative influence of recovery on softening is dependent on the relation between subgrain size and flow stress. The best fit for available experimental data on substructural strengthening (Thompson [1977], Gil Sevillano *et al.* [1980]) is obtained through a Hall–Petch type of relationship(s):

$$\sigma = \sigma_0 + k_5 d^{-c} \tag{5.6}$$

where σ is the yield stress, d the subgrain size, σ_0, k_5 and c are constants. The exponent c is equivalent to the classical Hall–Petch exponent; $c = -0.5$ for subgrains, which behave like grains, and $c = 1$ for low-energy dislocation substructure. The constant k_5 is also proportional to $(\theta)^{1/2}$, where θ is the average misorientation. It is, however, fair to point out that these generalized values or relationships are often not too consistent (Thompson [1977], Gil Sevillano *et al.* [1980]), making it difficult to relate recovery-induced structure with properties.

5.2.4 Extended recovery/continuous recrystallization

Standard recrystallization (Section 5.3) is a discontinuous process by which strain-free grains absorb deformed/recovered areas via high-angle grain boundary movement. As discussed in Section 5.1, this may severely limit recovery in the deformed

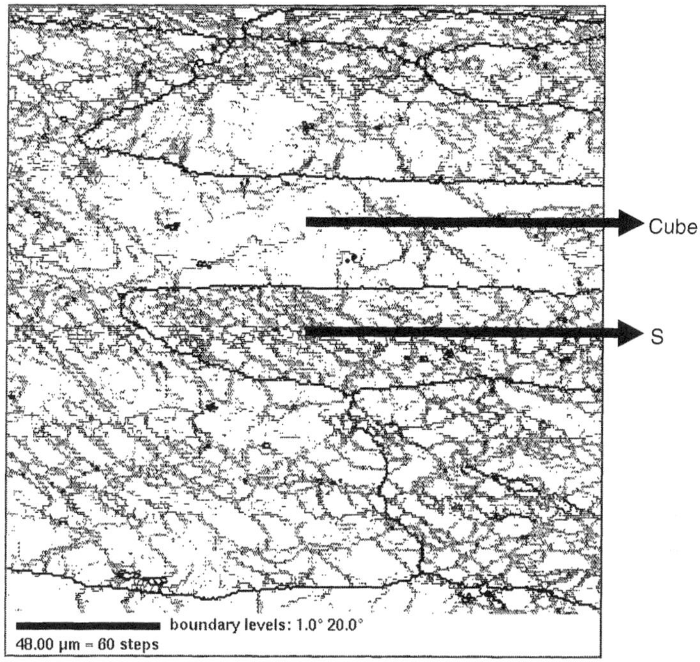

Figure 5.5. Difference in substructure between Cube {001}⟨100⟩ and S {231} ⟨346⟩ in aluminum – EBSD measurement on hot worked AA 1050 (Samajdar *et al.* [2001]).

areas. In certain circumstances, it is however possible to suppress discontinuous recrystallization, leading to a relatively fine and uniform grain size through 'homogeneous subgrain growth' – often termed 'extended recovery'. There is debate as to whether this is purely due to low-angle boundary movements or includes high-angle boundary migration. According to Humphreys and Hatherly [1995], the former is associated with extended recovey, while the latter leads to continuous recrystallization. These phenomena are observed locally (i.e. inside individual grains) as well globally (overall microstructural feature) and are discussed separately.

- *Local extended recovery*. This has been reported for aluminium (Hjelen *et al.* [1991]), low-carbon steel (Gawne and Higgins [1969], Samajdar *et al.* [1997b]) and Zr alloys (Vanitha [2006]) for grains of specific orientation families. These are often the grains/orientations free from strain localizations and hence subjected to homogeneous deformation.
- *Extended recovery/continuous recrystallization as an overall microstructural feature*. This has been reported in the following special circumstances.

- *Two-phase alloys.* Pinning by 2nd phase precipitates may stabilize subgrain structure and subsequent dissolution/coarsening of the precipitates may lead to 'homogeneous subgrain growth' or extended recovery (Humphreys and Hatherly [1995]), though the phenomena was originally identified in as *in situ* or continuous recrystallization (Köster and Hornbogen [1968]).
- Severe plastic deformation (Chapter 17.1) is often used to obtain ultrafine grain sizes through continuous recrystallization – static or dynamic. In such processes, severe plastic deformation and/or formation/dissolution of a 2nd phase are often used (Lee and McNelly [1987], Gudmundsson [1991], Humphreys and Bate [2001], Belyakov *et al.* [2002], Gholinia *et al.* [2002], Yu *et al.* [2004]). Such processes may involve extended recovery/continuous recrystallization as well as geometrical necessary dynamic recrystallization (Section 5.3.5.2). Recovery may also facilitate the formation of strain localizations, which, in turn, through developments of long-range misorientation may provide the appearance of grain-like structures under polarized light (which are indeed clusters of subgrains) (Samajdar *et al.* [1998d]).

5.3. RECRYSTALLIZATION

Primary recrystallization (also termed 'discontinuous recrystallization') is often 'viewed' as a nucleation and growth process. Though similar in name, and even partly in approach, to the nucleation and growth of phase transformation or solidification (see Section 2.2.1.1), there are significant differences. First of all, there is no 'classical' nucleation – the driving force of recrystallization being usually much smaller than that of 'classical' nucleation (Gottstein *et al.* [1985]). *The recrystallized nucleus is a part of the deformed matrix* (Humphreys and Hatherly [1995], Doherty *et al.* [1997]). In other words, any large subgrain or relatively well-recovered region of a deformed grain can be considered as a potential recrystallization nucleus – purely from the consideration of driving force, or relative differences in stored energy between the potential nucleus and its immediate surrounding. Whether such a potential nucleus is real or active will depend on its growth possibilities, particularly the presence of a growth-favourable[4] boundary. The velocity of such a boundary can be generalized as

$$V = MP \tag{5.7}$$

$$P = P_{SE} + P_C = (\alpha \rho G b^2) + \frac{2\gamma_b}{r} \tag{5.8}$$

[4] Recrystallization requires the presence and movement of high-angle boundaries. Their mobility is significantly larger than that of low-angle boundaries. Special boundaries (see Section 2.2.2.2), 27° ⟨110⟩ for bcc and 40° ⟨111⟩ for fcc, may have even higher mobility.

where V, M and P are respectively, the grain boundary velocity, mobility and the net pressure on the boundary. P has two components – P_{SE} or pressure due to the driving force or stored energy and P_C or pressure due to boundary curvature. The former can be given in terms of dislocation density[5] (α = constant close to 0.5, ρ = dislocation density, G = shear modulus and b = Burgers vector), while the latter can be obtained from the Gibbs–Thompson relation. γ_b and r are respectively the grain boundary energy and curvature, which is negative for expanding recrystallized grains (for more details on Gibbs–Thompson, the reader may refer Porter and Easterling [1986]).

A critically important aspect of this subject is the misorientaion dependence of boundary mobility (Humphreys [1995], Doherty [1997]). Low angle boundaries, such as those between sub-grains formed by deformation/recovery, have very low mobility, while boundaries with larger misorientations (>10–15°) have very high mobility. As a result, nucleation occurs by the rapid growth of a very small minority of sub-grains that evolve into growing new grains. The first necessary condition for this is that the sub-grain has, or quickly acquires, a local misorientation of more than 15°. The rapid growth of a very few sub-grains, compared to the slow growth of the remaining sub-grains, gives this common type of recrystallization its heterogeneous character – described as a "Nucleation and Growth" process.

Experimentally, it is impractical and impossible to exactly separate nucleation from growth. A region free from strain (e.g. without grain boundaries or misorientation) and exceeding a certain size (often based on the dimensions of the deformed grains) is considered as a recrystallized grain. To achieve this size, typically of the order of a micron, both nucleation as well as local or limited growth can be involved.

To identify the exact nuclei, or potential nuclei *just* turning active, is very difficult. In typical metallic systems, the size difference between the original subgrains (potential nuclei) and the final recrystallized grain is 10–100 times, making the probability of finding the exact nuclei in the deformed/recrystallizing matrix of the order of 10^{-3}–10^{-6}. Naturally, the focus of experimental research is often directed at identifying the recrystallization sources and also indexing and understanding their relative contributions to the recrystallized microstructure – size, shape and orientations of the recrystallized grains.

5.3.1 Sources of recrystallized grains

A summary of the typical, albeit generalized, recrystallization sources is given below.

[5] Strictly spoken, the difference in stored energy ($\rho - \rho_N$) should be considered, with ρ the dislocation density in the deformed zone and ρ_N the dislocation density in the growing nucleus, but in most cases one can consider $(\rho - \rho_N) \cong \rho$.

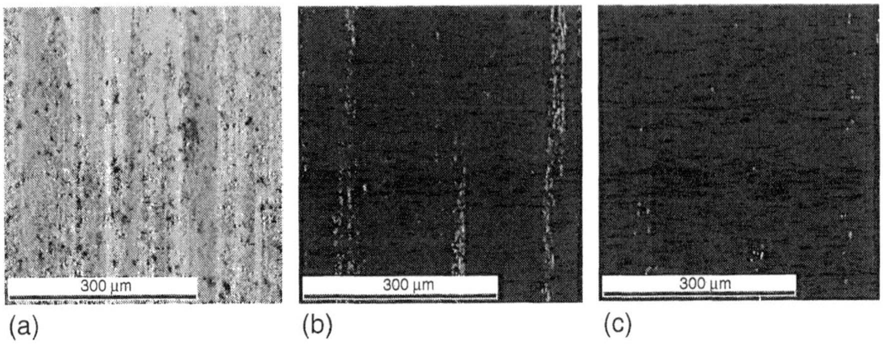

Figure 5.6. EBSD images of (a) hot deformed aluminium, (b) deformed Cube grain in (a), and (c) recrystallized Cube grain in (a). Distinction between (b) and (c) was established by considering grain orientation spread and grain size. A comparison of (b) and (c) indicates that the recrystallized Cube originate from deformed Cube. Courtesy Ravindra (IIT Bombay) – Research in Progress.

5.3.1.1 Deformed grains. As shown in Figure 5.6, a deformed grain may act as a source of recrystallized grains with more or less the same orientation as the 'mother' grain. Examples include Cube[6] recrystallization (Ridha and Hutchinsen [1982], Duggan *et al.* [1993], Samajdar *et al.* [1998d]) in fcc alloys and γ(ND//$\langle 111 \rangle$) fibre[7] recrystallization (Hutchinson [1984], Ray *et al.* [1994], Samajdar *et al.* [1999]) in low-carbon steel.

Cube recrystallized grains in fcc alloys originate from deformed Cube-oriented bands (see Figure 5.6). Such Cube bands, on the other hand, are suspected (Doherty *et al.* [1997], Samajdar *et al.* [1998e]) to be part of the original Cube-oriented grains in the undeformed structure. Cube grains nucleate favourable from the Cube bands, as the latter has lower stored energy (Ridha and Hutchinsen [1982], Doherty *et al.* [1997], Samajdar *et al.* [1998e]) and/or presence of growth favourable 40° $\langle 111 \rangle$ boundaries (Duggan *et al.* [1993]).

Recrystallized grains of γ(ND//$\langle 111 \rangle$) fibre in low-carbon steel also originate from deformed bands of similar orientation. Such bands are an effective recrystallization source due to extensive formation of grain interior strain localizations (Akbari *et al.* [1997]). The selective formation (Samajdar *et al.* [1999]) of strain localizations lead to fragmentation of the γ bands and correspondingly large variations in stored energy.

[6] Control of Cube recrystallization is the key issue in the TMP of both aluminium can stock (see Chapter 14.1), and of aluminium capacitor foils (see Chapter 14.2).

[7] Crucial for formablity of steel (see Chapter 11.10), and hence for the TMP of car body steel (see Chapter 15.1).

5.3.1.2 Shear bands. Shear bands cutting across several grains may also act as a source of recrystallized grains. The potency is often related to high stored energy and correspondingly large variations in relative misorientations and the possible presence of growth-favourable boundaries. Shear banding, briefly described in Section 4.2.1, is particularly strong in low-SFE metals and alloys. The formation of shear bands is often orientation sensitive (Dillamore *et al.* [1979], Gil Sevillano *et al.* [1980]); and hence, strong developments in recrystallization texture (preferred orientations of the recrystallized grains – for more details reader may refer Chapter 8) can be associated with recrystalliztion from shear bands.

5.3.1.3 Particle stimulated nucleation. Dislocations can be trapped by relatively large non-shearable particles (Ashby *et al.* [1970]). The dislocation density (ρ) around a particle is represented as

$$\rho = \frac{3F_v s}{rb} \tag{5.9}$$

where F_v, r, s and b are volume fraction and size of the of the 2nd phase particles, shear strain and dislocation Burgers vector. The entrapment and corresponding increase in dislocation density leads to the formation of a deformation zone – region around the 2nd phase particles with large misorientation developments. Though considerable efforts has been expended for modelling the deformed zones, a comprehensive physical model remains to be developed (Humphreys and Hatherly [1995]).

Recrystallization is favoured from particle-deformed zones due to large differences in the stored energies. Recrystallization of this type is referred as particle stimulated nucleation (PSN). The particle-deformed zones, and correspondingly the PSN grains, are of randomized orientations (Van Houtte [1995]).

In particle-containing commercial alloys, the annealing behaviour, including recrystallization texture developments, is strongly influenced by annealing temperature. Low-temperature annealing is often related to stronger randomization (Ørsund and Nes [1988], Samajdar *et al.* [1998c]). To explain such behaviour, two approaches have been proposed. The first approach (Ørsund and Nes [1988]) assumes the presence of inner and outer deformed zones – the former being more randomized. The annealing temperature is expected to determine the relative contributions from the zones and in turn decides the recrystallization behaviour. An alternative approach (Samajdar *et al.* [1998c]) proposes that the relative contributions from PSN and deformation bands are responsible for the overall recrystallization behaviour, including recrystallization texture developments.

5.3.1.4 Recrystallization twins. This is more valid for lower-SFE metals and alloys – e.g. copper, austenitic stainless steel (see Chapter 17.2), etc. Multiple twinning in the early recrystallization stages has been reported (Berger *et al.* [1988]) to play an important role in determining the orientations of the recrystallized grains. In the real sense this is not a true recrystallization source, but the only mechanism which may form recrystallized grains with new orientations – orientations otherwise absent in the deformed matrix.

5.3.2 Recrystallization mechanisms

Developments in recrystallized microstructures depend on the nucleation and growth of the recrystallized grains. While nucleation can be generalized as relative activation of different recrystallization sources, strain-free nuclei consume the deformed matrix during the growth process. The patterns of recrystallization, including recrystallized microstructure, are decided by the relative advantange/disadvantage of nucleation and growth.

The advantage/disadvantage relations of nucleation and growth are often generalized as ON (oriented nucleation) and OG (oriented growth) respectively, representing preferred nucleation and growth (Doherty [1985], Humphreys and Hatherly [1995], Doherty *et al.* [1997]). This distinction has one major problem – it is difficult to make a precise demarcation between the nucleation and growth stages of recrystallization. An alternative is to describe the developments in recrystallized microstructure, especially recrystallization texture, in terms of frequency and size advantage of recrystallized grains of different orientations (Doherty [1985]). Thus, if the volume fraction of a particular component i increases during recrystallization, then such an increase has to be due to larger number of i recrystallized grains and/or their bigger sizes. Frequency (α) and size (β^3) advantage factors are described as

$$\alpha = \frac{i_{\text{rex}}}{i_{\text{random}}} \tag{5.10}$$

$$\beta^3 = \left(\frac{D_i}{D_{\text{average}}} \right)^3 \tag{5.11}$$

where i_{rex} and i_{random} are the respective number fraction of i grains in the recrystallized material and in a random texture; D_{i} and D_{average} are the mean grain sizes

Table 5.3. Summary of the micromechanisms responsible for frequency/size advantage.

Frequency advantage	*Stored energy advantage*: Nucleation of a particular recrystallization source and/or orientation may be preferred. The preference is typically caused by large variation of stored energy within the source (example: recrystallization from deformed γ fibre bands in bcc steel (Hutchinson [1984], Ray *et al.* [1994], Samajdar *et al.* [1999])) or between the source and the surrounding (example: recrystallization from Cube bands in fcc alloys (Ridha and Hutchinsen [1982], Duggan *et al.* [1993], Samajdar *et al.* [1998d]))
	Micro growth selection: Presence of growth-favourable boundaries to intiate preferred nucleation (example: presence of 40° $\langle 111 \rangle$ boundary or S next to deformed Cube in fcc alloys (Duggan *et al.* [1993]))
Size advantage	*Growth advantage/disadvantage*: Unlike single crystals, in heavily deformed polycrystalline material, a global growth advantage is controversial (Doherty *et al.* [1997]) – as the nature of the grain boundary between a recrystallized grain and the deformed matrix is expected to change many times. Growth disadvantage (Section 5.3.4) through particle pinning and/or orientation pinning may, however, lead to a size disadvantage
	Early nucleation: Early preferential nucleation can also lead to size advan tage (example: large Cube grain in fcc aluminium was reported to be caused by early nucleation (Samajdar *et al.* [1998e]) – the estimated growth velocity of Cube and non-Cube being similar)

for i and average (i.e. irrespective of any crystallographic orientation) grains. Assuming spherical grain shape, the size advantage factor was taken as β^3 (Doherty [1985]). Evidently, the recrystallization texture will be strengthened when α and/or β^3 are larger than 1.

Such an approach relates directly to recrystallized microstructures and stays clear from the ON and OG models, a subject of nearly half century old debates and discussions (for more details on ON and OG, the reader may refer to Section 8.4.2). Table 5.3 summarizes the micromechanisms responsible for the frequency and size advantages. As shown in the table, the frequency advantage can be caused by both preferred nucleation and/or micro growth advantage – the latter can be considered as part of the early nucleation process. The size advantage, on the other hand, has been attributed to growth advantage/disadvantage and/or early nucleation.

Any discussion of recrystallization micromechanisms remains incomplete without describing further the mechanisms behind stored energy and growth advantage/disadvantage. The stored energy advantage/disadvantage, or differences in stored energy in different grains or orientations, is caused by deformation and orientation-dependent recovery – a topic described earlier in Section 5.2.3. The growth advantage/disadvantage, on the other hand, depends on the effects of 'pinning' on different types of grain boundary. While differences in grain boundary nature have been covered in Section 2.2.2.2, the present section tries to summarize three different types of grain boundary 'pinning'.

- *Particle Pinning or Zener Drag*. The pinning pressure (P) is given as

$$P = \frac{3F_v\gamma}{2r} \tag{5.12}$$

  where F_v, r and γ are particle volume fraction, particle size and grain boundary energy, respectively. Low-energy boundaries, such as low coincidence site lattice (CSL) boundaries, have low-Zener drag.
- *Solute drag*. The presence of solute is also expected to have a retarding effect on growth – the so-called solute drag (Lücke and Stüwe [1963], Gordon and Vandermeer [1966], Dimitrov *et al.* [1978]). The solute drag is lower for low-CSL boundaries.
- *Orientation pinning*. A growing recrystallized grain can be pinned by a deformed region of similar orientation (Juul Jensen [1995]) – i.e. by the presence/occurance of a low-angle boundary. Orientation pinning has been reported to have strong influence in both fcc Cube recrystallization (Doherty *et al.* [1997]) and in γ fibre recrystallization of low-carbon steel (Samajdar *et al.* [1999]).

5.3.3 Recrystallization kinetics

The analysis of recrystallization kinetics is often based on one of two tools – JMAK (Johnson, Mehl, Avrami, Komogorov) analysis and the microstructural path methodology (MPA). JMAK[8] relates the fraction recrystallized with time (t) as

$$1 - \exp(-kt^n)$$

where k is a constant and n the JMAK exponent. The JMAK exponent can be used to get a feel for the overall recrystallization kinetics. For example in a three-dimensional (3D) structure, theoretical n values of 3 and 4, respectively correspond to site saturation (nucleation happening at one instant) and to a constant nucleation rate (Humphreys and Hatherly [1995]). n can also be used to index sluggish growth or delayed nucleation. Random nucleation is critical for the validity of JMAK analysis. For example, fine-grained copper with random nucleation follows ideal JMAK, but the coarse-grained material does not (Hutchinson [1989]).

[8] Chapter 10 (Section 10.2.1) describes the JMAK analysis in details.

In fine-grained material, less than one successful nucleation from each deformed grain is expected, effectively randomizing the spatial distribution of the nuclei.

MPA attempts to extract the nucleation and growth rates from the experimental data. The primary problem of estimating nucleation rates from two-dimensional (2D) microstructural observations is circumvented by using the Laplace transform (Gokhale and Dehoff [1989], Vandermeer and Rath [1989]), and nucleation and growth rates are estimated as direct output.[9] For isothermal annealing without any shape change for the recrystallizing grains, the nucleation rate (N_τ, nucleation rate at a time instant τ) and the grain size (D – major axis of spheroidal grains) can be expressed as

$$N_\tau = N_1 t^{\delta-1} \tag{5.13}$$

$$D_{(t-\tau)} = G(t-\tau)^r \tag{5.14}$$

where N_1, δ, G and r are constants, t is the time and G is the growth rate. Though MPA is capable of separating out the nucleation and growth from experimental data, both JMAK and MPA are usually restricted by the basic assumption of random nucleation.

5.3.4 Role of second phase

Table 5.4 outlines the possible effects of different types of 2nd phase particles on the recrystallization behaviour. As shown in the table, the effects of 2nd phase particles can be generalized as (i) nucleation advantage[10] through PSN and/or nucleation on particle-induced strain localizations and (ii) growth disadvantage through particle pinning or Zener drag. The latter can be used effectively to control the grain size in a particle-containing alloy. The 'pinning' limited grain size (D_{Zener}) is often approximated as

$$D_{\text{Zener}} = \frac{4r}{3F_v} \tag{5.15}$$

where r and F_v are the particle size and volume fraction, respectively.

[9] It is to be noted that the original methodology was proposed by Gokhale [1985] for estimating nucleation rates during classical phase transformation. This was adopted for recrystallization by Vandermeer [1989].

[10] Nucleation disadvantage may come indirectly through inhibition of recovery during simultaneous precipitation.

Table 5.4. Effects of different types of 2nd phase particles on recrystallization.

Nature of the particle	Nucleation	Growth
Non-shearable particles	*Nucleation advantage*: For large particles – PSN; also possible is nucleation on particle-induced strain localizations	Particle pinning or Zener drag – growth disadvantage
Fractured particles	Small plate-shaped carbides may undergo (Kamma and Hornbogen [1976]) fracture and initiation of strain localizations. The latter may aid nucleation	Particle pinning or Zener drag – growth disadvantage
Coherent particles	No effect – small particles	Zener drag and growth inhibition – example: coherent precipitates in Cu–3%Co alloys (Phillips [1966])
Semi-shearable particles	No effect – soft 2nd phase does not form deformed zones or strain localizations	Zener drag and growth inhibition, especially dominant in regions with 2nd phase continuity
Shearable particles	No effect	Zener drag – example: pores and gas bubbles
Simultaneous precipitation	Depending on annealing conditions and precipitation kinetics, recovery and hence nucleation can be hindered	Depending on annealing condi tions and precipitation kinetics, Zener drag and growth inhibition

5.3.5 Dynamic recrystallization

Under certain conditions, the structure can recrystallize during deformation giving rise to dynamic recrystallization. In principle, this form of recrystallization can also occur during cold deformation, but in practice, this is only exceptionally observed, e.g. in very pure metals. In this section, dynamic recrystallization is classified as either discontinuous dynamic recrystallization (Derby [1987, 1991]) or as one of two other types. The latter are geometric dynamic recrystallization (McQueen *et al.* [1989]) and dynamic recrystallization through progressive subgrain rotation (Gardner and Grimes [1979]), and both involve strain-induced phenomenon with limited or no movements of high-angle boundaries. Following usual practice, these other types have been included here as part of the present section on dynamic recrystallization.

5.3.5.1 Discontinuous dynamic recrystallization. Figure 5.7 shows typical flow curves during cold and hot deformation. During hot deformation, the shape of the flow curve can be 'restricted', or work hardening rates counterbalanced, by

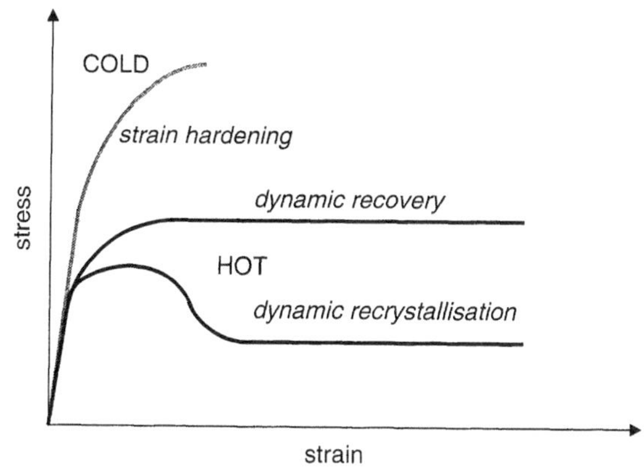

Figure 5.7. Typical flow curves during cold and hot deformation.

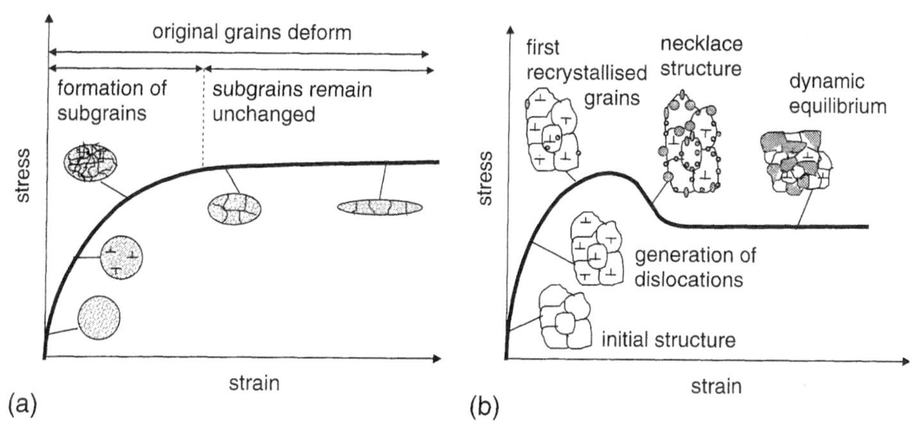

Figure 5.8. Evolution of the microstructure during (a) hot deformation of a material showing recovery and (b) dynamic recrystallization or discontinuous dynamic recrystallization.

dynamic recovery or by dynamic recrystallization (i.e. discontinuous dynamic recrystallization). Dynamic recovery is typical of high-SFE metals (e.g. aluminium, low-carbon ferritic steel, etc.), where the flow stress saturates after an initial period of work hardening. This saturation value depends on temperature, strain rate and composition. On the other hand, as shown in Figure 5.7, a broad peak (or multiple peaks) typically accompany dynamic recrystallization.

Figure 5.8 illustrates schematically the microstructure developments during dynamic recovery and dynamic recrystallization. During dynamic recovery, the original grains get increasingly strained, but the sub-boundaries remain more or less equiaxed. This implies that the substructure is 'dynamic' and re-adapts continuously to the increasing strain. In low-SFE metals (e.g. austenitic stainless steel, copper, etc.), the process of recovery is slower and this, in turn, may allow sufficient stored energy build-up. At a critical strain, and correspondingly at a value/variation in driving force, dynamically recrystallized grains appear at the original grain boundaries – resulting in the so-called 'necklace structure'. With further deformation, more and more potential nuclei are activated and new recrystallized grains appear. At the same time, the grains, which had already recrystallized in a previous stage, are deformed again. After a certain amount of strain, saturation/equilibrium sets in[11](see Figure 5.8b). Typically equilibrium is reached between the hardening due to dislocation accumulation and the softening due to dynamic recrystallization. At this stage, the flow curve reaches a plateau and the microstructure consist of a dynamic mixture of grains with various dislocation densities.

It is important, at this stage, to bring out further the structural developments and structure–property correlation accompanying dynamic recovery and dynamic recrystallization respectively. Both the subgrain size ($d_{subgrain}$, from dynamic recovery) and grain size (D_{rex}, dynamic recrystallization) are increasing functions of temperature and of inverse strain rate. Both follow a Hall–Petch-type (Eq. (4.3)) relationship.

$$\sigma = \sigma_1 + k_1(\mathrm{d}_{subgrain})^{n_1} \tag{5.16}$$

$$\sigma = \sigma_2 + k_2(\mathrm{D}_{rex})^{n_2} \tag{5.17}$$

where σ is flow stress and σ_1, k_1, n_1, σ_2, k_2 and n_2 are constants. Corresponding to dynamic recovery (Eq. (5.16)), σ_1 has a low value and n_1 is close to 1, while k_1 depends on alloy composition (being higher at higher solute content). For dynamic recrystallization (Eq. (5.17)), n_2 typically falls within 0.5–0.8.

Unlike static and dynamic recovery, recrystallization includes another classification – a sort of grey area between dynamic and static: meta-dynamic recrystallization. In this situation, the recrystallized nuclei form or nucleate dynamically, during hot deformation, but growth takes place during subsequent static annealing.

5.3.5.2 **Other types.**

- *Geometric dynamic recrystallization*. Grains with serrated boundaries (see Figure 5.9), a by-product of dynamic recovery, may pinch each other when subgrain size approaches the grain thickness. Such a mechanism (McQueen *et al.* [1989]) can

[11] Because of the dynamic recrystallization, the material becomes softer, with corresponding appearance of a broad peak in the flow stress (see Figure 5.7).

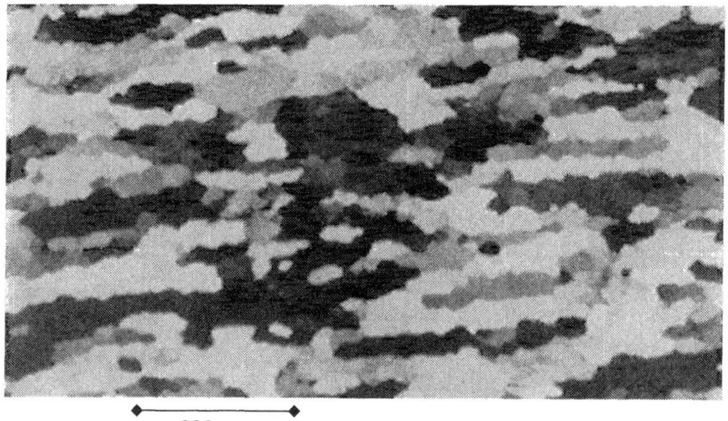

Figure 5.9. Polarized light optical image of hot deformed aluminium showing serrated grain boundaries (Samajdar *et al.* [2001]).

produce microstructures appearing 'recrystallized' (e.g. dynamically recrystallized), but this structure is more a product of deformation and recovery.

- *Dynamic recrystallization through progressive subgrain rotation.* This was originally reported for geological formations as rotational recrystallization, and was subsequent observed in high solute metals (Gardner and Grimes [1979]). During deformation, subgrains in the neighbourhood of pre-existing boundaries may undergo significant higher rotations than the grain-interior subgrains. At large strain, high-angle boundaries may develop. Though the exact mechanism is not fully understood, it is normally believed (Humphreys and Hatherly [1995]) to involve a combination of inhomogeneous plastic deformation, accelerated dynamic recovery in near-grain boundary regions and grain boundary sliding.

5.4. GRAIN COARSENING

Post recrystallization grain growth is driven by the surface energy or the grain boundary energy and this driving force is about two orders of magnitude lower than that of recrystallization. Though often called grain growth, the term grain coarsening is preferable – parity with interfacial energy driven precipitate coarsening[12].

Grain coarsening can occur in two forms: normal and abnormal[13]. In normal grain coarsening, the central mechanism is the loss of the smallest grains, while

[12] As in Chapter 7, interfacial energy driven precipitate coarsening (or competitive growth) is distinguished from supersaturation driven precipitate growth.

[13] also called secondary recrystallization

maintaining a nearly constant grain size distribution. In abnormal grain coarsening a few grains grow into a pinned grain structure. The pinning occurs usually by particles or by a high frequency of low angle, and thus low mobility, boundaries. In the present section on grain coarsening, some of the theories/models are highlighted first, followed by a brief summary of the factors affecting grain coarsening.

5.4.1 Theories of Grain Coarsening

Table 5.5 makes an attempt to collate some of the important theories of grain coarsening. As shown in the table, the theories and models range from fairly simple early statistical theories (Burke [1952]) to more complex statistical (Hillert [1965], Rhines [1974], Abbruzzese [1986], Mullins [1989a, 1989b]) and deterministic (Atkinson [1988], Anderson [1989]) models available today. The list (Table 5.5) is not an exhaustive one. For more details, the reader may refer to Atkinson [1988], Humphreys [1995] and Martin [1997].

It is important to point out that the statistical theories need to take simplifying assumptions, while the deterministic models are often too computation intensive. Even in primarily single phase metals, the model predictions and the experimental results often may not tally[14] – such differences are attributed to different factors: solute drag, Zener pinning, differences in grain–grain boundary mobilities and initial grain structure being different form the ideal one (i.e. non-equiaxed grains and/or different grain size distribution).

5.4.2 Factors affecting grain growth

The factors affecting grain growth are given below.

- *Specimen size.* The driving force is significantly reduced when grain size approaches ½ thickness of the specimen. The effect is due to reduced curvature of the grain boundary, since at the equilibrium grain boundaries must fall normally onto the sample surface. In addition, at grain boundary intersection with the surface, grooves form, further hindering boundary migration.
- *Temperature.* Grain boundary mobility depends strongly on the temperature (a Arrhenius-type relationship) and in turn affects grain growth kinetics. Another indirect effect of temperature is through dissolution/coarsening of 2nd phase particles pinning grain boundaries.

[14] And it is impossible to pin-point the best theory or model. The basic model assumptions, for some of the earlier model (see Martin [1997]) can also be debated. Mullins W.W., Acta Metall., 37 (1986b), 2979.

Table 5.5. Summary of some of the important theories/models of grain growth.

General classification	Model details	Remarks
Statistical models	*Burke [1952]:* Driving force = $P=\{(2\gamma)/R\}$ where γ and R are the grain boundary energy and radius of curvature	Offers only average grain size and not a distribution
	Average grain size at time $t = R_{av} = ct^{1/n}$. c is a constant and n is the grain growth exponent	Simplistic, yet relevant
	Deviation from an ideal exponent of 2 is observed even in pure metals – often explained from variations in boundary mobility (due to solute drag) and the presence of 'limiting' grain size (Zener pinning)	
	Feltham [1957] and Hillert [1965]: $dR_c/dt = (c_1M\gamma)/(4R_c)$. R_c is a critical radius below and above of which grains would respectively, shrink or grow. c_1 is a shape factor (0.1 and 1 for 2D and 3D) and M is the boundary mobility	Parabolic growth kinetics
	Rhines [1974]: Considers the role of topology. Modified Rhines analysis: $V_{av} = V_0 + ct$; where V_0 and V_{av} are the mean grain volume at time instance 0 and t	Growth exponent of 3 is predicted
	Abbruzzese [1986]: 2D topological model using an experimental relationship between grain sides/facets and grain size	Parabolic growth kinetics
	Mullins [1989a, 1989b]: Parabolic coarsening kinetic was established, provided (i) rate controlling boundary mobility and (ii) grain size distribution normalized by the time dependent average grain size remain constant	Perhaps the most important analysis on the statistical approach.
Deterministic models	*Equation of motion – Atkinson [1988]:* Intial grain structure is allowed to stabilize/equilibrate (based on some governing equations) by several iterations	Parabolic kinetics; sensitive to starting structure
	Monte Carlo – Anderson [1989]: Use of Monte Carlo algorithm in grain growth simulation is a relatively mature subject	Growth exponent approaching 2

- *Initial structure.* The initial structure, especially the shape (topological effects) and the grain size distribution (Eqs. (5.18) and (5.19)) can have very strong influence on the grain growth. The grain size distribution (represented by a distribution function $f(D)$) is often generalized as log-normal or Rayleigh type, the former being more relevant to experimental data. Changes in the distribution at the initial stages or during grain growth can affect the growth significantly:

$$\text{Log normal distribution: } f(D) = k_1 \exp(-(\ln D)^2) \tag{5.18}$$

$$\text{Rayleigh distribution: } f(D) = k_2 \exp(-k_3 D^2) \tag{5.19}$$

where k_1, k_2 and k_3 are constants.

- *Pinning*. Primarily the pinning is caused by solute drag, Zener pinning and orientation pinning (see Section 5.3.2). Whatever the source, the extent of pinning often differs between different types of grain boundaries. Whatever might be source, extent of pinning often gives rise to different mobilities of different types of grain boundaries. As a result, the relative presence of growth-favourable (low-CSL boundaries) and non-favourable (low-angle boundaries) boundaries may have significant effects on grain coarsening. The problem, however, is a complex one – as the relative presence of different types of boundaries may change significantly during grain growth.

As a result of these complications, the actual kinetic of grain coarsening cannot yet be reliably predicted in any given microstructure. Nor is it known how, why or after what time of annealing abnormal coarsening can occur in a pinned grain structure.

LITERATURE

Doherty R.D., Prog. Mater. Sci., 42 (1997), 39.
Doherty R.D., Hughes D.A., Humphreys F.J., Jonas J.J., Juul Jensen D., Kassner M.E., King W.E., McNelley T.R., McQueen H.J. and Rollet A.D., Mater. Sci. Eng., A238 (1997), 219.
Humphreys F.J. and Hatherly M., "Recrystallization and Related Annealing Phenomena", Elsevier, UK (1995).

Chapter 6

Alternative Deformation Mechanisms

Chapter 6
Alternative Deformation Mechanisms

6.1. INTRODUCTION

In Chapter 4, the main deformation mechanism for metallic materials (dislocation glide) has been discussed, both for cold and hot deformation. However, under certain conditions, other deformation mechanisms, such as twinning, creep, grain boundary sliding (GBS) and strain associated with phase transformations may occur. In the present chapter some of these alternative deformation mechanisms will be described. Plasticity (assisted) by a phase transformation (mainly the transformation of a parent phase into martensite) will not be treated here because the martensitic transformation is discussed in Chapter 7 and transformation-induced plasticity (TRIP) is illustrated in Section 15.2.

Before starting the discussion on individual deformation mechanisms, the influence of the major external parameters – stress, temperature and strain rate – and of microstructural characteristics, e.g. grain size, on the activation of a particular deformation mechanism will be illustrated with the help of deformation mechanism maps.

6.2. DEFORMATION MECHANISM MAPS

During (hot) deformation, several deformation mechanisms can occur, dependent on temperature and applied stress. In most cases, one mechanism is dominant in the sense that it gives the largest contribution to the strain rate (*rate controlling*). It is possible to indicate, for a given material, for which conditions of temperature and stress a particular mechanism is dominant. Such diagrams are called '*deformation mechanism maps*'. An example of such a map is given in Figure 6.1.

The upper boundary of the diagram is formed by the 'theoretical shear stress'. It is the stress that should be applied to deform a perfect, defectless crystal by a collective displacement of lattice planes over each other. It is known that this stress should be around $G/20$ with G the shear modulus. The value of G falls somewhat with increased temperature. When dislocations are present, deformation is possible with much lower stress. The deformation will now take place by movement of dislocations in their slip planes, aided when the temperature is high, by dynamic recovery caused by cross-slip and climb. In the lower part of the diagram, we can find the creep region. The creep region can be subdivided in different sub-regions, each characterized by a particular creep mechanism. Creep can be dominated by

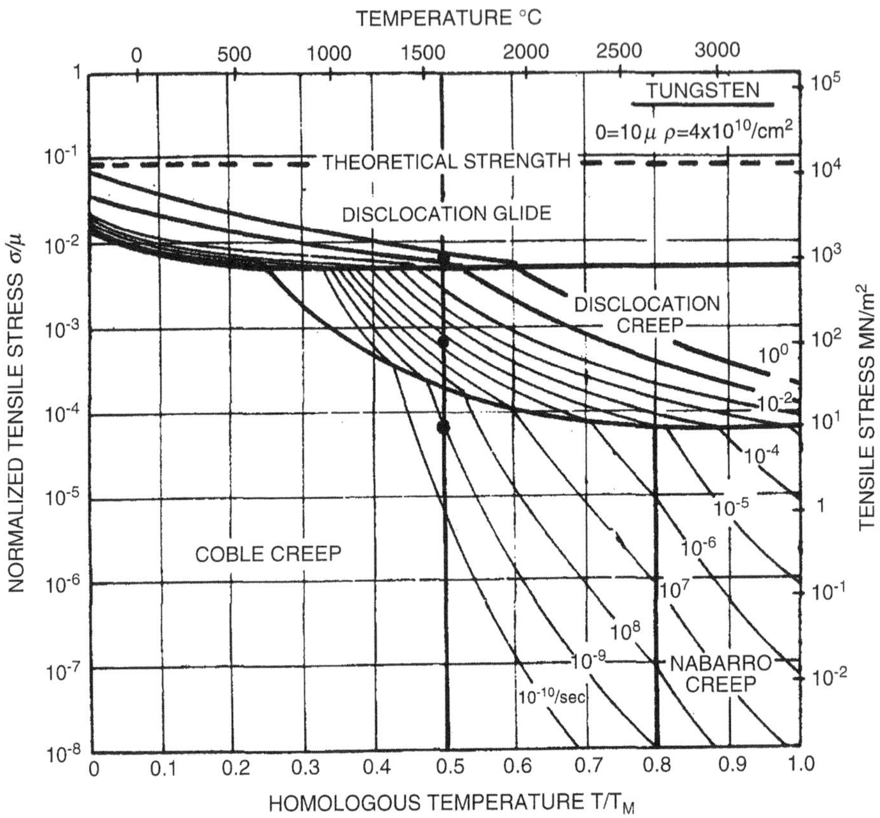

Figure 6.1. A deformation mechanism map for Tungsten (grain size 10 μm). Reprinted from Ashby [1972], Copyright (2007), with permission from Elsevier.

dislocation movement or by diffusion. For the latter, two mechanisms can be discerned: Nabarro creep and Coble creep. These creep mechanisms will be further discussed in the next paragraph. In principle, a material can also be deformed by twinning, but this mechanism is only relevant for some materials and is not considered in Figure 6.1. Another deformation mechanism that appears absent in Figure 6.1 is GBS. In a bicrystal such sliding can occur by itself, but for it to occur in a polycrystal, it must be accommodated, usually by diffusion creep. Furthermore, during diffusion creep itself, GBS is required and so GBS is occurring within the diffusion creep regions in the deformation maps. GBS is known to take place during superplastic forming. This occurs readily in materials with a very fine grain size at elevated temperatures. It will be discussed in Section 6.4.

For each of these mechanisms a specific relation (constitutive equation) between the stress, the temperature and the strain rate can be found. It should be noted that these maps represent a mixture of theoretical models, for example for the theoretical strength and for diffusion creep mechanisms, and mainly experimental results – for dislocation glide and creep – although these are supported by theoretical analysis.

- When a certain stress level and temperature have been chosen, the strain rate and the dominant deformation mechanism are automatically fixed. Consider, for example the case of Tungsten (Figure 6.1), for a temperature of 1560 °C (half of its melting point in Kelvin). When the material is subjected to a very low stress (e.g. $\sigma = 10$ MPa), the material will deform by diffusion creep at a strain rate of 10^{-9}/s. Under these circumstances it would take about three years to reach 10% deformation. This may look insignificant, but for certain applications this cannot be tolerated.
- At a somewhat higher load, e.g. $\sigma = 100$ MPa, the material will deform by dislocation creep at a rate of 10^{-6}–10^{-5}/s. At a stress level of $\sigma = 1000$ MPa, the material will deform by classical dislocation glide. The strain rate will be 0.1–10/s. When the stress level is raised in the dislocation glide region, a very much higher strain rate results. Note that in a conventional hot deformation process like hot rolling, this relation will be seen from a different point of view: a strain rate will be imposed by the rotation speed of the rolls. To perform the deformation a certain stress level (a certain rolling force) will be necessary. The higher the imposed strain rate, the higher the stress required (cf. relation 4.16).

Although some controversies about the exact creep mechanisms may still exist, many authors now seems to think that the mechanisms of high-temperature dislocation creep are not different from those occurring during conventional hot deformation processes. Both depend on the balance of dislocation generation by slip and annihilation by dynamic recovery. Both can be described by the same constitutive equation and result in the same type of dislocation substructure (see, e.g. McQueen *et al.* [1994]).

6.3. CREEP

6.3.1 The creep curve

Creep can be described as 'the strain response of a material under constant load over a long period of time'. The response is a gradual elongation followed by fracture. Creep mainly occurs at high temperatures. Deformation by creep takes place

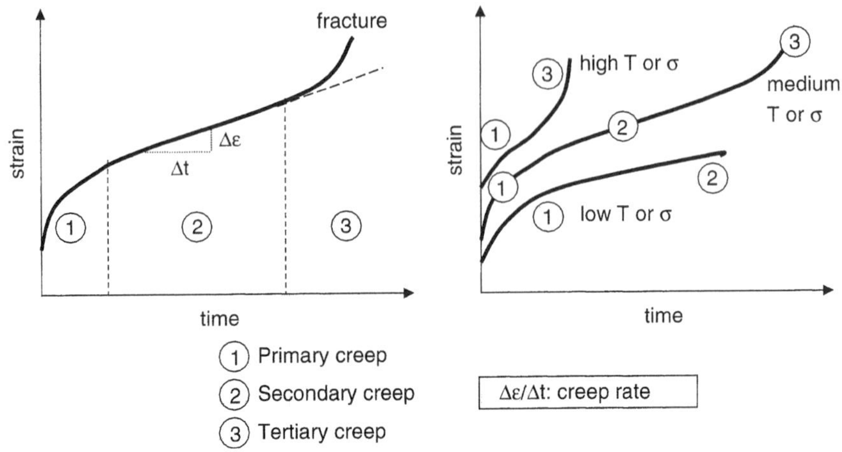

Figure 6.2. Illustration of some typical creep curves.

in service at very low strain rate and relatively low stress and is, in most cases, not desired. The elongation of a material under constant load is shown in Figure 6.2 as a function of time. Such a curve is called a 'creep curve'. It can be divided into three regions:

- Region 1 is the region of primary creep. When the load is applied, the material starts deforming. During deformation new dislocations are generated. They hinder the movement of each other and this causes a decrease of the creep rate ($\varepsilon^{o}=d\varepsilon/dt$).
- Region 2 is the region with constant creep rate; it is called 'secondary creep' or 'steady-state creep'. In this region strain hardening and dynamic recovery are in equilibrium. This situation is similar to the constant flow stress in a conventional hot deformation test (see Figure 4.10). In the conventional test, however, a constant strain rate is imposed, and after a certain time a constant flow stress is reached. In a creep test, a constant flow stress is applied and after a certain period a constant strain rate is observed. In both cases the deformation mechanisms are the same.
- Region 3 is called 'tertiary creep'. Owing to the development of voids, the effective stress increases and this increases the creep rate. The void growth soon leads to fracture.

Experimental creep curves are often 'less typical' as the one shown in Figure 6.2. At low temperature, the region of strain hardening dominates and the steady-state region is hardly reached. At high temperature or high stress, some void formation can already occur after Region 1 and the steady-state region is small or absent.

6.3.2 Creep mechanisms

6.3.2.1 Diffusion creep. When a load is applied to a sample, the material can try to achieve the imposed deformation by transport of vacancies through the crystal or along the grain boundaries. The deformation is achieved by the preferential formation of vacancies on those grain boundaries under tension and the destruction of vacancies at boundaries under compression. According to Lagneborg [1981], the constitutive equation for this type of diffusion creep can be expressed by

$$\varepsilon^{\circ} = 14\left(\frac{\sigma\Omega}{kTD^2}\right)D_{\mathrm{v}}\left(\frac{1+\pi\delta D_{\mathrm{b}}}{D\bullet D_{\mathrm{v}}}\right) \tag{6.1}$$

with Ω the atomic volume, D the grain size, D_{B} and D_{V}, respectively the diffusion coefficient for bulk diffusion and for grain boundary diffusion, δ the effective area of the grain boundary for grain boundary diffusion (see also Ray and Ashby [1971]). Equation (6.1) assumes that the ancillary processes of the formation and destruction rates at the boundaries and the required GBS, both require negligible energy dissipation and thus stress. If these assumptions are not valid, then the strain rate will be less than that given by Eq. (6.1) and the predicted linear dependence of stress may not be valid. When bulk diffusion is dominant, this type of creep is called Nabarro or *Nabarro–Herring creep*, and when grain boundary diffusion is dominant (at somewhat lower temperature), it is called *Coble creep*.

6.3.2.2 Dislocation creep. When in the deformation mechanism map, a stress is chosen above the region of diffusion creep, another deformation mechanism becomes active. Although the deformation is realized by dislocation glide and climb, the rate-controlling step is diffusion, which is needed to allow recovery of the dislocations by climb. Depending on temperature and stress, four different subclasses of dislocation creep mechanisms can be found: glide and bulk diffusion-controlled climb; glide and pipe diffusion-controlled climb; Harper–Dorn creep and power law breakdown. The first two mechanisms are characterized by the following constitutive equation (Mukherjee *et al.* [1969]):

$$\varepsilon^{\circ} = A\left(\frac{D_{\mathrm{B}}Gb}{kT}\right)\left(\frac{\sigma}{G}\right)^{n} \tag{6.2}$$

This is called the 'power law'. It is basically the same equation as Eq. (4.15), used to describe classical hot deformation. A and n are material-dependent constants, with A known experimentally, for fcc metals, to rise as the stacking fault energy (SFE) falls. D_B is the diffusion coefficient, b the Burgers vector and G the shear modulus. At higher stress, the power law relation is no longer followed and an exponential relation is found. It is called the 'power law breakdown'

$$\varepsilon^{\circ} = K\exp(\beta\sigma) \tag{6.3}$$

with K and β constant. Both regimes can also be described by the sinushyperbolic equation, defined in Section 4.16.

For some materials under certain conditions, a linear relation between stress and strain rate is found. This is called 'Harper–Dorn' creep.

6.3.3 Influence of the microstructure

An important aspect of creep is the 'time to rupture'. During creep small voids are generated, most commonly at grain boundaries. These voids will link up and form a crack that propagates and finally causes fracture. Certain microstructural features like grain size, precipitates, etc. have an influence on both the deformation mechanism and on the creep rate. This gives us a means to improve the creep resistance of a material. Figure 6.3 shows two deformation mechanisms diagrams for a MAR–M200 alloy used for turbine blades (9Cr; 10Co; 12.5W; 5Al; bal Ni, in wt%). An increase of the grain size from 100 μm to 1cm expands the domain of dislocation creep, but shifts also the iso-strain rate lines.

The normal working area of turbine blades (shaded area in Figure 6.3) falls for materials with a grain size of 100 μm clearly in the area of diffusion flow. For a homologous temperature of 0.5 and a load of 100 MPa, the creep rate is about 10^{-9}/s, which is about 3% per year. Using the same alloy but with a grain size of 1 cm, no deformation is detectable. For this reason, most high-temperature turbine blades are now directionally solidified single crystals. Besides the grain size, precipitates, particularly the coherent ordered $Ni_3(Al,Ti)$ in nickel-based 'superalloys' have a large influence on the creep rate, by impeding dislocation glide.

6.4. GRAIN BOUNDARY SLIDING

6.4.1 GBS and superplasticity

Superplasticity can be described as 'the capability of crystalline materials to undergo very large plastic deformations under tensile loading'. This means strains of typically several hundred per cent. In most cases, superplasticity is obtained at high temperature and small strain rate. The stress necessary to deform the material superplastically is rather low (in the order of 1–20 MPa). By around 1920, the first observations of 'abnormally high strains' were reported. Systematic investigations

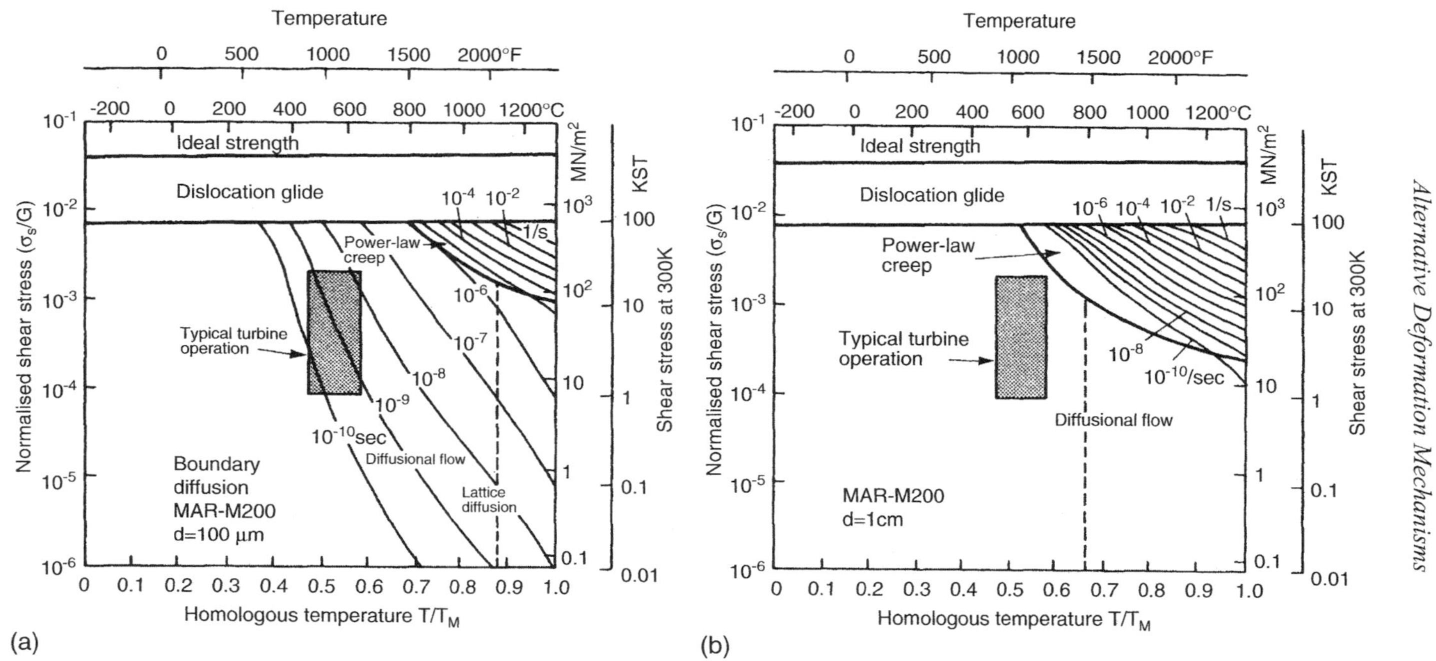

Figure 6.3. Influence of the grain size on the creep behaviour of a MAR–M200 Nickel alloy (a) 100 μm grain size, (b) 1 cm grain size (Gittus [1975]). (Courtesy Applied Science Publishers, London).

of the phenomenon only started in the 1960s and has in the meanwhile strongly increased. Today, the scientific interest for the subject is still very strong, despite the relatively few commercial applications of superplastic forming. The process is limited to some specific applications (such as components for the aeronautical industry) mainly because of the low forming velocities. Currently, it is most suited to form complex shapes where only a small number of such shapes are required.

The experimentally demonstrated mechanism for superplasticity is GBS. Sometimes it is also called 'fine structure superplasticity', because a fine grain is a necessary condition for this type of superplasticity. There also exists other types of superplasticity, e.g. 'transformation superplasticity' and 'internal stress-induced superplasticity' and superplasticity at very high strain rate, also called 'hyperplasticity'. Those kinds of superplasticity will not be discussed here. During GBS two grains slide over each other due to resolved shear stresses. In contrast to the mechanism of dislocation glide where grains are extended in the direction of deformation, grains more or less keep their aspect ratios after extensive GBS. At low strain rates, GBS gains importance over the dislocation glide mechanism and during superplastic forming it certainly is the prevalent and dominant deformation mechanism. It is however easy to show that GBS alone would result immediately in the formation of cavities in between the grains where grains are sliding apart and an overlap of material where they come together. It is clear that during superplastic forming an 'accommodation mechanism' has to be active which is able to postpone for some time the formation of cavities. To allow sliding to continue, material must be transported, usually by diffusion creep from regions where the grains are trying to overlap (regions under compression) and be deposited in regions where voids are trying to form (boundaries under tensile stress). Given the fine grain size and high diffusivity of superplastic forming processes, Eq. (6.1) indicates that the required creep should occur at the required rates even at the low stress of superplasticity.

6.4.2 Conditions for superplasticity

Experimentally it is found that one can obtain superplastic behaviour in a fine-structured material if certain conditions, which are not independent, are fulfilled. They can be considered as the framework in which superplasticity can occur.

(1) The material should have a high m-value at the deformation temperature (m is the strain rate sensitivity coefficient defined as $\partial \log \sigma / \partial \log \varepsilon^{\circ}|_{T,\varepsilon}$). A high strain rate sensitivity minimizes the tensile neck formation (localized strain) expected in the absence of strain hardening. With GBS accommodated by diffusion creep, the only mechanism leading to strain hardening is expected to be grain coarsening (Eq. (6.1)).

(2) The testing temperature should be sufficiently elevated. As a rule of thumb, one can state that the testing temperature should be higher than half the melting temperature (in Kelvin). This condition is necessary to allow the activation of the accommodating diffusion processes that go hand in hand with GBS.

(3) In most cases, superplasticity takes place at very low strain rates ($\varepsilon^\circ < 10^{-4}$–$10^{-2}$/s).

This can be understood considering Figure 6.4 and the simplest form of the constitutive equation:

$$\sigma = k(\varepsilon^\circ)^m \tag{6.4}$$

Suppose that two deformation mechanisms are possible: classical glide of dislocations (DISM) and GBS accommodated by diffusion creep. Note that m of Eq. (6.4) is the reciprocal of the n in Eq. (6.2). In Eq. (6.1), $m = 1$. Since the strain rate sensitivity coefficient for GBS is higher than for a DISM ($m > 1/n$), the $\sigma - \varepsilon^\circ$ curve (on a log scale) will have a higher slope for GBS than for dislocation creep. At higher strain rates a DISM will require a lower stress and will be dominant. At lower strain rates GBS will be dominant.[1]

(4) The material is required to be fine grained (e.g. 1–10 μm). This condition however goes together with the previous ones: the smaller the grain size, the higher

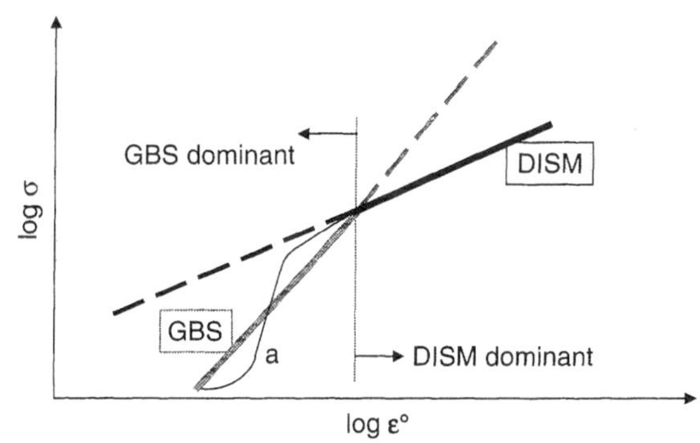

Figure 6.4. Scheme of the relation (6.4) with $m^{DISM} < m^{GBS}$.

[1] In the domain of GBS accommodated by diffusion creep, m is a function of the strain rate (curve a in Figure 6.4). At the lowest strain rates the value of m is usually reported as again falling below 0.5. This region is obviously of little industrial interest but scientifically it is a fascinating problem. The usual explanation is that grain boundaries are for some reason either not perfect sources and sinks for vacancies or they are unable to slide readily at very low stresses. Some form of threshold stress for these processes is postulated.

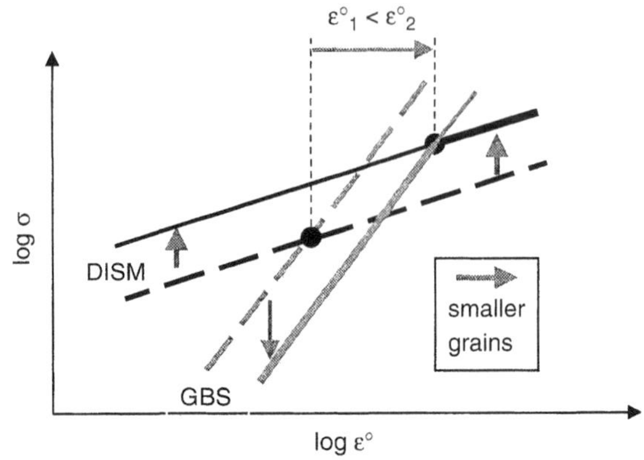

Figure 6.5. Influence of grain size on the relation between stress and strain rate for grain boundary sliding (GBS) and a dislocation mechanism (DISM).

the strain rate at which superplasticity is observed. To understand the influence of grain size, Eq. (6.4) needs some further explicitation.

For GBS, one can find a constitutive equation of the form:

$$\sigma^{\mathrm{GBS}} = kf'(\text{structure}, T)(D/b)^2 \varepsilon^{\circ m} \quad \text{with } m = 0.5 \tag{6.5}$$

For a classical DISM,

$$\sigma^{\mathrm{DISM}} = kf''(\text{structure}, T)(b/D)^2 \varepsilon^{\circ m} \quad \text{with } m < 0.5 \tag{6.6}$$

The influence of grain size is different for both mechanisms (Figure 6.5). A decrease of the grain size shifts the intersection point towards larger strain rates and thus extends the strain rate domain of GBS and superplasticity. The current interest in ultra-fine grained materials (Section 17.1) is partially due to the expectation that the smaller grain size will allow higher production rates (high strain rate superplasticity).

(5) A finely distributed 2nd phase, with the length scale of the grain size, is needed in order to prevent grain growth at the imposed low strain rates and high deformation temperatures. Several studies have shown significantly accelerated rates of grain coarsening in materials undergoing superplastic strain. The grain size is found, after superplastic deformation, to be coarser in the deformed gauge section than in the undeformed grip regions even though both regions have undergone the same times at high temperature.

(6) Superplasticity requires grain boundaries which can slide easily and also act as ready sources and sinks of vacancies. Unfortunately, the structure and

nature of grain boundaries has till now not been studied extensively and consequently it is not exactly known how they interfere with the phenomenon of superplasticity. It was found that GBS is favoured by the presence of high-angle grain boundaries rather than by low-angle grain boundaries. On the other hand, it seems that cavities originate more often at high-angle grain boundaries. If those cavities are linked with each other to form a network, they can easily coalesce into cracks. Obviously, it is not only the frequency of high- and low-angle grain boundaries which is important, but also their distribution in the material. This leads to the introduction of the notion 'grain boundary character distribution'. Furthermore it seems that, as for the case of recrystallization and grain growth, certain CSL boundaries (Coincidence Site Lattice; cf. Section 2.2.2.2) play an important role in a sense that they tolerate less GBS than a random grain boundary.

Failure during superplastic forming is most often caused by the formation of cavities (cavitation). Small cavities form at grain boundaries (triple points, precipitates) and when they grow towards each other during further deformation finally result in macroscopic cracks. Here again the nature of the grain boundaries plays a vital role.

A lot of materials can undergo superplastic deformation. The most important are a number of eutectic alloys, some iron alloys, titanium alloys, ceramics and several aluminium alloys. In the last series, one finds amongst others Al–Cu–Zr alloys (Supral), the AA7475 alloy (Al–Zn–Mg) and Al–Mg alloys such as AA5083 (Al–4.5Mg–0.7Mn) eventually alloyed with Cu. Most of them are commercially available. In Section 11.10, some technological aspects of superplastic forming are discussed.

6.5. TWINNING

6.5.1 Introduction

A twin is a region of a crystal in which the lattice orientation is a mirror image of the orientation of the rest of the crystal. Twins may form during crystal growth (annealing twins) and during deformation (mechanical or deformation twins, cf. Hosford [1993]).

In the present chapter we will mainly be concerned with mechanical twinning. It is the general belief that mechanical twinning occurs in most cases together with classical dislocation slip. In crystals with lower symmetry, like hexagonal metals, where the five independent slip systems, required to generate an imposed shape change may not be available, twinning can be extensive. Under shock loading, twinning does become more important, especially at low temperatures.

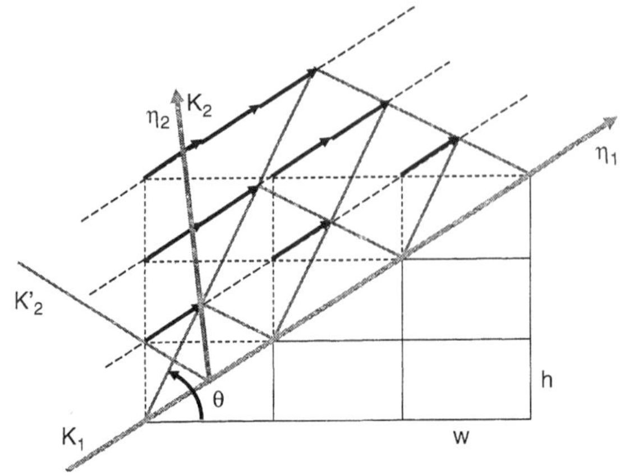

Figure 6.6. Formation of a twin by shearing over parallel atom planes.

Table 6.1. Twinning elements for rational twins: K_1 and K_2 are the two undistorted planes, η_1 the shear direction and η_2 a direction in K_2 perpendicular to the intersection of K_1 and K_2; h and w are defined in Figure 6.6 (Hosford [1993]).

Structure	K_1	K_2	η_1	η_2	h	w
fcc	{111}	{11−1}	⟨11−2⟩	⟨112⟩	a	a√2
bcc	{112}	{11−2}	⟨11−1⟩	⟨111⟩	a	a/√2
hcp	{10−12}	{−1012}	⟨10−11⟩	⟨−1011⟩	c	a√3

6.5.2 Twinning mechanism

Figure 6.6 illustrates the atom displacements (shearing) that leads to the formation of a twin. Similar to the dislocation slip mechanism, twinning is characterized by a specific plane (the 'mirror' or 'twinning' plane K_1) and a specific direction (the shear direction η_1), shown in Figure 6.6. K_1 is an invariant unrotated plane. Twinning can be described by a formal notation system. Besides K_1 and η_1, a second invariant plane K_2 and a second characteristic direction η_2 describe the twin. Both are indicated in Figure 6.6, with K'_2 the position of plane K_2 before twinning. Based on this notation, twins can be classified into three different types. The first type consists of twins for which only K_1 and η_2 are rational; for the second type only K_2 and η_1 are rational and for the third type (compound or rational twins) both planes and directions are rational. Table 6.1 gives an overview of the twinning elements for rational twins in fcc, bcc and hcp materials. It is important to notice that twinning is polarized: reversal of the η_1 direction will not produce a twin. This is an important difference between twinning and deformation by dislocation glide.

The amount of twinning shear (γ) and the angle of lattice reorientation (θ) (Figure 6.6) can be calculated from (Hosford [1993])

$$\gamma = \frac{w}{h} - \frac{h}{w} \quad \text{and} \quad \theta = 2\arctan\left(\frac{\mathrm{h}}{\mathrm{w}}\right) \tag{6.7}$$

Twinning occurs by a process of nucleation and growth. According to Christian and Mahajan [1995], there is no theoretical or experimental evidence to support a concept of homogeneous nucleation, so twins are thought to form only when a suitable defect configuration is present. The twin nuclei consist of faults that are bounded by partial dislocations (twinning partials) and grow normal to the twinning plane. For more details about these twinning mechanisms the reader is referred to overviews of Hull [1963], Christian and Mahajan [1995] and Mahajan [2002].

6.5.3 Influence of some parameters on twinning

A large number of external parameters may have an influence on twinning, e.g. temperature, strain rate, alloy composition, grain size, precipitates, crystal orientation, etc. Only the first four parameters will be discussed. More information can be found in Christian and Mahajan [1995].

In most bcc, fcc and hcp metals the importance of twinning will increase as the deformation temperature is lowered. It is thought that the twinning stress rises less steeply with decreasing temperature than the flow stress, so a transition from dislocation glide to twinning as the main deformation mode can be expected. This is especially true for bcc materials where the increase in flow stress with decreasing temperature is large. Twinning in fcc materials is less sensitive to temperature.

Strain rate has often the opposite effect as temperature. In general, the contribution of twinning to deformation will increase with increasing strain rate.

The influence of composition on twinning is rather complex, since compositional changes will in most cases simultaneously change the SFE and the amount of solutes or interstitials. It is often believed that a low value of the SFE will promote twinning. This is supported by the fact that low SFE materials such as stainless steel and brass show extensive twinning, while medium-to-high SFE materals like Cu and Al alloys do not twin under 'normal' strain rates. As the SFE decreases, the stacking faults become wider and cross-slip more difficult and mechanical twinning is favoured. For fcc materials, Venables found a parabolic relationship between increasing SFE and increasing stress needed to induce twins (Venables [1963]). Hirth and Lothe [1968] proposed a dimensionless SFE parameter normalized by the shear modulus (G) and the Burgers vector (b) as the base for a twinning criterion:

$$q = \frac{\text{SFE}}{Gb} \tag{6.8}$$

The lower the q, the higher the stress necessary to cross-slip and the lower the propensity for twinning. In bcc metals, interstitials seem to reduce or even inhibit the tendency for twinning, while substitutional solutes increase the tendency to deform by twinning. This can be explained by the hypothesis that the substitutional solutes affect the core structure and hence the mobility of the screw dislocations (Christian and Mahajan [1995]).

Twinning, like the related process of martensite formation, is well known to be greatly inhibited by reduced grain size (El-Danaf *et al.* [1999]).

6.5.4 Twinning and deformation

Twinning is a deformation mechanism that will be activated when material parameters (such as SFE and grain size) and processing parameters (such as temperature and strain rate) fulfil certain conditions, as discussed in the previous paragraph. In many cases twinning and dislocation glide can operate simultaneously and hence it can be expected that some interaction will occur. Twinning has an influence on the yield locus, strain hardening and texture development. The shape of the yield locus can substantially be affected by twinning. In most cases the yield locus becomes asymmetric. This is a direct consequence of the fact that the twinning shear is directional: shear in one direction, e.g. (111) [11−2] in fcc crystals, can lead to twinning, but shear in the opposite direction, (111)[−1−12], will not.

A good example to illustrate this, is given by Hosford [1993]. A cold-rolled Mg sheet has a strong basal plane texture, with the c-axis parallel to the normal direction (cf. Section 8.5.3). In a compression test (with sample axis parallel to the rolling direction), yielding occurs by twinning. In a tensile test, twinning is not possible, and deformation by slip is difficult because of the unfavourable orientation of the slip systems. As a consequence, the yield strength in tension is much larger than the yield strength in compression.

The influence of twinning on strain hardening has recently been analysed by Asgari *et al.* [1997] and El-Danaf *et al.* [1999]. They found a clear difference in strain-hardening behaviour between fcc metals with medium-to-high SFE and metals with low SFE. Polycrystalline materials deforming by slip alone, show mainly Stage III and Stage IV work hardening, as discussed in Chapter 4. Low SFE fcc materials exhibit a four-stage strain-hardening regime, as schematically

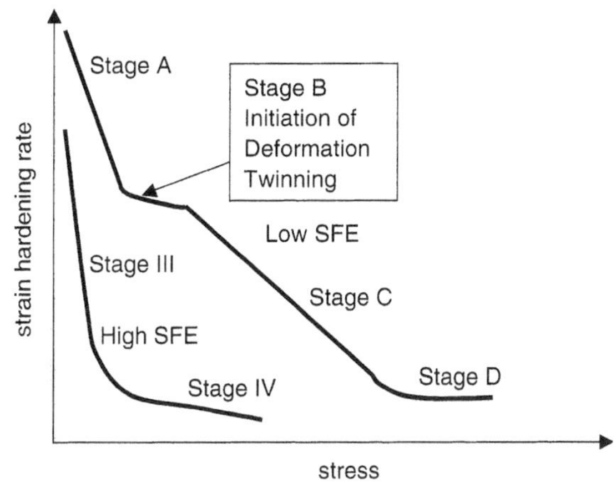

Figure 6.7. Schematic illustration of the strain hardening in polycrystalline fcc metals. After Kalidindi [1998a].

shown in Figure 6.7. The first stage (Stage A) is the same as Stage III hardening in conventional hardening (dynamic recovery regime). The initiation of the first deformation twins was found to coincide with a first plateau in the hardening curve (Stage B). Stage C corresponds to a decreased rate of primary twin formation and finally Stage D appeared to correlate with the onset of secondary twins that intersected the primary twins. Twin boundaries obstruct the motion of dislocations, as ordinary grain boundaries do. This is known as the Hall–Petch effect of twins.

The high strain hardening observed in many materials that deform (partially) by twinning, leads to interesting combinations of high strength and high ductility (large uniform elongation). Although a low SFE clearly promotes twinning in fcc alloys, careful measurements of the stress for the onset of twinning (El-Danaf *et al.* [1999]) proved that this twinning stress was independent of SFE for fcc alloys with a constant grain size. However, above a critical value of SFE, in fcc alloys with the same grain size, the extensive twinning leading to Stage B ceased. By increasing the grain size, twinning was found in alloys with somewhat higher SFE. The authors reported that SFE appeared to have only *indirect* effects on twinning. First, low SFE inhibits dynamic recovery allowing higher strain hardening, as seen in Stage A (which is identical to Stage III). Second, low SFE also inhibited slip-induced grain fragmentation. Allowing the initial grain size to remain essentially unchanged by the strain at which the critical stress for twinning was reached appeared to be required to obtain the extensive twinning and thus Stage B. This model accounted for the results reported and appears to describe for fcc metals how the enhanced strain hardening of extensive twinning can be achieved.

It is obvious that the local lattice reorientation that occurs during twinning, should affect the deformation textures of a material. However, in fcc metals, the volume fraction of twins is usually only a few per cent, so the direct effect of twinning is unlikely to account for the texture changes reported in fcc metals that twin. Low SFE fcc materials like Brass that readily twin have, after high strains, very different deformation textures than are seen for medium-to-high SFE fcc materials (Section 8.3.3). Recent studies have suggested that the texture difference seen in rolling for brass as against copper correlate not only with the onset of twinning (Stage B), but with the later onset of shear banding that is promoted by the twinning (El-Danaf *et al.* [2000]). So it appears that the strain localization associated with the heavily strain hardened twinned microstructures is responsible for the texture transition.

It has however been for many years an outstanding problem to incorporate mechanical twinning as a mode of plastic deformation into classical crystal plasticity models. One of the main problems is to handle the large number of possible crystal orientations created by twinning. A number of approaches have been reported in literature and the reader is referred to these papers for more information (Chin *et al.* [1969], Van Houtte [1978], Tomé *et al.* [1991], Kalidindi [1998b]). Through a combination of experiments and modelling the full effect of twinning on deformation microstructures may in the near future become more clearly understood.

LITERATURE

Boyer E. (ed.), "Atlas of Creep and Stress-Rupture Curves", ASM-International, Metals Park, Ohio, USA (1988).
Bressers J. (ed.), "Creep and Fatigue in High Temperature Alloys", Applied Science Publishers Ltd, London (1981).
Chandra N., "Constitutive Behavior of Superplastic Materials", Int. J. Non-Linear Mech., 37 (2002), 461–484.
Mahajan S., 'Deformation twinning', in "Encyclopedia of Materials: Science and Technology", Elsevier (2002).
Padmanabhan K.A. and Davies G.J., "Superplasticity", MRE-serie Vol 2, Springer, Berlin (1980).
Pilling J. and Ridley N., "Superplasticity in Crystalline Solids", The Institute of Metals, London (1989).
"Superplasticity", Agard lecture series N 168 (ISBN 92-835-0525-5) (1989).

Chapter 7

Phase Transformations

Chapter 7

Phase Transformations

7.1. INTRODUCTION

Phase transformations control many of the useful properties of practical metallic alloys. They occur during the standard heat treatments of metals which undergo allotropic transformations – such as steels and Ti and Zr alloys. In metallic systems with only one principal crystal structure alloying is widely practised to facilitate hardening by 2nd phase precipitation (e.g. Al alloys and Ni base superalloys). Major efforts have therefore been made to understand and control the wide variety of phase transformations that can occur in metallic systems. Given the importance of the subject many excellent textbooks on classical phase transformations are available (Christian [1975], Porter and Easterling [1986], Reed-Hill and Abbaschian [1992], Honeycombe and Bhadeshia [1995], Martin *et al.* [1997], Smallman and Bishop [1999], Bhadeshia [2001]) and the reader is advised to consult them for more details.

Most standard heat treatments, and associated phase transformations, are carried out on the final or near-final component to obtain the required properties. However, controlled thermo-mechanical processing (TMP) is also carried out in close 'co-ordination' with the phase transformations that can occur both at the deformation processing temperatures and during subsequent treatments. Practical examples described later on in Part III include the development of Dual phase (Section 15.2) and High-strength low-alloy (HSLA) steels (Section 15.3), Patented steel wires (Section 15.5), Zr alloys (Section 16.1), Ti alloy forgings (Section 16.2) and Al aerospace components (Section 14.4). Modern TMP requires a clear understanding of the basic types of phase transformations, i.e. the underlying thermodynamic principles and the rate-controlling mechanisms. This chapter aims to provide the reader with a summary of the general concepts of phase transformations; for more specialized analyses, this should be complemented by one or more of the above textbooks.

Phase transformations are generally categorized based on two transformation mechanisms: (I) Nucleation and growth and (II) Spinodal decomposition. The latter is usually limited to some very solute-rich alloys (more than 10 or 20 at% solute) under specific thermodynamic conditions (see, e.g. Cahn [1956, 1968], Christian [1975], Porter and Easterling [1986], Martin *et al.* [1997]). Nucleation and growth transformations comprise the immense majority of phase transformations in current materials. They are characterized by the requirements of creating a critical nuclei size and overcoming an activation energy barrier to enable the transformation to a more stable state. Type (I) transformations are often further classified as civilian (or diffusion-controlled) and military (or diffusionless). The diffusion-controlled transformations are typically 2nd phase precipitation or dissolution, eutectoid transformations and allotropic transformations involving solute partitioning. They imply solute atom diffusion over a period of time by atom movements in a concentration gradient. Military transformations on the other hand require all atoms of a transforming element to displace simultaneously by the same quantity – a well-known example is the martensitic transformation. To do this the interface between the new and parent phases has to be highly mobile, glissile in fact.

These standard classifications are not always perfectly well defined; for example, a bainitic transformation is a military transformation considered to involve diffusion (Bhadeshia [2001]), while a massive transformation is a diffusionless civilian transformation (Porter and Easterling [1986]). To overcome this classification difficulty, the terms reconstructive and displacive transformations have been suggested recently (Bhadeshia [2001]). A change in crystal structure has to be accompanied by a change in shape. In a transformation, where the change in shape is compensated, fully or in part, by mass transport, the transformation is termed reconstructive. Absence of mass transport would define the transformation as displacive. A summary, along with the basic definitions, of all these is included in Figure 7.1.

7.2. THE THERMODYNAMIC BASICS

The thermodynamic basics of phase transformations can be understood from the mixing behaviour of two, initially pure components (e.g. A and B) in a binary system. As shown in Figure 7.2, mixing two sets of atoms leads to new relative atom positions and therefore an energy of mixing is involved. The relative equilibrium positions of the atoms is determined by the enthalpy of mixing, ΔH_{mix}, whose sign controls the type of preferred configurations; zero, negative or positive values of ΔH_{mix} favour, respectively, randomness, ordering or clustering (hence phase separation). The effects are illustrated by plotting the Gibbs free energy G as a function of composition (X) in $G(X)$ diagrams (see Figure 7.3). In this figure, G is taken as $\Delta G_{mix} = \Delta H_{mix} - T\Delta S_{mix}$, recalling that ΔS_{mix} $(= -R[X_A \ln X_A + X_B \ln X_B])$ is positive, X_A and X_B being the respective mole fractions of A and B (≤ 1).

With negative ΔH_{mix} values, G–X has a typical cup-shaped appearance, see Figure 7.3a, as a consequence of the predominant negative energy contribution from the entropy of mixing, $T\Delta S_{mix}$. The only exception to the cup shape of G–X occurs when $\Delta H_{mix} > 0$ (and hence H has an inverse cup shape) and low T. Two

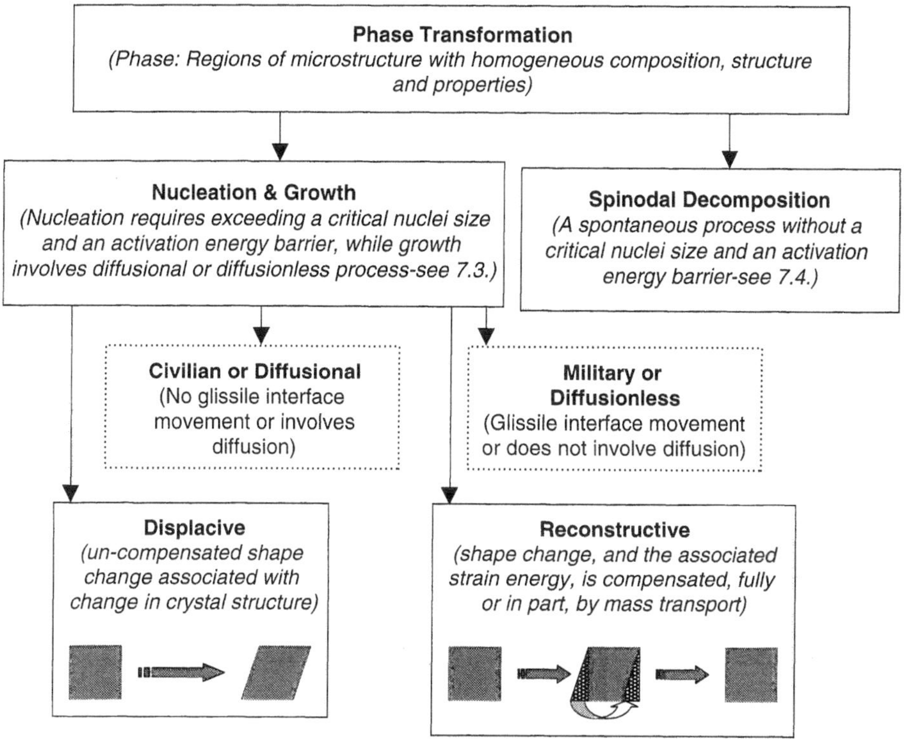

Figure 7.1. Classification of phase transformations based on the transformations mechanisms.

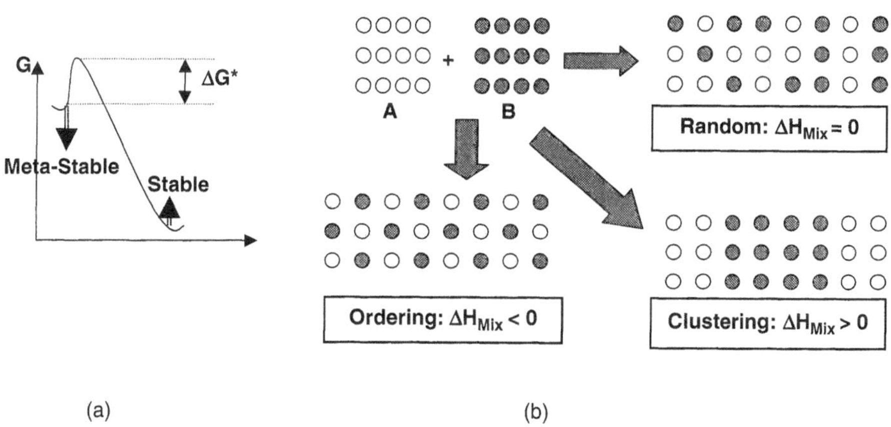

Figure 7.2. Mixing of pure components A and B atoms in a binary system. (a) The total energy determines the metastable or stable (lowest energy) nature of the phases – the metastable phase is 'retained' by ΔG^*, the activation energy barrier. (b) The relative equilibrium positions of the atoms are determined by the enthalpy of mixing.

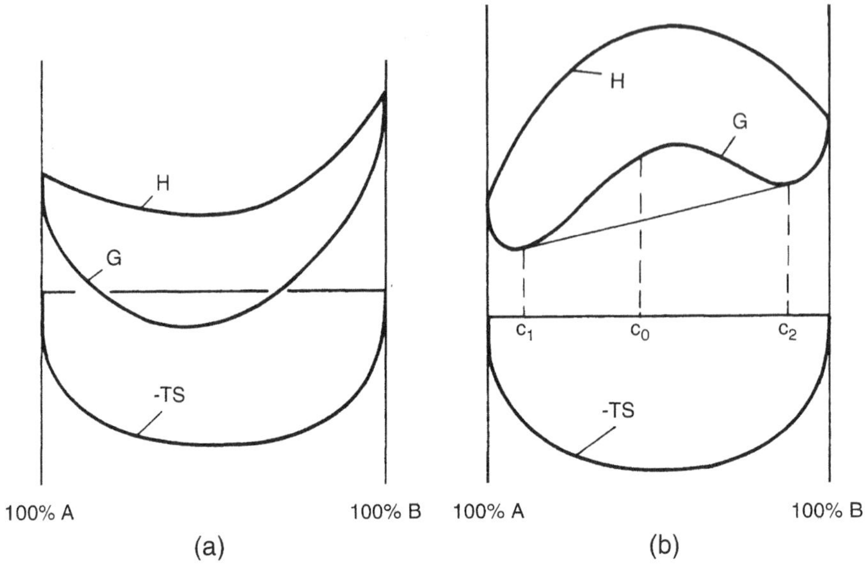

Figure 7.3. Schematic Gibbs free energy of mixing-composition (*G–X*) plots of binary solutions. (a) Normally, *G–X* appears cup-shaped, except for (b) $\Delta H_{mix} > 0$ and low *T*. In (b), a composition c_0 would tend to decompose into A-rich c_1 and B-rich c_2.

minima develop in the *G–X* plot leading to potential decomposition of a random solution into A-rich and B-rich phases (see Figure 7.3b).

The *G–X* plot gives the energy of a phase at a given composition *X*. In the case of several phases, the common tangent construction between the minima of the *G–X* plots provides the compositional ranges for phase stability (see the *G–X* curves in Figure 7.4). Phase diagrams are therefore the output of many *G–X* plots at different temperatures. It is important to appreciate, at this stage, what phase diagrams can and cannot offer. The phase diagrams provide information on the types of phases and their relative amounts and compositions. The other microstructural details are, however, a product of transformation mechanisms and kinetics and phase diagrams are incapable of bringing them out.

7.3. NUCLEATION AND GROWTH-TYPE TRANSFORMATION

In this section, the basics of nucleation and growth in phase transformations are covered. The section also tries to summarize a minimum knowledge base on kinetics, the formation of metastable phases and the mechanisms of invariant plane strain (IPS). Finally, a few specific nucleation and growth-type transformations,

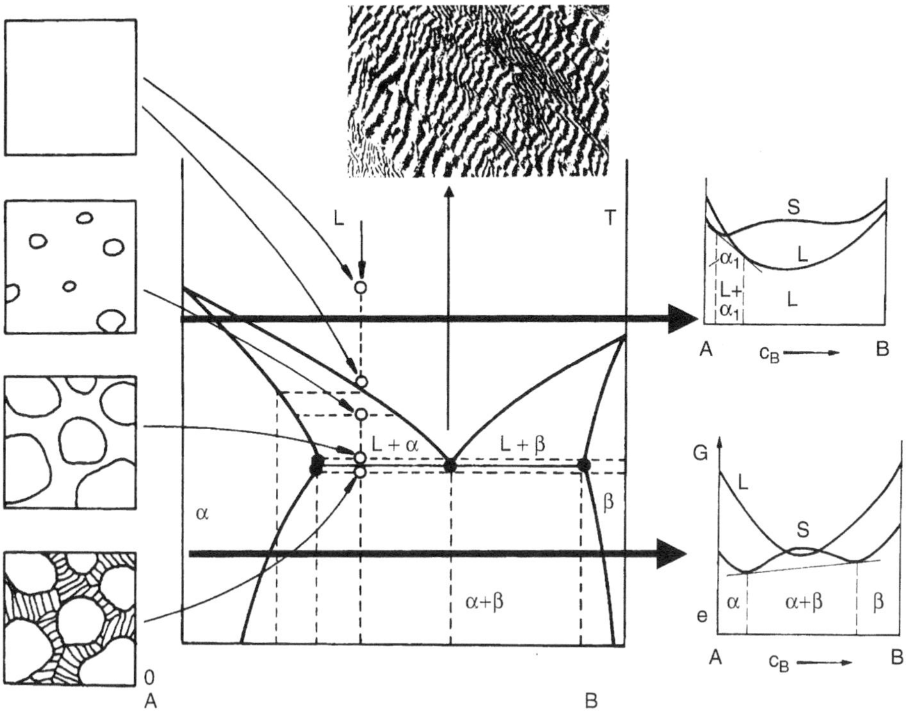

Figure 7.4. Schematic of a eutectic phase diagram, along with the constituent *G–X* plots for two temperatures. Possible microstructures at different locations of the phase diagram are also given.

both displacive and reconstructive – with interesting differences in mechanisms and/or structure, are given – these include eutectoid vs. discontinuous, IPS vs. massive, martensite vs. bainite and order–disorder transformations.

7.3.1 Nucleation

The energetics of nucleation in solid-state phase transformations is similar to that of solidification (see Section 2.2.1.1). The energy balance, Eq. (2.2), is simply modified by the addition of a (positive) term ΔG_S for the misfit strain energy between matrix and 2nd phase – per unit volume. Lattice strains (both plastic and elastic) are almost invariably associated with nucleation or the early stages of phase transformation. Incorporating ΔG_S into Eq. (2.2), the energy balance for homogeneous nucleation of a spherical particle of radius r can be obtained as

$$\Delta G_{\text{hom}} = \frac{4}{3}\pi r^3 \Delta G + \frac{4}{3}\pi r^3 \Delta G_S + 4\pi r^2 \gamma \tag{7.1}$$

$\Delta G = G_{II} - G_I$ represents the free energy difference per unit volume between the parent (I) and new (II) phases; for typical transformation conditions where the new phase is more stable, ΔG *is negative*. It should be recalled that for many phase transformations the order of magnitude values of ΔG are -100 MJ/m^3 (or -1 kJ/mole).

Correspondingly, the critical nucleus radius r^* and activation energy ΔG^* can be estimated as

$$r^*_{\text{hom}} = \frac{-2\gamma}{(\Delta G + \Delta G_S)} \tag{7.2}$$

$$\Delta G^*_{\text{hom}} = \frac{16\pi\gamma^3}{3(\Delta G + \Delta G_S)^2} \tag{7.3}$$

For heterogeneous nucleation (and ignoring the ΔG_S term) r^* and ΔG^* are

$$r^*_{\text{het}} = \frac{-2\gamma}{\Delta G} \tag{7.4}$$

$$\Delta G^*_{\text{het}} = \Delta G^*_{\text{hom}} S(\theta) \tag{7.5}$$

where the shape factor $S(\theta) = \frac{1}{2}(2 + \cos\theta)(1 - \cos\theta)^2$.

The nucleation rate is proportional to two exponential terms: $\exp(-\Delta G^*/kT)$ and $\exp(-\Delta G_M/kT)$, where ΔG_M is the activation energy barrier for diffusion and k the Boltzman constant:

$$N = c_1 C^* \left[\exp\left(\frac{-\Delta G^*}{kT}\right)\right]\left[\exp\left(\frac{-\Delta G_M}{kT}\right)\right] \tag{7.6}$$

where c_1 is a constant and C^* the number density of nucleation sites. Using Eq. (7.6), the behaviour of the nucleation rate (N) with T can be established (see Figure 7.5). As shown here, the two exponential terms define the trend in nucleation rate. The variation of N with T can be 'shifted' by changes in ΔG^* (see Figure 7.5d). For example, an increase in ΔG^* is expected in the following order: free surface → grain/phase boundaries → dislocations → vacancies → homogeneous. N would therefore follow an exactly opposite pattern.

It is to be noted that, with one or two rare exceptions, e.g. Cu–Co (Servi and Turnbull [1966]), homogeneous nucleation is not reported for the stable or equilibrium 2nd phase. The reason is the strong influence of the surface energy term

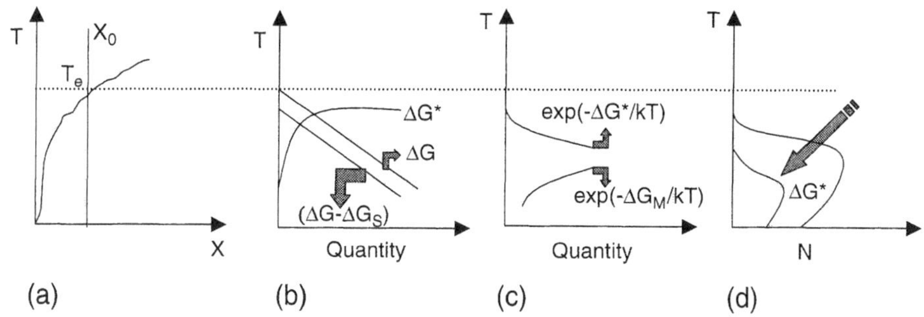

Figure 7.5. Establishing the trends in nucleation rate. After Porter and Easterling [1986]. (a) Section of a phase diagram indicating X_0 as alloy composition and T_e as equilibrium temperature. (b) Variation of ΔG and ΔG^* with T. (c) $\exp(-\Delta G^*/kT)$ and $\exp(-\Delta G_M/kT)$ vs. T. (d) N vs. T also showing its strong dependence on ΔG^*.

γ on ΔG^*; a large γ effectively makes N insignificant. Large γ values then favour the nucleation of metastable rather than stable phases (see Section 7.3.4).

During TMP the same basic thermodynamic and kinetic relations hold but the terms can be modified to allow for the changes in microstructure due to deformation processing. Very broadly, the thermodynamic relations for r^* and ΔG^* are not strongly affected as the energy terms γ and ΔG_S are very similar and since the chemical driving force ΔG is usually at least an order of magnitude greater than the dislocation energies introduced by plastic deformation. However, plastic deformation can strongly influence the kinetics via its influence both on the number density of sites for heterogeneous nucleation (along grain/phase boundaries and dislocations) and on diffusion rates (by vacancies and dislocations).

7.3.2 Growth

The growth of a 2nd phase takes place beyond r^*. As shown in Figure 7.6a, growth with composition change involves both diffusion and interface movement, while growth without any change of composition requires only interface movement. In the former case, the slower of the two steps would control the rate of growth. As shown in Figure 7.6b, the signature of a diffusion-controlled process is the local 'dip' in the composition profile near the interface.

Table 7.1 summarizes different types of growth. The growth associated with typical diffusional phase transformations may range from ordinary solute diffusion-controlled growth to ledge growth. The former, valid for non-coherent interfaces, follows a parabolic growth law ($r \propto \sqrt{t}$) for both spherical and planar 2nd phases (Aaron *et al.* [1970]); while the growth velocity (v) depends on $\sqrt{D}$ and ΔX_0, representing respectively the diffusivity and a term ($X_0 - X_e$), which is

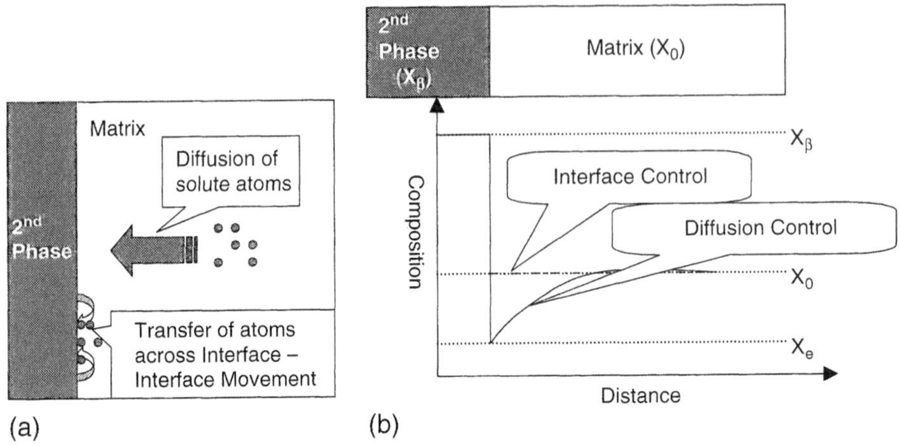

Figure 7.6. (a) Growth with change in composition between matrix and 2nd phase requires diffusion (or transfer of atoms to the interface) and interface movements (or transfer of the atoms across interface). (b) Schematic composition profiles for diffusion control (i.e. slowest step). X_β and X_0, respectively represent the concentrations of solute atoms in the 2nd phase and in the matrix. X_e represents the equilibrium solute concentration in the matrix phase.

Table 7.1. Generalizing the growth mechanisms associated with phase transformations.

Growth mechanism	Description
Diffusion controlled (Aaron *et al.* [1970], Martin *et al.* [1997])	Parabolic growth law: 2nd phase size $\propto \sqrt{Dt}$; $v \propto \sqrt{D/t}$ and $v \propto \Delta X_0$, where v and ΔX_0 are respectively the growth velocity and compositional undercooling
Ledge controlled (Laird and Aaronson [1969], Martin *et al.* [1997])	Valid for coherent and semi-coherent interfaces (and also smooth solid–liquid interfaces). The thickening is intermittent, possibly by ledge nucleation
Glissile interface movement (Porter and Easterling [1986], Bhadeshia [2001])	Interfacial dislocations glide in a co-ordinated manner, resulting in interface movement (also see Figure 7.8a, b). This is associated with IPS (invariant plane strain – Section 7.3.5) transformations
Competitive growth (Porter and Easterling [1986])	Diffusion-controlled coarsening of 2nd phase particles. The origin (see Figure 7.8) is the increased solubility of fine particles (Gibbs–Thomson effect). The thickening rate is proportional to D, γ and X_e (equilibrium solubility)

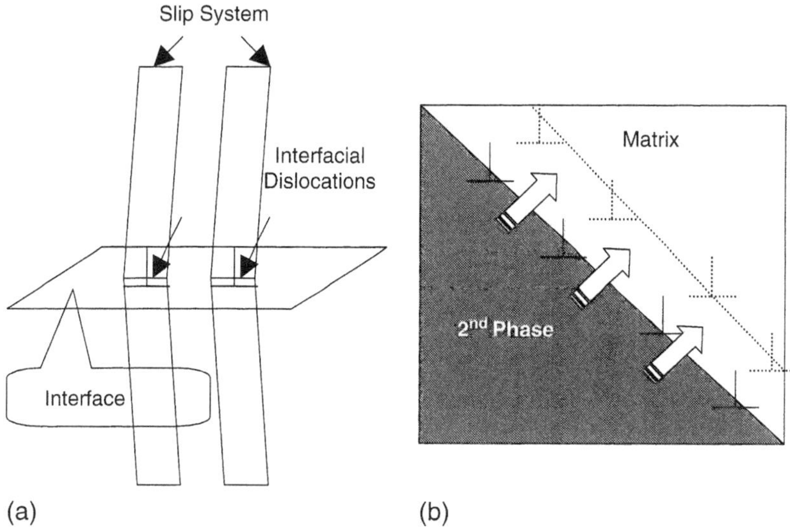

Figure 7.7. (a) Schematic of a glissile interface. The slip systems for the interfacial dislocations are shown. (b) A co-ordinated movement of the interfacial dislocations along the slip systems can move the interface.

proportional to the compositional undercooling ($T_0 - T_e$) (for more details, see Porter and Easterling [1986]). Ledge growth, on the other hand, is 'different' from standard diffusion-controlled growth (Porter and Easterling [1986], Martin *et al.* [1997]) – the thickening is typically intermittent (Laird and Aaronson [1969]), the pauses in time being normally attributed to the nucleation of ledges.

The co-ordinated movement of glissile dislocations, as in Figure 7.7a, b, can move the glissile interface (Porter and Easterling [1986], Smallman and Bishop [1999]). Such an interface movement is associated with IPS (see Section 7.3.5) transformations and involves shape change – a change that is accommodated locally through slip and/or twinning.

Competitive growth, or Ostwald ripening of particles, is not strictly associated with a phase transformation, but can determine the size distribution for the 2nd phase. The growth rate is proportional to D, γ and X_e and inversely proportional to the square of the particle radius.

7.3.3 Kinetics

The kinetics of a phase transformation would involve the kinetics of nucleation and growth. They are usually described by TTT (time–temperature–transformation) diagrams and/or by a JMAK (Johnson–Mehl–Avrami–Kolmogorov kinetic) analysis

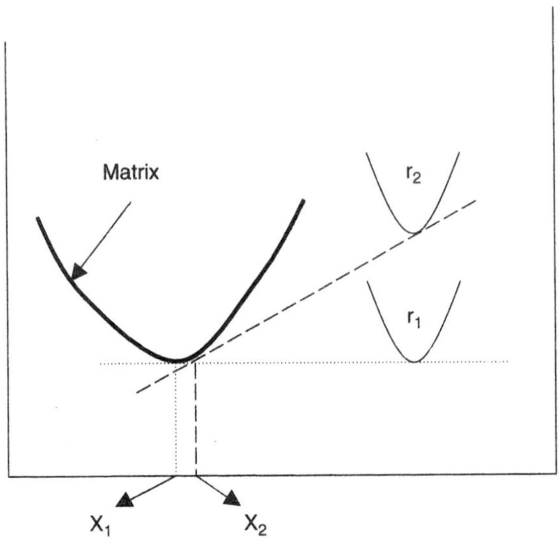

Figure 7.8. Thermodynamic origin of competitive growth. r_1 and r_2 represent the $(G–X)$ curves for the 2nd phase – size of $r_1 > r_2$.[1] As given by the common tangent construction, X_1 and X_2 are the maximum solute solubility, at equilibrium, between matrix and r_1 and r_2. X_2 being larger than X_1, a diffusion flux of solute from r_2 to r_1 is expected. This is the thermodynamic rationale for competitive growth.

(for more details on JMAK, the reader may refer to Sections 5.3.3 and 10.3.1). Figure 7.9 shows an example of the basics of a TTT diagram. Phase transformation data, from several (sigmoidal) curves of fraction transformed vs. time at different temperatures, are combined to provide C-shaped TTT curves representing per cent transformation as a function of time and temperature. The basic diagram does not, however, describe the scale of the microstructure. As shown in the example of Figure 7.9, the difference in cooling rate (FC and AC) is not only reflected in per cent of bcc β, but also in the relative morphology and continuity of the β lamellae.

7.3.4 Formation of metastable phases

One of the important issues in phase transformations is that of metastable phase formation. The formation of stable or equilibrium phase is often preceded by a metastable phase or a sequence of metastable phases (Christian [1975], Porter and Easterling [1986], Reed-Hill and Abbaschian [1992], Martin *et al.* [1997]). The latter are able to form since, unlike most stable phases, they can have high homogeneous

[1] Finer particle has more free energy because of Gibbs–Thomson effect – for more details, the reader may refer to Porter and Easterling [1986].

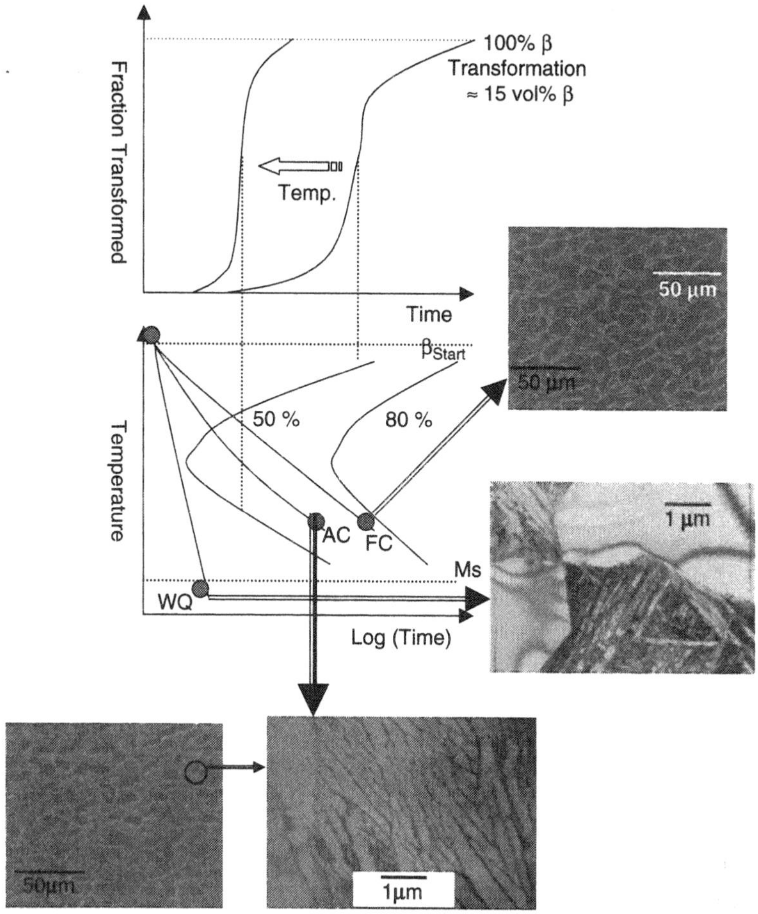

Figure 7.9. Schematic TTT diagram for Zr–2.5 wt% Nb (from bcc β to hcp α) as derived from individual C-plots for transformation percentages. The figure shows three different cooling rates – FC (furnace cooled), AC (air cooled) and WQ (water quenched). Though the formation of martensite in WQ and the lamellar $\alpha-\beta$ structures in FC and AC are expected, the TTT diagram does not predict the relative fineness of $\alpha-\beta$ lamellae between FC and AC. After Kiran Kumar [2005].

nucleation rates. Their interfacial energy is usually significantly lower than that of the stable phase and this energy difference is sufficient to compensate for the lower driving force, so their ΔG^* is decreased, Eq. (7.6), enabling rapid formation of metastable phases. Figure 7.10 illustrates the schematics of possible *G–X* and phase diagrams and TTT diagrams for a hypothetical transformation sequence: $\beta'' \rightarrow \beta' \rightarrow \beta$, in the increasing order of interfacial energy and stability.

A large number of precipitation reactions in common alloys give rise to metastable phases such as the hardening phases in Al and Ni base alloys.

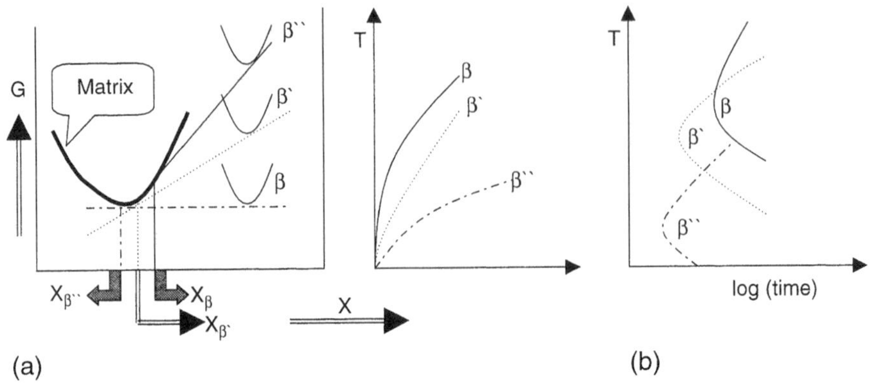

Figure 7.10. (a) Schematics of *G–X* and phase diagram for a 'sequence' of phase transformation: $\beta^{``} \to \beta^{`} \to \beta$, in the increasing order of interfacial energy and stability. (b) Schematics of TTT diagram for the same sequence of phase transformations.

7.3.5 Invariant Plane Strain transformation

The formation of martensite is often viewed from a TTT diagram. In steel or in Zr (see Figure 7.9), a fast cooling rate may 'suppress' diffusion-controlled transformations ($\alpha-\beta$ lamellar structures in Zr or classical pearlitic structure in steel) and form martensite. To a beginner in metallurgy, martensite is often associated with a crystal structure (e.g. bcc to bct in carbon steel) or with the athermal nature (again associate with Fe–C system). It needs to be pointed out, at this stage, that such definitions are incorrect – for example, between different alloy system the crystal structure of martensite can vary (e.g. hexagonal closed packed (hcp) martensite in Zr) and isothermal martensitic transformation also exists (Honeycombe and Bhadeshia [1995]).

The best way to define martensitic transformation is to classify it as a diffusionless IPS transformation. IPS represents homogeneous shear, in which (i) each atom of the parent unit cell retains its position in the product (though interatomic distances and angles may change (Honeycombe and Bhadeshia [1995])) and (ii) the interfacial plane between parent and product (the so-called habit plane) remains invariant, or does not rotate away. It is to be noted that though IPS is often identified with martensitic transformation, it is also valid for bainitic transformation and formation of acicular ferrite (Honeycombe and Bhadeshia [1995], Bhadeshia [2001]). However, formation of bainite and acicular ferrite also involves diffusion.

The theory of IPS transformation is phenomenological and the present section makes an attempt to summarize this theory. For further details on IPS transformation, the readers may refer to Honeycombe and Bhadeshia [1995] and Bhadeshia [2001].

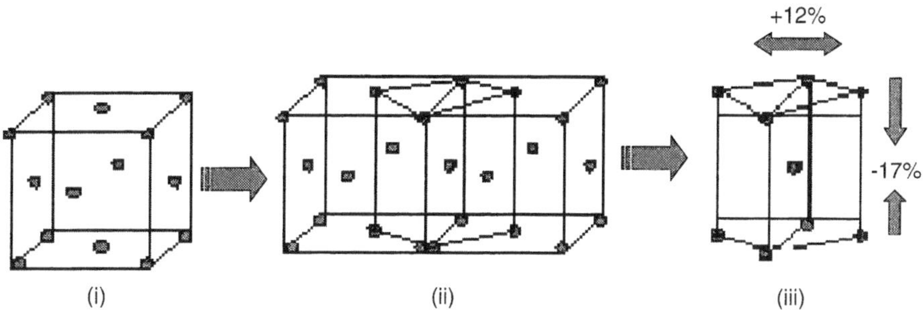

Figure 7.11. (i) Schematics of austenite unit cell – changing into martensite unit cell. After Honeycombe and Bhadeshia [1995]. (ii) A bct unit cell can be 'curved' from two fcc austenite unit cells. (iii) The bct unit cell would require approximately 17% compression and 12% expansion for conversion into bcc. This is B – the Bain strain.

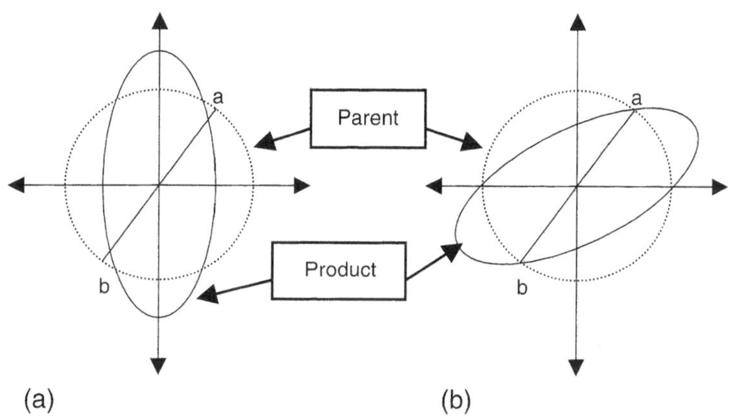

Figure 7.12. Introducing the concept of R to maintain invariant line strain (ILS). After Bhadeshia [2001]. (a) Shows the schematic of a parent sphere changing into product ellipsoid by pure B. In such a case, maintaining an ILS would not be possible – as any line common to parent and product (e.g. 'ab') would develop discontinuity. (b) To maintain continuity, a rigid body rotation (R) is necessary.

As shown in Figure 7.11, conversion of parent (austenite) into product (martensite) would involve transformation strains – generalized as Bain strain (B). B alone is, however, incapable of providing invariant line strain (ILS), the latter being necessary in the habit plane to ensure a glissile interface (Bhadeshia [2001]). Combining B with rigid body rotation (R) would 'fix' this problem (see Figure 7.12). No rotation can, however, convert ILS into IPS.

Observations, on the other hand, have confirmed the IPS nature of the transformation. This was mainly established from apparent continuity of scratches – if the habit plane was 'variant', then discontinuity of scratches on the parent–product interface must take place.

The theory of IPS transformation assumes that $BR = P_1P_2$, where P_1 is the IPS (also observed macroscopically) and P_2 a second homogeneous shear. In other words, P_1 and P_2 combined can equate BR, while only P_1 is observed macroscopically. To resolve this apparent dilemma, it is considered that the effect of P_2 is cancelled macroscopically by inhomogeneous lattice-invariant deformation, slip and/or twinning.

The IPS transformations can be stress assisted or strain induced (Patel and Cohen [1953], Christian [1982], Honeycombe and Bhadeshia [1995], Bhadeshia [2001]). These have strong relevance in TMP, e.g. – the transformation-induced plasticity (TRIP) in Section 15.2, and also to TMP failures (Chapter 15). The stress-assisted and strain-induced IPS are caused by the enhancement of ΔG (driving force of transformation) by $\Delta G_{\text{mechanical}}$ (free energy change provided by the external stress system) – the enhancement being reflected on the increase in IPS start temperature (i.e. M_S or B_S representing respectively the martensite or bainite start temperatures, Bhadeshia [2001]). If the domain of temperature increase corresponds to an energy state (or temperature) below yielding, then it is called stress assisted. If it exceeds yielding, then the IPS is referred as strain induced. It needs to be noted that the increase in IPS start is possible till a critical temperature (Patel and Cohen [1953], Christian [1982]), M_d or B_d, respectively for martensite and bainite.

7.3.6 Selected examples on different mechanisms and/or structures

As shown in Table 7.2, the microstructures of the same alloy can differ remarkably through the influence of heat-treatment conditions on nucleation and growth mechanisms. This section summarizes some of these classically different structures and mechanisms.

7.3.6.1 Eutectoid vs. discontinuous precipitation. An eutectoid or near-eutectoid transformation ($\alpha \rightarrow \beta + \gamma$) can form a lamellar structure. Examples of such lamellar structure are the so-called pearlitic structure in the Fe–C system or the lamellar structures of two-phase Zr alloys (see Figure 7.9). Though the so-called pearlitic transformation was originally referred to (Mehl and Hagel [1956]) as 'edgewise nucleation and sidewise growth' (see Table 7.2), subsequent studies (Honeycombe and Bhadeshia [1995]) have indicated three types of nucleation sites – clean austenite boundaries and nucleation on cementite and ferrite. Branching of lamella during growth has also been observed.

Table 7.2. Schematics of different possible microstructures in the Fe–C system.

Name	Schematic of the microstructure	Comments
Grain boundary allitromorphs	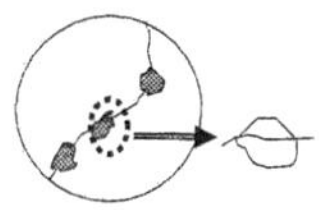	Heterogeneous nucleation at grain boundaries. The grain boundary 2nd phase may have faceted interface (coherent or semi-coherent) with one grain and non-faceted with the other
Grain interior idimorph	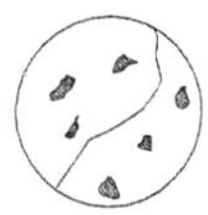	Grain interior heterogeneous nucleation – sufficient undercool ing is required. The shape of the 2nd phase would depend on the combined effects of nucleation and growth, interfacial energy and misfit strain energy
Lamellar	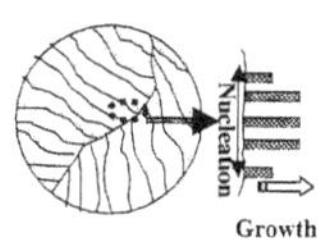	The so-called eutectoid or pearlitic structure. Originally referred as 'sidewise nucleation and edgewise growth' – branching of the lamellae may also play a strong role on the morphological appearance of the nodules
Widmanstätten		Based on nucleation site – can be both 'intra'- and 'inter'-granular. Alternate theories were proposed on the growth of Widmanstätten plates and no general agreement has yet been reached
IPS	Figures 7.15 and 7.17	The microstructural differences are based on involvement of diffusion and carbide precipitation (between martensite and bainite) and nucleation sites (between bainite and acicular ferrite)

A classical example of 'exploiting' eutectoid transformation is in patented steel wires (Section 15.5). Fine pearlite colony (through eutectoid transformation) is further refined to form a real 'nano-composite', a technology perfected earlier. The technology of patenting is based on an 'optimized' undercooling – the inter-lamellar spacing is inversely proportional to undercooling ΔT (this relationship has been observed experimentally – see Bogers and Burgers [1964]), while growth of pearlite colony is proportional to $(\Delta T)^2$.

[2] Discontinuous solute concentration change, between α and α', across the cell front.

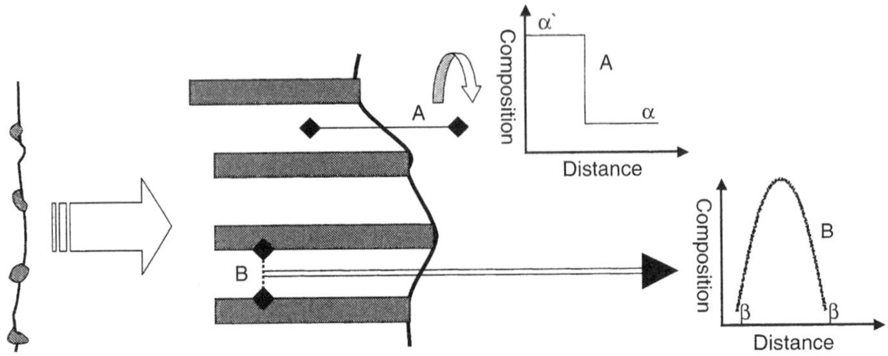

Figure 7.13. Schematic of a discontinuous or cellular precipitation – $\alpha' \rightarrow \alpha + \beta$, where α' is a supersaturated solid solution, α the same phase with lower solute content and β the solute-rich equilibrium 2nd phase. Schematics of the compositional profiles along the lines 'A' and 'B' are also given.

A discontinuous precipitation, see Figure 7.13, provides a similar microstructure as that of a eutectoid transformation, but the mechanisms are quite different (Porter and Easterling [1986]). As shown in Figure 7.13, the transformation takes place 'discontinuously'[2] behind the moving cell front, e.g. $Mg_{17}Al_{12}$ (β) precipitation in Mg–9 at% Al (α).

7.3.6.2 IPS vs. massive transformation. IPS, basic mechanism is discussed in Section 7.3.5, does not require diffusion – though specific IPS transformations (e.g. bainite – Section 7.3.6.3) may also involve diffusion. The massive transformation (Porter and Easterling [1986]), on the other hand, is a diffusionless transformation, without IPS or glissile interface movement. As shown in Figure 7.14a, Cu–38 at% Zn is stable as β at about 800°C, $\alpha + \beta$ between 500°C and 800°C and as α below 500°C. A fast cooling from β can suppress Cu-rich α precipitation, resulting in retention of β below 500°C. It is then possible for β to transform into α *without any long-range diffusion – diffusionless civilian transformation.*

7.3.6.3 IPS – martensite vs. bainite. Martensite and bainite both involve IPS (Honeycombe and Bhadeshia [1995], Bhadeshia [2001]). Microstructurally both may show the main features of Figure 7.15 – surface relief and internal twinning (and/or slip). The structures of martensite and bainite cannot be distinguished from crystallography, both may exhibit similar orientation relationships and self-accommodating nature of the crystallographic variants (see Figure 7.17). The only way to distinguish between martensite and bainite is from the presence of 2nd phase (see Figures 7.16 and 7.17). The upper bainite, with low ΔT, has presence

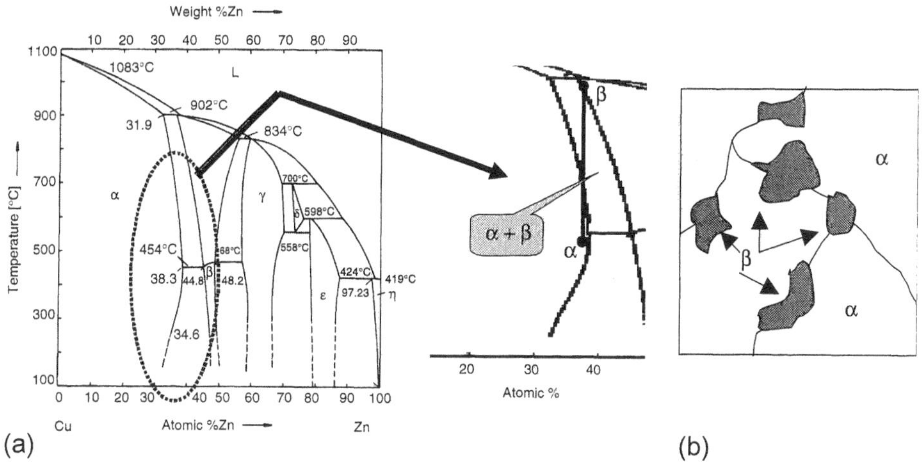

Figure 7.14. (a) Cu–Zn phase diagram – showing massive transformation at approximately 38 at% Zn. (b) Schematic of the massive β grain forming at $\alpha-\alpha$ boundary.

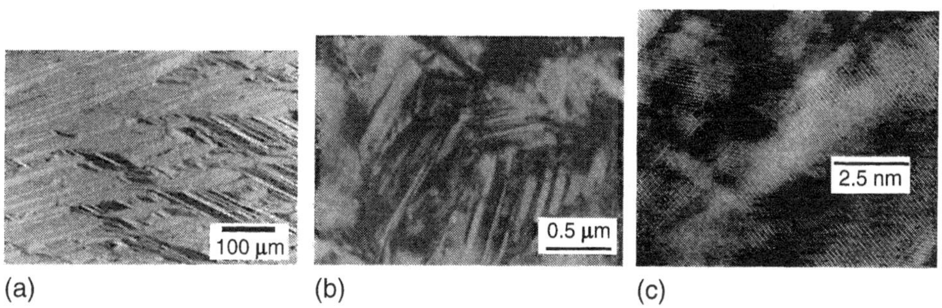

Figure 7.15. Martensite in Ni–Ti. (a) Optical micrograph under polarized light, (b) TEM and (c) HREM image of internal twinning. Courtesy of Madangopal and G.K. Dey, BARC, India.

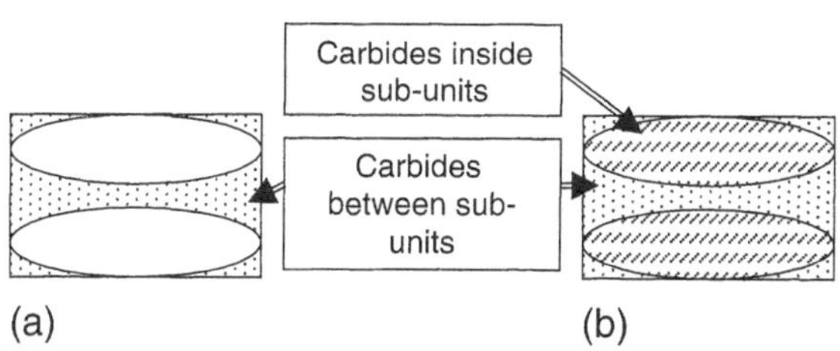

Figure 7.16. Schematics of (a) upper and (b) lower bainite. After Honeycombe and Bhadeshia [1995]. In (a) ferrite sub-units are separated by 2nd phase (typically carbides), while in (b) carbides inside the sub-units are observed at an approximate angle of 60° with the axis of bainite plate(s).

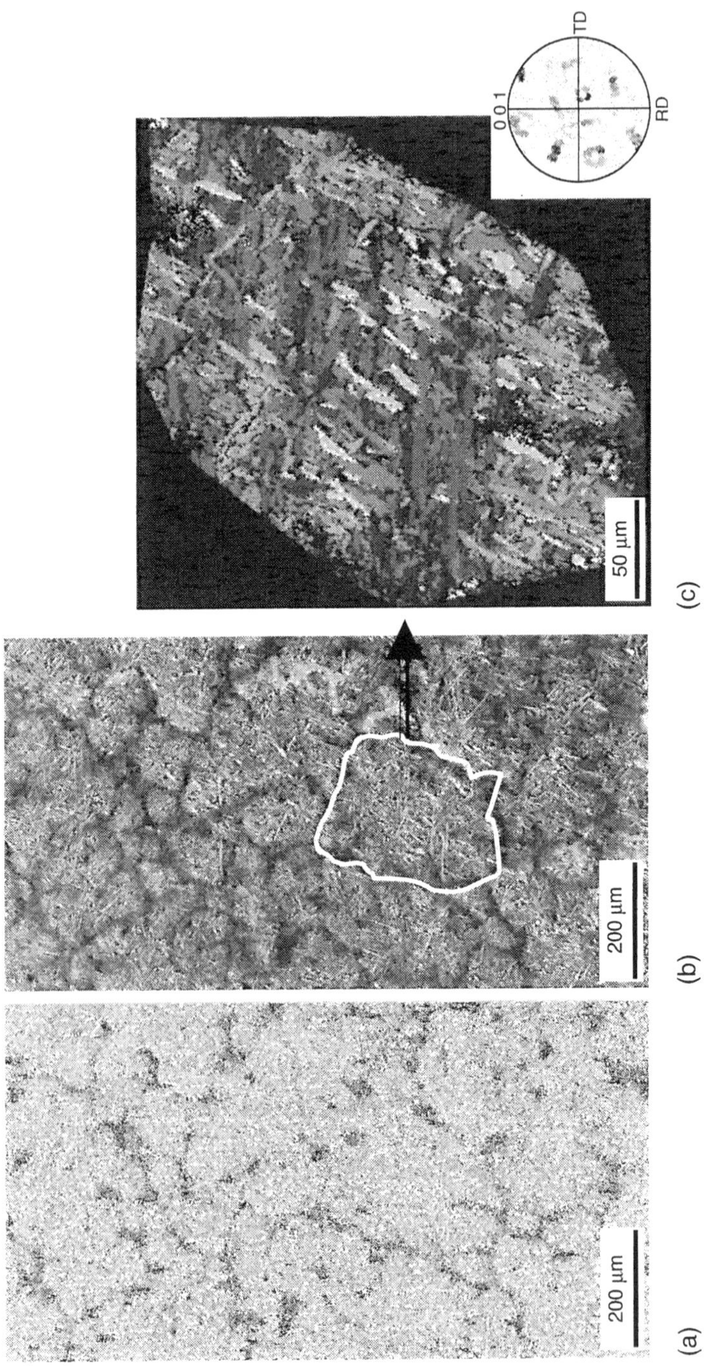

Figure 7.17. EBSD images of upper bainite in medium carbon steel. (a) Phase map – white ferrite, yellow carbide and black non-indexed points. (b) Image quality map of the same region. (c) A zoomed up region from (b) is shown in different colours – colours corresponding to different crystallographic variants. The variants and the relative locations of the sub-units can be explained from self-accommodation or minimization of strain energy. Courtesy of V. Pancholi and Madangopal, *Research in Progress*.

of carbides between the ferrite sub-units (as in Figure 7.17c, the ferrite sub-units are approximately single crystals); while in lower bainite, higher ΔT or more driving force (high ΔT or more carbon content) for carbide formation, sub-unit interior carbides are also observed. The bainitic transformation has been a subject of large controversies, with different schools proposing independent or simultaneous carbide precipitation. Though direct experimental evidence in support of either of the mechanisms would be difficult to acquire, indirect evidences (and also the majority opinion) seem to be tilted more towards IPS with simultaneous carbide precipitation.

7.3.6.4 Order–disorder transformation. As shown earlier in Figure 7.2, $\Delta H_{mix} < 0$ would make a preference for dissimilar bonding (A being preferred next to B) – this is the basis of the so-called ordering or ordered intermetallics (Porter and Easterling [1986]). Three points need to be brought out for any discussion on ordering:

1. Ordering, and the nature of the ordered unit cell, needs a specific stoichiometry (Figure 7.18a).
2. Ordering takes place below a critical temperature (T_C). Above T_C the entropy factor ($-T\Delta S$) dominates (Figure 7.18a, b).
3. An ordered structure has another type of boundary – a boundary between different ordered domains (see Figure 7.18c). These APBs (anti-phase boundaries) can have significant effects on both the mechanical properties and order–disorder transformations of intermetallics (Christian [1975], Porter and Easterling [1986], Reed-Hill and Abbaschian [1992], Smallman and Bishop [1999]).

The LRO or long-range order (as in Figure 7.18b) can be generalized as

$$\text{LRO} = \frac{P_{AB} - P_{AB}^{\text{random}}}{P_{AB}^{\text{maximum}} - P_{AB}^{\text{random}}} \tag{7.7}$$

where P_{AB}, P_{AB}^{random} and P_{AB}^{maximum}, respectively represent number of AB bonds in the specimen, number of AB bonds in a random structure and maximum number of AB bonds possible. LRO may change suddenly or more gradually with temperature (see Figure 7.18b). This represents the so-called order–disorder transition – a transition which is of practical importance to any TMP of intermetallics. Plastic deformation and also recovery/recrystallization are extremely difficult below T_C and hence a typical TMP of intermetallics always involves temperatures above T_C.

(a)

(b)

(c)

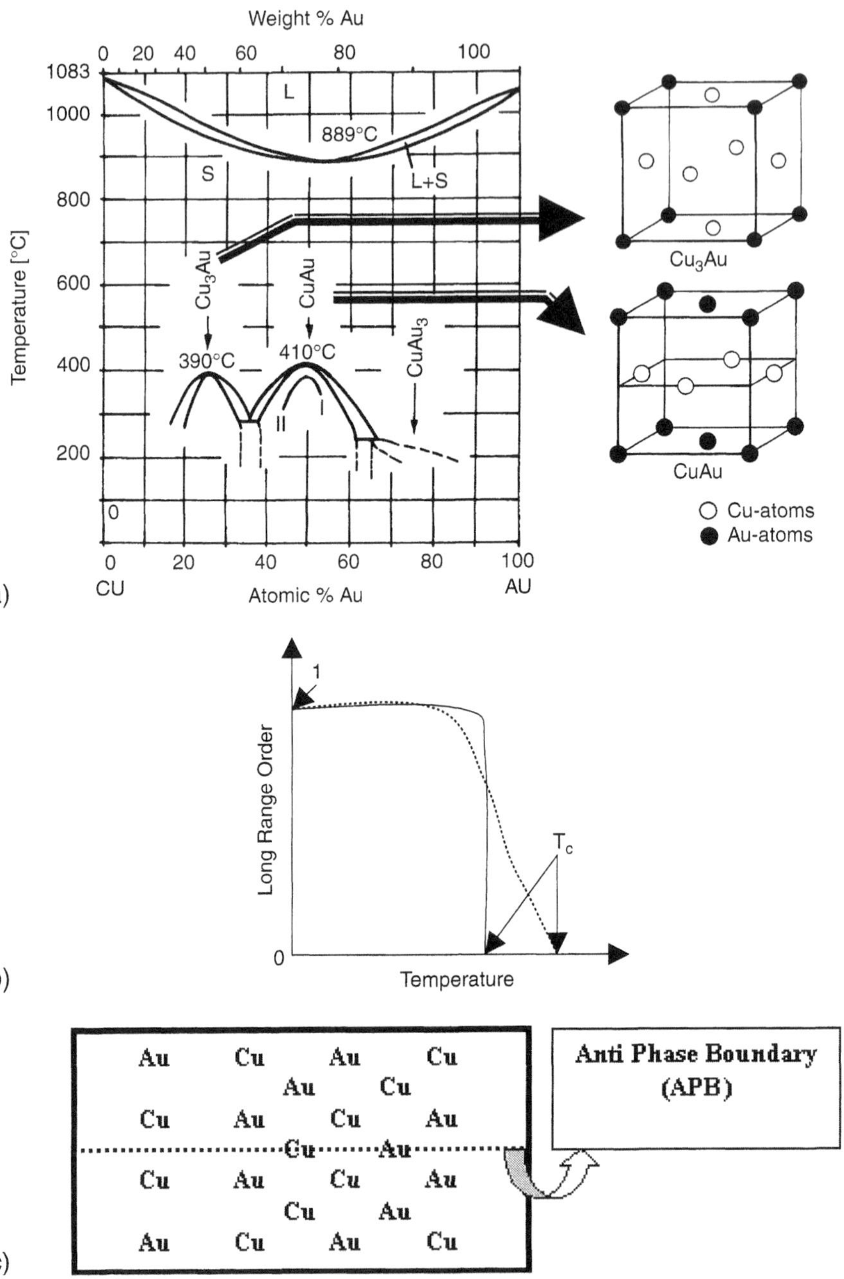

Figure 7.18. (a) Cu–Au binary phase diagram showing the presence of Cu_3Au and CuAu-ordered intermetallic phases – the unit cells of the respective phases are also shown. (b) Schematic of the LRO vs. temperature. (c) Schematic of 2D CuAu lattice showing presence of anti-phase boundary (APB).

7.4. SPINODAL DECOMPOSITION

As pointed out earlier in Section 7.3.1, TMP does not violate the thermodynamic basis of phase transformation. TMP, especially plastic and elastic stresses, can, however, affect the kinetics of transformation and the nucleation and growth. These, in turn, can generate significantly different microstructures – an issue which is often exploited in product and process developments. Almost all the case studies, in Part III, are examples of such exploitation.

The different mechanisms of phase transformations described so far are often witnessed in commercial alloys and in turn are relevant to the TMP. The last section of this chapter, spinodal decomposition, is, however, relatively exotic for TMP. Even then a 'minimum' on spinodal decomposition is included in the present section, as an overall understanding of phase transformation remains incomplete without an understanding of spinodal decomposition. For more details on spinodal decomposition, interested readers may refer classical textbooks on phase transformation.

A phase transformation may take place 'spontaneously' (Cahn [1956, 1968], Porter and Easterling [1986], Martin *et al.* [1997]) – without a critical nuclei size (r^*) and the need to exceed an activation energy barrier (ΔG^*). This is the so-called spinodal decomposition. As shown in Figure 7.19, the miscibility gap in a binary phase diagram typically corresponds to a tendency of phase separation,

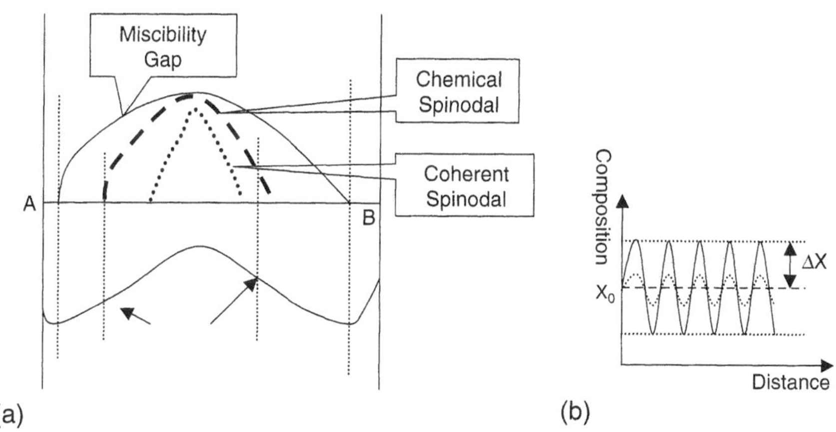

Figure 7.19. (a) Binary phase diagram (A–B) showing miscibility gap and corresponding *G*–*X* at room temperature. The chemical spinodal is defined by the region of the *G*–*X* with $\frac{d^2G}{dX^2} < 0$. The coherent spinodal lies within chemical spinodal – for infinite wavelength of compositional fluctuations. (b) Development of compositional fluctuation (dotted to solid line – as a function of time) during spinodal decomposition.

as discuss earlier in Figure 7.3. The spinodal decomposition would, however, require spontaneous up-hill diffusion (A going into A-rich region and B going into B rich – spontaneously) – diffusivity $D < 0$ or $\frac{d^2G}{dX^2} < 0$. As shown in Figure 7.19a, the region of a phase diagram satisfying this criterion is called chemical spinodal. A number of alloy systems show signatures of spinodal decomposition: Cu–Ni–Fe, Cu–Ni–Si, Al–Zn, Fe–45%Cr, etc. The signature of spinodal decomposition is typically characterized by small angle X-ray scattering, electron diffraction and analytical electron microscopy (Martin *et al.* [1997], Smallman and Bishop [1999]).

Chapter 8

Textural Developments during Thermo-Mechanical Processing

Chapter 8

Textural Developments during Thermo-Mechanical Processing

8.1. INTRODUCTION

A polycrystalline, single phase material consists of grains with an identical crystal structure, but with a different crystallographic orientations. When these orientations are statistically distributed at random, the material is crystallographic isotropic. However, in most materials, the grains are not randomly oriented. In that case, the material has a crystallographic texture. Besides a crystallographic texture, materials can also show a morphological texture, e.g. when all grains are elongated in a particular direction or when the material contains precipitates that are oriented in one or more preferential directions (Figure 8.1). In the chapter on plasticity (Chapter 3), there is a brief explanation of why a crystallographic texture occurs during deformation (concept of lattice spin). This is developed in more detail in Chapter 10. Another important source of texture generation (or texture change) is the process of recrystallization and grain growth, which has been described in Chapter 5 (softening mechanisms).

The present chapter will be devoted to the description and interpretation of crystallographic textures. In the next paragraph, an introduction to the world of

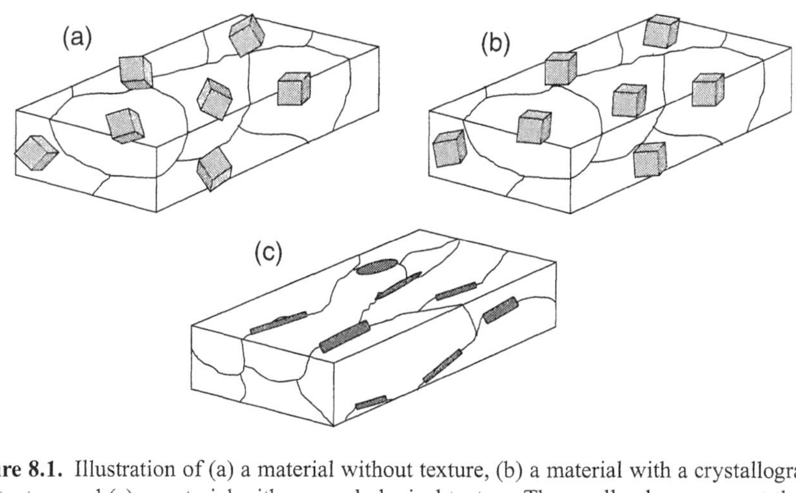

Figure 8.1. Illustration of (a) a material without texture, (b) a material with a crystallographic texture and (c) a material with a morphological texture. The small cubes represent the crystallographic orientations of the grains.

graphical representation of texture data will be provided (stereographic projections, pole figures and orientation distribution functions (ODFs)). After that, a short survey of the most important cold deformation textures and of recrystallization textures will be presented. The industrial relevance of crystallographic textures will be illustrated in several parts of this book, but in particular, in the Sections 11.7 (sheet metal forming), 14.1 (beverage cans), 14.2 (capacitor foils), 15.1 (steel for car bodies) and 17.2 (grain boundary engineering).

8.2. GRAPHICAL REPRESENTATION OF TEXTURE DATA

8.2.1 Grain orientation

The orientation of a grain is always expressed relative to an external co-ordinate system. In flat products (plates, sheets), this external reference frame traditionally consists of the rolling direction (RD), the normal direction (ND) and the transverse direction (TD). Any crystal orientation can be expressed with the help of Miller indices and is written as: (*hkl*)[*uvw*]. This means that a plane (*hkl*) is parallel with the rolling plane and the direction [*uvw*] is parallel with the RD. For example, the orientation of the crystal in Figure 8.2a should be written as (001)[1–10]. When all the crystallographic equivalent orientations are considered, the Miller indices are expressed as {*hkl*}<*uvw*>. In axisymmetric products (wires, extruded bars), the crystal orientation is described by one set of Miller indices [*uvw*], indicating that this crystallographic direction is parallel with the sample axis, e.g. [111] in Figure 8.2b. All rotations around [*uvw*] are crystallographic equivalent.

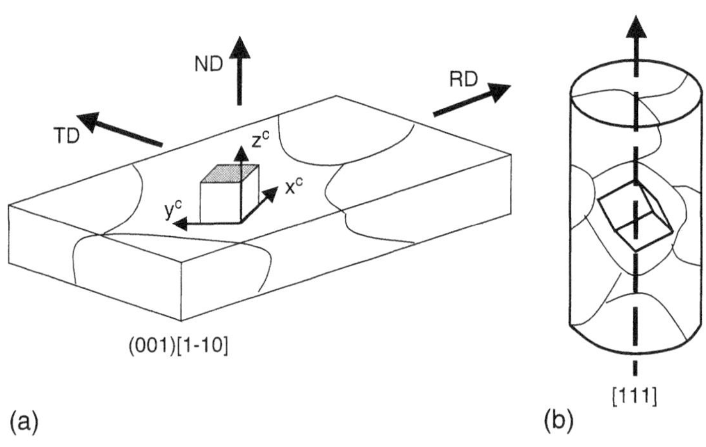

Figure 8.2. Examples of crystal orientations in sheet and wire, expressed with Miller indices. The three cube axes are shown as x^c, y^c and z^c.

8.2.2 Pole figures[1]

For many years, pole figures have been the standard tool for the representation of texture data although more and more scientists (encouraged by increased computer power) prefer to use ODFs. Both representations will be discussed in the following paragraphs.

In a pole figure, a particular set of equivalent crystal orientations (e.g. all the <100> orientations) is shown, following the principles of the stereographic projection.[2] The sample is placed in the centre of the stereographic sphere. For a rolled sheet, the rolling plane normal (ND) is pointed towards the north pole of the sphere (Figure 8.3a). For samples with axial symmetry (wires, extruded bars), the drawing direction is placed like the RD of a sheet. For each grain in the sample, a set of crystallographic equivalent directions is chosen (e.g. <100>) and projected on the equatorial plane as shown in Figure 8.3a. Only the points on the northern hemisphere are taken into account. This equatorial plane, with all intersecting points of the <100> directions, is called the (100) 'pole figure'. For a polycrystal without crystallographic texture, the projection points are scattered all over the projection plane

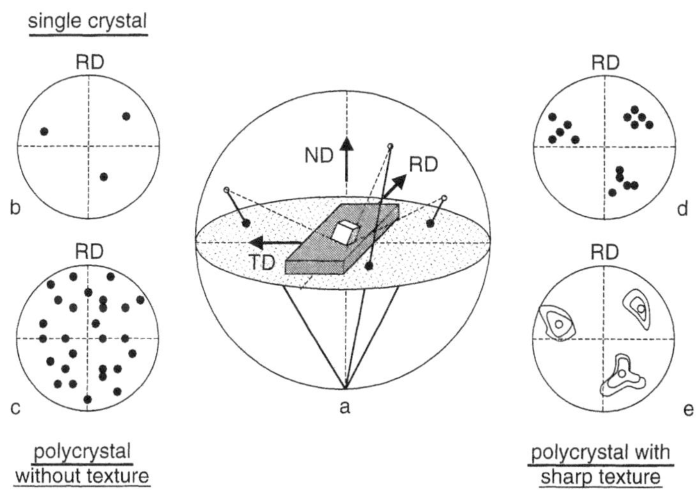

Figure 8.3. Construction of a (100) pole figure. (a) Stereographic projection of the (100) poles; (b) projection of the (100) poles of one grain on the equatorial plane; (c) projection of the (100) poles of a polycrystal; (d) projection of the (100) poles of a textured polycrystal; (e) contour map of the (100) pole density distribution.

[1] For a more detailed description of pole figures and inverse pole figures, the reader is referred to Hatherly and Hutchinson [1979], Kocks *et al.* [1998] and Randle and Engler [2000].
[2] For an introduction to stereographic projections, the reader is referred to Johari and Thomas [1969].

(Figure 8.3c). When the material has a crystallographic texture, the projections will cluster together (Figure 8.3d). In most cases, it is useful not to show individual points but contour lines of equal pole density distribution as shown in Figure 8.3e.

Of course, not only <100> orientations, but all crystallographic directions can be used to construct pole figures. Figure 8.4 shows three stereographic projections of the same grain using respectively the <100>, <110> and <111> directions of the grain. This grain has an orientation {110}<001>.

Although a pole figure of a crystalline material contains information regarding the crystallographic orientations of all the grains, in many cases it is not so simple to extract texture information from an experimentally determined pole figure. This is illustrated by the pole figures of cold-rolled interstitial free (IF) steel in Figure 8.5. A complete analysis of pole figures can be done by comparing them with standard projections (see Section 8.2.3). The 'art of reading pole figures' will, however, not be further discussed in this text.

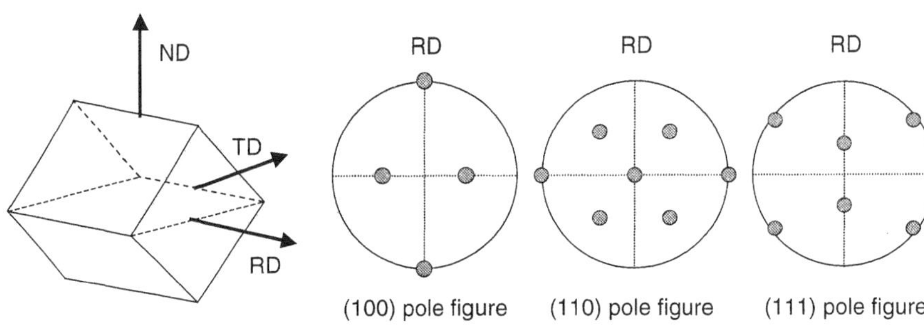

Figure 8.4. (100), (110) and (111) pole figures of the grain on the left.

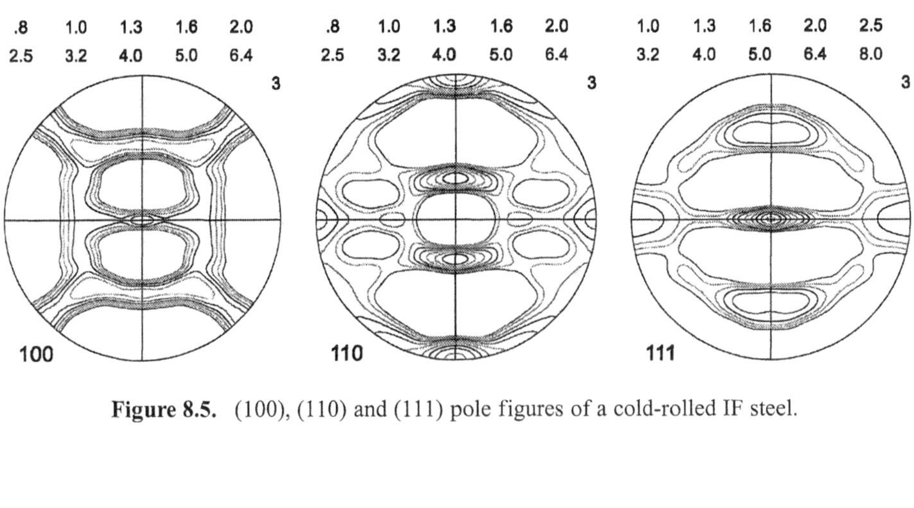

Figure 8.5. (100), (110) and (111) pole figures of a cold-rolled IF steel.

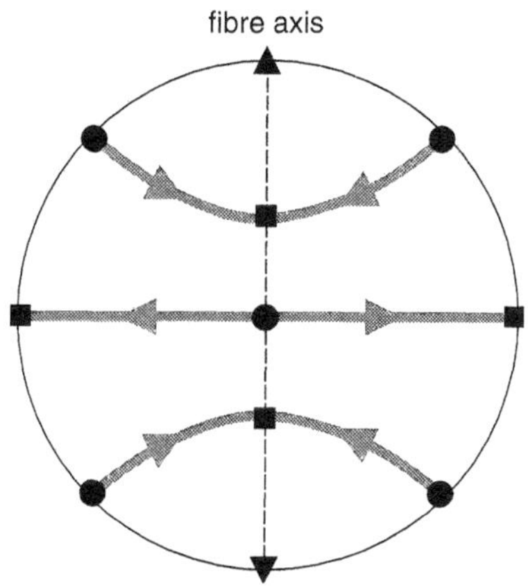

Figure 8.6. (100) Pole figure of a <110> fibre texture.

In wires or extruded products, the grains have a tendency to align themselves with one (sometimes more) particular crystallographic direction parallel with the sample axis. This is called a 'fibre texture'. Fibre textures are often represented by inverse pole figures, which will be discussed in Section 8.2.3, but they can, of course, also be represented in conventional pole figures. Consider, for example, a wire with a perfect <110> fibre texture (all grains have a <110> direction parallel with the wire axis). The wire is positioned in such a way that the stereographic projection of the axis is located at the upper and lower position in a (100) pole figure (Figure 8.6). The <100> poles of a grain with a {100} plane parallel with the plane of projection are indicated with the black dots. In a fibre texture, a grain is allowed to make rotations around the fibre axis <110>. The arrows indicate the result of such a rotation on the location of the <100> poles. In conventional pole figures, the fibre texture is characterized by long, small bands of orientations around a common symmetry point.

8.2.3 Inverse pole figures

For rolled samples, pole figures (or ODFs; see below) are commonly used to represent texture data. For wires or other samples with axial symmetry, the so-called 'inverse pole figures' are more appropriate. Inverse pole figures make use of the so-called 'unit triangle', which is part of a 'standard projection'. In a standard projection, one single crystal with a particular (chosen) orientation (e.g. (001)[100] in Figure 8.7a) is considered and all major directions of that crystal are projected onto the equatorial plane using the principles of the stereographic projection

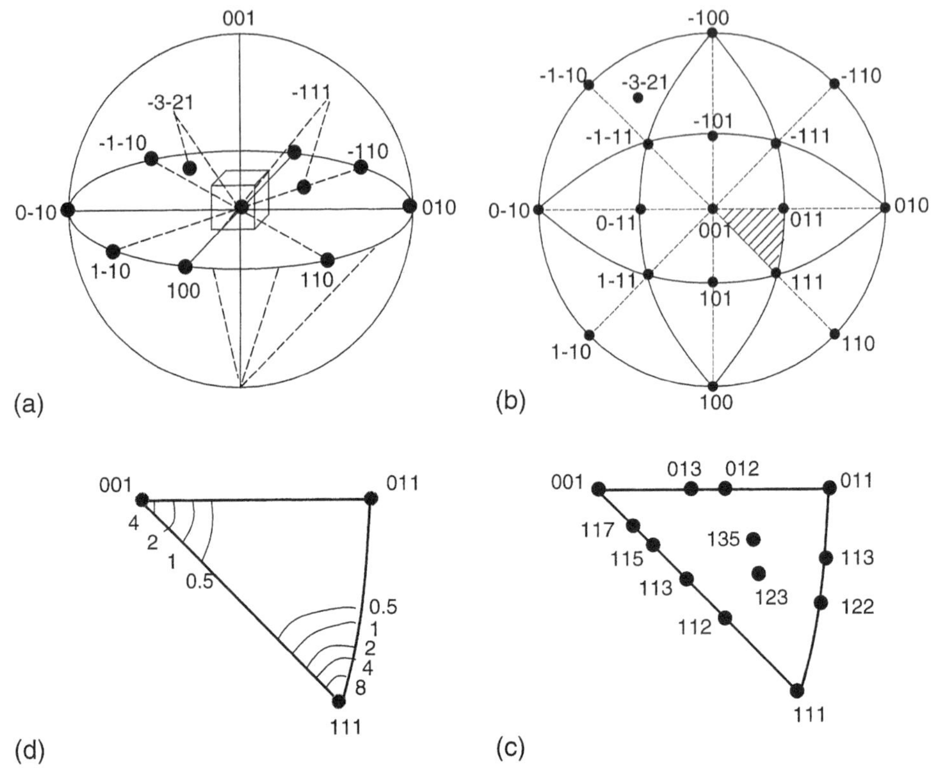

Figure 8.7. (a) Stereographic projection of a (001)[100] oriented crystal; (b) (001) standard projection: equator plane of (a) showing the projection of some major crystal orientations; (c) the [001][011][111] unit triangle; and (d) inverse pole figure using the unit triangle to indicate with the help of contour plots how intense a particular crystallographic direction is oriented parallel with an external reference direction (e.g. the wire axis).

(Figure 8.7b).[3] The projection in Figure 8.7b is called a '(001) standard projection'. Because of symmetry considerations (an axisymmetric process makes different parts or 'unit triangles' of the standard projection identical), it is often sufficient to show only part of this standard projection, e.g. the quarter between the [001], [010] and [100] directions or only the 'unit triangle' [001], [011] and [111] (Figure 8.7c). To construct an inverse pole figure, a unit triangle is used to indicate – with the help of discrete or contour plots – how intense each crystallographic direction is oriented parallel with the external reference direction like the

[3] In Figure 8.7b, only a very limited number of orientations are shown; more detailed standard projections can be find in several textbooks, e.g. Taylor [1961] or Johari and Thomas [1969].

wire axis or the axis of an extruded bar. The (schematic) inverse pole figure in Figure 8.7d, for example, shows that the volume fraction of grains oriented with a <111> direction parallel to the wire axis, is more than 8 times random. The grains with a <001> direction parallel to the wire axis, are present 4 times above random. The inverse pole figure does not give any information about the rotational spread of each orientation around the wire axis.

8.2.4 Orientation distribution functions

8.2.4.1 Introduction. In the preceding paragraphs, it has been demonstrated how pole figures can be used to represent information about the texture of a material. It has, however, also been mentioned that 'reading' a pole figure is not so easy. In fact, it is necessary to analyse all intensities separately, and this leads in practice to a situation where only the main intensities (main texture components) are looked at.

With the help of the so-called 'orientation distribution function' (ODF), it is possible to describe the complete texture information of a sample. Suppose that the orientation of a grain can be represented by a parameter 'g'. An ODF is a mathematical expression, for which we will use the symbol f that describes the volume fraction of grains in all intervals $g \pm \mathrm{d}g$.

$$\frac{\mathrm{d}V}{V} = f(g)\mathrm{d}g \qquad (8.1)$$

The integral of the ODF over all possible orientations should be equal to 1. The ODF of a sample without texture is a constant. If the sample has a texture, the ODF has maxima and minima.

8.2.4.2 Euler angles. In order to give a graphical representation of an ODF, a method must be found to define the orientation g of a grain. This is done with the help of the so-called 'Euler angles'. Two different co-ordinate systems are defined. The first is connected to the sample (sample axes system X_i) and the second to the crystal of a grain (crystal axes system X_i^c). Both systems are Cartesian and right handed (Figure 8.8).

The sample system is related to the shape of the sample. For example, for a rolled sheet, the axis X_1 is taken in the RD, X_2 in the transverse and X_3 in the ND of the sheet. The orientation of the crystal axes system can now be expressed in the reference frame of the sample axes system by three rotations (Figure 8.9). These three rotations bring both systems together. In the literature, several conventions have been proposed to perform these rotations. The most widely used system (which will also be followed in this book) is the system of Bunge [1965, 1985].

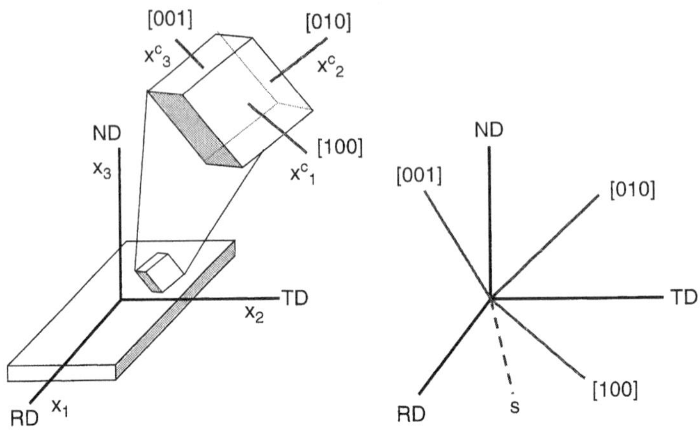

Figure 8.8. Orientation of the crystal axis system $\{X_i^c\}$ and the sample axis system {RD, TD, ND}; s is the intersection of the planes (RD–TD) and ([100]–[010]).

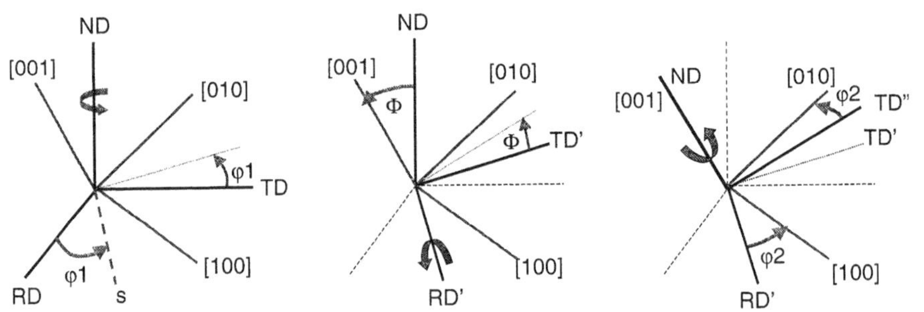

Figure 8.9. Definition of the Euler angles φ_1, Φ en φ_2 in the Bunge convention.

- First a rotation φ_1 around ND is performed; this brings RD in the position s, with s the intersection of the planes (RD–TD) and ([100]–[010]). The new positions of RD and TD are now RD′ and TD′.
- A rotation Φ around RD′; this brings ND together with [001]; TD′ will now get the position TD′′.
- A rotation φ_2 around the ND axis (which is now equal to [001]); due to this rotation, RD′ falls on [100] and TD′′ comes together with [010].

The angles (φ_1, Φ, φ_2) are called 'Euler angles'. The equations to calculate Euler angles from Miller indices or vice versa for cubic materials can be found in Humphreys and Hatherly [1996].

8.2.4.3 Euler space. When the three Euler angles are plotted in Cartesian co-ordinates, we get the so-called 'Euler space'. This space is limited for φ_1 and φ_2 between 0° and 360°, and for Φ between 0° and 180° (Figure 8.10). Each crystal orientation can be represented in this Euler space. In this representation, each family of orientations will be found at several locations of the Euler space. This has two main reasons.

First of all, each random orientation {*hkl*}<*uvw*> has one or more crystallographic related, but physically different variants. This is illustrated in the Figure 8.11 for the so-called 'Brass orientation' {110}<1−12>. When a (110) plane is

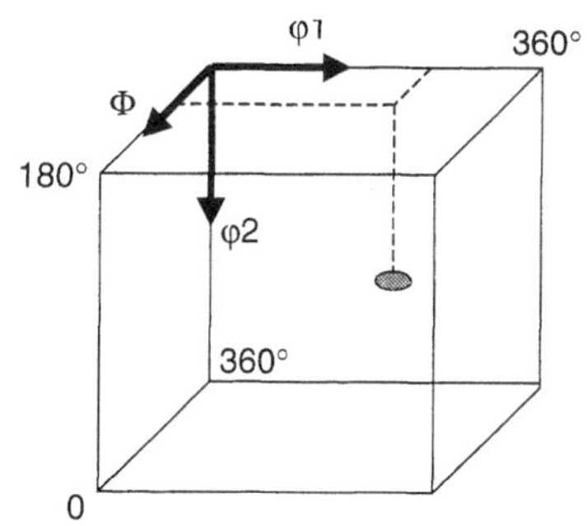

Figure 8.10. Graphical representation of a crystallographic orientation with Euler angles: φ_1=270°, Φ=60° en φ_2=180° (Miller indices: (0–74)[0–4–7]).

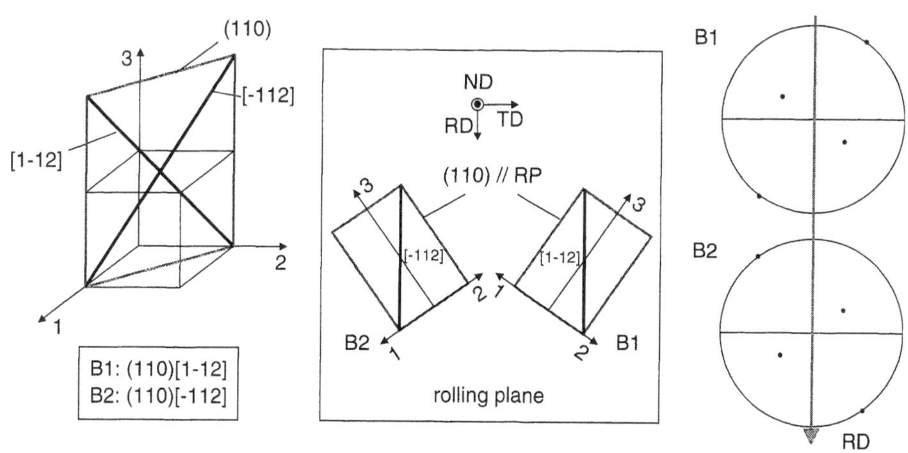

Figure 8.11. Two crystallographic related, but physically different variants of the brass orientation {110}<1–12>. The (100) pole figure of both variants are included.

Table 8.1. Miller indices and Euler angles for the crystallographic variants of the S {123}<63–4>, brass {110}<1–12> and H {001}<110> orientations.

Variant	Miller indices		Euler angles		
	(*hkl*)	[*uvw*]	φ_1	Φ	φ_2
S1	(123)	[63–4]	301	36.7	26.6
S2	(123)	[–6–34]	121	36.7	26.6
S3	(213)	[36–4]	239	36.7	63.4
S4	(213)	[–3–64]	59	36.7	63.4
B1	(110)	[1–12]	54.7	90	45
B2	(110)	[–112]	125	90	45
H	(001)	[110]	315	0	0

chosen parallel with the rolling plane, than both a [1–12] and a [–112] direction can be oriented parallel with the RD. As illustrated in Figure 8.11, showing the rolling plane seen from above the variants B1 and B2 can physically be distinguished from each other, although they are equivalent from a crystallographic point of view. For example, under the same stress state, the deformation or shape change can be different for the two variants. In general, a random orientation {*hkl*}<*uvw*> in a cubic crystal lattice has four different variants. When {*hkl*} is a plane of symmetry of the crystal, as for example (110), than the number of variants is reduced to two. When on top of that, also [*uvw*] is a symmetry axis, for example [100] or [110], only one variant is possible. Table 8.1 shows some examples. For all these variants, it is now possible to write the Miller indices in 24 crystallographic equivalent ways. This is due to the fact that for a particular crystallographic orientation, the crystal axes system can be defined in different ways. This is illustrated in Figure 8.12 for the brass variants B1 and B2.

In the left-hand figure, the situation of Figure 8.11 is considered again. In the right-hand figure, the same orientation is drawn, but with another choice of the crystal axes. The Miller indices of the variants B1 and B2 have now changed, although the crystal has not been moved.

In general, each variant can be written in 24 different ways. Because of symmetry in the crystal structure or in the sample geometry, it is often not necessary to show all these solutions in the whole Euler space. In most cases, only a part of this space will be considered. Table 8.2 gives a short survey of some possibilities.

In the case of a cubic crystal (fcc or bcc), and an orthotropic sample geometry (e.g. a rolled sheet without shear strain in the surface region), all orientations will still appear three times in the (90–90–90°) Euler space. In principle, this space could be further reduced, but this would give a too complicated construction and it is not done.

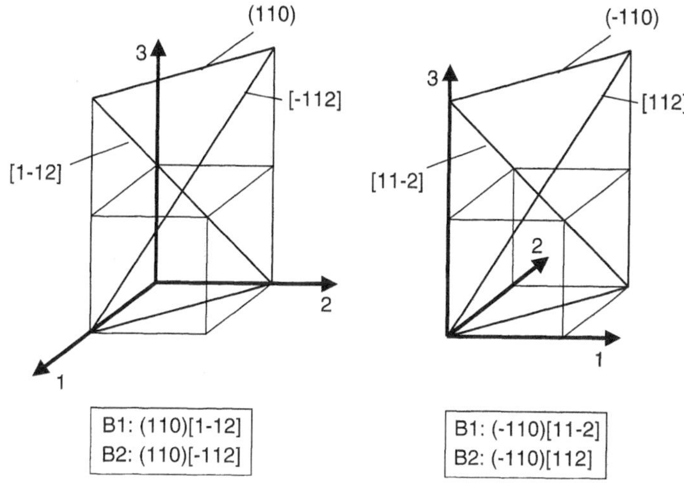

Figure 8.12. Illustration of two crystallographic equivalent descriptions of the same crystal orientation.

Table 8.2. Size of the Euler space for different crystal structures and sample geometry's (Randle and Engler [2000]).

Crystal structure	Crystal symmetry		Sample geometry		
			No	Monoclinic	Orthotropic
	Φ (°)	φ_2 (°)	φ_1 (°)	φ_1 (°)	φ_1 (°)
Triclinic	180	360	360	180	90
Monoclinic	90	360	360	180	90
Trigonal	90	120	360	180	90
Hexagonal	90	60	360	180	90
Orthorhombic	90	180	360	180	90
Tetragonal	90	90	360	180	90
Cubic	90	90	360	180	90

8.2.4.4 Two-dimensional representation of an ODF. For practical reasons, ODFs are not represented in the 3D Euler space, but in 2D sections of this space. For fcc metals such as Cu and Al, the Euler space is usually divided in sections of constant φ_2, for example, sections with φ_2=0°, φ_2=5°, φ_2=10°, etc. till φ_2=90°. Figure 8.13 shows an example of such an ODF for two orientations from Table 8.3.

Table 8.3. Euler angles for two texture components.

Component	φ_1	Φ	φ_2
Brass			
(011)[2–11]	35	45	0
(110)[1–12]	55	90	45
(101)[−1–21]	35	45	90
Goss			
(110)[001]	90	90	45
(011)[100]	0	45	0
(101)[0–10]	0	45	90

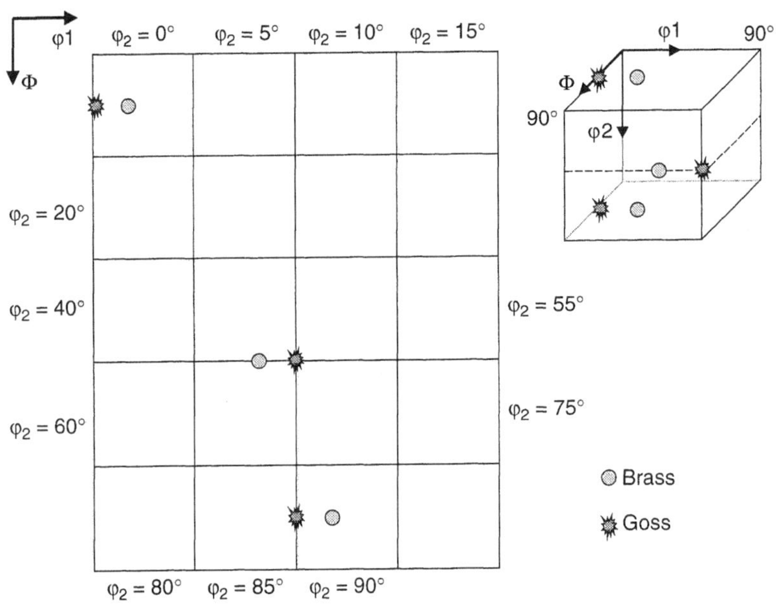

Figure 8.13. ODF sections for constant φ_2 with Euler angles from Table 8.3.

The representation described above, is know as the 'Bunge' notation (Bunge [1969]). This notation will be used in this course. In some publications, the notation of Roe and Williams is used (Roe [1965]). In this notation, three Euler angles are also used (Ψ, Θ, φ), but another rotation is used to bring the crystal axes together with the sample axes. The relation between the notation of Roe–Williams and Bunge is given by the formula:

$$\Psi = \varphi_1 - \frac{\pi}{2} \quad \text{and} \quad \Theta = \Phi \quad \text{and} \quad \phi = \varphi_2 + \frac{\pi}{2}$$

Another possibility is to use the notation of Kocks (Ψ, Θ, φ); the relation between the Kocks and the Bunge notation is given by:

$$\Psi = \varphi_1 - \frac{\pi}{2} \quad \text{and} \quad \Theta = \Phi \quad \text{and} \quad \phi = \frac{\pi}{2} - \varphi_2$$

Still another method to represent orientation distributions is the 'Frank–Rodrigues' space. For more information, we refer to the literature, e.g. Kocks *et al.* [1998].

In practical situations it would be rather unique to have an assembly of grains, all pointing exactly in the same direction. In most cases, we rather see a clustering of points around a particular orientation. According to the 'sharpness' of the texture, the intensity peak in an ODF will be more or less smeared out. This spread of the intensity around a mean value has a consequence: a component is not only visible at the exact location of the Euler angles, e.g. $\varphi_2 = 0°$, 45° and 90° for brass, but also in the adjoining sections. This minimizes the risk that an important texture component would be missed, because it falls in between two plotted φ_1 or φ_2 sections.

For certain Miller indices, it will not be possible to derive a unique value for the Euler angles. An example is the so-called 'cubic' orientation {001}<100>. This component is of course found on the corners of the Euler space, but also for all Φ=0 and $\varphi_1 + \varphi_2$=90°

The ODF of a textured polycrystalline material will thus be represented by some 'intensity peaks' around typical preferred orientations, superimposed on the background intensity from grains without preferred orientation. The major advantage of an ODF is that it contains the complete texture information of a sample. A possible drawback is that one needs some experience to interpret the 2D graphical representation. ODFs can also be used to obtain additional information. This include volume fraction estimates for major orientations, skeleton lines representing intensity/location of fibres and texture-related properties like normal anisotropy ($\bar{R}$ – values). Such representations are not only 'user-friendly', but provides direct information on the nature/extent of texturing.

8.3. SOME IMPORTANT COLD DEFORMATION TEXTURES

8.3.1 Introduction

The most important cold deformation textures are those developed by cold rolling (in many cases, approximated by plain strain compression) and by axisymmetric deformation (cold extrusion, wire drawing). The development of a crystallographic texture depends on the crystal structure of the material and on the deformation mode, but also the starting texture and some microstructural developments, like twinning or grain splitting, can have an influence. Hence, it cannot be expected that the texture of two materials with the same crystal structure will be exactly the same after an identical deformation. Nevertheless, some main lines can be discerned and will be illustrated below. In the following paragraphs, the cold deformation texture of fcc, bcc and hexagonal metals after plain strain and axisymmetric deformation will be discussed.

8.3.2 Axisymmetric deformation

During axisymmetric deformation, a 'fibre texture' is generated. Cold-drawn fcc materials have in general the so-called 'double fibre texture': <100> + <111>. It means that most of the grains either have a <100> or a <111> direction parallel with the wire axis. An easy way to illustrate fibre textures is to use an inverse pole figure (see Figure 8.7). The intensity ratio between the <100> and <111> components has been reported to be a function of the reduced stacking fault energy (SFE) of the material, as illustrated in Figure 8.14 (English and Chin [1965]). The reduced SFE is dimensionless, and is the SFE divided by the shear modulus (*G*) and the Burgers vector (*b*). The highest fraction of <100> component is found in heavily (>98%) drawn Ag wire. However, the schematic description given above, must be considered with some care. Measurements reported by Stout *et al.* [1988], carried out on wires with smaller reductions (ε=2), did not find such a strong influence of SFE. It was suggested that the results of English could perhaps be influenced by the presence of a strong texture before wire drawing. Another fact is that in many commercially drawn wires the high drawing speed may cause heating of the deformation zone and partial recrystallization of the wire, resulting in a mixture of deformation

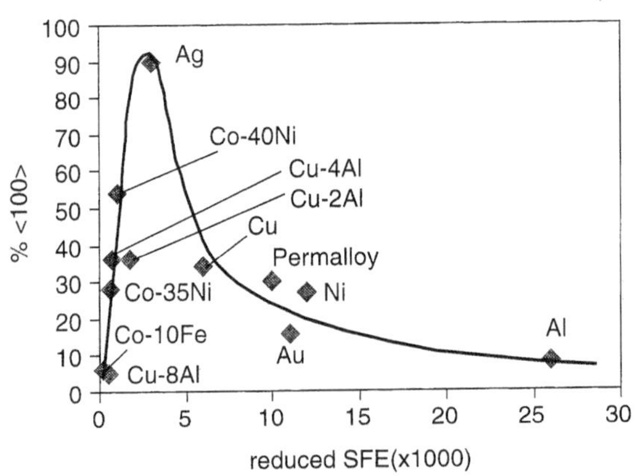

Figure 8.14. Correlation between the fraction of the <100> texture component in fcc metals and the reduced SFE (SFE/*Gb*); all reductions are between 98.5 and 99.8%. Data from English and Chin [1965].

and recrystallization components (Inakazu *et al.* [1994]). Besides the drawing speed, other drawing parameters may influence the texture. In electrolytic tough pitch (ETP) Cu wire, it has been observed that an increase in die angle (and a related increase in redundant deformation, cf. Section 11.4) reduces the sharpness of both the <111> and <100> components, but increases the <100> to <111> ratio (Kraft *et al.* [1996]).

Cold-drawn *bcc materials* have in general a simple <110> fibre texture. This texture is, for example, found in cold-drawn pearlitic wire or in drawn W-wire (Figure 8.15). The latter may be drawn at 800–1000°C, but because of the high melting point of W (3410°C) this can still be considered as 'cold' drawing. The strong <110> texture gives rise to a peculiar microstructural development, called the 'curling effect', which is described in Section 11.4.

In hexagonal metals, a lot of confusion on fibre texture has been reported; in a number of cases a <10–10> fibre texture is found.

In some cases an additional symmetry is present, as illustrated in Figure 8.16. In the left-hand figure, a <110> fibre texture is illustrated: all grains have this direction parallel with the fibre axis. The right-hand figure shows the same grains, but now with a <001> direction parallel with the wire surface. This is called a 'cylindrical texture'. In a classical pole figure, no difference between both textures can be detected, because no correlation exists between the location of a grain in the wire and the intensity in the pole figure. A cylindrical texture can be detected with a local orientation measurement in TEM or with EBSD.

In uniaxial compression of metals, the fibre textures are often more or less the inverse of the situation in drawing. In fcc metals, in general, a texture close to an <110> fibre texture is found after uniaxial compression. In brass, a spread towards <111> is however observed (Stout *et al.* [1988]).

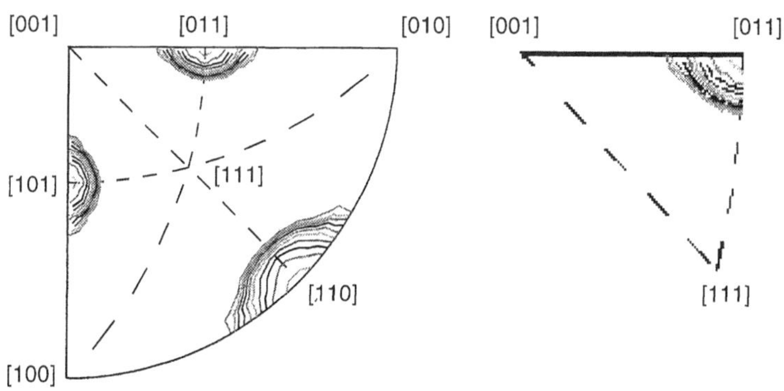

Figure 8.15. <110> Fibre texture in a drawn W-wire in the [100][010][001] quadrant and in a simple unit triangle (contour levels: 1.6, 2.0, 2.5, 3.2, 4.0, 5.0, 6.4, 8.0, 10, 13).

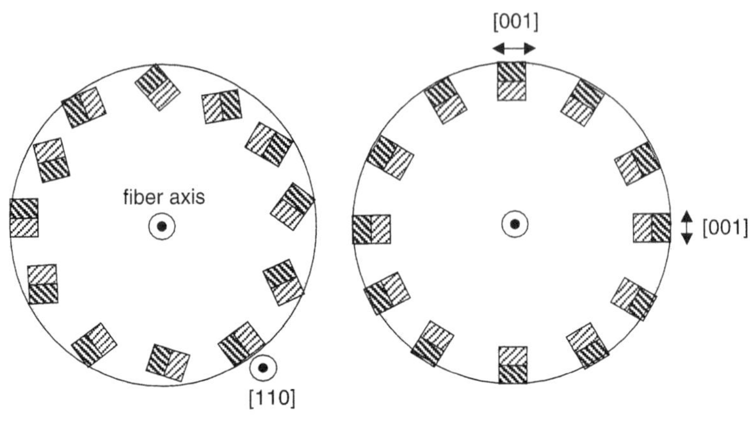

Figure 8.16. Difference between a fibre texture and a cylindrical texture.

In bcc metals, a double fibre texture is often found after uniaxial compression: a strong <111> and a weak <100> component.

8.3.3 Plane strain deformation of fcc materials

The crystallographic texture after plane strain deformation (cold rolling) of an fcc material can be represented in a somewhat idealized form as shown in Figure 8.17.

The texture consists of two 'families' of orientations, called respectively, the α- and the β-fibre. These fibres form a line (or a tube) through the Euler space. The α-fibre consist of orientations that all have a <011> axis parallel with the rolling plane normal, starting from the 'Goss' orientation (G:{011}<100>) and ending at the 'brass' orientation (B:{011}<211>). The β-fibre starts at the brass orientation and goes over through 'S' (S:{123}<634>), sometimes also denoted as S{123}<412> and ends in 'Copper' (Cu:{112}<111>). Both fibres can be found three times in the (90°, 90°, 90°) part of the Euler space. The Table 8.4 gives the co-ordinates of the main fibre components. In reality, the experimentally observed textures can be more complicated than shown in the schematic representation of Figure 8.17 and can be different from one material to another. Moreover, the fcc rolling texture will evolve during rolling. This will be further illustrated with two typical textures. The first one is the rolling texture of copper, which is representative of fcc metals with medium to high SFE, like Cu (SFE: 78 mJ/m^2) and Al (SFE: 166 mJ/m^2). The second is called the 'Brass texture' and is representative of low-stacking fault materials, like Cu–30 wt% Zn (SFE: 20 mJ/m^2) and Ag (SFE: 22 mJ/m^2).

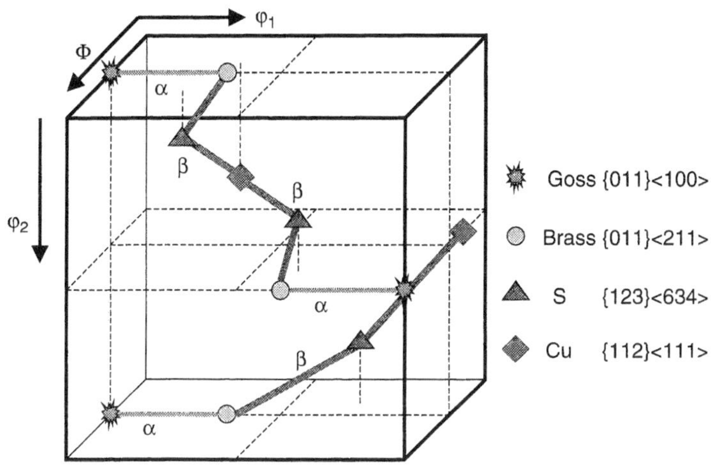

Figure 8.17. Schematic representation of the cold rolling texture of fcc materials.

Table 8.4. Co-ordinates of the main α- and β-fibre components in the (90°, 90°, 90°) part of the Euler space.

Name	*(hkl) [uvw]*	φ_1	Φ	φ_2
Goss	(011)[100]	0	45.0	0
Brass	(011)[2–11]	35.3	45.0	0
S	(132)[6–43]	27.0	57.7	18.4
Copper	(112)[1–11]	37.5	42.6	25.5
Copper	(112)[1–11]	37.5	42.6	25.5
S	(231)[3–46]	52.9	74.5	33.7
Brass	(110)[1–12]	54.7	90.0	45.0
Goss	(110)[001]	90.0	90.0	45.0
Goss	(101)[0–10]	0	45.0	90.0
Brass	(101)[−1–21]	35.3	45.0	90.0
S	(231)[−3–64]	59.0	36.7	63.4
Copper	(112)[−1–11]	90.0	35.3	45.0

In Figure 8.18, the evolution of the texture with increasing rolling reduction is given for both materials with the help of 'skeleton lines'. These lines are obtained by connecting all the maxima of the β-tube in the respective φ_2 sections of the ODF and plotting them as a function of φ_2.

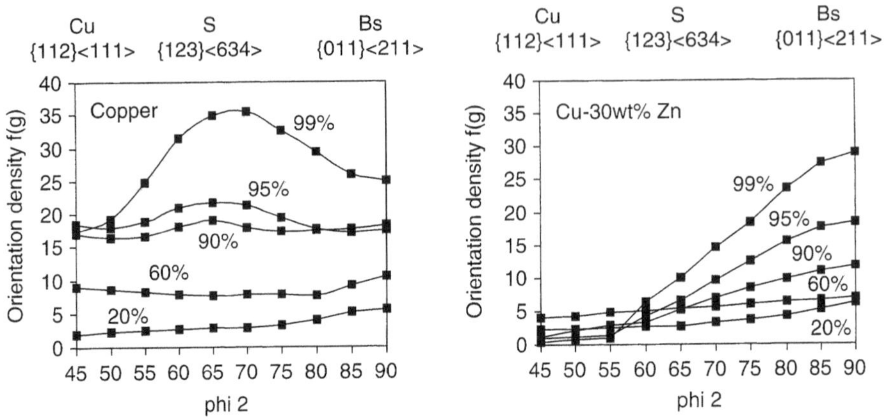

Figure 8.18. Orientation density $f(g)$ along the β-fibre in cold-rolled Cu and Cu–30 wt% Zn, as function of the cold rolling reduction. Adapted from Hirsch and Lücke [1988].

The figure shows that for Cu, the texture intensity increases over the whole β-fibre in a more or less homogeneous way with increasing rolling reduction. Only above 90% reduction some asymmetry in the β-fibre develops, with a peak in the S component. For moderate strains up to 60%, the texture evolution in brass is more or less comparable; but with larger reductions, the texture evolution is completely different and a strong asymmetry in the development of the β-fibre becomes obvious. The Cu component decreases and the brass component sharply increases. More information about these typical textures can be found in Hirsch and Lücke [1988].

The cold rolling texture of materials such as aluminium (alloys), austenitic Fe–Ni steel and nickel are similar to the pure Cu-type texture, while examples of the brass-type are found in alloys with low SFE, like austenitic stainless steel and silver. Some alloys show a transition between the two types. It has to be mentioned that not only the composition (or SFE) of a material affects the type of texture, but also the deformation temperature. Some examples of textures developed during hot deformation will be discussed in Section 8.5.

Figure 8.19 shows the ODF of an Al–Mg–Mn alloy (type AA5182), measured after hot and cold rolling. The cold rolling reduction was 70%. The β-fibre components (brass, S and Cu) are clearly present, as well as a Goss component. An additional cube {100}<001> component is however also visible. This is not a traditional component after cold rolling, but rather a component that originates during recrystallization (cf. Section 8.4). Its presence in the cold rolling texture can be understood from the fact that this material has previously been hot rolled; and during hot rolling, a number of recrystallization events took place. The cube grains

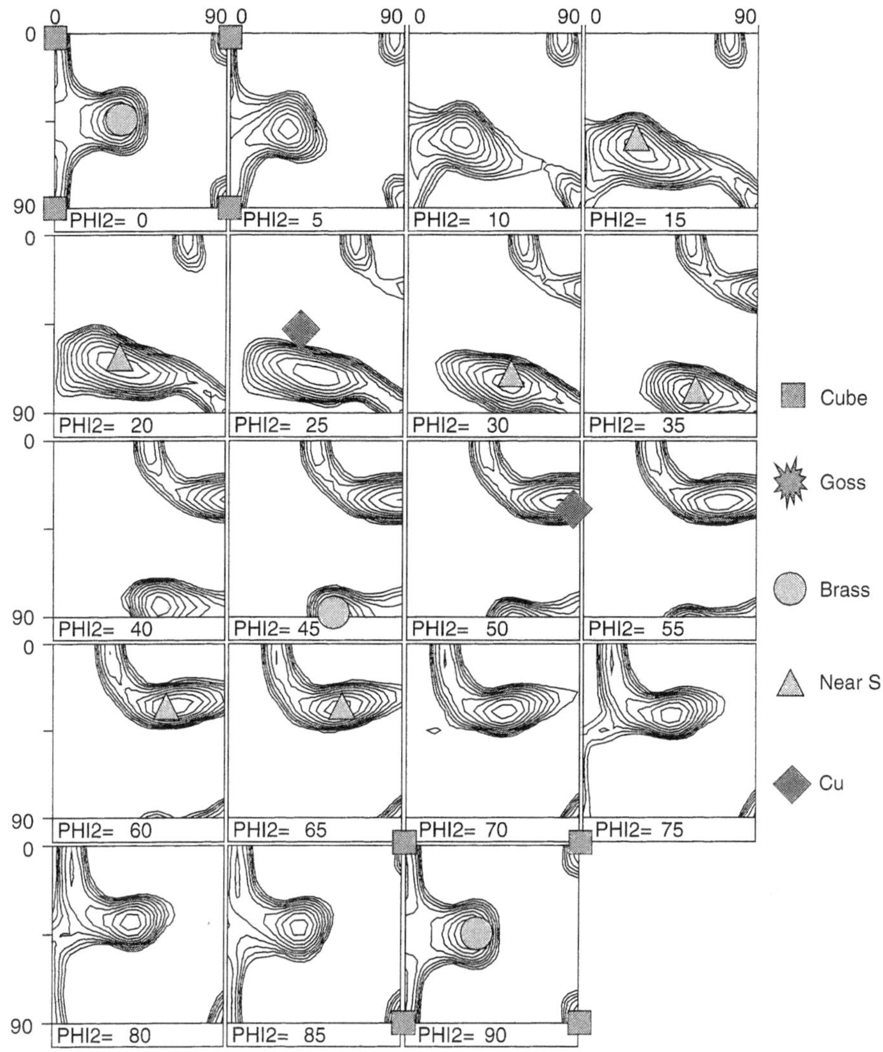

Figure 8.19. ODF of a cold-rolled Al–4.5Mg–0.4Mn alloy (constant φ_2 sections) with indication of some important texture components (not all cube positions have been indicated). Contour levels between 1 and 8 times random.

that have been generated do not disappear entirely during further cold rolling. Parts of the cube grains have been re-oriented towards new orientations, but many of them resisted and are still present after cold deformation. This example shows that also the starting texture (reflecting the pre-history of the material) will influence the texture after cold rolling.

8.3.4 Plane strain deformation of bcc materials

The typical cold rolling texture of a bcc material (like ferritic low-carbon steel) consists of two 'fibres' through Euler space. The alpha fibre consists of grains with a <110> axis parallel to the RD. The most important components are traditionally indicated with the letters H, J, I and E as indicated in Figure 8.20.

The gamma fibre contains orientations with a {111} plane parallel with the rolling plane (or a <111> direction parallel with the rolling plane normal). The most important components are indicated with E, F, E′ and F′. E is thus the cross point of both fibres.

The complete ODF of low-carbon steel is shown in Figure 8.20, with constant φ_1 sections. During cold rolling, the intensities along the gamma fibre will evolve and the details of the ODF will change, but the general picture (alpha and gamma fibre) is clearly present in most cold rolling textures of steel.

8.3.5 Plane strain deformation of hexagonal materials

The information on cold rolling textures of hexagonal metals is less abundant than that of fcc and bcc metals. Nevertheless, the textures of hexagonal materials are of interest in a number of applications like the use of Ti alloys in the aeronautic industry, Mg for automotive applications or Zr alloys in the nuclear sector (cf. Chapter 16).

The cold rolling texture of hexagonal materials can be classified in three categories, depending on the value of their c/a ratio.

Materials with a c/a ratio close to the ideal value of 1.633 (magnesium, cobalt), have a natural tendency to align the principal slip plane (the {0001} basal plane), parallel with the rolling plane; or in other words, the *c*-axis parallel with the rolling plane normal. The scatter around this ideal position is often equal in all directions or has sometimes a slight preference towards the RD (Barrett and Massalski [1980]) as shown in the first (0002) pole figure in Figure 8.21. According to Wang and Huang [2003], the texture can be fairly well described as a {0001} fibre texture; there is no preferred orientation in the rolling plane.

In materials with a large c/a ratio, like Zn with c/a = 1.856 and Cd with c/a = 1.885, the *c*-axis is tilted 15–25° away from the ND towards the RD. The <11–20> directions, which are the primary slip directions, are aligned parallel with the RD. The <10–10> directions are parallel with the TD of the sample. This texture development is a result of the combined action of basal slip and twinning (Wang and Huang [2003]).

Materials with a small c/a ratio, like Zr with c/a = 1.589 and Ti with c/a = 1.587, have a rolling texture that can be described as a rotation of the basal plane of 20–40° towards TD and the <10–10> poles are parallel with the RD.

For more information about textures in hexagonal materials, the reader is referred to Philippe [1994], Kocks *et al.* [1998] and Wang and Huang [2003].

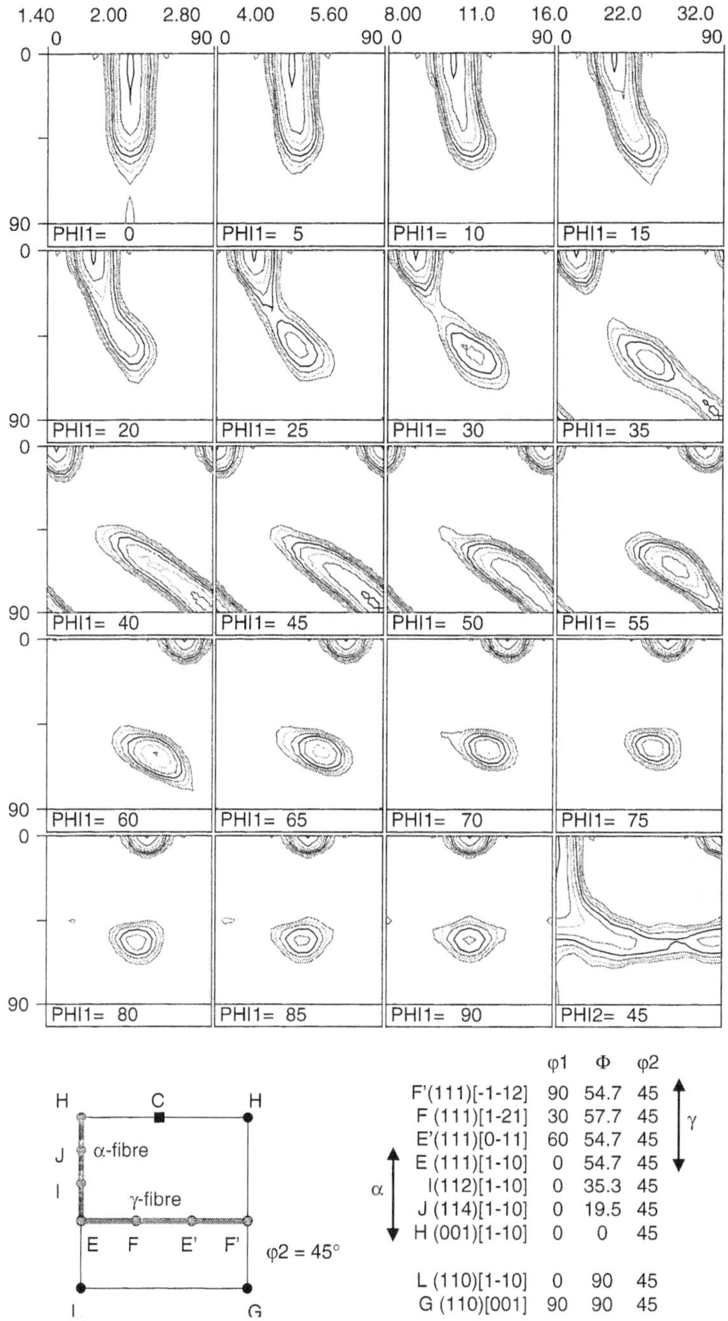

Figure 8.20. ODF of an IF steel after hot and cold rolling constant φ_1 sections and a φ_2=45° section showing the α- and γ-fibre contour levels from 1.4–32.

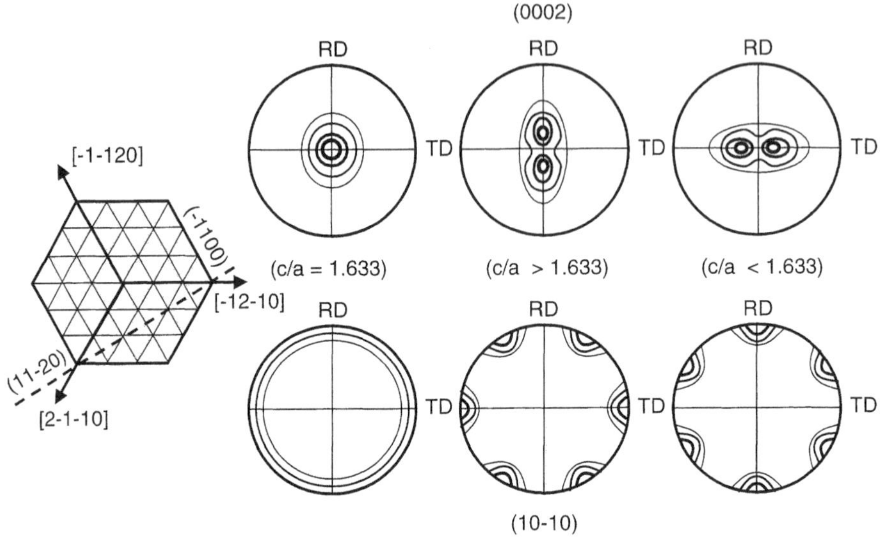

Figure 8.21. Schematic (2000) pole figures of cold-rolled hexagonal materials.

8.4. RECRYSTALLIZATION TEXTURES

8.4.1 Introduction

When a cold deformed material is annealed, the initial cold deformation texture will change during recrystallization and eventually during subsequent grain growth. We call this new crystallographic texture a 'recrystallization texture' or an 'annealing texture'.[4] Since many industrial materials will receive their final forming operation in annealed condition, the understanding of recrystallization or annealing textures is very important. This texture is to a large extend responsible for the anisotropy in the mechanical properties of the material and will in many cases determine the properties of the end product. After more than 50 years of intensive research, no general accepted mechanism of texture formation during annealing exists. This is related to our incomplete understanding of the cold deformation structures (and textures) of metallic materials and with the experimental difficulties in visualizing the initial stages (nucleation) of the recrystallization process. When the first recrystallized

[4] The term 'annealing texture' is more appropriate in most practical circumstances because in most industrial annealing processes, a certain amount of grain growth cannot be excluded or is in some cases even required. Since in this chapter we want to look at the effect of a recrystallization process on the crystallographic texture, we will use the term 'recrystallization texture'.

grains become visible in a sample, the nucleation stage is already wiped out and we are looking at the so-called 'lost evidence'.

In the following paragraphs, some brief summary of two competing mechanisms of texture formation will be given. After that a short review of some important recrystallization textures is provided.

8.4.2 Oriented nucleation or oriented growth: . . . half a century of discussions

In Chapter 5, it has been shown that the origin of a recrystallized grain can be found in the deformed structure. Although the precise mechanism of recrystallization is not yet completely understood, it seems clear that small zones in the deformed matrix form the nuclei of the recrystallized grains. This implies that the nucleus of a recrystallized grain will often have a similar orientation as the deformed region where it originates (but since grains can 'break up' into regions with small misorientations during deformation, the recrystallized grains will not have exactly the same orientation as the 'mother grain'). If all grains should produce the same amount of nuclei and if all nuclei would grow in the same manner, the recrystallization texture would just be a replica of the deformed texture, which is clearly not the case in practice. This means that some 'filter' must be active: either some mother grains produce more potent nuclei than others, or some nuclei grow faster than others. These two possibilities have triggered two competing recrystallization theories: the 'oriented nucleation' theory and the 'oriented growth' theory.

In the oriented nucleation theory, it is assumed that in a deformed matrix some grains or zones have 'more potent' nuclei than others. This suggestion is based on the observation that certain sites are preferred sites for nucleation of recrystallized grains, for example, persistent cube bands, zones around non-shearable 2nd phase particles, shear bands, etc. It is also observed that in most of these preferred sites a predominance of recrystallized grains with a particular orientation can be found, e.g. Goss {011}<100> oriented grains in transition bands and in shear bands in steel or cube {100}<001> oriented grains in persistent cube bands, S {123}<63–4> grains in shear bands and randomly oriented grains around particles in aluminium alloys, etc. From such observations it has been concluded that not all potential nuclei (all orientations present in the deformed matrix) are really 'activated', but that some kind of preferred nucleation exist: oriented nucleation.

In the oriented growth theory it is assumed that all potential nuclei are activated, without any preference; but that those nuclei which have a 'fast growing' grain boundary nature with their neighbours will develop much faster and will determine the recrystallization texture. This theory is based on the observation that the components of a recrystallization texture can often be deduced from the deformation texture, by some particular rotations around some simple crystal orientations. Some well-known examples are:

30–40° rotation around a common [111] axis in fcc metals.
25–30° rotation around a common [110] axis in bcc metals.
30° rotation around a common [0001] axis or 90° around [1010] in hexagonal metals.

All these orientation relations are 'misorientations with a high mobility'. The role of a 'global growth advantage' or an overall oriented growth has been accepted in single crystals. However, controversies do exist on the contributions of such a mechanism in heavily deformed polycrystalline material. In such cases, a 'selective growth advantage' at the early nucleation stages is often considered (cf. Chapter 5). Both the theories have undergone modifications over the years, including formulations of the so-called 'compromise theories' – combining the principles of both theories. Even then a general agreement does not exist on the relative effectiveness of oriented nucleation and oriented growth in defining the developments in recrystallization texture. More details can be found in Chapter 5 and in Humphreys and Hatherly [1996].

8.4.3 Some important recrystallization textures in rolled fcc metals

Recrystallization textures of fcc materials are influenced by several parameters, like the amount of preceding cold deformation, the starting texture and microstructure, the purity of the metal, the amount of grain growth overlapping with recrystallization, etc. In that respect, it is difficult to define 'typical' textures.

In very pure fcc metals and under favourable processing conditions (large rolling reduction, small initial grain size and sufficiently high annealing temperature), the main annealing texture component is the cube orientation {100}<001>. For low-SFE materials, the twin orientation of cube can also be present. However, as illustrated in Figure 8.22, larger initial (hot rolled) grain size or lower cold rolling reductions can considerably reduce the amount of the cube component. In industrial practice, the 'ideal' conditions are often not met and the texture after recrystallization of cold-rolled Cu is a 'retained' rolling texture: the cube component has increased during recrystallization and the β-fibre components have decreased, but are still present (see Figure 8.23).

Small amounts of impurities like As, P or Sn in Cu, can also significantly weaken the cube texture (Barrett and Massalski [1980]). Alloying elements like Zn can drastically change the texture. The cube component decreases with increasing Zn content and at around 8 wt% Zn a {236}<385> component begins to develop and reaches a maximum around 15 wt% Zn (Virnich and Lücke [1978]). The dominant recrystallization texture component in a Cu–30 wt% Zn alloy (like the one used in Figure 8.18) is {236}<385>.

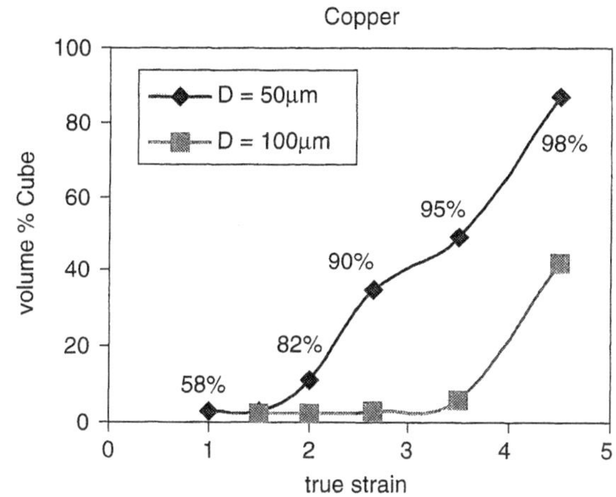

Figure 8.22. Volume per cent cube in 99.995 wt% Cu in fully recrystallized condition, as function of hot-rolled grain size and cold rolling reduction. After Necker *et al.* [1997].

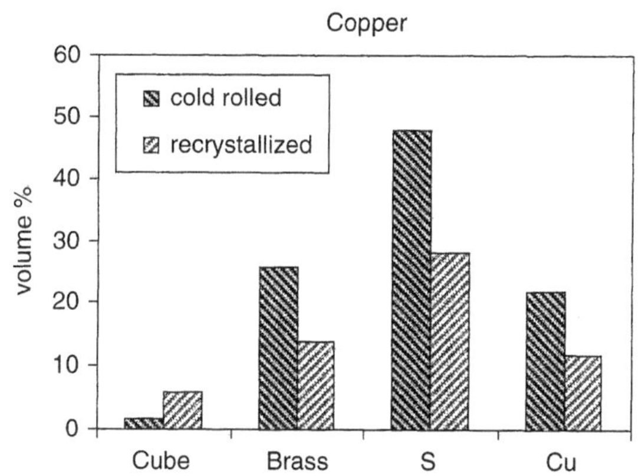

Figure 8.23. Some important texture components in 99.995 wt% Cu after 93% cold rolling and after recrystallization at 200°C.

In very pure Al, the same cube texture as in pure Cu can be found. As for the case of Cu, the processing conditions must be well chosen to maximize the amount of cube texture. This will be illustrated in Chapter 14.2 (aluminium sheets for capacitor foils). Most industrial aluminium alloys contain, however, a number of impurities and alloying elements. When Mg is added to aluminium (most Mg is in solute solution), recovery will slow down and shear bands will be formed during cold deformation. Recrystallization will now take place in these shear bands and orientations, like Goss {011}<100>, P {011}<112> and Q {013}<231> will emerge.

Most commercial aluminium alloys contain Fe and Si and these atoms will form precipitates (eventually together with other alloying elements). These precipitates can further complicate the texture evolution during recrystallization. In general, it is observed that 'large' particles (roughly above 1 μm) which induce particle stimulated nucleation (PSN) will have a tendency to weaken the recrystallization texture, because the grains that develop around the particles have a near random orientation. This randomness is due to the very wide range of orientations that are present in the cold deformed matrix around the particles. The effect of PSN seems to be larger after short annealing at high temperature than after batch annealing (long anneal at moderate temperature). This is illustrated in Figure 8.24 for an AA5182 alloy. The cold rolling texture consists mainly of deformation components (brass, S, Cu). After fast annealing, the texture is very weak which is due to the randomizing effect of the large particles. After batch annealing, part of the cold rolling texture has been retained. This means that beside PSN, nuclei have also been activated in the cold deformed brass, S and Cu grains and recrystallized grains of these orientations have grown out. Very fine precipitates or dispersoids will have an influence on the recrystallization kinetics; in the sense that the recrystallization will be slowed down and in extreme cases completely suppressed. In that case a sort of extended recovery will take place during annealing, and the annealing texture will be nearly identical to the deformation texture.

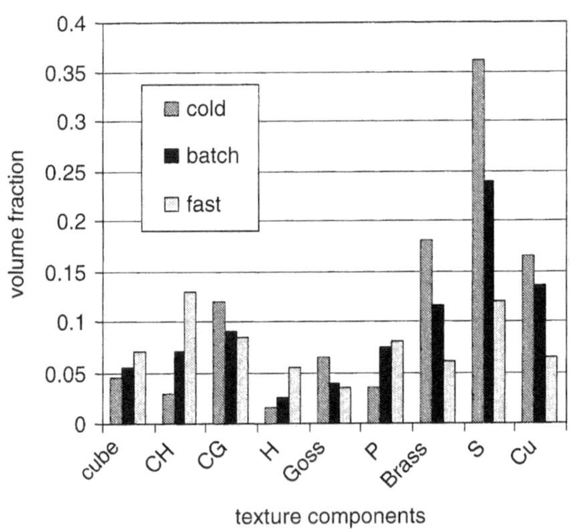

Figure 8.24. Influence of annealing treatment on the recrystallization texture of an AA5182 alloy (Al–4.3 wt% Mg–0.29 wt% Mn); the samples were cold rolled (70% reduction or $\varepsilon = 1.2$) and subsequently batch annealed (24 h/360°C) or fast annealed (500°C/10 s).

These examples show that during recrystallization, a competition between potential nucleation sites takes place and that the outcome is largely influenced by processing parameters. This makes it difficult to give a simple classification of fcc recrystallization textures (although a valuable attempt has been made by Engler and Lücke [1992]), but on the other hand, it provides us (in principle) with a possibility to control the recrystallization texture, within certain limits. A careful design of rolling and heat treating schedules are of critical importance in most aluminium alloys. An example of earing control in the production of sheets for can stock, by careful steering of the texture development during hot and cold rolling, will be given in the Chapter 14.1.

8.4.4 Some important recrystallization textures in rolled bcc metals

The most relevant bcc metal from an industrial point of view is steel. The cold deformation texture of cold-rolled low-carbon steel has been described in Figure 8.20. It consist of the so-called 'γ- and α-fibre'. During recrystallization of the sheets, it is observed that the deformed grains belonging to the γ-fibre (like F{111}<−211> and E{111}<1–10>) recrystallize first. It is thought that this is due to the high stored energy, combined with a large spread of this stored energy inside the grains (Figure 8.25). The recrystallized grains are mainly of F and E orientations. Deformed grains of the α-fibre go through a long period of recovery and start recrystallizing at the end of the recrystallization process. In the mean time, part of them have already been consumed by the growing γ-grains. The result is that the recrystallization texture resembles the cold deformation texture, but the fraction of γ-grains has increased and the fraction of α-grains has decreased. This

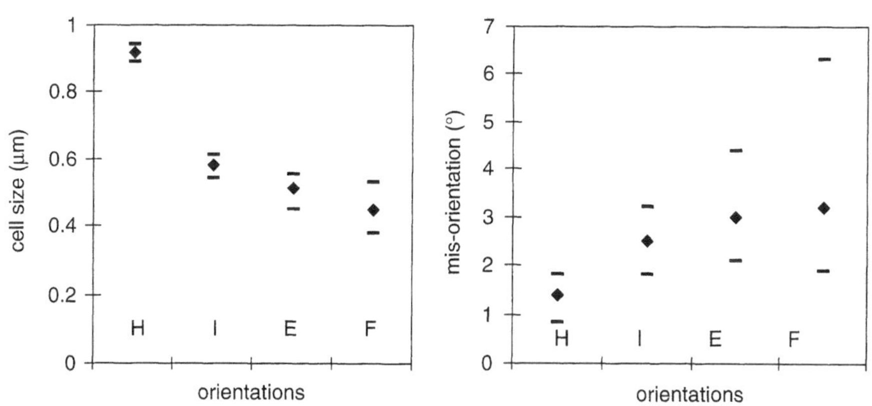

Figure 8.25. Stored energy (misorientation/cell size) and spread on this stored energy for four important texture components in low-carbon sheets. Data from Samajdar *et al.* [1998e].

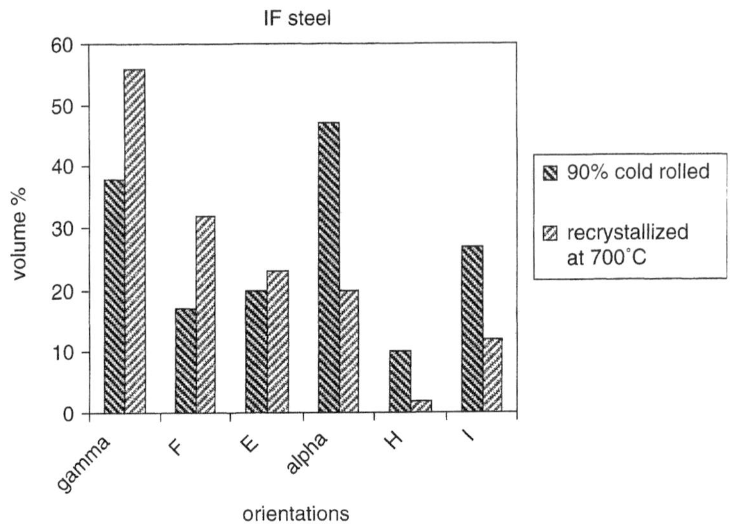

Figure 8.26. Main texture components of IF steel sheet after cold rolling and after recrystallization. Adapted from Samajdar *et al.* [1997a].

is very important for industrial applications (mainly in car body manufacturing) because it has been shown that the deep drawability of sheets improves with an increase of the ratio of γ-grains over α-grains (see Figure 8.26).

8.5. TEXTURES IN HOT DEFORMED MATERIALS

8.5.1 Introduction

Texture formation during cold deformation, as described in Section 8.3, is relatively well understood and can be fairly well predicted (cf. Chapter 10). The texture formed during hot deformation and inherited (eventually in modified form) at room temperature, is more complicated. A first issue is that the structure at room temperature (e.g. ferrite in low-carbon steel) is not always the same as the structure that has been deformed at high temperature (e.g. austenite). The texture measured at room temperature is in that case called a 'transformation texture'. Transformation textures are observed in many alloys, such as steel, Ti alloys, Fe–Ni alloys, etc. Another difficulty can arise when the deformed, high-temperature phase dynamically recrystallizes. The texture recorded at room temperature is the result of a simultaneous deformation and a recrystallization process. This is observed in Cu, Ni, austenitic stainless steel and many other materials. But even when no phase transformations and no dynamic recrystallization occur, the texture formed at high temperature may differ substantially from the cold deformation texture. This is observed, e.g. in many Al alloys. In the following paragraphs, two cases mentioned above will be illustrated with an example. However, it has to be mentioned that in many multi-pass industrial forming operations, the situation may be more complicated because after each high-temperature-forming pass, the structure and texture may be (partially) wiped out by classical static recrystallization.

8.5.2 Hot deformation textures in Al alloys

During plane strain compression at room temperature, most Al alloys (and other high-SFE materials) develop a rather uniform β-fibre texture, as discussed in Section 8.3.3. Plane strain channel die compression tests with an Al–1Mn–1Mg alloy at high temperature (200–400°C) have shown a systematic increase of the brass and the cube components with increasing temperature and decreasing strain rate, at the expense of the Cu and S components (Maurice and Driver [1997]). Careful analysis of slip traces in single crystals (Maurice and Driver [1993]) led to the conclusion that non-octahedral slip, i.e. the activation of slip systems other than {111}<110> (mainly on {110} planes), is responsible for the presence of a significant cube component. The occurrence of non-octahedral slip was also demonstrated in Al–1% Mn crystals deformed by hot plane strain compression at strain rates of 10/s (Perocheau and Driver [2002]).

The development of the brass component is more difficult to explain. Recently, a tentative explanation for the strength of the brass component in a Al–6Cu–0.4Zr alloy (Supral) was proposed by Bate *et al.* [2004]. Channel die experiments showed that at strains up to 2, a rather uniform β-fibre developed and that at higher strains Cu and S decreased and brass became dominant. This increase in brass intensity was associated with growth of the brass grains. This differential dynamic grain growth was explained by different substructural energy densities in grains with different orientations. However, it seems that further investigations on this subject are required to fully understand the problem.

In industrially hot-rolled Al alloys, a traditional β-fibre texture, albeit with dominant brass component, is generally observed in the central section of the plate where plane strain conditions prevail. Near the surface the Cu component is stronger because this component developed during tandem rolling from the {001}<110> shear component introduced during the first rolling passes in the reverse rolling mill (Driver *et al.* [2000]).

8.5.3 Transformation textures in steel

During hot deformation of steel, the parent phase (austenite) develops a crystallographic texture, which is inherited by the transformation product (ferrite, bainite or martensite). The transformation textures found in ferrite after hot rolling of austenite have been extensively reviewed by Ray and Jonas [1990, 1994a]. In plain carbon steels, it is not straightforward to derive the relations between the austenite and ferrite texture from experiments, because it is generally not possible to measure the texture of the hot-rolled product at high temperature. One way to overcome this problem consists of using another fcc material with the same SFE but with the deformed fcc phase retained at room temperature, assuming that this texture is representative of the austenite texture. Another method is to use known 'orientation relations' between parent and product phase (e.g. Bain, Kurdjumov–Sachs, . . . for steel) to recalculate the product phase. A drawback of this method is that variant selection may occur during transformation (not all variants are produced to the same extent) or that selective growth of α grains of certain orientations takes place. Another problem is that some orientations in ferrite can be derived from more than one orientation present in austenite. Hence, it is sometimes impossible to derive unambiguously the origin of a particular ferrite orientation in austenite. In some steels (e.g. in TRIP steel; cf. Section 15.2), it is possible to retain part of the austenite phase at room temperature and to determine experimentally the texture of the deformed austenite (Verlinden *et al.* [2001]).

In spite of all these difficulties, the transformation textures in plain carbon steels are fairly well understood. As long as austenite is deformed above the recrystallization temperature (also called the 'no-recrystallization temperature', T_{NR}), a cube component is formed, the strength of which depends of the amount of strain accumulated before recrystallization. The transformation texture of recrystallized austenite consists mainly of a rotated cube component {001}<110>.

When austenite is rolled below the recrystallization temperature, a sharp β-fibre texture develops, together with a weaker Goss component. It has been pointed out that the brass component {110}<112> transforms into a {332}<113> component, the Cu component into a {113}<110> component and several transformed S variants are also close to one of these two transformation components. Of course, also the Goss component and the other components of the β-fibre will transform. A detailed overview can be found in Ray *et al.* [1994a].

For deep-drawing applications (see also Sections 11.11 and 15.1), the {332}<113> component is the most beneficial because during further cold rolling of the hot-rolled plate, this component rotates towards {554}<225> and then to {111}<112> and {111}<110>. The latter two orientations belong to the γ-fibre, which enhances the deep drawability. Hence after hot rolling, a strong brass component is desirable. In general, the intensity of the transformation texture increases with decreasing finishing temperature, because (in the absence of recrystallization) more strain can be accumulated before the transformation starts. In simple plain carbon steels, T_{NR} is close to the transformation temperature, so only a fairly weak transformation texture can be expected. Alloying elements like

Nb, V and Ti lower the transformation temperature, so more deformation can be applied in austenite before the transformation to ferrite starts and a much sharper transformation texture is observed. Substitutional solutes like Mn, Cr, Ni and Mo, but also a finer austenite grain size and faster cooling, have been observed to increase the desired {332}<113> component and leave the {113}<110> component unaffected. Although neither a widely accepted model for selective grain growth during transformation nor an accurate model for variant selection exists, reasonable good agreements between predicted and experimentally observed transformation textures are obtained.

LITERATURE

Aernoudt E., Van Houtte P. and Leffers T., 'Deformation and textures of metals at large strain', in "Materials Science ad Technology", R.W. Cahn (ed.), vol. 6, VCH, New York (1993), pp. 89–136.

Barrett C.S. and Massalski T.B., "Structure of Metals: Crystallographic Methods, Principles and Data", Pergamon Press, Oxford (1980).

Bunge H.J., "Texture Analysis in Materials Science – Mathematical Methods", Butterworths, London (1982).

Hatherly M. and Hutchinson B., "Introduction to Textures in Metals", vol. 5, Institution of Metallurgists, Chameleon Press (1979).

Hirsch J. and Lücke K., Acta Metall., 36 (1988), 2863–2882.

Kocks U.F., Tomé C.N. and Wenk H.-R., "Texture and Anisotropy: Preferred Orientations in Polycrystals and their Effect on Materials Properties", Cambridge University Press (1998).

Randle V. and Engler O., "Texture Analysis. Macrotexture, Microtexture and Orientation Mapping", Gordon and Breach (2000).

Taylor A., "X-ray Metallography", Wiley, New York (1961).

Wasserman G. and Grewen J., "Texturen Metallischer Werkstoffe", Springer, Berlin (1962).

Wenk H.-R., "Preferred Orientation in Deformed Metals and Rocks: An Introduction to Modern Texture Analysis", Academic Press, Orlando (1985).

Chapter 9

Residual Stress

Chapter 9

Residual Stress

9.1. INTRODUCTION

The subject of residual stress has been well covered in many[1] books and review articles (Heindlhofer [1948], Osgood [1954], Almen and Black [1963], Macherauch and Hauk [1986], Van Houtte and De Buyser [1993], Withers and Bhadeshia [2001a, b], Welzel *et al.* [2005]). The first issue, which needs to be clarified, is the exact definition of residual stress. A general definition is usually coined as – 'the stress in a body, when the body is unacted on by any external agency' (Heindlhofer [1948], Osgood [1954], Almen and Black [1963]). A similar, though more scientific, definition can be formulated as – 'the stress in a body which is stationary and at equilibrium with its surrounding' (Withers and Bhadeshia [2001a]). Perhaps both these definitions are slightly complicated for a beginner and an effort is made to provide a simpler picture.

Figure 9.1 illustrates schematically the creation of residual stress through deformation of a 'composite' two-phase material – one phase deforms elastically whereas the other undergoes an elasto-plastic deformation. On unloading to zero applied stress, the 'harder' phase is extended by the 'softer' phase, which in turn is compressed by its neighbour. At the level of the atomic planes, the spacings of a and b would respectively be increased and decreased. This effect can occur at all scales – metres (aircraft wing spans – e.g. 30 m) to nanometres (thin films in semiconductors – e.g. 10 nm).

Figure 9.2 tries to describe the atomistic origin of residual stress in further details. The process of deformation[2] can be classified in terms of its atomistic effects – deformation which may change interplanar spacing or may incorporate lattice defects. As in Figure 9.2, the former, or the elastic deformation, may remain in the body under suitable constraints. This is the residual stress. The constraints may range from constraints provided by a substrate in the case of thin films (Welzel *et al.* [2005]) to heterogeneous deformation involved in bulk-forming processes. For example, the surface of a bent bar containing more heavily deformed layers may provide the constraint necessary for retaining the interplanar spacing change in the sub-surface regions. Almost any manufacturing process would develop residual stresses (Heindlhofer [1948], Osgood [1954], Almen and

[1] Several recent review articles are cited in the literature list.

[2] The origin of deformation or stresses can be mechanical, thermal, phase transformation related, etc.

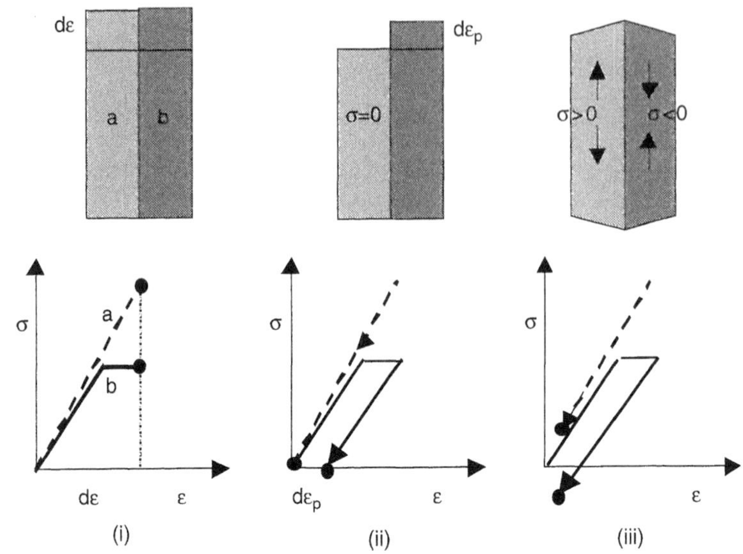

Figure 9.1. Schematics describing the creation of residual stress by deformation of a two-phase system (a+b); (i) both phases are extended in tension by dε, a being harder deforms elastically and b deforms elastically then plastically; (ii) represents the situation on unloading if the two phases were separate, b would be plastically deformed by $d\varepsilon_p$; (iii) the real situation on unloading when the two are joined so that a is extended by b and conversely b is compressed by a. The total resulting stress of (a+b) is zero.

Black [1963], Macherauch and Hauk [1986], Withers and Bhadeshia [2001b], Welzel *et al.* [2005]).

9.2. TYPES OF RESIDUAL STRESS

Residual stress can be microscopic or macroscopic as illustrated schematically in Figure 9.3. As shown in Figure 9.3a, the residual stress can vary inside a grain due to the presence of inclusions, dislocations, stacking faults, etc. The residual stress can also vary between the grains or crystallites (see Figure 9.3b), since each grain orientation possesses specific elastic and plastic properties. Finally, these two micro-stresses would combine to provide macro-stress resulting from inhomogeneous macroscopic strains. Though typically covering several crystallites or grains, the macro-stresses can also vary along the component dimensions and can be direction sensitive or anisotropic[3] (see Figure 9.3c).

[3] Micro-stresses can also be anisotropic in nature.

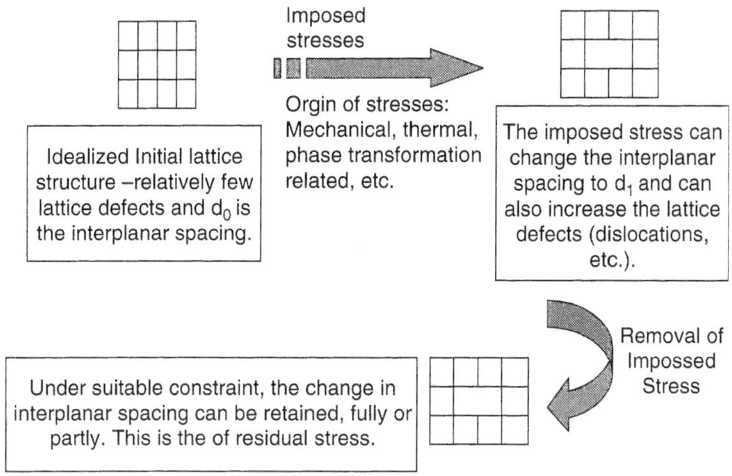

Figure 9.2. An undeformed idealized material with interplanar spacing d_0 is subjected to imposed stresses. This could lead to a change in interplanar spacing and to an increase in lattice defects (dislocations, stacking faults, etc.). The former (which can be classified as elastic strain) if retained, fully or partly, is the residual stress.

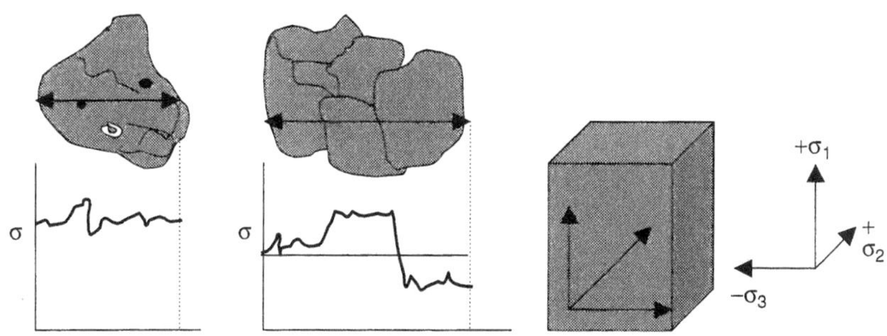

Figure 9.3. (a) Grain interior micro-stresses. The stress variation is caused by the presence of inclusions, dislocations, etc. (b) Micro-stresses varying between different grains. (c) The macro-stress is the average of local or micro-stresses. Though smaller in dimension than the overall length-scale of a component, this would cover significant fractions of the component. The micro- and macro-stresses can be anisotropic in nature.

9.3. CONTINUUM APPROACH TO RESIDUAL STRESS

As reviewed by Van Houtte and De Buyser [1993] and Welzel *et al.* [2005], the elastic deformation, and the distribution of elastic stress–strain, is governed by

- Hook's law: $\sigma_{ij} = E_{ijkl}\,\varepsilon_{kl}$, where σ_{ij}, ε_{kl} and E_{ijkl} are the tensorial quantities representing elastic stress, strain and modulus, respectively.

- Stress equilibrium equations and strain compatibility equations.
- Stress and strain boundary conditions.

Simplifying assumptions of homogeneous plastic stress and strain are often used in the continuum approach to plasticity.[4] For residual stress, similar simplifying assumptions are also adopted. For example (Van Houtte and De Buyser [1993], Welzel *et al.* [2005]):

- Reuss model: Stress homogeneity in the polycrystal; as elastic modulus would depend on grain orientation and elastic strain would vary between grains.
- Voigt model: Strain homogeneity.
- Hill model: Intermediate solution between Reuss and Voigt.
- Eshelby–Kröner model: An elastic self-consistent model – also with intermediate solution between Reuss and Voigt.

9.4. ORIGIN OF RESIDUAL STRESS

The origin of residual stress of any kind can be generalized in two steps: (i) 'misfit' through differences in interplanar spacing and (ii) constraint from immediate surroundings in 'retaining' the 'misfit'. The dimensions of the 'misfit' and the constraints involved in the macroscopic stresses, Figures 9.4a–c, are significantly larger than the microscopic stresses, Figures 9.4d and e, – but the generalized two-step origin exists in both cases. The macro-stresses are typically located within different parts of a component. The 'misfit'(s) or the difference(s) in interplanar spacing are chemically, thermally or plastically induced, with constraint being provided by the different parts. As in Figure 9.4a, plastically induced 'misfit' between plates is caused by the external clamping stresses of riveting – different parts of the assembly providing the necessary constraint. Examples of chemically and thermally induced misfits may range from simple carburizing or nitriding of steel to thermal mismatch[5] between different parts of an engineering component. In microscopic stresses, the scale of the 'misfit' and constraint is smaller. An example is provided in Figure 9.5. As in Figure 9.5a, two parts of an engineering component, metal and ceramic, can have different coefficient of thermal expansion and hence any thermal cycle would lead to stresses – macro residual stresses. In a metal–matrix composite, on the other hand, a similar situation will lead to micro residual stresses.

[4] Taylor model, see Chapter 10, treats plasticity from strain homogeneity, while Sachs model assumes stress homogeneity.

[5] Two different components of the same body or two different phases of the same alloy may have different coefficients of thermal expansion. Such body or alloy will have differences in strain during heating or cooling – strains created by the so-called thermal mismatch.

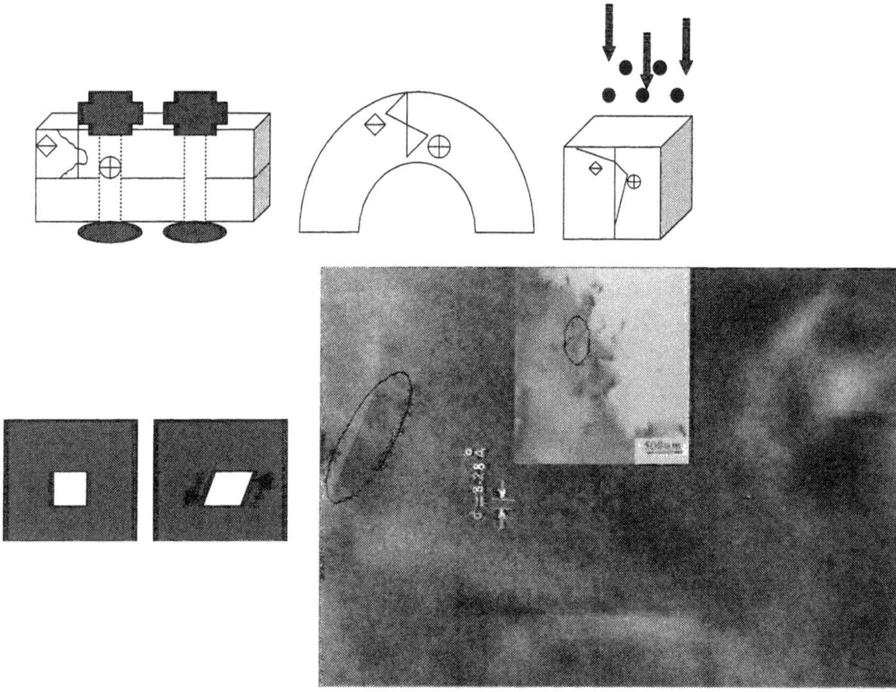

Figure 9.4. Residual stresses at different dimensions – (a)–(c) macroscopic stresses and (d)–(e) microscopic stresses. (a) Riveting of two plates, (b) bending of a bar and (c) shot peening – all bring plastically induced changes in interplanar spacing or elastic deformation retained through local constraints. The compressive and tensile nature of the elastic stresses are indicated in (a)–(c). (d) Displacive phase transformation (e.g. – martensitic transformation in iron-based alloys) and (e) coherent precipitation in austenitic stainless steel (Wasnik *et al.* [2003]). (d)–(e) also would generate residual stresses, though these are microscopic in nature.

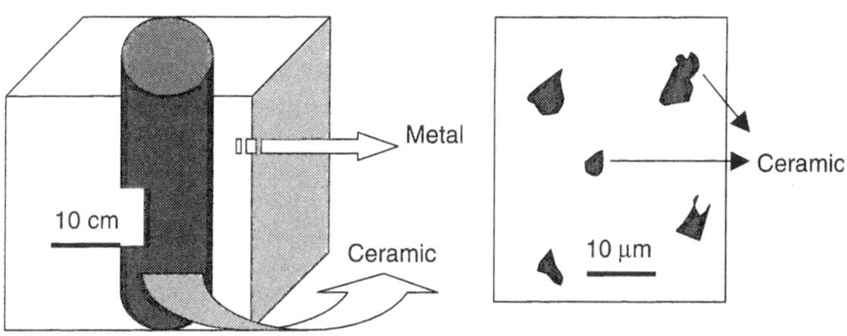

Figure 9.5. (a) An assembly of ceramic and metal parts in an engineering component. (b) Metal–matrix composite with ceramic reinforcement. Due to differences in the coefficients of thermal expansion, a thermal cycle would lead to 'misfit' and residual stresses in both (a) and (b) – dimensionally such stresses would, however, be termed respectively as macroscopic and microscopic.

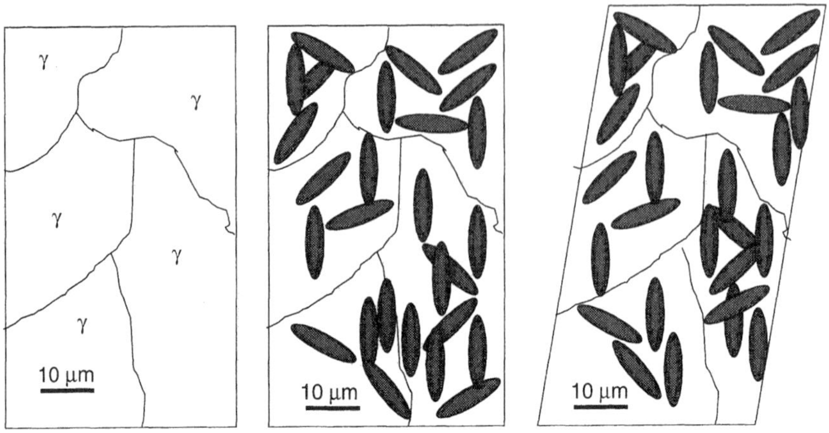

Figure 9.6. (a) Pre-transformation austenitic (γ) polycrystalline structure. (b) Austenite (γ) undergoing reconstructive transformation to ferrite platelets. (c) Austenite (γ) undergoing displacive transformation to martensite platelets. In (b) only microscopic in-grain residual stresses can form, while in (c) the physical distortion of the displacive (invariant plane strain) transformation can result in both micro- and macro-stresses.

It is important to point out that phase transformations[6] can also induce stresses, both macro and micro, as indicated in Figure 9.6. The stresses in reconstructive transformations will be typically microscopic, while displacive transformations can create both micro- and macro-stresses.

9.5. RESIDUAL STRESS MEASUREMENTS

The importance of residual stress is widely acknowledged. This includes both detrimental and beneficial aspects. For example, the relative nature (tensile or compressive) and the magnitude of residual stresses can enhance or retard fatigue crack growth (Almen and Black [1963], Macherauch and Hauk [1986], Withers and Bhadeshia [2001a]). Though treatments/processes to 'control'[7] residual stresses do exist, such control is often approximate and qualitative at best. One of the reasons for this can be illustrated indirectly with Figure 9.7.

[6] For more details on phase transformation, the reader may refer Chapter 7.

[7] These may include steps to create surface compressive stresses (e.g. nitriding, shot peening, etc.) or alternatively steps to reduce the residual stresses (tensile straining of Al thick plate (see Section14.4) compression of forged components and low-temperature recovery annealing).

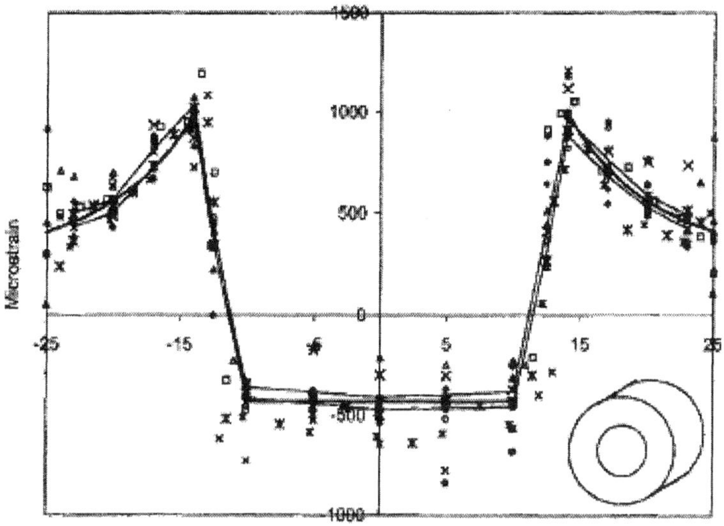

Figure 9.7. Results of 'round-robin' tests on microstrain measurements in an aluminium ring-plug. Courtesy of Withers [2001b] (copyright (2007), with permission from Maney Publishing). The measurements were obtained in 10 different neutron diffraction facilities. Analytical solution is shown by the central solid line.

Even using relatively sophisticated equipment by 'established' neutron diffraction facilities (Withers [2001b]), a large scatter in experimental internal stress data is not uncommon. Though a trend is evident in Figure 9.7, less sophisticated or calibrated measurements would offer even larger scatter and can indeed be misleading. Analytical or numerical solutions, however sophisticated they might be, cannot bridge the gap left by the absence of meaningful experimental data. For a practicing engineer or scientist, it is important to have a clear picture of the different measurement techniques for residual stress. It is also important to comprehend the relative advantages and resolution of such measurement techniques. Table 9.1 provides a summary of residual stress measurement techniques. As shown in the table, the methods for residual stress measurements can be classified into three categories (Macherauch and Hauk [1986], Van Houtte and De Buyser [1993], Withers and Bhadeshia [2001a], Welzel *et al.* [2005]) – mechanical methods, diffraction methods and methods involving physical property changes. All these methods are indirect and they estimate strain or property changes. A conversion of such measurements to residual stress estimates often leads to loss of information/accuracy.

It is perhaps imprudent to 'grade' the methods in terms of relative sophistication or effectiveness. For example, relatively simple and 'vintage' mechanical methods

Table 9.1. A summary of the residual stress measurement techniques.

Method	Details	Remarks
Mechanical	*Hole drilling*: Relaxation of residual stresses (or 'misfit') around a drilled hole is estimated by an assembly of strain gauges, interferometry or holography (Macherauch and Hauk [1986], Nelson *et al.* [1997], Withers and Bhadeshia [2001a])	Less sensitive with depth or with localized yielding
	Compliance: Stress relaxation around a crack by strain gauge interferometry (Macherauch and Hauk [1986], Withers and Bhadeshia [2001a])	Often coupled with progressive material removal
	Curvature: Curvature development of a substrate, during addition/removal of material, is used to estimate stresses. Contact (e.g. profilometry) or non-contact (e.g. video) measurements are possible (Withers and Bhadeshia [2001a])	Coatings and layers. Accuracy is limited by the available curvature
Diffraction	*Electron*: Convergent beam electron diffraction for micro-stresses between phases (Withers and Bhadeshia [2001a])	Best spatial resolution, but limited statistics
	X-ray: Slope of the interplanar spacing with specimen/detector tilt – ψ/Ω, *d* vs. $\sin^2\psi$ or $\sin^2\Omega$ (Van Houtte [1993], Welzel [2005]).	Possible to estimate the stress tensor
	Neutron: Conventional or continuous source (reactor source) – in principle same as X-ray, but with more penetration. Also time-of-flight measurements with pulsed beam (spallation source). The respective techniques estimate residual stress from shifts in diffraction angle and wavelength (Withers and Bhadeshia [2001a])	Normally peak shifts are estimated. Access to neutron diffraction is often restricted
Property changes	Changes in physical properties can be used to estimate residual stress. These can involve changes in electrical/magnetic properties, in ultrasound velocities, in thermoelastic/photoelastic effects, etc. Measurements of both macro- and micro-stresses are possible (Withers and Bhadeshia [2001a])	These are evolving techniques – often non-standardized, i.e. relative effects of microstructure are not fully understood

like hole drilling are increasing coupled with sophisticated arrays of lasers and sensors (Nelson *et al.* [1997]), which may offer significant improvements in their effectiveness. A detailed description of each technique is beyond the scope of the present chapter, and only a concise description of the X-ray diffraction is included.[8]

The 'internal strain' can be expressed as changes in interplanar spacing

$$\varepsilon = \frac{d_1 - d_0}{d_0} \tag{9.1}$$

where d_1 and d_0 are the interplanar spacing of the material with and without strain.

The exact value of d_0 is impractical to estimate theoretically, being sensitive to the exact composition, and typical X-ray stress measurements involve Ω and ψ configuration of the goniometer (Macherauch and Hauk [1986], Van Houtte and De Buyser [1993] – Figure 9.8). Normally, the d_ψ^{hkl} is measured for a range of ψ (0 to $\pm\psi_{\max}$) and then plotted[9] as a function of $\sin^2\psi$ (see Figure 9.11a). The measurements are obtained for constant ϕ (e.g. $\phi = 0°, 45°, 90°$, etc.), but can be repeated for other ϕ's for an anisotropic residual stress estimation. The X-ray lattice strain along L_3 (see Figure 9.8) is

$$\varepsilon_{33}^{XL} = \frac{d_{hkl}(\psi,\phi) - d_0}{d_0} \tag{9.2}$$

The superscripts X and L stand respectively 'as seen by X-rays' and 'expressed in the instrument/goniometer directions'. ε_{33}^{XL} represents the average strain – average among the grains favourably oriented for diffraction. A more general expression for ε_{33}^{XL} is

$$\varepsilon_{33}^{XL} = \frac{\int_{2\Pi}^{0} \varepsilon_{33}^{L}\left[g_{\phi\varphi}(\lambda)\right] f\left[g_{\phi\varphi}(\lambda)\right] \mathrm{d}\lambda}{\int_{2\Pi}^{0} f\left[g_{\phi\varphi}(\lambda)\right] \mathrm{d}\lambda} \tag{9.3}$$

where $\varepsilon_{33}^{L}[g_{\phi\varphi}(\lambda)]$ is the elastic strain in the direction L_3 of crystallite(s) with orientation $g_{\phi\varphi}(\lambda), f(g)$ representing the ODF (orientation distribution function – see

[8] This was considered necessary for the continuity with next Sections – 9.6 and 9.7. It is also to be noted that X-ray and neutron diffraction, the latter being far less accessible, are the only measurement techniques capable of measuring the stress tensor.

[9] Such plots are used to estimate d_0 values.

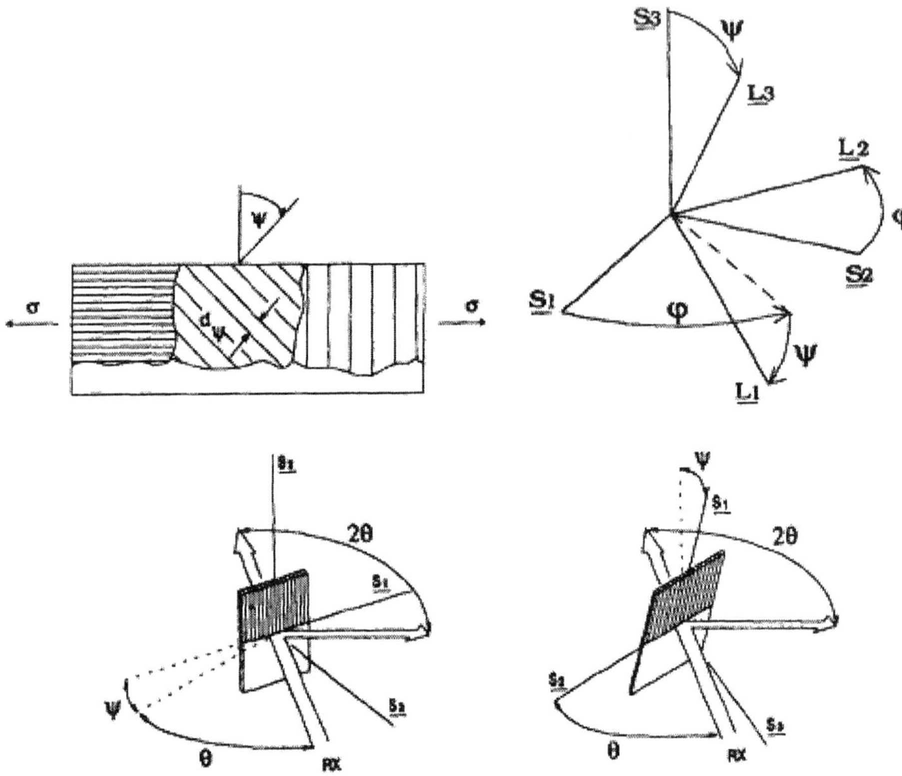

Figure 9.8. (a) Measurements of lattice spacing d_ψ of a $\{hkl\}$ plane. (b) Definition of ψ and ϕ angles. S_1, S_2 and S_3 – sample reference system. L_1, L_2 and L_3 – instrument/goniometer directions, L_3 is normal to the $\{hkl\}$ plane being measured. Configurations for (c) Ω and (d) ψ goniometers. To minimize errors in experimental data, high 2θ $\{hkl\}$ planes are typically used for residual stress measurements. Courtesy of Van Houtte and De Buyser [1993] (copyright (2007), with permission from Elsevier).

Chapter 8). To convert ε_{33}^{XL} to residual stress, a simplifying model or assumption (such as Reuss – as in Eq. (9.3)) (Van Houtte and De Buyser [1993]) is used

$$\varepsilon_{33}^{XL}[g_{\phi\varphi}(\lambda)] = r_{33ij}^{RL}\sigma_{ij}^{L} \tag{9.4}$$

r_{33ij}^{RL}, X-ray elastic compliances (XEC), can be expressed/calculated as

$$r_{33ij}^{RL} = \frac{\int_{2\Pi}^{0} s_{33ij}^{L}(\lambda) f[g_{\phi\varphi}(\lambda)]\mathrm{d}\lambda}{\int_{2\Pi}^{0} f[g_{\phi\varphi}(\lambda)]\mathrm{d}\lambda} \tag{9.5}$$

in which $s^L_{33ij}(\lambda)$ represents the expression for elastic compliances in the goniometer system. Extending the treatment (Van Houtte and De Buyser [1993]) would provide

$$d_{hkl}(\psi,\phi) = d_0[1 + s_1(\sigma_{11} + \sigma_{12}) + \tfrac{1}{2}\, s_2\, \sigma_\phi \sin^2\psi] \tag{9.6}$$

where $s_1 = r^{RL}_{3311}$ and $\tfrac{1}{2}s_2 = r^{RL}_{3333} - r^{RL}_{3311}$. If s_1 and s_2 are independent of ψ, a linear relationship of $d_{hkl}(\psi,\phi)$ with $\sin^2\psi$ would emerge. Similar approaches are also possible with other models (Voigt, Kröner, etc.) (Van Houtte and De Buyser [1993]).

9.6. MICRO-STRESS ANALYSIS – A TOOL FOR ESTIMATING DISLOCATION DENSITIES

As outlined in the earlier section, X-ray diffraction of internally stressed crystalline materials gives rise to shifts in the positions of the diffracted lines that are used to evaluate the degree of internal strain at the scales of grains or polycrystalline aggregates. However, internal stresses at the scales of nanometres, as occurs in the vicinity of dislocations, give rise to significant broadening of the X-ray peaks since, very locally, the lattice is heavily strained in both tension and compression. The form of this broadening can be used to evaluate the density and sizes of the nano-diffracting domains (e.g. the dislocation densities).

The classical analyses of X-ray line broadening were first developed by Warren and Averbach and Williamson and Hall. However, these methods suffered from problems related to separating out the respective influences of domain size and density and, particularly for dislocation density measurements, have been superseded over the last 10 years by new analyses. The latter are based on the Wilken's model of a restrictedly random distribution of dislocations (Wilkens [1970]) and adopt new methods for de-convoluting the broadening contributions due to strain and domain size. This brief survey describes some pertinent features of this emerging approach.

Figure 9.9 illustrates typical X-ray line profiles of a deformed metal (in this case hot-rolled Al–Mg) after different strains. It is worth noting the scales; the abscissa covers about $\pm 1°$ and the ordinate up to 4 orders of magnitude so that line broadening analysis requires very high angular resolution and accurate intensity measurements. The broadening occurs both in the tails of the distribution and the widths of the peak. Both types of broadening can be used for the analysis: widths of several peaks in a modified Williamson and Hall analysis (Ungar and Borbely

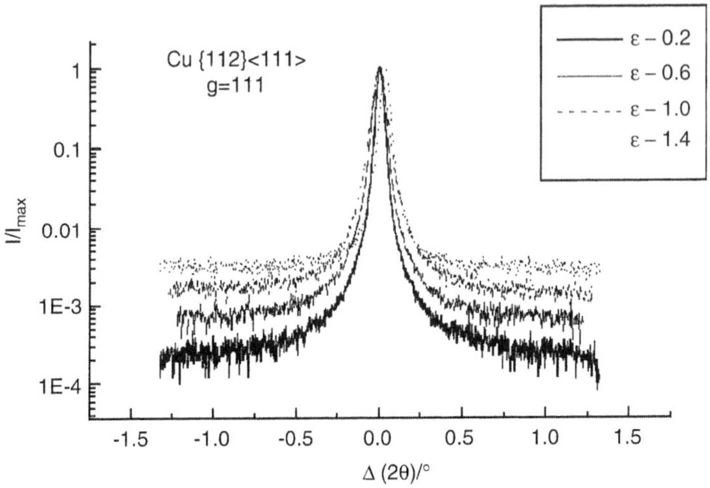

Figure 9.9. X-ray line profiles in hot-rolled Al–Mg (Guiglionda [2003]).

[1996], Guiglionda *et al.* [2004]) and the tails of any peak for a momentum analysis (Groma [1998]).

The intensity distribution in X-ray line profiles can be written as $I(q)$, where $q = 2/\lambda(\sin\theta - \sin\theta_0)$ – representing the angular distance from the ideal Bragg position θ_0 for wavelength λ and diffraction angle θ. As an example of the momentum method, the 2nd order moment can be written in terms that separate out the domain size ε_F from the dislocation density ρ as follows:

$$M_2(q) = \frac{1}{\pi^2 \varepsilon_F} q - \frac{L}{4\pi^2 K^2 \varepsilon_F^2} + \frac{\Lambda\langle\rho\rangle \ln(q/q_0)}{2\pi^2} \tag{9.7}$$

From Eq. (9.7), and without going into a description of all the terms, it is apparent that the size effect (ε_F) should be roughly linear in q, whereas the strain effect due to dislocations should have a logarithmic dependency. Figure 9.10 illustrates three practical examples from three types of microstructures: (i) a pure size broadening in nanoparticles of Fe (with a linear relation), (ii) a pure strain, or dislocation broadening, from cold-rolled Al–Mg (logarithmic relation) and (iii) mixed broadening in hot-rolled Al–Mg, which possesses a dislocation subgrain structure, i.e. both domains and dislocations. The slopes and the intercepts of these plots, as in Figure 9.10a, are used to determine the respective domain sizes and dislocation densities. An advantage of this method is that, by a suitable choice of the goniometer angles, peaks from individual texture components can be analysed to give these

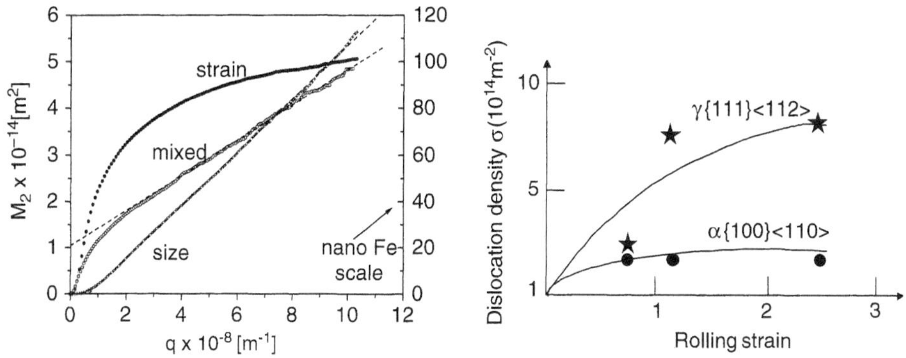

Figure 9.10. (a) Different plots of the 2nd order restricted moment. Courtesy of Borbely A. and Driver J.H. Effects of size, strain and a combination (mixed) on the 2nd order restricted moment, M_2 vs. q as in Eq. (9.7). (b) Dislocation densities determined by the momentum method in two texture components of high-purity cold-rolled iron (γ which accumulates high dislocation densities and α which does not) (Borbely *et al.* [2000], Borbely and Driver [2005]).

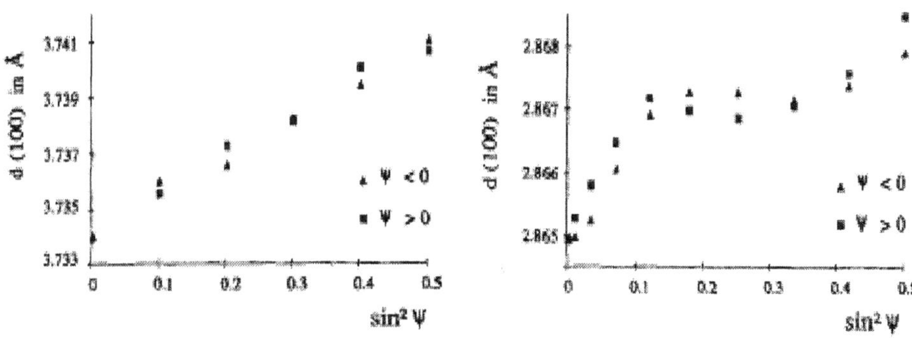

Figure 9.11. (a) d vs. $\sin^2\psi$ for gold coating. A tensile residual stress of 201 MPa is estimated from the slope of the linear plot, using Reuss model for isotropic material. (b) Non-linear d vs. $\sin^2\psi$ for cold-rolled steel, the non-linear behaviour being attributed to the presence of crystallographic texture. Courtesy of Van Houtte and De Buyser [1993] (copyright (2007), with permission from Elsevier).

parameters as a function of grain orientation. An example of the application of the method to the dislocation densities in two different texture components of cold-rolled iron is given in Figure 9.10b.

9.7. RESIDUAL STRESS AND CRYSTALLOGRAPHIC TEXTURE

A non-linear relationship of d vs. $\sin^2\psi$ can be related to the presence of strong crystallographic texture (see Figure 9.11b). Different methods, as summarized in

Table 9.2. A summary of the methods typically adopted to analyse residual stress data for materials with strong crystallographic texture.

Details	Remarks
Using {h00} or {hhh} planes: Linear d vs. $\sin^2\psi$ is expected (Evenschor and Hauk [1975]) in cubic materials. Isotropic XEC values can be used for linear curves	Uses elastic models. Main issue is to find suitable plane of sufficient intensity and high 2θ
Using ODF: XEC r_{33ij}^{RL} is estimated (using ODF) as a function of ϕ and ψ. Using a suitable model (e.g. Eq. (9.6)), non-linear experimental d vs. $\sin^2\psi$ is fitted (e.g. least-square procedure) by adjusting σ_{ij}	Indirectly uses elastic models. Different procedures have been adopted (see Van Houtte and De Buyser [1993])
Using ideal orientations: $\varepsilon_{33}^{XL}(\phi,\psi)$ are measured (Hauk and Vaessen [1985]) for individual ideal orientation(s), as represented by suitable stereographic projections – 2θ, ϕ and ψ. Several measurements for each ideal orientation are obtained and then converted into elastic or residual stresses	No model assumptions or usage – but labour intensive. Often more than one ideal orientations contribute to the intensity at any particular direction – this is in contradiction to the basic assumption of the method
Using strain plots: Instead of measuring $f(g)$, as in an ODF, the residual stress is measured at different 2θ, ϕ and ψ. The strain pole figures (Hauk and Oudelhoven [1988]), see Figure 9.12, can represent residual stress as a function of orientations/texture. $\varepsilon_{ij}(g)$ or $\sigma_{ij}(g)$ functions can also be formulated and stress/strain distribution functions can be worked out (Wang *et al.* [2001])	No model assumptions or usage. Extremely labour intensive, but perhaps the only way to bring out completely the effects of texture on residual stress

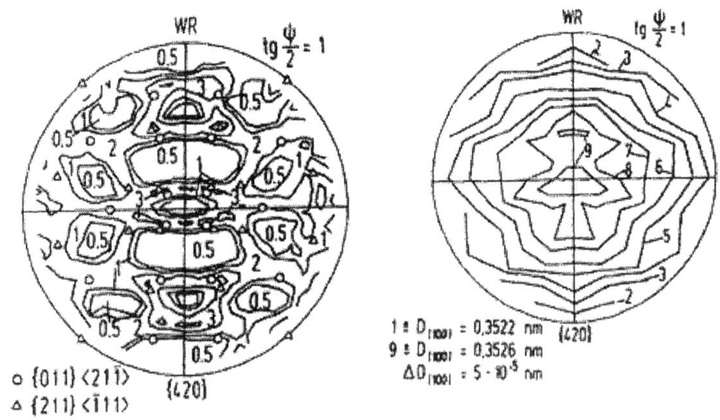

Figure 9.12. (420) Projection representing (a) pole figure or crystallographic texture and (b) strain plot in cold-rolled nickel. WR represents the rolling direction. Courtesy of Hauk and Oudelhoven [1988] (copyright (2007), with permission from Max-Planck-Institut für Metallforschung).

Table 9.2, are typically (Van Houtte and De Buyser [1993]) adopted to analyse residual stress data for 'textured' materials. As shown in the table, even in such cases elastic models can be adopted to estimate residual stress – namely by restricting measurements to suitable planes or by using suitable models in combination with ODF. The other alternative is to measure residual stress for ideal orientation(s) or to estimate strain plots, both strain pole figures (Hauk and Oudelhoven [1988]) as well as stress/strain distribution functions (Wang *et al.* [2001]). Though extremely labour intensive, the latter is the only way to effectively bring out the effects of crystallographic texture on residual stress.

LITERATURE

Van Houtte P. and De Buyser L., Acta Metall. Mater., 41(2), (1993), 323.
Welzel U., Ligot J., Lamparter P., Vermeulen A.C. and Mittemeijer E.J., J. Appl. Crystal., 38, (2005), 1.
Withers P.J. and Bhadeshia H.K.D.H., Mater Sci. Tech, 17, (2001a), 355.
Withers P.J. and Bhadeshia H.K.D.H., Mater Sci. Tech, 17, (2001b), 366.

Chapter 10

Modelling

Chapter 10

Modelling

10.1. INTRODUCTION

Mathematical and numerical modelling of thermo mechanical processing (TMP) has become an important tool of scientists and engineers over the last 20 years. The complexity of many of the coupled phenomena that take place during even quite simple operations has led to the development of mathematical techniques for simulating the behaviour of materials during processing. The obvious advantage of modelling is the capacity to predict what will happen when processing parameters and/or alloy compositions are changed, in terms of microstructure evolution and therefore material properties. In theory, this 'through process modelling' will enable manufacturers to dispense with costly, time-consuming, experiments each time they change processing parameters or modify compositions to optimise them for new applications.

However, it should be recognized that for many applications this 'material scientists dream' is some way from practical realization. Although some of the basic physical concepts of say, phase transformations or work hardening and softening have been available for some time, modelling processing tools are presently in a state of continous development. This is because of the large numbers of parameters that enter into the models and their multi-scale requirements due, in part, to the heterogeneous nature of many deformation processes. As an example, take a typical extrusion process in which a metallic alloy is extruded through a die to form a long profile. The process has been practised for decades, but successful modelling of the properties of a new extruded alloy would require a quantitative analysis of the material behaviour during extrusion through

- a detailed description of the entire history of the strain paths and flow stresses of typical material elements during the extrusion process;
- the thermal cycle of the above elements;
- a reliable model for the evolution of the grain orientations and deformation, i.e. dislocation substructures, during the process as a function of the above parameters; and
- good models for recovery, and eventually recrystallization, allotropic transformations and precipitation as a function alloy composition, thermal cycle and deformation substructure (or stored energies).

In most cases, the entire set of these ingredients are not available; so, modelling of many practical processes is limited to subsets of the above. This can, however, be very useful, and quite sufficient for many less demanding requirements, for example, predicting the loads to be applied during the shaping process or predicting grain sizes in hot-rolled steels. Microstructural modelling of processes involving phase transformations without concurrent plastic deformation is quite well developed (see, e.g. the recent review by Grong and Shercliff [2002]). The aim of current modelling developments is to extend these specific cases to a wide variety of thermo-mechanical treatments.

To model microstructure evolution in greater detail for relatively simple deformation paths, it is now becoming almost routine to use finite element (FE) or finite difference (FD) methods to determine the strain and thermal histories of the material. The FE and FD methods will not be developed here since there are many good descriptions of them, e.g. Zienkiewicz *et al.* [2005]. Given the macroscopic strain and thermal history, a combination of physical and/or semi-empirical models is then often used to estimate the change of texture, substructure and 2nd phase distributions for each of the representative material elements. In the following, we shall give examples of modelling developments in some thermo mechanical processes starting with deformation texture evolution.

10.2. DEFORMATION TEXTURES

All large plastic strains lead to the formation of a deformation texture due to the fact that, during deformation, the orientations of most individual grains change, usually to stable orientations which can more easily accommodate the imposed strain. The case of a single crystal deformed in a tensile test is well known (see Figure 10.1); the orientation defining the tensile axis of a randomly oriented fcc crystal rotates progressively to the [100]–[111] border and then, by the activation of other slip systems, towards [112].

The case of a grain in a polycrystal in tensile extension is similar but slightly different at medium to high strains since the straining conditions are different; a grain is constrained by the entire set of surrounding grains. In fact, it continues towards either [100] or [111] as shown in Figure 10.1d. At high strains, this gives a crystallographic texture defined by two components, which have the [100] or [111] crystal directions aligned with the extension axis.

During shaping operations, the strains are often much higher than in a tensile test so the resulting textures can be very pronounced. These crystallographic textures lead to anisotropic mechanical (and physical or chemical) properties, which can be predicted from the basic theory of crystal plasticity. A practical example of the

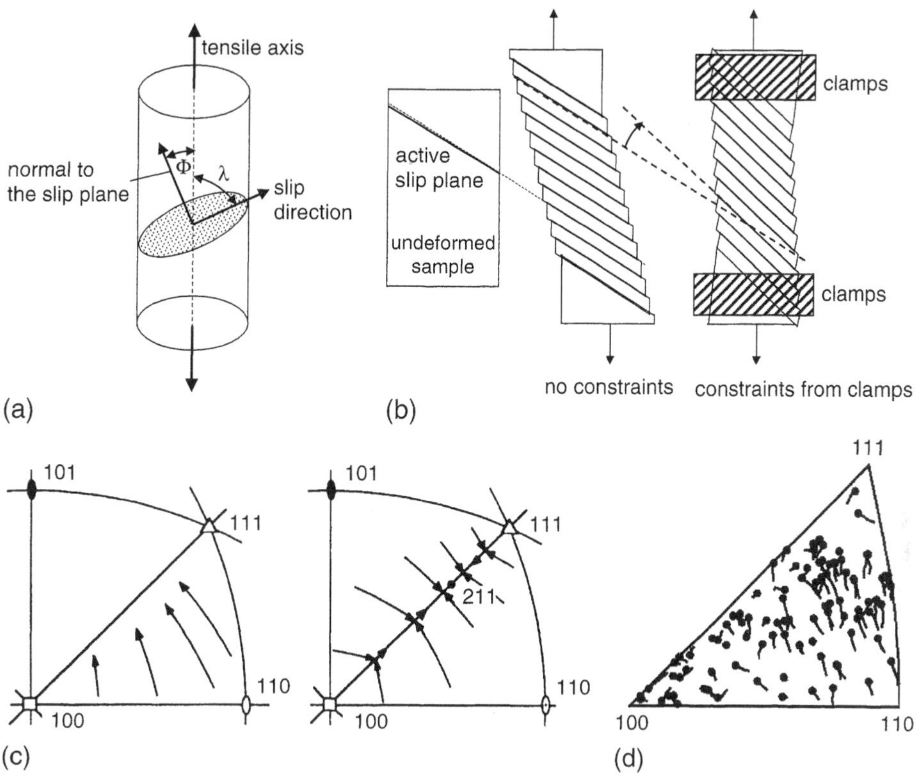

Figure 10.1. Lattice rotations due to slip in a tensile deformed fcc crystals (a) slip system in a single crystal, (b) the effect of grip constraints on the slip plane orientation of the crystal, (c) theoretical single crystal reorientations as depicted in a stereographic plot from Hosford [1993] and (d) experimental lattice rotations of a set of grains measured inside a polycrystalline Al sample deformed 6% in tension (the end orientations are indicated by filled circles). *In situ* measurements by high energy, Synchrotron 3D X-ray diffraction (Poulsen *et al.* [2003]). Permission obtained from Elsevier.

strength anisotropy of an aeronautical alloy is given in a Figure 3.6 and the well-known case of earing in drawn cups in Figure 3.7. In most cases, as in can drawing, the deformation is multiaxial, often close to biaxial, so that the flow stresses under combined stress states are required. These are often characterized by an anisotropic yield surface as described in Section 11.7 on sheet metal forming.

10.2.1 Principles of deformation texture formation

A grain, or a crystal, usually changes orientation during plastic deformation since the activated slip/twinning systems are rarely 'perfectly oriented' to accommodate the imposed plastic strain. For example, as shown in Figure 10.1 a single crystal elongated in a tensile test usually starts deforming on one slip system, and thereby develops a shear deformation; the difference between the shear due to slip and the imposed elongation is accommodated by a simultaneous rotation of the crystal lattice. The rate of change of orientation (shortened to lattice rotation rate) then depends upon both the slip systems activated in the grain and its current deformation mode as imposed by the macroscopic shaping operation.

Methods used to evaluate the slip systems in crystals undergoing a strain rate tensor $\dot{\varepsilon}$ are described below. This requires solving the relations between the slip amplitudes or rates $\dot{\gamma}$ and the $\dot{\varepsilon}$ components. The latter values are often prescribed by the deformation process, but the slip systems and their rates have to be determined from the yield surface of the grain. This first involves calculating for each strain increment the current yield surface of the crystal. At low temperatures, when rigid-plastic slip predominates the crystal yield surface is an n-dimensional polyhedron ($n = n°$ of imposed strain components) with facets corresponding to each potential slip system of critical resolved shear stress τ_c^S, in accordance with the Schmidt law. The facets are defined by the linear relations between the stress deviator **s** and the orientation factors M_{ij}^S of the slip system;

$$\tau_c^S \geq \sum_{ij} S_{ij} M_{ij}^S \qquad (10.1)$$

The matrix of orientation factors M_{ij}^S is simply composed of linear functions of the direction cosines of the slip plane normals **n**, and the slip directions **b**:
so that for system s

$$M_{ij}^S = \sum_S \frac{1}{2}\left(\mathbf{b}_i^S \mathbf{n}_j^S + \mathbf{b}_j^S \mathbf{n}_i^S\right) \qquad (10.2)$$

they describe the current slip system orientation with respect to a suitable reference frame. Figure 10.2 illustrates an example of a 2D yield surface of an fcc single crystal.

Since the current stress state of the grain is not known *a priori* from the current yield surface, a slip system selection criterion is then used to determine the set of critically stressed systems. The selection criterion are standard continuum mechanics principles, e.g. minimum internal work done by the crystal: $\Sigma_S \tau_c^S \dot{\gamma}_S$Min as first suggested by Taylor [1938], or maximum external work carried out by the machine: $\Sigma_{ij} \sigma_{ij} \dot{\varepsilon}_{ij}$Max as proposed by Bishop and Hill [1951].

Adopting the maximum work method, the stress state corresponds to the vertex of the yield surface that yields the highest value of the plastic work and the

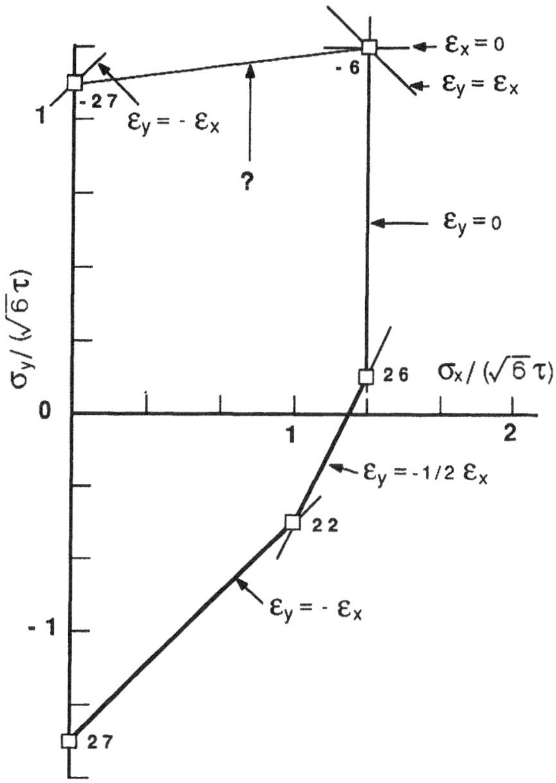

Figure 10.2. 2D yield surface of an fcc single crystal with $x = [1-12]$ and $y = [-111]$ (from Hosford [1993]). The facets correspond to the inner envelope of the linear relations (Eq. (10.1)) for each potential slip system. The vertices are stress states, which activate two or more potential slip systems.

critically stressed systems are then given by Eq. (10.1). Given these critically stressed slip systems, e.g. 6 or 8 {111}<110> systems in fcc crystals, their slip rates can be determined by the set of linear relations between the $\dot{\gamma}$ and the $\dot{\varepsilon}$ components:

$$\dot{\varepsilon}_{ij} = \sum_{S} M_{ij}^{S} \dot{\gamma}_{S} \tag{10.3}$$

The above relation – due to Taylor [1938] – cannot always be solved unambiguously since, as in the case of fcc slip on {111}<110>, the number of potential slip systems $\dot{\gamma}$ is often higher than the number of prescribed strain rate components. Various methods have been devised to get around this problem, see, for example, Van Houtte [1988]. However, when the number of slip systems increases, the ambiguity problem tends to become less acute so that bcc crystals which can slip on several planes along the <111> directions do not give much ambiguity.

When deformation occurs by twinning a similar relation to 10.3 between the amount of twinning and the strain can be written, but since it is the twinning volume fraction that changes during a strain increment this becomes:

$$d\varepsilon_{ij} = M_{ij}^{t} \cdot \gamma^{t} \cdot df\gamma^{t} \tag{10.4}$$

where γ^t is the twinning shear of system t and $df\gamma^t$ the change in volume fraction of its twins during a strain increment $d\varepsilon_{ij}$.

At high temperatures, when slip becomes thermally activated, this 'yield surface' is rounded off according to the value of the strain rate sensitivity parameter $m = (d\ln\sigma/d\ln\dot{\varepsilon})_T$ $0 \le m \le 1$. The slip rates can then be determined by numerical resolution of the relation for the stress state of the grain:

$$\dot{\varepsilon}_{ij} = \sum_{\mathrm{S}} M_{ij}^{\mathrm{S}} \cdot \dot{\gamma}_0 \left(\frac{\sigma_{kl} M_{kl}^{\mathrm{S}}}{\tau_0} \right)^{1/m} \tag{10.5}$$

where τ° and $\dot{\gamma}_0$ are reference quantities for the shear stresses and strain rates as used in the standard relation $\tau_{\mathrm{s}} = \tau_{\circ}^{\mathrm{S}}(\dot{\gamma}^{\mathrm{S}}/\dot{\gamma}_{\circ}^{\mathrm{S}})^m$ (no summation over s).

Given the current slip rates in the crystal, the lattice rotation rate (with respect to the external refernce frame) is obtained as the difference between the (macroscopic) velocity gradient tensor **d** imposed by the external boundary conditions and the velocity gradient tensor **g** due to glide on the slip systems. The relation between the lattice rotation, the prescribed displacements and the displacements due to glide is readily shown by an analogy with the 'kinematics' of slip in a pack of cards (Figure 10.3).

This fundamental difference of the velocity gradients is the rotation rate of the crystal lattice, which with respect to the X_1, X_2, X_3 co-ordinate system, is described by an antisymmetric tensor $\dot{\mathbf{r}}$:

$$\dot{\mathbf{r}} = \dot{\mathbf{d}} - \dot{\mathbf{g}}, \quad \dot{r}_{ij} = \dot{d}_{ij} - \dot{g}_{ij} \tag{10.6}$$

In the above example, the X_1, X_2 reference frame is fixed; but in many practical shaping operations, this reference frame can also rotate by rigid body rotations as the material flows round corners, etc. The basic analysis is then written

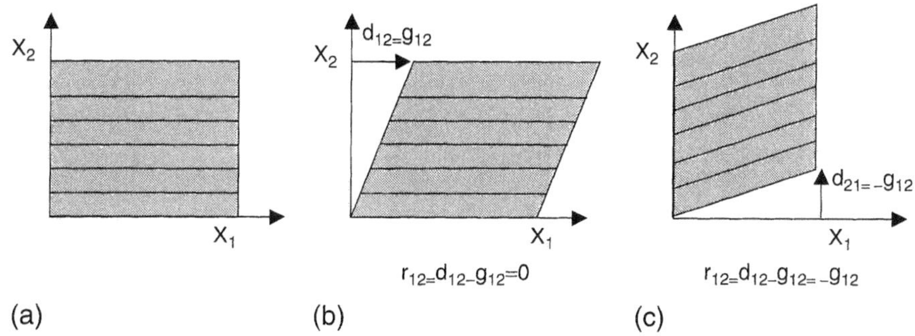

Figure 10.3. The lattice rotations due to imposed displacements on a pack of cards (a) initial card set with slip planes represented by lines along X_1; in (b) the cards are sheared by a quantity d_{12} along the X_1 direction corresponding to the usual shear of the cards (g_{12}) and so the "lattice" rotation is zero; in (c) the cards are sheared along the X_2 direction which, since they have only one "slip system" enabling the g_{12} displacement, can only be accomplished by g_{12} and a simultaneous lattice rotation $=-g_{12}$.

in a form which separates out the symmetrical and antisymmetrical parts of the two velocity gradient tensors, and is often to be found in the mechanics literature as:

$$\dot{r}_{ij} = \dot{\omega}_{ij} - \frac{1}{2}\left(\dot{g}_{ij} - \dot{g}_{ji}\right) \tag{10.7}$$

where $\dot{\omega}$ is the rigid body rotation rate, or the antisymmetrical part of the imposed velocity gradient tensor.

The components of the velocity gradient tensor **g** due to glide on the slip systems are defined by

$$\dot{g}_{ij} = \sum_{S} \mathbf{b}_i^S \mathbf{n}_j^S \dot{\gamma}_S \tag{10.8}$$

The rotation matrix can also be written as a 3 component vector $\mathbf{r} = (-r_{23}, r_{13}, -r_{12})$.

After an infinitesimal strain $\mathrm{d}\varepsilon = \dot{\varepsilon}\mathrm{d}t$, there is an infinitesimal rotation $\mathbf{dr}$, which can be expressed as the product of a unit rotation vector $\mathbf{R}$ and an infinitesimal angle $\mathrm{d}\theta$ ($\mathbf{dr} = \mathbf{R}\,\mathrm{d}\theta$).

The rotation angle $\mathrm{d}\theta = \sqrt{(\mathrm{d}r_1^2 + \mathrm{d}r_2^2 + \mathrm{d}r_3^2)}$ and the rotation axis $\mathbf{R} = (\mathrm{d}r_1/\mathrm{d}\theta)$, $(\mathrm{d}r_2/\mathrm{d}\theta)$, $(\mathrm{d}r_3/\mathrm{d}\theta)$.

The new orientation of any vector $\mathbf{V}_1$ after rotation is $\mathbf{V}_2$ given by $\mathbf{V}_2 = \tilde{\tilde{\mathbf{R}}}\mathbf{V}_1$, where the rotation matrix is defined for infinitesimal rotations as

$$\tilde{\tilde{\mathbf{R}}} = \mathbf{I} + \mathrm{d}\mathbf{r} \tag{10.9}$$

Rigourously one should use the finite rotation matrice, which is obtained from **dr** by relations given, for example, by Bunge [1982] and Kocks *et al.* [1998].

Equation (10.9) then enables one to calculate the new orientation of any grain after a small strain increment and then repeat the entire calculation for the following increment(s). For a polycrystalline aggregate, this calculation is performed for a set of orientations which represent the crystallographic texture of the aggregate. It turns out that the calculated grain orientations then rotate progressively towards orientations, which are practically stable (i.e. $\dot{r} \to 0$); these are the theoretically stable texture components for the deformation mode. In axisymmetric elongation, e.g. wire drawing, of fcc polycrystals they would be the <111> and <100> orientations aligned with the tensile direction (see Figure 10.4). The case of axisymmetric deformation of bcc metals was first treated by Chin *et al.* [1967].

There are some further subtle points to be considered; in particular, what is the relation between the (incremental) strain of the grain and the imposed macroscopic strain? To a first approximation, these are taken equal (the Taylor or Full

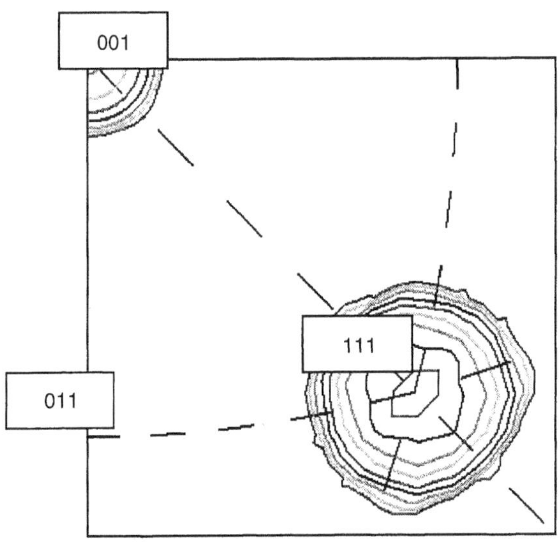

Figure 10.4. Simulated inverse pole figure of the elongation directions for the axisymmetric deformation of an fcc polycrystal after a strain of 3.5. Courtesy of D. Piot, Ecole des Mines de Saint-Etienne.

Figure 10.5. Schematic of grain deformation in the RC lath model for rolling.

Constraints model described above for which five independent strain components are applied to the grain). This gives deformation texture simulations which are similar but usually not quite the same as experimental textures. To a second approximation, following Honeff and Mecking [1978] it is often considered that the highly flattened and elongated grains typical of strongly deformed metals can undergo some localized in-plane shears without creating major plastic 'incompatibilities'; this is the origin of the Relaxed Constraints (RC) model since the corresponding stress components are relaxed (usually to zero). In most cases, for strongly rolled sheet, one relaxes the ε_{xz} grain shear (where x is the rolling direction and z the sheet normal) corresponding to the 'lath' model (see Figure 10.5). In the 'pancake' model, both the ε_{xz} and ε_{yz} shears are relaxed. The other, imposed, strain components are then set equal to the macroscopic values as in the Taylor model. This means that the single crystal yield surface for the non-imposed stresses is modified to become 4D (lath) or 3D (pancake). The principal of solving for the slip rates is similar to the FC method described above, but the equations are modified to allow for the mixed boundary conditions as first proposed by Renouard and Wintenberger [1976].

There have also been major efforts to improve on these relatively simple models of grain deformation. One is the self-consistent method in which the set of grain orientations are treated simultaneously to obtain local stresses and strains which sum up, consistently, to the macroscopic values. The Lebensohm and Tomé [1993] model for viscoplastic slip is particularly useful for texture simulations. Note, however, that these models assume implicitly that the deformation is always homogeneous at the grain level. Both the RC and the self-consistent models usually give somewhat better agreement with experiment than the FC model.

Other models attempt to account for the possibility of grain break-up during plastic deformation, i.e. grain strain rates which can vary within the grain and be significantly different from the macroscopic values. As described in Chapter 4, many grains do not deform homogeneously during plastic strain and often separate out into bands of different orientations separated by transition bands. This occurs extensively in many alloys at low temperatures but less so at high temperatures. The texture models developed recently for this behaviour are those of Lee and Duggan [1993], Crumbach *et al.* [2001] and Van Houtte *et al.* [2005]. Most of them assume some degree of co-ordination between the deformation bands to retain compatibility with the macroscopic strain.

Standard crystal plasticity analysis also gives the relations between microscopic (crystal) and macroscopic (polycrystal sample) flow stresses as follows.

The sum of the slip rates $\Sigma\dot{\gamma}^s$ in a grain g is used to define the Taylor factor of the grain $M(g)$. This factor relates the microscopic and macroscopic flow properties by the plastic work rate $\dot{W}(g)$:

$$\dot{W}(g) = \sum_{S} \dot{\gamma}^{S} \tau_{c}^{S} = \sigma(g)\dot{\varepsilon} \tag{10.10}$$

If one assumes that the critical shear stresses of the different slip systems are identical (termed 'isotropic latent hardening' and probably quite reasonable for medium to high-SFE metals) then:

$$M(g) = \frac{\dot{W}(g)}{\tau_C \cdot \dot{\varepsilon}} = \frac{\Sigma \dot{\gamma}_S(g)}{\dot{\varepsilon}} = \frac{\sigma(g)}{\tau_C} \tag{10.11}$$

For a polycrystalline aggregate, the average Taylor factor over all grains $<M>$ can be calculated for a given distribution of orientations. In the case of the tensile deformation of an fcc metal with a random orientation distribution $<M> = 3.07$. However, non-random distributions as in textured polycrystals lead to other $<M>$ values; typically they lie in the range 2–4.

The Taylor factor is also important for the work hardening behaviour. To first order for small strains $\sigma = <M> \tau_c = (<M>/\mathrm{d}\varepsilon)$ and therefore $(\mathrm{d}\sigma/\mathrm{d}\varepsilon) \approx <M> \cdot (\mathrm{d}\tau/\mathrm{d}\varepsilon)$ so that

$$\text{if } \mathrm{d}\varepsilon = \frac{\Sigma \mathrm{d}\gamma}{<M>} \quad \text{then} \quad \frac{\mathrm{d}\sigma}{\mathrm{d}\varepsilon} \approx <M>^2 \cdot \frac{\mathrm{d}\tau}{\Sigma \mathrm{d}\gamma} \tag{10.12}$$

i.e. the microscopic and macroscopic work hardening rates are related by the square of the Taylor factor (see Table 10.1).

10.2.2 Simulated deformation textures

As described in Chapter 8, textures are commonly represented by pole figures or, more quantitatively by ODFs, i.e. intensity plots of the crystallographic orientation densities in Euler space. In Figure 10.6, a typical simulation of an fcc cold rolling

Table 10.1. Some typical Taylor factors, *M* and <*M*>, for axisymmetric deformation of fcc and bcc metals.

M/Orientation	{111}<110> Slip (fcc)	{110, 112, 123} <111> Slip (bcc)
M <100>	2.449	2.121
M <110>	3.674	3.182
M <111>	3.674	3.182
<*M*> Random	3.067	2.754

texture is shown as an ODF indicating a set of orientations which spread along the β fibre axis from Bs (brass) through Copper to S. Note that the simulated Bs component is weak compared with the usual experimental values.

During reversible hot rolling, and as described in Section 11.2, there are significant alternating shears near the surface regions; the latter develop a {001}<110> shear texture, while the centre develops the standard β fibre. Figure 10.7 shows examples of these texture gradients in a hot-rolled Al alloy after 95% reduction, both measured and simulated using FEM streamline strain paths and the mixed slip model.

Bcc rolling texture simulations have been the object of much research because of their importance for sheet steel anisotropy and formability. Figure 10.8 below compares some simulations and experimental rolling textures in the form of the constant $\varphi_2 = 45°$ sections (see Chapter 8). They include the α and γ fibres which are quite well predicted for the more recent grain interaction models.

10.3. RECOVERY AND RECRYSTALLIZATION

Recrystallization models cover a wide scale from the semi-empirical to the fundamental, but complete models which can satisfactorily describe the entire set of nucleation and growth reactions in terms of kinetics and quantitative microstructure evolution are unavailable at the present time. This is because recrystallization is a very localized process that occurs in highly unstable and heterogeneous systems resulting from the plastic flow of polycrystals. The variations in deformation microstructure at the level of microns give rise, in many cases, to recrystallization nucleation but these local variations are difficult to predict quantitatively.

The general kinetics of transformations will be treated first using the 'JMAK' formulation since this is what is used most often in TMP, in combination with some empirical parameters. We shall then examine some more microstructure-based models.

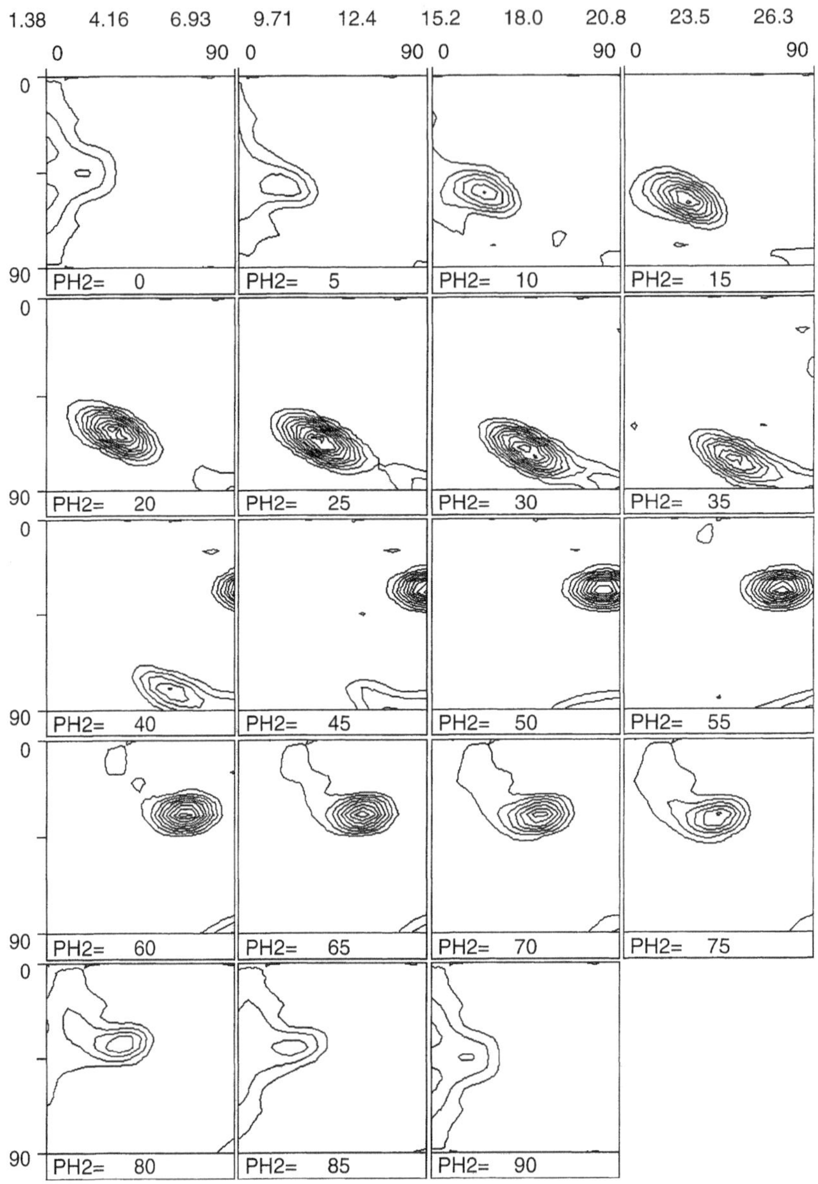

Figure 10.6. Simulated ODF of an fcc metal cold rolled to a strain of 1.5 assuming the RC model. From Perocheau [1999]. Permission obtained from Elsevier.

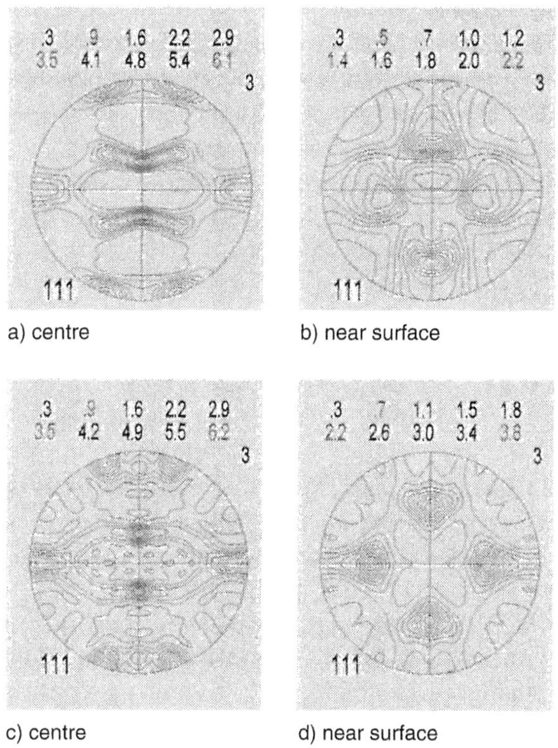

Figure 10.7. Simulated (a, b) and experimental (c, d) {111} pole figures for hot-rolled Al AA 3103 (Al–1%Mn). From Perocheau and Driver [2000]. Permission obtained from Elsevier.

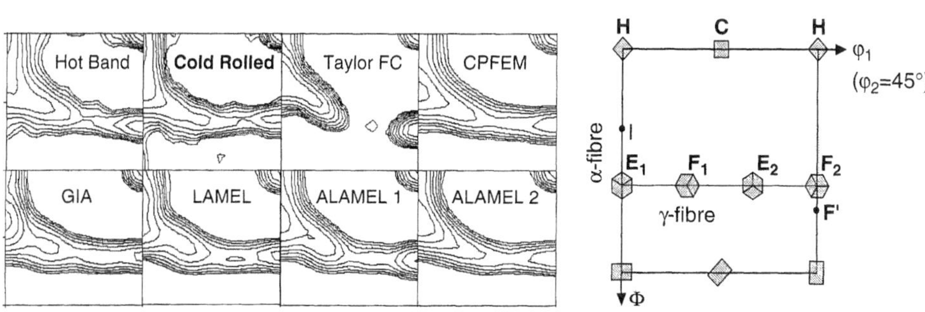

Figure 10.8. Rolled IF steel textures (ODF at $\varphi_2 = 45°$) for hot band, cold rolled and simulations by the Taylor FC, crystal plasticity FEM and four other recent models, which allow some localized strain relaxations in the grains (Van Houtte *et al.* [2005]). Permission obtained from Elsevier.

10.3.1 General transformation kinetics

Most solid-state phase transformations take place by a set of reactions involving nucleation of the new phase and its growth into the (old) parent phase. The kinetics of these transformations are important issues in TMP since many processes involve concurrent phase and mechanical transformations (or alternatively are timed to avoid them). Modelling phase transformation kinetics is a wide field but one of the common basic ingredients is the JMAK formulation of transformation kinetics. JMAK refers to Johnson and Mehl [1939], Avrami [1939, 1940] and Kolmogorov [1937] who, virtually independently, developed the following concepts and expressions for the kinetics of reactions in which nucleation and growth (by interface movement) occur simultaneously. It is mostly used for allotropic transformations and recrystallization, but is perfectly general.

Transformation kinetics usually follow a sigmoidal plot of fraction transformed vs. time (the latter often best plotted on a logarithmic scale) (see Figure 10.9).

The fraction transformed X is the ratio of the volume fraction of new phase to the total volume (V_n/V_T), but the nucleation rate $\dot{N} = dN_v/dt$ is defined by the number of germs formed per unit time/volume of non-transformed phase V_o (o for old phase).

During a time interval dt, the number of germs created $= \dot{N} \cdot V_o \cdot dt$, but the volume of the old phase V_o decreases at a rate controlled by the transformation kinetics (which is the required quantity). It is mathematically easier to treat the kinetics in terms of an imaginary, **extended volume V_e** in which nucleation and growth occur freely at all points without hindrance or impingement (e.g. 'phantom' nucleation can occur in volumes already transformed). In this extended

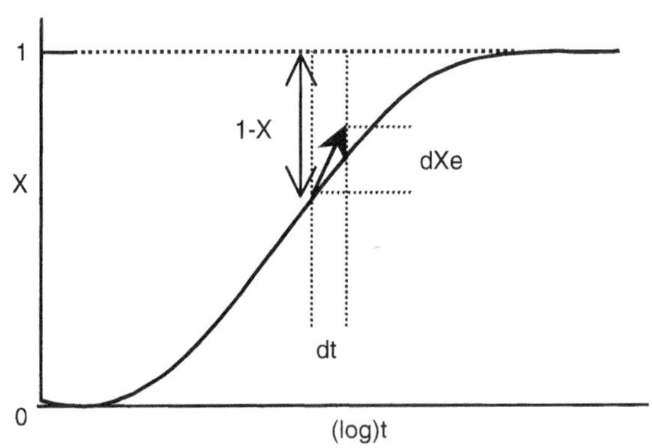

Figure 10.9. Schematic growth kinetics $X(t)$; the arrow indicates the rate of increase of the extended volume fraction dXe/dt compared with the true rate dX/dt of the solid line.

volume, the number of nuclei formed (i.e. real and phantom) during a time increment is $\dot{N} \cdot V_T \cdot dt$. Their growth leads to an increase of the extended volume which can be directly related to the kinetics. Thus, the change in extended volume resulting from the growth of nuclei formed in a time interval $t = \tau$ to $t = \tau + d\tau$ is

$$dV_e = V_g(t - \tau) \cdot \dot{N}(\tau) V_T d\tau \quad (t > \tau) \tag{10.13}$$

where $V_g(t - \tau)$ is the volume, at time t, of a germ previously formed at $t = \tau$. The extended volume V_e is then given by the time integral of Eq. (10.13) and the extended volume fraction by dividing with V_T:

$$X_e = \frac{V_e}{V_T} = \int_0^t V_g(t - \tau) \cdot \dot{N}(\tau) \cdot d\tau \tag{10.14}$$

During a small time interval, the extended volume fraction increases by dX_e which is clearly more than the real volume fraction increase (dX) since the latter can only grow into the remaining fraction $(1 - X)$ (see Figure 10.9); it follows that $dX = (1 - X)\, dX_e$ and so by integration $X_e = -\ln(1 - X)$. Equating the two expressions for X_e gives the fundamental relation for the kinetics of nucleation and growth:

$$\mathbf{X} = 1 - \exp\left\{ -\int_0^t V_g(t - \tau) \cdot \dot{N}(\tau) \cdot d\tau \right\} \tag{10.15}$$

As an example, take the case of a transformation involving 3D growth. At some time t, the volume of a spherical germ which nucleated at the instant τ is

$$V_g(t - \tau) = \left(\frac{4}{3}\right) \pi r^3 = \left(\frac{4}{3}\right) \pi G^3 (t - \tau)^3$$

where G is the interface velocity. If the growth and nucleation rates G and $\dot{N}$ are assumed constant and homogenous during the transformation, direct integration of Eq. (10.15) after inserting for $V_g(t - \tau)$ gives the Johnson–Mehl formula:

$$X = 1 - \exp\left[\left(-\frac{\pi}{3} \right) \cdot G^3 \cdot \dot{N} \cdot t^4 \right] \tag{10.16}$$

If all the nuclei are present initially, with a density N_0, and their number does not change (termed site saturation) then

$$V_g = \frac{4\pi}{3} G^3 t^3 N^3 N_0 V_T \Rightarrow X_e = \frac{4\pi}{3} G^3 N^3 N_0 t^3$$
$$\text{and} \quad X = 1 - \exp\left[\left(-\frac{4\pi}{3}\right) G^3 N_0 t^3\right] \qquad (10.17)$$

Similarly it is easily shown that growth in 2D at constant $\dot{N}$ leads to a time exponent of 3. The above examples assume homogeneous nucleation and unhindered growth in 2D or 3D which is not always respected. Clearly, the time exponent depends upon the detailed mechanisms of the transformation and since G and $\dot{N}$ are not always known, one often writes:

$$X = 1 - \exp(-kt^n) \quad 2 < n < 5$$
$$\text{or} \quad X = 1 - \exp\left\{-\left(\frac{t}{t_r}\right)^n\right\} \qquad (10.18)$$

where t_r is the time to $X = 0.63$. Examples of this treatment for recrystallization are given in Chapter 5. It is worth noting that the experimental values of G and $\dot{N}$ are rarely constant so that Eq. (10.15) has to be integrated numerically using their current values. The time exponent n will then not take a constant, integer value as in the above elementary examples. It turns out that in practice for recrystallization $1 < n < 3$ since nucleation and growth take place at **heterogeneous** sites in the deformed substructure. The application of the generalized 'Avrami' method to inhomogeneous structures has been treated in some detail by Furu *et al.* [1990] by assuming volume elements with different nucleation rates. This work confirms the dependency of the time exponent on time, G and $\dot{N}$ for heterogeneous rates. An extension to recrystallization modelling of hot-deformed Al alloys using three types of nucleation sites (PSN, cube bands and grain boundaries) has been proposed by Vatne *et al.* [1996a].

The JMAK treatment is a statistical treatment of the kinetics, generally without much microstructure input, but it does give a useful estimate of the recrystallized grain size D_{rex}:

$$D_{rex} \approx 2Gt_r \approx 2\sqrt[n]{\frac{G}{\dot{N}}} \qquad (10.19)$$

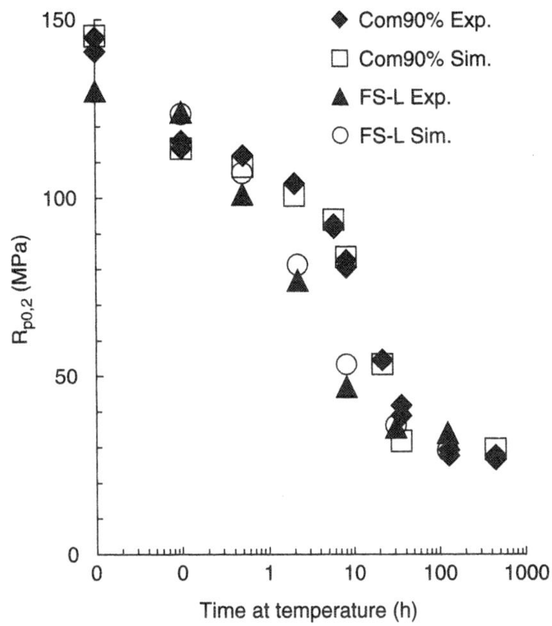

Figure 10.10. Kinetics of recovery softening of 90% cold-rolled Al in terms of yield stress $=f(\log t)$ from Ekstrom *et al.* [2004] (Com: AA 1200 and FS-L is a lab. casting).

10.3.2 Recovery kinetics

Recovery kinetics (see Chapter 5 for mechanisms) are often characterized by a flow stress, after strain hardening, which decreases with the logarithm of recovery time. The rate of decrease is sensitive to the temperature and alloy content. Experimental examples of this characteristic time dependence and a review of some models have been treated in detail by Nes [1995] and Humphreys and Hatherly [2004] (see Figure 10.10).

The kinetics analysis often uses, and develops, an original idea proposed by Kuhlmann *et al.* [1949] according to which the effective activation energy varies with internal stress and therefore with recovery time. In their first model dislocations, of number N, were supposed to anneal out of pile-ups of length L at a rate given by:

$$\frac{\mathrm{d}N}{\mathrm{d}t} = -k' \exp\left\{-\left(\frac{Q - \beta GN/L}{kT}\right)\right\} \tag{10.20}$$

where Q is the standard activation energy for dislocation climb, G the shear modulus and k' and β are constants. As the number of dislocations decreases the effective activation energy (in round brackets) increases, so the process becomes increasingly difficult and slows down. Integrating equation gives a kinetic law which is clearly logarithmic:

$$N = N^{\circ}\left\{1-\left(\frac{kT}{A}\right)\ln(1+Bt)\right\} \tag{10.21}$$

with A and B as constants.

In the more recent analysis of Nes [1995], the dislocations are considered to be arranged in cells, essentially as low-angle boundaries, which is more in keeping with the typical work hardened and recovered structures of medium to high-SFE metals. Also, since recovery rates are very sensitive to solute content, solute diffusion is considered to be the rate limiting process. In this model, recovery takes place by subgrain growth driven by dislocation line tension, decrease but the rate of movement of the subgrain boundaries is controlled by solute diffusion at dislocation pinning points (jogs). It is then shown that for dislocation cells of radius r, the driving force (per unit length of dislocation) for growth is $F \approx \alpha G b^2/r$, where α is a constant close to unity.

The average speed of dislocations is controlled by the motion of jogs pinned by solute atoms with a jog spacing l giving:

$$V = b\nu \exp-\left\{\frac{U_{\mathrm{d}} - Fbl}{kT}\right\} \tag{10.22}$$

where ν is the Debye frequency and U_{d} the activation energy for solute diffusion. Clearly, this equation is of the same form is the Kuhlmann expression with an effective activation energy that increases as the cell size increases (and the driving force or dislocation density decreases).

Assuming that the average rate of growth of the cells ($\mathrm{d}r/\mathrm{d}t$) is proportional to V and substituting for F leads to a growth rate law:

$$\frac{\mathrm{d}r}{\mathrm{d}t} = \left(\frac{b\nu}{c}\right)\exp-\left\{\frac{\left(U_{\mathrm{d}} - \alpha G b^3 l\right)/r}{kT}\right\} \tag{10.23}$$

Separating out the term equivalent to the solute diffusion coefficient $B_{\mathrm{d}} = (b\nu/c)$ $\exp-\{U_{\mathrm{d}}/kT\}$ gives for the sub-grain growth rates $dr/dt = B_{\mathrm{d}}\exp\{(U_{\mathrm{d}} - \alpha G b^3 l/r)kT\}$.

If the spacing of the pinning points l is taken proportional to the initial cell size $2r_{0,}$ then an approximate solution by integration leads to a logarithmic time dependency of the cell size evolution:

$$\frac{r_0}{r} = 1-\left(\frac{kT}{A}\right)\ln\left\{\frac{1+t}{\tau}\right\} \tag{10.24}$$

where the constant is proportional to αGb^3 and τ is a characteristic diffusion time.

Given the equivalence of the residual strain hardening to (r_0/r), this analysis can be integrated into a thermo-mechanical process routine to estimate recovery softening during annealing and also during interpass rolling schedules.

10.3.3 Recrystallization

The fundamental mechanisms of recrystallization are described in Chapter 5. Since it is basically a heterogeneous process, modelling recrystallization requires a knowledge of the nucleation and growth rates at different sites in the heterogeneous, as-deformed substructure, i.e. the microstructure varies strongly in space and with time. This is one of the most difficult metallurgical processes to model and treatments are now largely based on numerical simulations using high-speed computers. A good review is to be found in the book by Humphreys and Hatherly [2004]. This section is limited to giving the basics of current methods and some examples.

10.3.3.1 Discrete models. To describe space and time-resolved microstructure evolution during recrystallization, two basic discrete methods are used – the Monte Carlo and Cellular automata models.

The *Monte-Carlo* method was first applied to grain growth by Anderson *et al.* [1984] and then extended to recrystallization by Srolovitz *et al.* [1986]. The method consists of setting up a 2D or 3D microstructure of elements, or cells, on a regular lattice. Each cell represents a basic building block of the microstructure; and for grain growth simulations, a grain typically consists of several identically numbered cells (see Figure 10.11). A grain boundary is then characterized by the

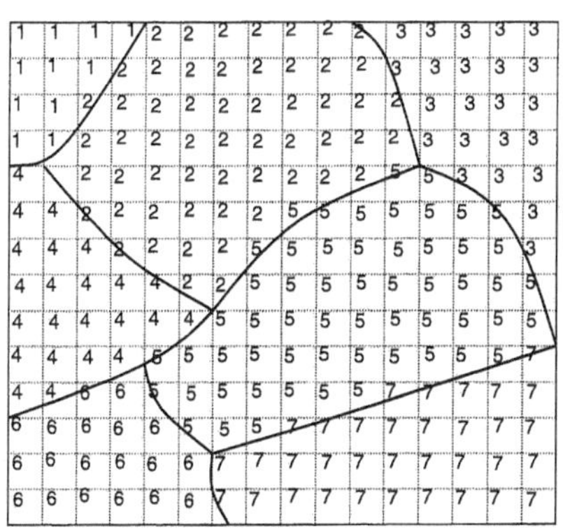

Figure 10.11. Schema of a 2D Monte Carlo grain structure.

relative numbers of cells with different neighbours, e.g. 2/5 pairs in 2D, and an energy is attributed to this pairing. The microstructure evolution is controlled by an energetic criterion for the transformation probability, P, of the pairs. According to the Metropolis algorithm this is

$$P = P_0 \quad \text{for } \Delta E \leq 0$$

$$P = P_0 \exp\left(\frac{-\Delta E}{kT}\right) \quad \text{for } \Delta E > 0$$

where P_0 is often set $= 1$ in early versions.

A random generator picks a cell, changes it with one of the neighbours, then evaluates the energy of the new state. According to the energy change, the transition is accepted or rejected. Like pairs do not have much chance of changing, and transitions are favoured at microstructural features with higher densities of unlike pairs, e.g. curved grain boundaries. This then gives curvature-driven grain growth. The algorithm for transition attempts is repeated a large number of times. The timescale is defined by the number of Monte Carlo time Steps (MCS), where each MCS corresponds to a number of attempted transitions equal to the number of cells or lattice sites in the model.

Recrystallization can be treated using a grain structure similar to that of Figure 10.11, but a stored energy H is attributed to all sites within a grain and can be varied to simulate local differences of stored energy. The number of Q different cell states is often taken at about 48 to limit computer times, while retaining equivalence with the physical processes. More recently, using powerful computers a $Q = 999$ version has been applied to the simulation of nucleation in recrystallization (Holm *et al.* [2003]).

An example of such a Potts model of recrystallization is given in Figure 10.12.

Other recent developments in MC modelling include the 3D interactions of moving boundaries with particles (the Zener drag problem) treated by Miodownik *et al.* [2000] (see also below). It should be emphasized that Zener drag is a 3D problem and cannot be correctly treated in 2D, so that previous attempts at 2D simulations of Zener drag are of doubtful use.

The MC method possesses the advantage of being relatively easy to programme but suffers from the difficulty of assigning physical significance to the cells which are the basic elements of the microstucture. For example, during recrystallization in deformed metals, the crucial scale for nucleation is much less than a micron so that the cell size should probably correspond to less than 0.1 μm. In the largest simulations, the number of cells does not exceed $400 \times 400 \times 400$, which would then be of the order of magnitude of the volume of one grain. Another problem

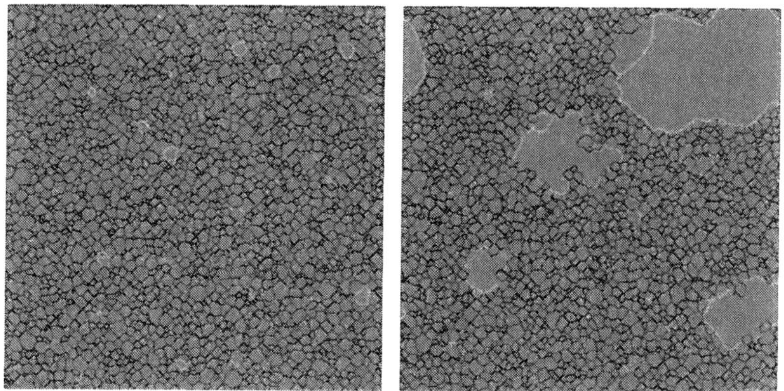

(a) (b)

Figure 10.12. MC model simulation of recrystallization nucleation, from Holm *et al.* [2003]. Permission obtained from Elsevier.

concerns the true timescale of the simulations as required for kinetics analyses. The merits and difficulties of the MC model have been analysed in more detail by Humphreys and Hatherly [2004]. It appears that the method is being superceded by the more flexible cellular automaton algorithms.

The *Cellular Automaton (CA)* method was first proposed for recrystallization by Hesselbarth and Göbel [1991] and subsequently developed by Marx *et al.* [1999] and Raabe [1999]. It also discretizes space and time like the MC models but the transformation rules are different; a cell transition occurs in a deterministic way according to the state of its neighbours (either nearest, or more usually, up to second nearest neighbours). It intrinsically assumes that the energetic requirements of the transformation are respected, which is largely true of recrystallization where growth of the recrystallization front only occurs in one 'thermodynamic sense'. By setting more flexible transition rules, the transformation of an element can become dependent on stored energy, misorientation and temperature. These parameters can then change with time, so approaching the real physical processes of recovery and recrystallization.

As pointed out by Humphreys and Hatherly [2004] the CA method consists of defining a spatial framework into which analytical or empirical equations can be inserted to describe specific mechanisms. They can then enable microstructure and texture inhomogeneities to be treated more satisfactorily than with the MC method.

10.3.3.2 Vertex or cellular models. The cellular models are essentially based upon the dynamics of boundary movements in cell assemblies as controlled by the

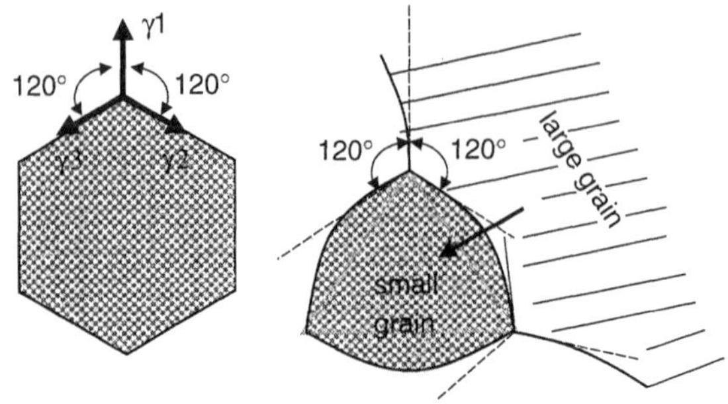

Figure 10.13. Schematic of surface tensions and boundary movements in six-sided and three-sided grains.

contradictory requirements of space filling and minimizing boundary energy. The simplest example is that of standard grain growth (see Figure 10.13 for a 2D illustration), where the boundary tensions at a triple point, and along the entire boundaries, are only in equilibrium (120°) for the particular case of a six-sided grain.

The smaller grains with, say three neighbours, must have curved boundaries to maintain local equilibrium at the triple point and this geometrically induced curvature then drives boundary motion towards the centre of the small grain (Figure 10.13) in accordance with the Gibbs–Thomson law. In contrast, large grains with more than six neighbours and a radius of curvature of opposite sense will then expand; *in fine* the large grains grow and the small grains shrivel and disappear so the average grain size increases and the overall interface energy decreases. Hillert [1965] developed an analytical mean field model for this type of process:

$$\frac{dR}{dt} = M\gamma\left(\frac{1}{\bar{R}} - \frac{1}{R}\right)$$

where R is the radius of a growing grain in a 3D assembly of grains of average radius $\bar{R}$ and M and γ their respective mobilities and boundary energies.

Modelling the evolution of cellular structures has been carried out at two levels: (i) analytical or semi-analytical models of mean field type, which give useful estimates of the overall behaviour of cellular structures and (ii) numerical vertex simulations which attempt to describe the true local processes.

The *mean field approach* has mostly been developed by John Humphreys [1997, 1999a]. As above, it consists of comparing the rate of growth of an individual cell (grain or subgrain of radius R) with that of the average cell (radius $\bar{R}$). By means of a local energy balance and the standard $V = MP$ law (interface velocity=mobility x driving pressure), the growth rate of a spherical cell of radius R is readily shown to be:

$$\frac{dR}{dt} = M\gamma\left(\left(\frac{1.5\beta\bar{\gamma}}{\bar{R}}\right) - \frac{2\alpha\gamma}{R}\right) \tag{10.25}$$

where α and β are geometric constants of ordre unity. γ and $\bar{\gamma}$ are the boundary energies of the particular cell and the average value of the assembly. Growth becomes unstable when the rates of growth of individual and average grains are such that:

$$\bar{R}\left(\frac{dR}{dt}\right) - R\left(\frac{d\bar{R}}{dt}\right) > 0 \tag{10.26}$$

giving an instability criterion in the form of the relation:

$$M\gamma - \left(\frac{\bar{R}M\gamma}{R}\right) - \left(\frac{R\bar{M}\bar{\gamma}}{4\bar{R}}\right) > 0 \tag{10.27}$$

which, when expressed in size, energy and mobility ratios ($R/\bar{R}$, etc.), becomes a quadratic with two roots. This is illustrated graphically in Figure 10.14a in terms of the critical size $R/\bar{R}$ and mobility $M/\bar{M}$ ratios for unstable (discontinuous) growth. The upper and lower branches of the discontinuous growth zone correspond to the two roots of Eq. (10.27). Clearly, unstable growth of a cellular structure is expected for large cells with boundaries of high mobility and small energies.

This approach can be extended to the case of unstable growth of a subgrain in a typical subgrain structure during annealing (equivalent to the late stages of recovery and eventually nucleation of recrystallization). Assuming reasonable values for sub-boundary and boundary mobilities, and energies based on the Read and Shockley analysis, Humphreys [1997] has derived the conditions for instability for the size ratio as a function of misorientation as shown in Figure 10.14b. If the sub-boundaries develop mean misorientations of say 10°, then a size ratio of about 2 is sufficient to develop faster growth of the subgrain. However, the upper size limit of this rapid growth stage is about 5× average which is not very high. True rapid nucleation to develop large, fast growing new grains requires subgrains in a field with lower average misorientation (e.g. 2–5°) but with much higher initial size ratios. This type of analysis highlights some of the difficulties of nucleating fast-growing grains in a subgrain assembly.

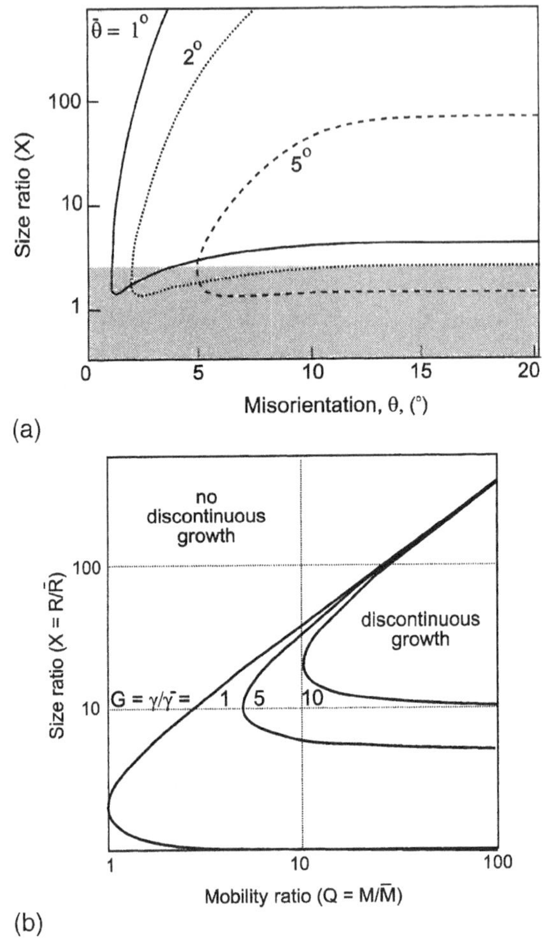

Figure 10.14. The conditions for instability of a cellular microstructure (a) in terms of the size and mobility ratios and (b) for sub-boundaries as a function of their size and misorientation from Humphreys and Hatherly [2004]. Permission obtained from Elsevier.

The above mean field method may also be extended to the case of growth in the presence of particles by the introduction, into the equation for the growth rate dr/dt, of a term corresponding to the average Zener drag pressure, i.e. proportional to $-F_V\gamma/d$. It has also been used for predicting nucleation by strain-induced boundary migration in Al alloys (Hurley and Humphreys 2003).

There have been many different types of *numerical 'network' or 'vertex' simulations* starting with Fullman [1952]. They all describe the cellular or granular microstucture as composed of interconnected boundaries whose movements are closely linked with those of the triple point vertices (or quadruple points in 3D). Since only the co-ordinates of the triple points are stored in the computer (plus a numbering scheme for the grains and their orientations), much larger cellular assemblies can be treated during an 'annealing' treatment. However, they have to include numerical treatments of the topological transformations such as grain switching and the disappearence of small grains – this is not too difficult in 2D but a major task in 3D. The various vertex models developed over the years differ by the way they handle the boundary dynamics. The most recent versions often adopt the Kawasaki *et al.* [1989] formulation for the coupled equations of boundary motion. According to this analysis, the change of boundary energy, due to an incremental area change, d*s*, is entirely dissipated by viscous motion of the boundary leading to:

$$\text{For vertex } i\text{:}\ \left(\frac{\partial D}{\partial \mathbf{v}_i}\right)+\left(\frac{\partial L}{\partial \mathbf{x}_i}\right) \quad \text{with } D=\int\left[\frac{(\mathbf{v}\cdot\mathbf{n})^2}{2m}\right]\mathrm{d}s \ \text{ and } \ L=\int_s \gamma\,\mathrm{d}s \tag{10.28}$$

$$\text{And for a set of } N \text{ vertices} \sum_{i=1}^{N}\left(\frac{\partial D}{\partial \mathbf{v}_i}+\frac{\partial L}{\partial \mathbf{x}_i}\right)\delta\mathbf{v}_i=0 \tag{10.29}$$

where m is the mobility, γ the boundary energy per unit area and **v** and **n** the velocity and the boundary normal vectors. This is a discretized version of the well-known relation $V = MP$ and its application leads to the movement of boundaries, and boundary interconnects, towards their centre of curvature at rates dependent on the local values of the above parameters. This method has the immense advantage that real time, space, energy and mobility parameters are used.

The model has been applied to growth in 2D (see review by Maurice [2001]) and 3D and, in particular to the behaviour of microstructural heterogeneities. Figure 10.15 gives an example of discontinuous subgrain growth in high-purity aluminium. From the whole subgrain size spectrum about 30 subgrains with a larger size than average (1.5 – 1.7 × average) and slightly higher misorientation (3° instead of the average 2°) were introduced. Their corresponding growth rates are given in Figure 10.15b. The basic similarity with the mean field method is self-evident, but it is worth noting that the growth rates of individual cells are dependent on their environment (Figure 10.15b), a result which cannot come out of the mean field treatment.

Figure 10.16 also shows the transformation of a heavily rolled structure of flattened grains into an *in situ* recrystallized structure as modelled by a 2D vertex simulation.

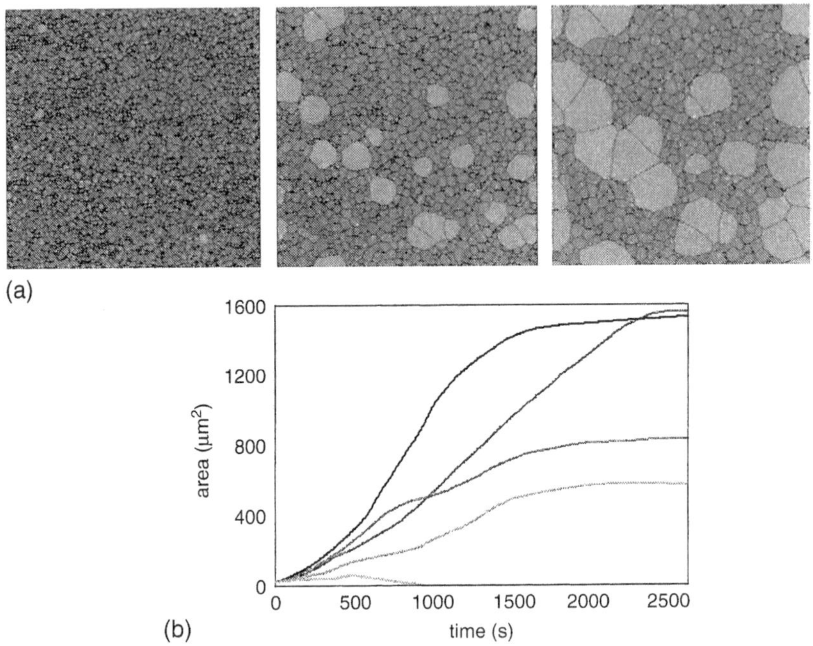

Figure 10.15. (a) 'Recrystallization' by discontinuous subgrain growth in Al at 473 K and (b) growth rates of abnormally growing subgrains (Maurice and Humphreys [1998]). Permission obtained from TMS.

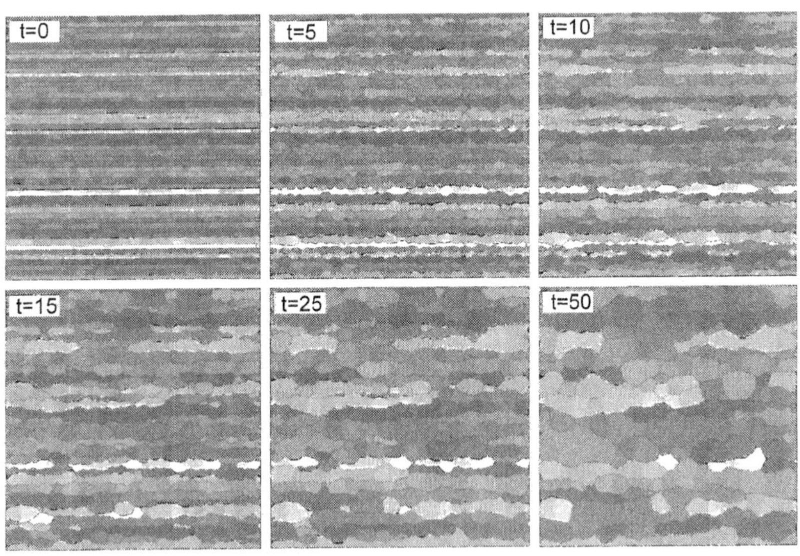

Figure 10.16. 2D Vertex simulation of *in situ* recrystallization of heavily deformed metal showing the transformation of lamellar to equi-axe grain structures (Hayes *et al.* [20). Permission obtained from Elsevier.

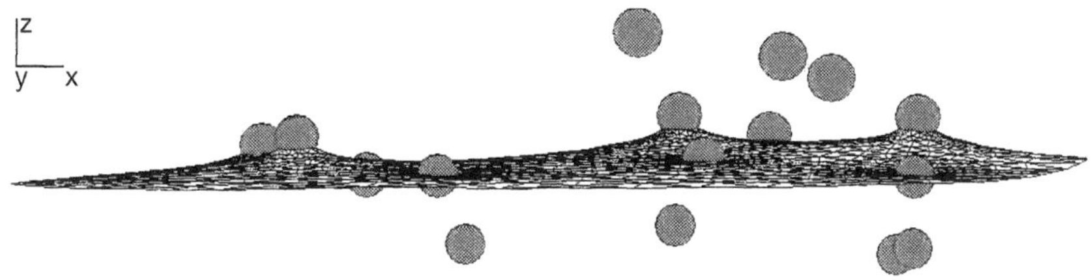

Figure 10.17. 3D Vertex simulation of the interaction of a moving boundary with a set of particles. The boundary is just pinned by particles of radius 0.05 Rd (the boundary radius) and a volume fraction of 0.22%, from Couturier *et al.* [2005]. Permission obtained from Elsevier.

Very recently, the method has been applied to the interaction of a moving boundary with a set of 2nd phase particles (Zener drag) by Couturier *et al.* [2003, 2005]. Figure 10.17 illustrates the form of the moving boundary as it is pulled through a field of particles. Using this FE vertex model to analyse the forces at the particles leads to the conclusion that the standard Zener drag pressure of $-3F_v\gamma/d$ should in fact be multiplied by a factor of 2.2.

LITERATURE

Anderson M.P. and Rollett A.D. (eds), "Simulation and Theory of Evolving Microstructures", TMS, Warrendale, PA (1989).

PART II: TECHNOLOGY

Chapter 11
Forming Techniques

Chapter 11
Forming Techniques

11.1. GENERAL INTRODUCTION

The first section (Chapters 1–10) of this book dealt with the metallurgical issues involved in thermo-mechanical processing (TMP). Of an equal[1] importance are the issues pertaining to production technology – the subject of this second section. Many of these issues are domains of mechanical/production engineers, but the basics should also be mastered by materials engineers.

This chapter about forming techniques, deals with techniques ranging from rolling to forging, from sheet metal forming to superplastic forming. The individual sub-chapters attempt to bring out details on the machines, die-tool design, process optimization, etc. All such issues are covered based on their relevance to the respective forming techniques. This general introduction, on the other hand, tries to outline two issues – 'friction and lubrication' and 'TMP furnaces'. The former, though extremely relevant to any forming technology, has not been specifically covered in the subsequent sub-chapters. The 'TMP furnaces' is not a forming technique, but was considered as a necessary input in the present section on TMP technology.

11.1.1 Friction and lubrication

The resistance that is encountered when two bodies are rubbed against each other is called *friction*. Friction is an important factor in metal forming. It dissipates energy and hence increases the force needed to deform a material. It generates heat, which complicates the control of the deformation temperature. It can affect the material flow during deformation and leads to inhomogeneous deformation. Last but not least it can degrade the quality and appearance of the surface. For all these reasons, it will be necessary in almost all forming operations to reduce the friction by adequate *lubrication*.

The first 'recorded' recognition of friction came from Leonardo da Vinci (Schey [1970]). The subject was rediscovered by Amontons (Amontons [1699], Schey [1970], Möller and Boor [1996]). Friction coefficient was postulated by Amontons and Coulomb (Coulomb [1785], Schey [1970], Möller and Boor [1996]) as

[1] Perhaps, of larger interest to actual production are the technological issues. For example, forging was the earliest forming techniques and still remains an attractive one in terms of metallurgical properties of the product, but dominance of other forming techniques (e.g. rolling and extrusion) is based purely on technological considerations and on cost economics.

proportional to the normal force and independent of contact area and relative speed of the moving surfaces. Subsequent theoretical developments, in particular in the19th century (Schey [1970]) and Hardy's theories (Hardy [1936]) on boundary lubrication, formulate today's understanding of friction in metal-working processes. The present section makes an attempt in bringing out issues relevant to friction and lubrication in metal forming – issues discussed under three brief sub-sections on friction during plastic deformation, friction measurement and lubrication.

11.1.1.1 Friction during plastic deformation. The friction between a die and a workpiece is an important factor in a shaping process. Nevertheless, this factor is often neglected (or first neglected and then 'corrected') for the sake of simplicity. The possible impact of friction on the forces in a simple metal-forming operation can be illustrated with the following example.

Consider the uniaxial compression of a flat disc. It is assumed that no barrelling occurs and that the vertical compression stresses are homogeneously distributed over the disc. It is also assumed that the (Coulomb) friction coefficient μ is constant and that p represents the stress perpendicular to the surface. During compression, the disc will expand sidewards (radial direction) and this will generate a shear stress τ towards the centre of the disc. One can write:

$$\mu = \frac{\tau}{p} \tag{11.1}$$

As a consequence of shear stress τ, a lateral compressive stress will be generated. It is possible to show that the pressure distribution over the surface of the disc is

$$p = \sigma_{\mathrm{y}} \exp\left[2\mu\left(\frac{r-a}{h}\right)\right] \tag{11.2}$$

where r, a and σ_{y} represent respectively the radius of the disc, distance from the centre and flow stress. A derivation of expression (11.2) can be found in several handbooks – example Dieter [1976]. The pressure distribution is illustrated in Figure 11.1. On the edge of the disc, the compressive stress has a minimum and is equal to the flow stress of the material. Because friction hinders the flow of the material away from the centre, the compressive stress increases towards the centre. This characteristic distribution is often called a *friction hill*.

The average compressive stress (p_{av}) can be calculated by integrating Eq. (11.2):

$$p_{\mathrm{av}} \int_0^r \frac{\sigma_{\mathrm{y}}}{r} \exp\left[2\mu \frac{r-a}{h}\right] \mathrm{d}a \sim \sigma_{\mathrm{y}}\left[1+\frac{\mu r}{h}\right] \tag{11.3}$$

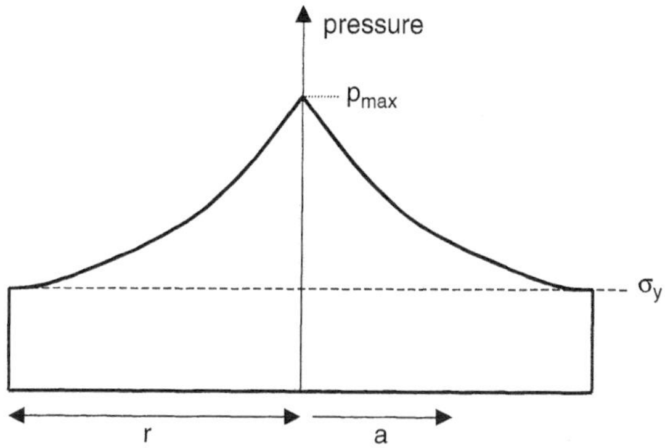

Figure 11.1. Distribution of the compressive stress during uniaxial compression of a flat disc.

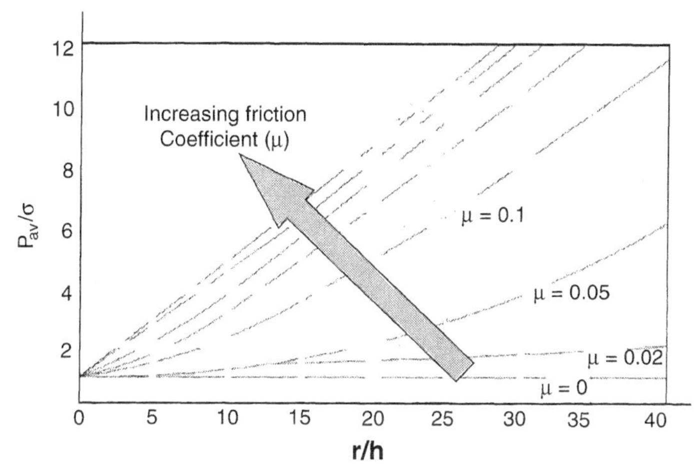

Figure 11.2. Effects of friction during compression tests. P_{av}/σ (average compressive stress/flow stress) vs. radius/height ratio (r/h – of the specimen undergoing compression) as a function of friction coefficient μ. After Dieter [1976].

Figure 11.2. tries to highlight the importance of friction in 'metal forming'. As shown in the figure, even at low friction coefficients ($\mu < 0.02$) the forming/compression pressures can be 2–4 times larger than the flow stress of the material. Friction being more relevant for flat discs (or high r/h).

11.1.1.2 Friction measurement. The friction coefficient must, in most cases, be determined by experiments. In principle, this could be done during a production process; but because of practical limitations and the risk of damage to production equipment, the friction coefficient is normally determined in lab-scale tests. The results of such tests may never be exactly representative of the production line, but are quite suitable in comparing the influence of materials and lubricants on friction. One of the most popular tests is the so-called 'ring compression test' (Kalpakjian [1995]) (see Figure 11.3).

A flat ring is compressed between two flat plates and expands outwards. Without friction, the inner and outer diameter of the ring would expand proportionally. With increasing friction between the ring and the plates, the decreasing height is less and less compensated by the outward expansion of the outer diameter and more and more by a lesser expansion of the inner diameter. With very high friction, the inner diameter can even expands inwards (see Figure 11.3).

For a standard ring with a ratio of outer diameter to inner diameter to height of 6/3/2, calibration curves have been determined (see Figure 11.3b) (Kalpakjian [1995]). With such chart, it is possible to estimate the friction coefficient directly from the results of a ring compression test. For example, if the height is reduced by 40% and the inner diameter becomes 10% smaller, the friction coefficient of

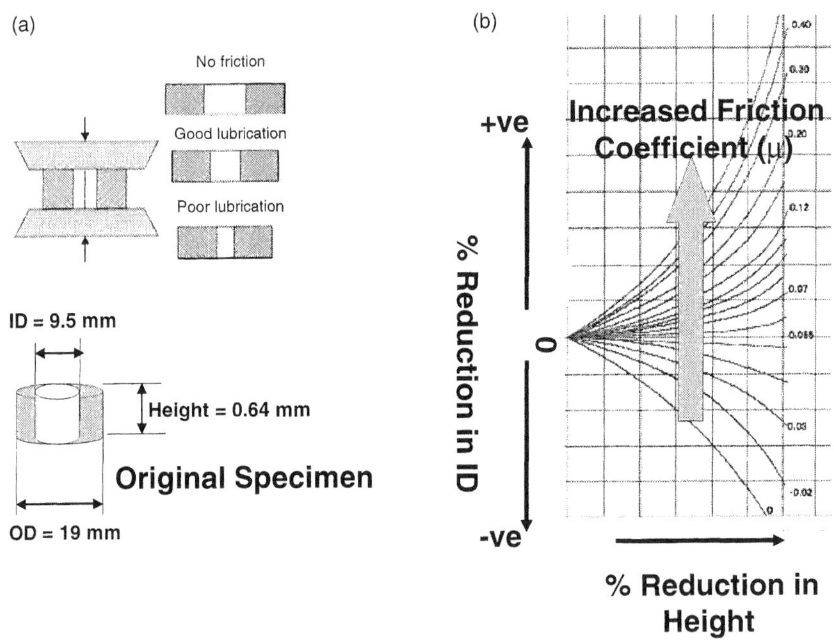

Figure 11.3. (a) Illustrations of 'ring compression test'. (b) Friction coefficient from the data of the ring compression tests. After Kalpakjian [1995].

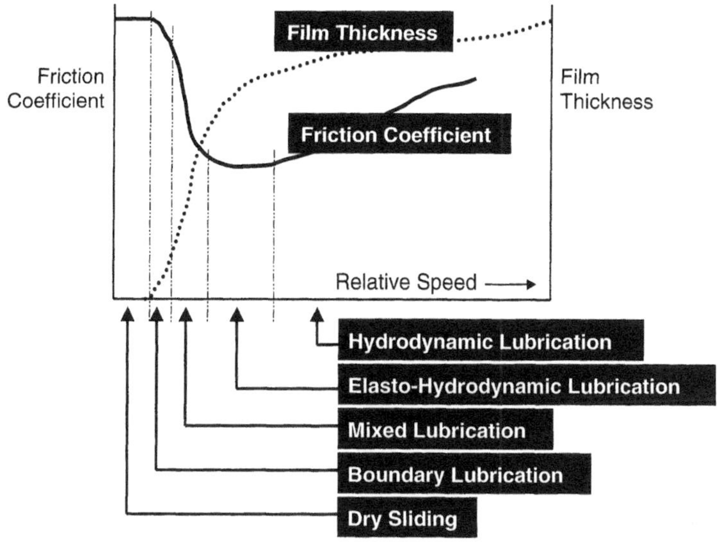

Figure 11.4. Stribeck curve – schematic diagram representing different frictional regimes in forming operations. After Möller [1996].

the tested system (material for the ring, material for the platens and lubricant) is equal to 0.1.

11.1.1.3 Lubrication. Friction accounts for 10–20% of the force in typical metal-forming operation (Blazynski [1976]). To counter frictional effects, including tribological effects, appropriate use of lubrication is required (Möller and Boor [1996]). Though no historical record of 'intentional' lubricant use exist in primitive metal forming (as in the archaeological evidences of hammering operations in the Middle East (Singer [1954])), unintentional usage like use of greasy fingers and hammering on asphalt, however, did possibly exist (Schey [1970]). Subsequent societies used lubricants as a common tool and except for two isolated developments in the present century,[2] basic ingredients of today's lubricants were known for some time.

Figure 11.4, the so-called 'Stribeck curve' or 'Reynolds–Sommerfield' curve (Möller and Boor [1996]), offers a glimpse at different lubrication regimes. Such

[2] Phosphate conversion lubrication for severe cold deformation (as in drawing and extrusion) in Germany and glass pad lubrication for hot extrusion in France (Schey [1970], Möller and Boor [1996]).

regimes are outlined based on relative speed of the forming operation and film thickness/friction coefficient.

- *Dry sliding*: Direct contact of the workpiece and die-tool surfaces – surface damage and even material removal.
- *Boundary lubrication*: Lubricants form surface layers, absorbed or chemically modified layers, and fill surface imperfections. The layers may undergo a dynamic process of 'rubbing off' and 're-generation'.
- *Mixed lubrication*: Small quantity of lubricant is allowed to form light shearing layers by physical or chemical reaction. This avoids direct contact and reduces friction and wear.
- *Fluid lubrication*: Complete separation of contact surfaces by thick fluid/lubricant layers, friction is due to the shear resistance offered by the lubricant. Based on thickness/stability of lubricant layer, these are sub-divided as elasto-hydrodynamic and hydrodynamic. Other than friction/tribology, the lubricant also plays an important role in heat transfer. For example, the selection of a mineral oil would not only depend of the viscosity and thermal stability of the oil, but also on its thermal conductivity and specific heat.

Table 11.1 provides a summary of common lubricants used in metal forming. The most important liquid lubricants are:

- *Oil*: Most oils stick well to metals and have excellent lubrication properties, but have a rather low thermal conductivity and specific heat. For this reason, oil is less suitable when frictional heat has to be conducted away. Moreover, it is difficult to remove oil from the surface after the deformation process.
- *Emulsions*: An emulsion is a 'composite' of two immiscible liquids, e.g. water with oil; due to the presence of water, most emulsions have reasonable cooling properties; emulsions are frequently used in high-speed machining (also in specific forming operations) of metals.
- *Foams and greases*: Most foams are reaction products of specific salts; they form a protective layer between two metal surfaces and prevent a direct contact between them. Grease is a viscous lubricant that frequently used in all types of machinery, but it has a rather limited use in manufacturing.

The most useful solid lubricants are:

- *Graphite*: In the presence of air or moisture, graphite has a low friction coefficient and is well suited as a lubricant – even at high temperatures. In vacuum or on inert gas atmosphere, however, graphite can not be used.

Table 11.1. A summary of common lubricants used in metal forming.

Type of lubricant	Details
Solid	Graphite {0.1–0.2; +450°C}, moly-di-sulphide (MoS_2) {0.04–0.09; +400°C} and polymer (PTTE) {0.04–0.09; +250°C}. Respective friction coefficients and operating temperatures are mentioned within '{}'
Semi-solid	*Greases*: Thickened mineral oils and synthetic oils with soaps (containing different metal ions – Na, Ca, Li, Al). Adhesive/tacky: bitumen is the predominant substance
Mixed or partial lubrication	For mixed lubrication – as in Figure 11.4. Depending on relative stability of the boundary layer mild or strong anti-wear additives are used *Mild*: Saturated/unsaturated fatty acids and primary/secondary alcohols *Strong*: Compounds of chlorine, sulphur or phosphorous
Fluid	From liquefied gasses to different types oils and water. Most common are hydrocarbon-based mineral oils (80–85 wt%C and 10–15 wt%H) and synthetic lubricants. The former can be classified as paraffins, napthalenes and aromatics; while a range of esters, polyglycols, silicone oils, etc. are possible as synthetic fluid lubricants. Different additives are also used for improved performance and life

- *Molybdenum disulphide*: Molybdenum disulphide (MoS_2) is somewhat similar to graphite, but is only useful at high temperature – example, MoS_2 mixed with oil is frequently used in warm-forming operations.
- *Glass*: Glass becomes viscous at high temperature and is suitable as lubricant; it has a low thermal conductivity and forms a thermal barrier between the workpiece and the tools; it is frequently used as lubricant in extrusion and in forging.
- *Metallic films and polymeric films*: Polymers (e.g. teflon and polyethylene) and soft metals (e.g. lead, tin; and in some cases, copper) can be useful as lubricant – example, during drawing of high-carbon wire (to be used in auto tires) a thin layer of Cu is electrolytically deposited on the wire and is used as lubricant.

The delivery of the lubricants, especially the automated delivery, is also of enormous technological importance. For example, in a modern rolling mill, an important aspect of the lay out and operations is the appropriate lubrication delivery. It involves lubrication reservoir, pumps, filters, gauges and controls, heating/cooling, automation and quality checks. The quality checks would typically involve checks for density, colour, flash point, aniline point, carbonization tendency, neutralization properties and saponification values (Möller and Boor [1996]). A more

Table 11.2. Broad classifications of industrial furnaces used in TMP.

Classification	Types	Remarks
Heat source	Electric Fuel-fired	The nature of electric power, e.g. resistance, induction, etc.; and the nature of fuel would decide the furnace design
Sub-classification of electric	Resistance Induction Microwave	Resistance heating is most common. Induction and microwave heating in TMP furnaces are still topics for R&D
Sub-classification of fuel-fired	Direct fired Indirect fired	Combustions products are circulated over work piece in direct fired, while in indirect fired the workpiece is protected. Examples of the latter include muffle (workpiece is enclosed in a muffle and then heated) and radiant tube (enclosing flame and combustion products)
Job type	Soaking pit Slot-forge	Furnaces are also classified based on its application. For example, soaking pits are ingot heating furnaces. Similarly, slot-forge, wire, carburizing, etc. are named for specific operations
Material handling	Batch Continuous	Batch: 'in-and-out' furnaces Continuous: material is moved while being heated

detailed description on lubricants in TMP can be found in Singer [1954], Schey [1970], Miller [1993] and Möller and Boor [1996].

11.1.2 TMP furnaces[3]

As shown in Table 11.2, furnaces can be classified using different indices. Of all the classifications shown in the table, the batch and continuous furnaces (and annealing practices) are particularly relevant to the TMP – the latter being more applicable for 'steady-state' operations. As shown in Figure 11.5, a range of batch furnace designs are possible - designs based on specific applications/loading. The schematic of a 'multi-zone' fuel-fired continuous furnace, typical application – steel reheat, is given in Figure 11.6. Table 11.3 outlines the basic information on several types of continuous furnaces. The classifications on batch/continuous furnaces is not exhaustive, for further details on industrial furnaces the reader may refer Trink [2004].

[3] The annealing practices related to TMP does not end with furnaces and furnace design. Of an equal importance are atmosphere and environment control, heat transfer, heating, cooling, etc. In any of these, several issues of large technological importance exist. For example, cooling would involve cooling/quenching media (controlled water jets to oil or polymer quenching), control and optimization of cooling – including automation, de-scaling during cooling, etc.

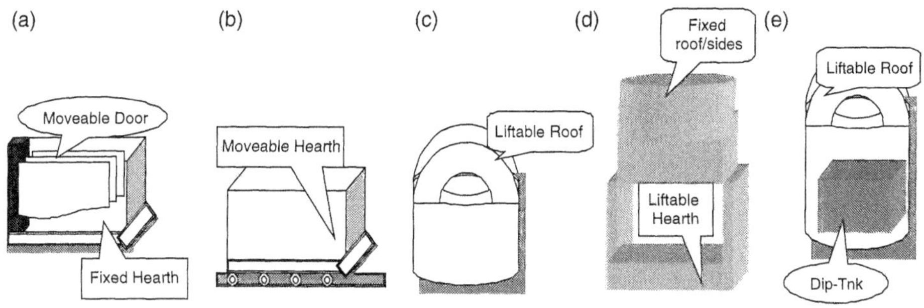

Figure 11.5. Schematics of different types of batch annealing furnaces. (a) Fixed hearth – this can be box or slot type. (b) Car or rollable hearth –hearth on steel wheels and rails for heavy loads (examples: car type, car bottom, lorry hearth, etc.). The furnace doors can be fixed to the car; or guillotine doors, to seal front/back ends, can be used. (c) Bell – fixed hearth with liftable roof/sides. (d) Elevator – liftable hearth, fixed roof/sides. (e) Dip-tank – for example, salt/lead-bath. The batch furnace types are not restricted to the five types mentioned in this figure, but can vary widely based on loading/applications.

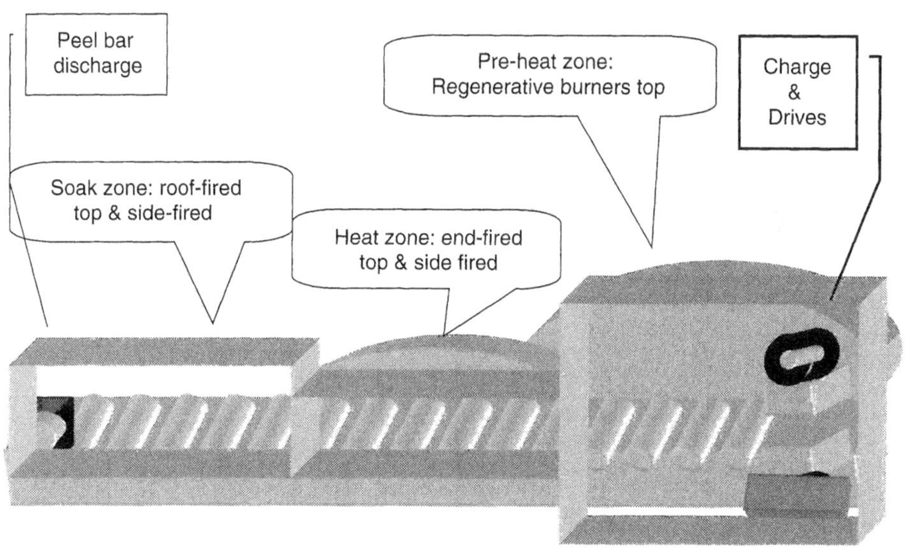

Figure 11.6. Schematic of a 'multi-zone' fuel-fired continuous furnace.

Table 11.3. Several types of continues furnaces, furnaces typically used in TMP.

Furnace type	Details
Tunnel	A tunnel furnace, as in Figure 11.7, typically has different heating zones and conveyorized material movement. The latter may range from conveyors to rollers, typically kept within the furnace to preserve heat and to prevent belt/roller failures by repeated heating/cooling. Heating may range from electric to fuel fired. Several design innovations are possible – including heat circulating 'plug-fans', radiant tube, etc.
Shuttle car-hearth	Hybrids between batch and continuous. Example: A box-type furnace with doors at both ends with two rolling hearths for quick loading/unloading
Saw-tooth walking beam	Saw-tooth walking beams provide rollover action for round objects (example: pipes). Cold material is picked up by the saw-tooth and at every step of the walking beam, the material rolls down – exposing different parts to heating. The furnace typically employs top and bottom firing
Rotary hearth	Donut or oval shaped – the charged material rotates through different zones and gets heated and then discharged. The major design issue is to obtain reduced fuel rates. Water seals may limit, but does not eliminate, 'air infiltration' – the major cause for energy inefficiency
Fluidized bed	Shaft furnace containing a thick bed of inert balls/pebbles – bubbled streams of combustions gasses rise through grate or perforated plate from a combustion chamber below. Rapid heat transfer and uniform heating of complex shapes are the typical advantages

LITERATURE

Kalpakjian S., "Manufacturing engineering and technology", Addison-Wesley (1995).
Möller U.J. and Boor U., "Lubricants in Operation" (Translated and Published from German), Ipswich Book Company, Suffolk, UK (1996).
Schey J.S., "Metal Deformation Processes: Friction and Lubrication", Mercel Dekker Inc., New York (1970).
Trink W., Mawhinney M.H., Shannon R.A., Reed R.J. and Garvev J.R., "Industrial Furnances", John Wiley & Sons, Hoboken, NJ (2004).

11.2. ROLLING

11.2.1 Introduction

Rolling has been used for about 500 years to form flat sections and sheet metal. In fact, it was probably first developed in the mid-16th century for the production of gold and silver strip of near-constant dimensions, to be used for coining and minting (and still is employed for this). It basically involves pushing a metal workpiece into the gap between two rotating rolls, which then simultaneously draw the workpiece into the rolls and compress it to reduce the thickness and increase the length. For large components this often requires significant power input, which was initially supplied by water mills, then steam power, before the advent of modern, electrically driven and highly automated rolling mills.

From an economic point of view, rolling is the most important metal working and shaping technique; it can be used to roll large ingots from half a meter thickness down to a few microns in the case of Al foil (of total length up to a hundred kilometres). Thus, 30 ton ingots are rolled down in a succession of rolls, often starting at high temperatures and finishing near room temperature. Using appropriately shaped rolls, hot rolling is also widely employed to form long profiles of more complex sections such as I beams and rails (known as 'shape rolling' as opposed to flat rolling). A very wide variety of forms, widths and thicknesses can therefore be manufactured, with high productivities, from semi-finished slabs through car-body sheet to packaging foil. Modern rolling mills are extremely efficient units capable of processing over a million tons of metal per year.

Rolling is carried out in a sequence of rolling passes during which the compressive strain can vary from a few percent to 50%. Since the deformation is only applied on the part of the workpiece between the rolls, i.e. a relatively small volume, the loads are reduced to moderate values even for very large ingots; this is of course the practical origin of the process. A reversible rolling mill is often used for the first stages of hot rolling of large sections such as ingots and slabs to reduce their thickness as rapidly as possible and avoid cooling down. Further reductions are achieved in a series of one-way mills known as 'stands'. Cold rolling of sheet is also usually carried out at high strain rates between roll stands; but for thin foil, reversible rolls are used with coilers at each end.

Both hot and cold rolling can lead to major improvements of the material properties by refining the microstructure. As-cast ingots are often characterized by large grain sizes, significant porosity and coarse 2nd phase particles. During hot rolling, porosity can be closed up, grain size reduced by recrystallization and coarse particles broken up leading to stronger, tougher alloys. Cold rolling can also be used to increase strength by work hardening; the latter is often sufficient to limit the amount of achievable rolling strains so that intermediate softening operations by annealing are necessary to continue rolling. If the property improvements by hot and cold working have been known for centuries, their microstructural origin have only been really understood over the last 50 years and the detailed process – structure–property relations, through TMP are the current object of intense research and development. Rolling lends itself to TMP because of the large number of variables during the process: temperature, strain and strain rate per pass and interpass periods, all of which need to be tailored to a specific alloy composition.

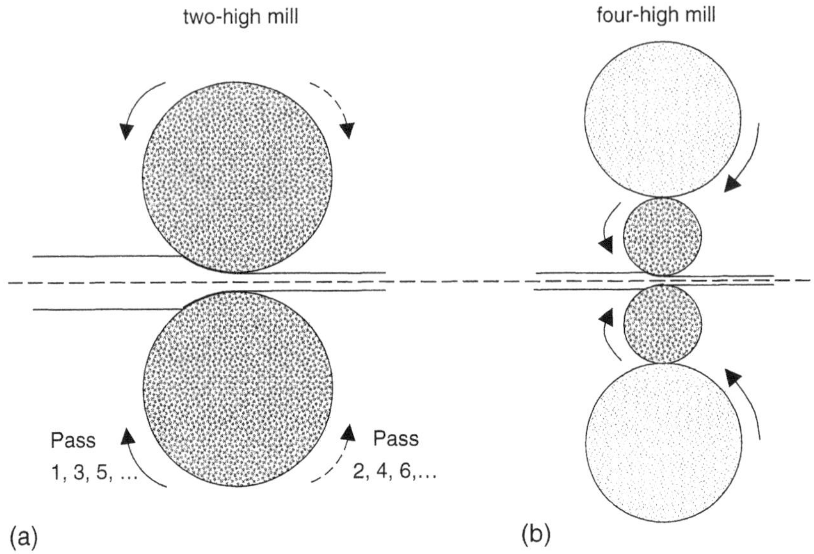

Figure 11.7. Schematic of a 2-high mill (a) and a 4-high mill (b).

11.2.2 Rolling equipment

11.2.2.1 Plate, sheet and foil. The first, and still a very common, type of rolling equipment is the two-high mill. Figure 11.7a illustrates a schematic two-high reversible mill in which the direction of rotation of the rolls is reversed after each pass to enable the workpiece to be passed successively backwords and forwards. This type of reversing mill with large diameter rolls is often used for the first stages of hot rolling ingots in the primary rolling mill (breakdown rolling in the blooming, slabbing or cogging mill). Typically, the ingot is reversibly hot rolled down from 500 to 30 mm (total average strain of 2.8) in a series of 10–20 passes.

Higher strains per pass are carried out during subsequent rolling operations down to sheet or foil using smaller diameter rolls to reduce the required power. However, smaller diameter rolls are less rigid than the large ones and therefore tend to bend significantly around the workpiece, producing camber in the strip particularly during cold rolling of hard metals. This is reduced, or eliminated, by using larger diameter *back up* rolls which support the smaller *work rolls*. Figure 11.7b shows a schematic of a four-high mill. The principle has been extended to the development of cluster mills, Figure 11.8, in which each roll is supported by two backing rolls. A Sendzimer mill is an example of such a cluster mill used to roll very thin sheet or foil.

High rates of production can be achieved in a continuous mill using a series of rolling mills often denoted tandem mills. Each set of rolls is placed in a stand and

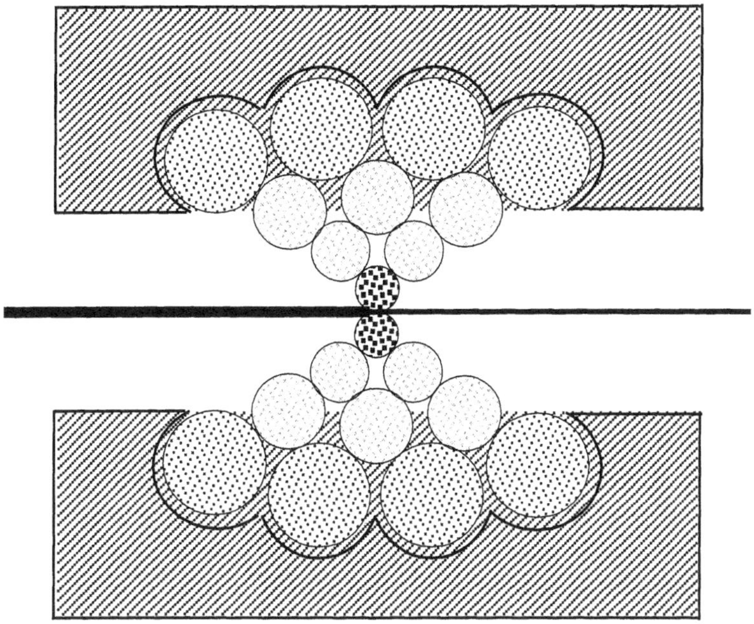

Figure 11.8. The Sendzimir mill, as an example of a cluster mill.

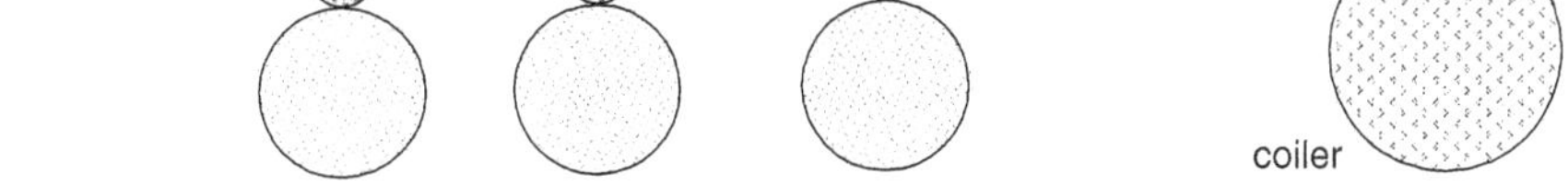

Figure 11.9. Schematic of a continuous mill.

since the input and output speeds of the strip at each stand are different, the strip between them moves at different (usually rapidly increasing) velocities. The rolling speeds of each stand are therefore synchronized so that the output speed of stand n equals the input speed of stand $n+1$, i.e. successive stands work in tandem. The final output sheet is usually coiled and both the coiler and the uncoiler can be adjusted to provide a back or a front tension. Continuous 4-high tandem mills are used for rolling

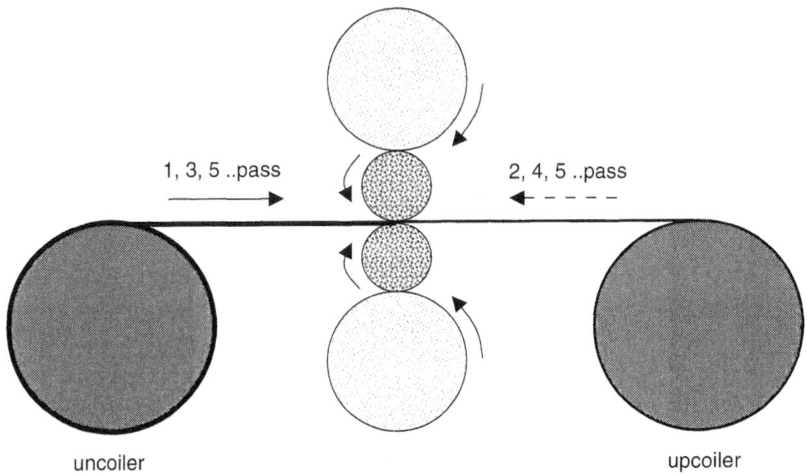

Figure 11.10. Illustration of a single-stand reversible cold rolling mill.

strip of typical thickness 30 mm down to a few mm in 3–5 stands; this can be done hot or cold (Figure 11.9). In the latter case, the production rates are very high but so are the capital costs and the product is very standardized (e.g. car-body sheet).

More flexible cold rolling is performed in 4-high single stand reversing mills with coilers at both ends (and which can also provide front and back tension) (see Figure 11.10).

A special type of mill for large reductions is the planetary mill which is made up of two large backing rolls surrounded by several small planetary rolls (Figure 11.11). Each of the latter gives a roughly constant reduction to the slab before it meets the next set of rolls. Thus, during a single pass (at high temperatures), the slab undergoes a large number of reductions so that it is, in effect, rolled down to strip in one pass.

11.2.2.2 Bars, rods and profiles. Long products such as beams, rails and wire rods are manufactured by rolling them through a series of work rolls of specific shapes, typically grooved rolls. The workpiece starts as an initial round or square bloom or billet which is then repeatedly passed through the calibrated work rolls, usually by reversible rolling. The pass geometry is generally established by empirical trials but some numerical methods are being developed.

During shape rolling, in contrast to flat rolling, the cross-section of the metal is reduced in two directions. In one pass, the metal is compressed in one direction and then rotated 90° for the following pass so that more 'equiaxed' sections can be achieved. Thus, a square billet is reduced to bar by alternate passes through oval and square-shaped grooves. The area of contact therefore changes continuously during

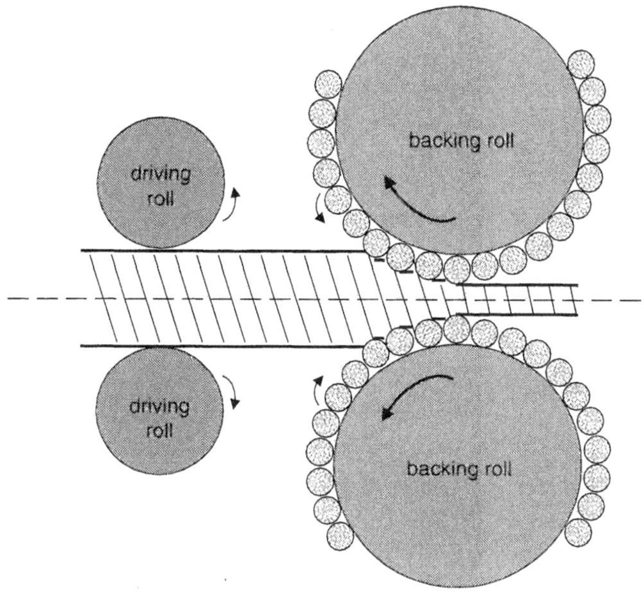

Figure 11.11. Schematic of a planetary mill.

the rolling process. The total reduction per pass is expressed in terms of the change in cross-sectional area, since there are both thickness reductions and width increase.

Shape rolling involves significant amounts of lateral spreading, which is difficult to control so the more complex shapes require very experienced designers.

11.2.3 Mechanics

Figure 11.12 illustrates the basic geometry of rolling (a) and the forces exerted during a flat rolling pass (b) under plane strain conditions (constant width w). A workpiece of initial thickness h_0 is passed between rolls of diameter R at an initial velocity v_0 and exits at thickness h_f and velocity v_f. The surface velocity of the rolls is v_c. At the first point of contact, the metal advances more slowly than the rolls but, by volume conservation during a thickness reduction Δh, it exits more quickly; only at the neutral (or no-slip) point N, defined here by the angle Ψ, are the surface roll and workpiece velocities identical.

The total reduction and total true strain in one pass are defined as:

$$r = \frac{h_0 - h_f}{h_0} \quad \text{and} \quad \bar{\varepsilon} = \ln\left(\frac{h_0}{h_f}\right) \tag{11.4}$$

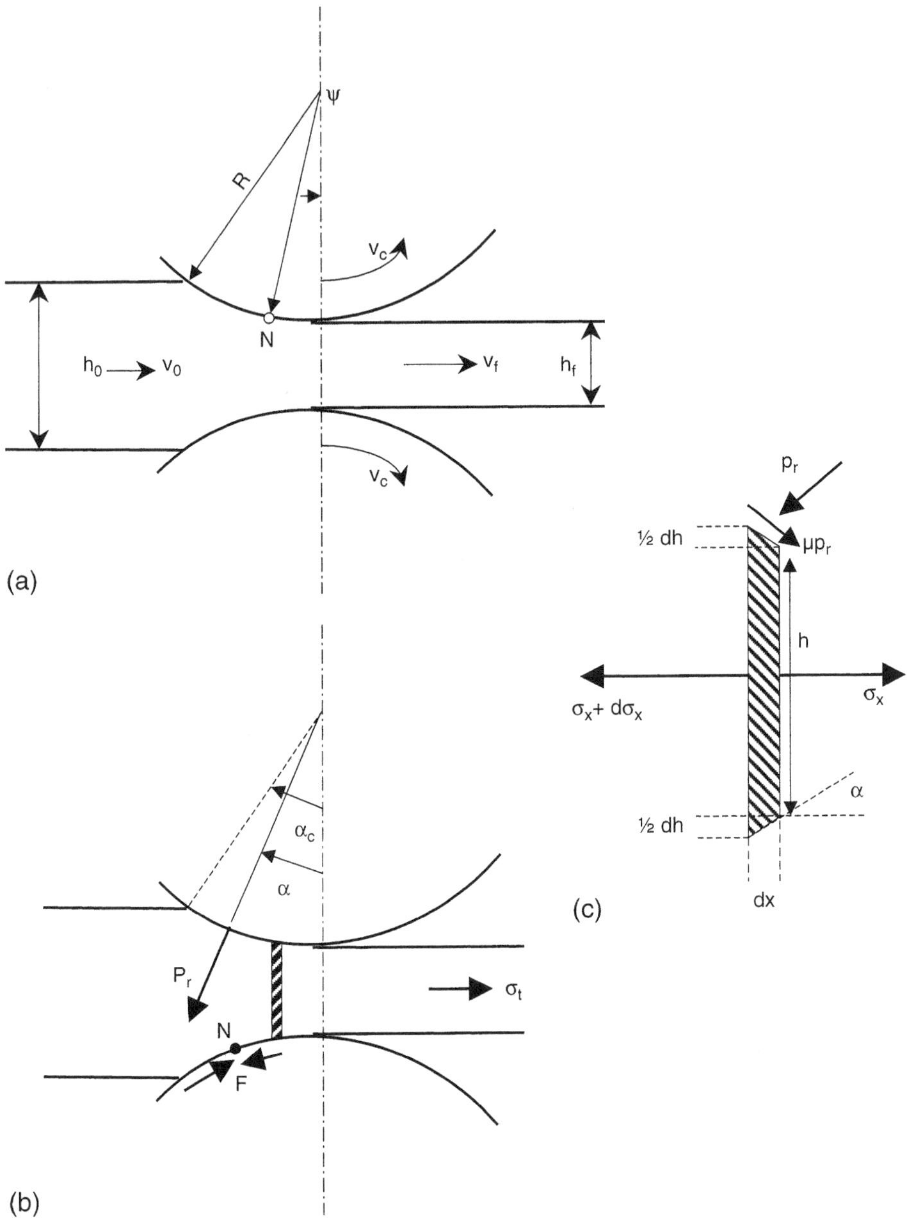

Figure 11.12. Basic geometry of flat rolling.

To feed the workpiece into the rolls, there has to be some friction between the two. The frictional force acting on the roll surfaces before the neutral point, N, pulls the strip between the rolls, while the friction force acting between the neutral point and the exit opposes the strip movement out of the rolls.

For the workpiece to enter the rolls, the horizontal component of the frictional force F, which acts towards the roll gap, has to be equal to or greater than the same component of the radial force P_r exerted by the rolls to compress the workpiece. The limiting condition at the entrance is expressed in the following way for the angle of contact α_c (half the total included angle for plane strain compression).

$$F \cos\alpha_c = P_r \sin\alpha_c \tag{11.5}$$

Also since the frictional force F is related to the radial force P_r by the friction coefficient $\mu : F = \mu P_r$. Combining these two equations gives for the lowest possible value of μ:

$$\mu = \tan\alpha_c \tag{11.6}$$

The workpiece will not be drawn into the rolls if the coefficient of friction is less than the tangent of the contact angle. The thicker the slab, the greater the friction coefficient required and if $\mu = 0$ rolling cannot occur.

The exact values of the friction coefficients during rolling are difficult to establish with precision since they are considered to vary along the contact arc of the rolls. However, for most analyses, a constant value is assumed; typically μ varys from 0.05 to 0.1 for cold rolling with lubrication to 0.2 or even above for hot rolling.

The vertical component of P_r is known as the rolling load P required to compress the metal. The specific roll pressure p is this force divided by the contact area, i.e. the product of the width and the projected length of arc of contact:

$$p \approx \left(\frac{P}{w[R(\Delta h)]} \right)^{1/2} \tag{11.7}$$

By simple geometry the draft Δh is related to the angle α_c by the length of the arc of contact $l_c = (R\Delta h)^{1/2}$ and the roll diameter R:

$$\tan\alpha = \frac{(R\Delta h)^{1/2}}{(R - \Delta h)/2} \approx \left(\frac{\Delta h}{R} \right)^{1/2}$$

But from Eq. (11.6)
$\mu \geqslant \tan \alpha_c$ so

$$(\Delta h)_{max} = \mu^2 \cdot R \tag{11.8}$$

This determines the maximum reduction that can be achieved in one pass for a given set of rolls.

The minimum exit thickness (h_{min}) that can be obtained in a strip has been estimated by Stone [1953]; it depends on the average equivalent flow stress of the material $\bar{\sigma}$ (averaged between the entrance and exit values), the friction coefficient μ and the elastic properties of the rolls (Young's modulus E_r and Poisson's ratio v_r):

$$h_{min} = \frac{A\mu R}{E_r}(1 - v_r^2)(\bar{\sigma} - \sigma_t) \tag{11.9}$$

A is a coefficient that takes values of 7 to 8 and σ_t is the tensile stress (if any) applied to the workpiece during rolling. In practice, it is possible to obtain thinner strip than expected from this relation by pressing the rolls together.

A rigorous calculation of the rolling force P (also known as 'the roll separation force') is not very easy to do. In a first very crude approximation, the effects of inhomogeneous deformation and of friction can be neglected and flat rolling is considered as simple plane-strain compression. The rolling force can than be calculated from Eq. (11.7) with the contact pressure p equal to the (plane strain) flow stress of the material:

$$P = \left(\frac{2}{\sqrt{3}}\right)\sigma_y w\left[R(\Delta h)\right]^{1/2} \tag{11.10}$$

A rough compensation for neglecting friction and inhomogeneous deformation consists in replacing the factor $(2/\sqrt{3}) = 1.155$ by a factor 1.5, leading to the following 'rule of thumb' for the estimation of the rolling force:

$$P = 1.5\sigma_y w\left[R(\Delta h)\right]^{1/2} \tag{11.11}$$

In reality, the roll pressure varies significantly along the arc of contact and a typical distribution is shown in Figure 11.13. The pressure goes through a maximum close to the neutral point and the general form of the curve is known as the 'friction hill'. The total area under this curve is proportional to the rolling load. The area under the dotted line represents the force required to deform the metal in plane

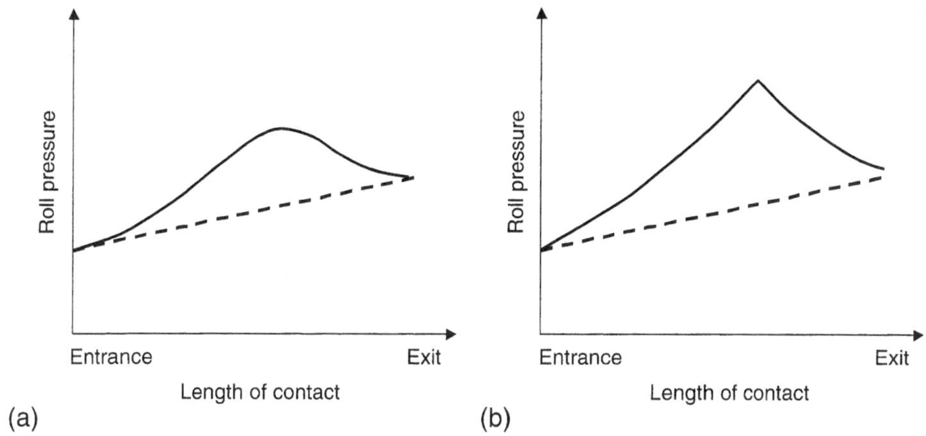

Figure 11.13. Schematic variation of roll pressure along the arc of contact (a) experimental and (b) calculated according to Eqs. (11.13)–(11.15).

strain compression. The area above this line is related to the force required to overcome the friction between roll and workpiece, hence the name of friction hill.

The height of the friction hill depends upon the value of the friction coefficient, but both the peak height and position can be shifted by the application of front or back tensions to the workpiece. A back tension will significantly reduce the rolling load and shift the peak towards the exit side so this is often applied industrially.

A more realistic roll pressure analysis follows from the standard theory of plastic working (see, e.g. Dieter [1988], Rowe [1977], Mielnik [1991]). The horizontal components of forces acting on an element of metal situated in the roll gap at a position described by the angle α (see Figure 11.12) are:

- $(\sigma_x + d\sigma_x)(h + dh) - h\sigma$: due to longitudinal stress.
- $2(p_r \sin\alpha)(dx/\cos\alpha)$: due to radial pressure on both rolls.
- $2\mu(p_r \cos\alpha)(dx/\cos\alpha)$:due to friction against both rolls.

The force balance gives:

$$h\,d\sigma_x + \sigma_x\,dh + 2\mu p_r\,dx \pm 2\mu p_r\,dx = 0 \qquad (11.12)$$

where, as before, p_r is the radial pressure, h is the current thickness and σ_x is the horizontal component of stress in the metal. The $\pm$ sign accounts for the change in sign of the frictional force at the neutral point. Equation (11.12) is often known as the 'von Kármán' [1925] equation, who first proposed it.

Rigourous solutions to this equation require numerical techniques, but an approximate analytical solution is given following Bland and Ford [1948] by taking the small angle approximations $\sin\alpha = \alpha$ (in rad), $\cos\alpha = 1$ and $Pr = P$ and assuming that the variation in flow stress is small compared with the variation in roll pressure so that one obtains:

$$\text{on the entrance side}\quad P = \frac{2\sigma_0 h}{\sqrt{3}h_0}\cdot\exp\left[\mu(H_0 - H)\right] \tag{11.13}$$

$$\text{and on the exit side}\quad P = \frac{2\sigma_0 \cdot h}{\sqrt{3}\cdot h_f}\cdot\exp\left[\mu H\right] \tag{11.14}$$

$$\text{where the quantity}\quad H = 2\sqrt{\frac{R}{h_f}}\cdot\tan^{-1}\left(\sqrt{\frac{R}{h_f}}\cdot\alpha\right) \tag{11.15}$$

If front tension (σ_f) and back tensions σ_t are applied to the strip, then these pressure are reduced by respective factors of $(1-(\sigma_f/\sigma_0))$ and $(1-(\sigma_t/\sigma_0))$.

These equations implicitly assume plane strain compression, homogeneous deformation and relatively low-friction coefficients; they apply essentially to the case of cold rolling strip, except for the case of very light reductions as in skin pass rolling, where the deformation takes place in the surface regions It should also be noted that due to the high loads, the radius of curvature of the rolls R is not necessarily constant; in fact, they tend to flatten significantly by elastic deformation and Hitchcock [1935] has given an estimate of the larger R value (R') to be used in Eq. (11.11).

$$R' = R\left[1 + \frac{CP}{(h_0 - h_f)w}\right] \tag{11.16}$$

with C the elastic constant of the roll ($C \sim 0.022$ mm^2/kN).

The mechanics of hot rolling is more complicated because of the higher frictions, the superimposed shear components in the surface regions for thicker slabs and the strain rate sensitivity of the flow stress. An early analysis of the roll stresses by Ekelund [1927] for this case gives:

$$\frac{P}{w} = \left\{\sigma_0 + \frac{2V\eta\sqrt{\Delta h/R}}{h_0 + h_f}\right\}\cdot\sqrt{R\Delta h}\left\{1 + \frac{1.6\mu\sqrt{R\Delta h} - 1.2\Delta h}{h_0 + h_f}\right\} \tag{11.17}$$

where σ_0 is the 'static' yield stress, V the peripheral speed of the rolls and η the material viscosity.

During hot rolling, the near-surface regions undergo additional (redundant) shears since the friction coefficient is significantly higher than for cold rolling and also because of the higher thickness to contact length ratio. There is often significant lateral spread.

The frictional forces lead to near-surface shears, whose sign changes at the neutral point. If x is along RD and z = ND, then from Figure 11.12b the displacement gradient component of the shear deformation $\mathrm{d}e_{xz} = \mathrm{d}u_x/\mathrm{d}z$ induced, by the friction stresses, is initially positive (in the upper half of the sheet), goes to zero at the neutral point then becomes negative afterwards. If the neutral point is near the middle of the arc of contact, then this shear component will roughly balance out to an accumulated friction-induced shear near zero. However, the geometrical shape change of the metal induced by what is the equivalent of a converging channel is described by the $\mathrm{d}e_{zx} = \mathrm{d}u_z/\mathrm{d}x$ component of the displacement gradient. For the same material element, this $\mathrm{d}e_{zx}$ component is clearly negative at first, decreases in amplitude during the pass and can also change sign but not necessarily at the neutral point. Consequently, the conventional shear strain $\mathrm{d}\varepsilon_{xz}$, which is half the sum of both components does not always give an adequate description of the true material deformation.

There is a rule of thumb for estimating the relative importance of these two types of shear from the ratio of the contact length l_c to average sheet thickness $\frac{1}{2}(h_0 - h)$:

- $l_c/\frac{1}{2}(h_0 - h) > 5$ means that the friction-induced effects at the surface, along the contact length, dominate; a thin sheet rolled with large diameter rolls undergoes very superficial surface shears $\mathrm{d}e_{xz}$ (as in a skin pass).
- $l_c/\frac{1}{2}(h_0 - h) < 0.5$ means that the shape change effect dominates and leads to large shear components inside the metal. This occurs during hot rolling, particularly when large draughts are applied to thick band. At ¼ thickness, there are strong shear effects $\mathrm{d}e_{zx}$, particularly during reversible rolling.

When $0.5 < l_c/\frac{1}{2}(h_0 - h) < 5$, the deformation is supposed to be homogeneous but homogeneous is clearly a relative term when applied to rolling deformation.

Hosford and Caddell [1983] also define a Δ-*factor* (thickness (or diameter)/contact length). The application to wire drawing is described in Chapter 11.4.

$$\text{For flat rolling, the }\Delta\text{-factor gives: } \Delta = \left[\frac{2-r}{2}\right]\left[\frac{h_0}{rR}\right]^{0.5} \tag{11.18}$$

In principle, one wants $\Delta < 1$–2; for $\Delta > 5$, most of the deformation is concentrated on the surface (thick specimen; small rolls); Δ can also be high for very small reductions.

The strain rate in rolling is related to the geometry of the mill and to the surface speed v of the rolls. If slipping friction between sheet and rolls is assumed, an approximate estimation of the mean strain rate in rolling can be obtained from Mielnik [1991]:

$$\dot{\varepsilon} = \frac{v}{h_0}\sqrt{\frac{h_0 - h_f}{R}} \tag{11.19}$$

There is a general tendancy to use finite element methods (FEMs) for the detailed analysis of hot rolling mechanics to derive both the roll forces and the local deformation modes of the material. Figure 11.14 indicates how the shear and compression strains of an aluminium alloy vary through the roll gap during two passes of breakdown rolling near the start of rolling (pass 4) and towards the end (pass 18).

In Figure 11.14a, the initial (small) negative ε_{xz} shear is reversed during the pass and reduced to almost zero, whereas during pass 18, Figure 11.14b the same, larger shear becomes positive and is retained at the exit.

11.2.4 Typical rolling schedules

11.2.4.1 Steel. Most steels are now produced by continuous casting with typical cross-sections of 0.2–0.3 m (thickness) and up to 2 or 3 m in width. In some cases, the as-cast steel is rolled hot directly after casting and solidification but usually it is cooled down and sectioned into slabs. The slabs are then reheated to temperatures around 1200°C then hot rolled down to thicknesses of the order of 50–30 mm in a roughing or slabbing mill before going into the finishing mill for reductions down to a few mm. Strain rates in the roughing mill are of the order of 10/s and then increase up to 100/s during finishing. Note that these conditions are only given as an indication. Individual mills can have quite different rolling schedules as a function of equipment, alloy composition and final application (see Chapter 15). The final sheet gauge is obtained by cold rolling down to thicknesses of about 1 mm for applications such as car-body sheet. In this case, the cold-rolled sheet is then annealed at temperatures of about 700°C to provide a 'soft' recrystallized sheet suitable for deep drawing.

In the case of plain carbon steels, the initial roughing starts at temperatures around 1150°C and ends at about 1000°C before the finishing in the temperature range 1000–910°C. For these steels, the final cooling after finishing is particularly important since the cooling rate controls the phase transformations that occur (e.g. austenite to perlite or bainite, etc.) and therefore the final properties. For special products such as the high-strength low alloys (HSLA) or multiphase steels, etc. the

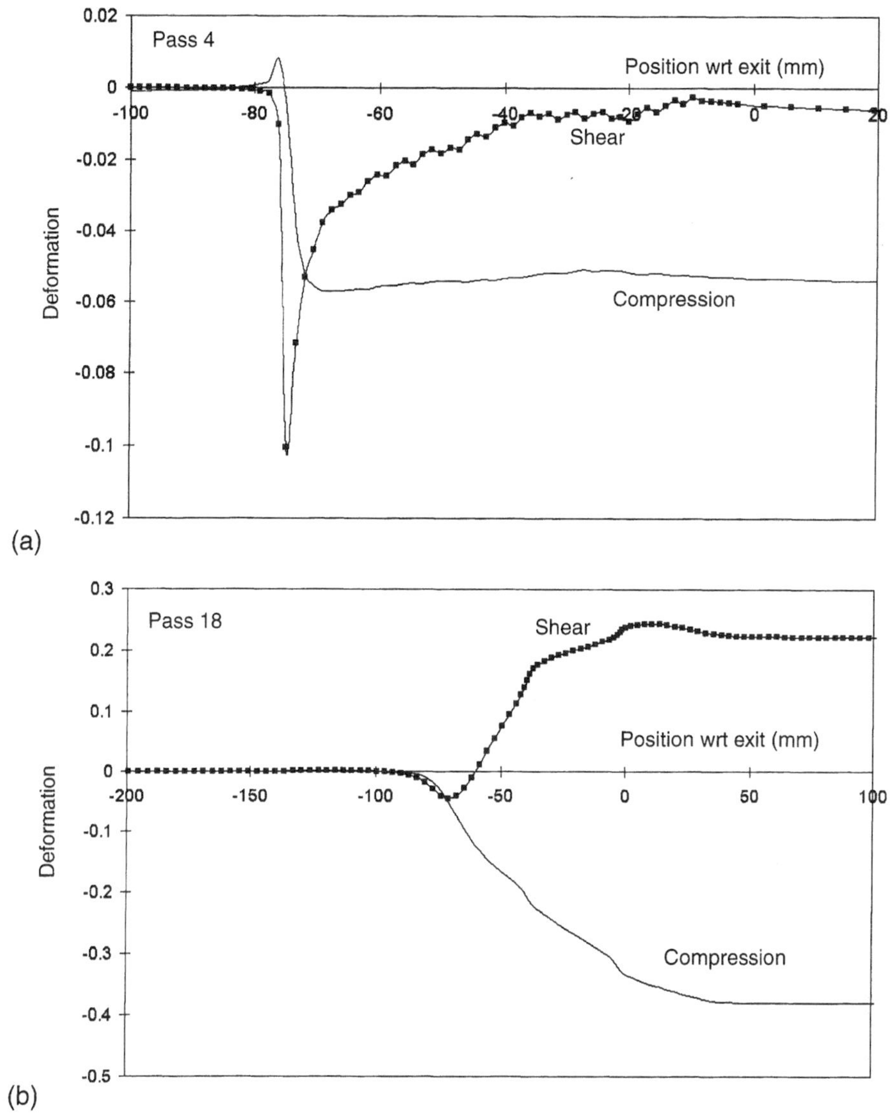

Figure 11.14. The cumulated shear (dotted) and compression (continuous) strain components through the roll gap of a breakdown mill, near the surface for two passes (strain rate and temperature fields calculated by finite element methods for an AA 3104 (Al–1Mn–1Mg) alloy in the stationary regime (Perocheau [1999]).

rolling schedule is closely linked to the austenite–ferrite phase transformations to develop fine-grained steels with improved properties (Chapter 15).

In the case of austenitic stainless steels, the slabs are reheated at 1280°C for about 3 h. then rolled down from 150–220 mm to 30–40 mm in the roughing mill. A common practice is to reheat after the first pass (about 25% reduction) then to continue roughing; the first pass and reheat facilitate homogenization of the as-cast structure. The finishing stage down to a few mm thickness takes place in a 5–7 stand tandem mill over the temperature range 1100–1000°C. These steels are often cold rolled to 5–0.5 mm thick sheet and then annealed.

11.2.4.2 Aluminium. Most aluminum alloy ingots are produced by direct chill (DC) semi-continuous casting with typical dimensions of 0.4–0.6 m thickness, 2 m width and up to 9 m length (weight 20–30 tons). These ingots are destined to be rolled down to plate, sheet, etc. by a rolling schedule comprising hot rolling, cold rolling and often intermediate anneals (see Figure 11.15 and Table 11.4).

The ingot is first homogenized, i.e. heated to a temperature in the range 500–600°C for relatively long time (at least a few hours) to reduce segregation and remove non-equilibrium, low melting point eutectics. This facilitates subsequent hot working and improves homogeneity. The homogenization treatment can also have a further effect on the final microstructures of some alloys in that precipitation reactions can occur during the treatment, in particular in the Cr, Mn or Zr-containing alloys that form dispersoids.

After homogenization, the ingot is usually hot rolled down to 30–10 mm thick strip in a reversible (breakdown) rolling mill. The number of passes varies from 9 to 25. The strip from the single stand breakdown mill is either coiled to await cold rolling or, in modern processing lines, further hot rolled in a multiple stand tandem mill. In current practice, the tandem mills have between 2 and 6 stands.

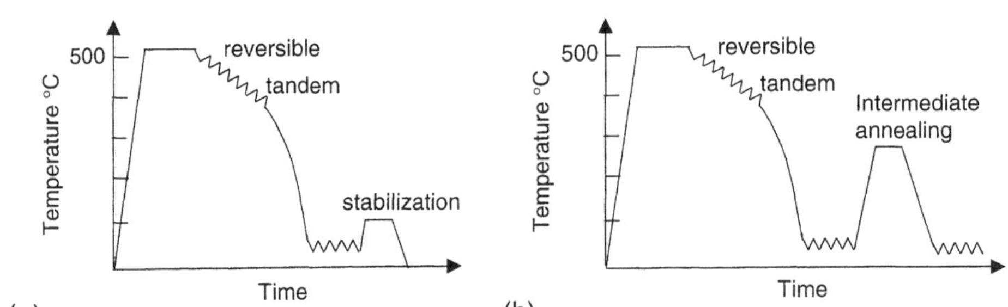

Figure 11.15. Schematic Al rolling schedules for the production of (a) can stock, (b) foil.

Table 11.4. Some typical rolling conditions for Al alloys.

	Reversible	Tandem	Cold rolling
Start temperature (°C)	500–600	400–500	20
Finish Temperature (°C)	400–500	250–350	100
$N°$ passes	9–25	2–5	2–10
Initial thickness (mm)	400–600	45–15	2–6
Final thickness (mm)	45–15	2–9	0.01–1
Strain per pass	0.1–0.5	0.7	0.3–0.7
Total strain	3.5	3	<5
Strain rates (s)	1–10	10–100	>50
Inter-stand times (s)	10–300	<3	

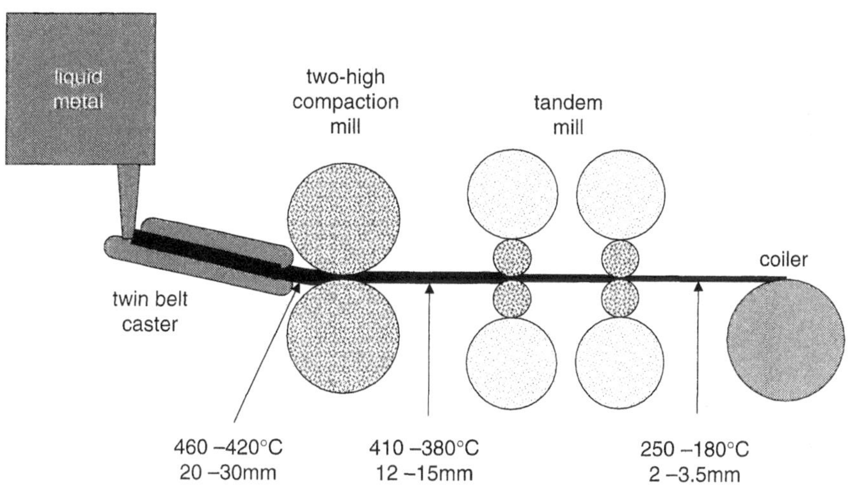

Figure 11.16. Schematic continuous strip casting line.

Cold rolling is usually carried out in a reversible 4-high cold mill between two coilers. When correctly set-up this equipment can be used to roll down the 'softer' alloys to a thickness of 15–20 μm. To obtain very thin packaging foil of about 6 μm thickness, the foils are then doubled up and re-rolled. Intermediate annealing is frequently necessary to achieve large cold rolling reductions. It is also worth noting that modern, high speed, cold rolling of medium strength alloys such as the 3xxx and 5xxx series for canning sheet (Chapter 14.1) generates a substantial temperature increase so that the output sheet can attain 120–150°C (or $0.45T_m$).

An increasing proportion of the less strongly alloyed sheet products are now produced by continuous strip casting methods. As shown in Figure 11.16, the hot metal is poured between rotating cylinders to produce 'thick' sheet (10–20 mm) which is then immediately rolled in a tandem mill.

LITERATURE

Backofen W.A., "Deformation Processing", Addison-Wesley, Reading, MA (1972).
Ginzburg V.B., "Steel Rolling Technology", Marcel Dekker, New York (ISBN 0-8247-8124-4), (1989).
Roberts W.L., "Cold Rolling of Steel", Marcel Dekker, New York (ISBN 0-8247-6780-2), (1978).
Rowe G.W., "Principles of Metal Working", 2nd edition, Edward Arnold, London (1977).
Singh R.V., "Aluminum Rolling, Process, Principles and Applications" TMS Warrendale, PA (2000).
Totten G.E., Funatani K. and Xie L., "Handbook of Metallurgical Process Design" Marcel Dekker, ISBN 0-8247-4106-4, 2004 in particular, chapter 3 "Design of Microstructures and Properties of Steel by Hot and Cold Rolling" by R. Colás, R. Petrov and Y. Houbert and chapter 4 "Design of Aluminum Rolling Processes for Foil, Sheet and Plate" by J. Driver and O. Engler.

11.3. EXTRUSION

11.3.1 Introduction

Extrusion is a process in which a billet of metal is first placed into a chamber with a die at one end and a ram on the other. The billet is then pushed through the relatively narrow die to form long profiles of constant section determined by the die geometry (rather like squeezing toothpaste out of a tube) (see Figure 11.17).

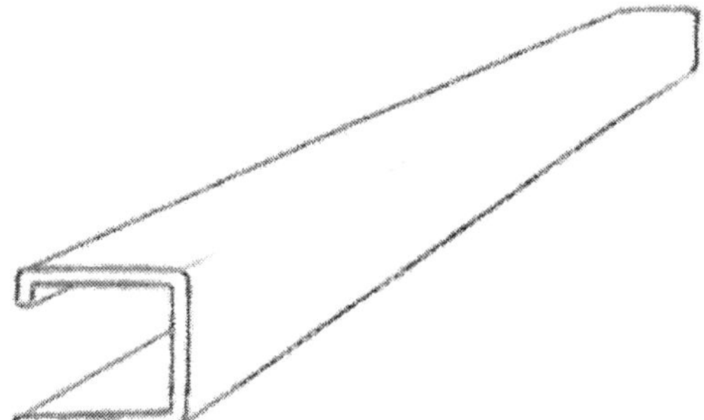

Figure 11.17. Example of an extrusion profile.

The process is usually carried out at high temperatures, of the order of 0.5–0.75T_m to reduce both the applied loads and wear on the dies, but on softer metals can be performed at room temperature. The loads are imposed by mechanical presses or hydraulic rams through a mandrel onto the metal billet, which is restrained by the chamber designed to withstand the large radial stresses generated during extrusion; Figure 11.18 is a schematic of the direct extrusion process. According to the die geometry, a very large variety of cross-sections can be produced varying from round or rectangular bars, through L, I or T shapes etc. to tubes and complex sections.

A second type of extrusion – known as 'indirect or back extrusion' – uses a hollow ram containing the die, while the other end of the billet is completely blocked off. As the ram is pushed into the billet, the metal is extruded out in the opposite direction through the die (back out along the hollow ram). This has the advantage that there is no relative movement between the container wall and the billet so that the friction forces and the required power level are smaller. However, the use of a hollow ram limits the loads that can be applied.

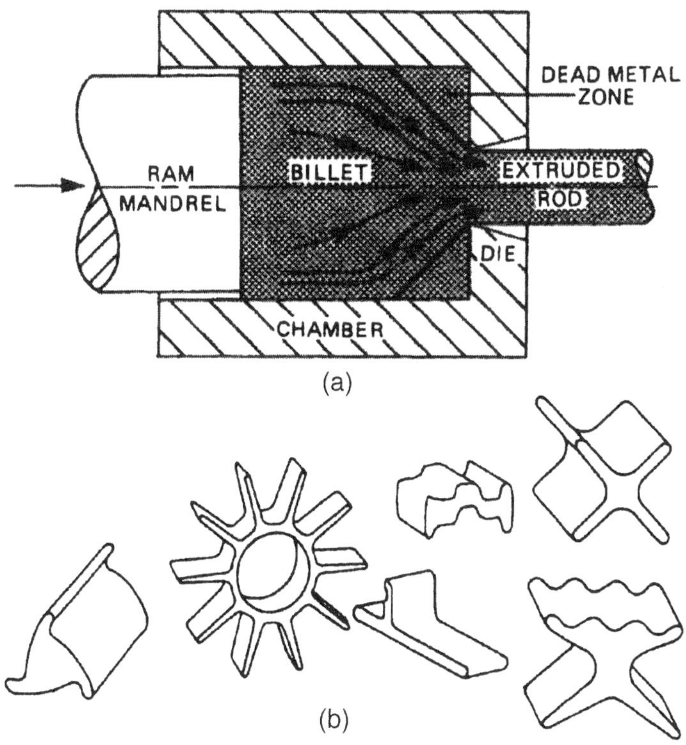

Figure 11.18. (a) Direct extrusion process and (b) some typical profiles – (Avitzur [1983]).

A variant of this is impact extrusion, which is used to produce short lengths of hollow shapes, e.g. toothpaste tubes, from solid rods or disks. A high-speed mechanical press pushes the ram into the disk placed above a female die so that the wall of the can or tube is then punched out between the ram and the die.

In general, extrusion tooling is inexpensive; lead times for custom shapes or prototypes are relatively brief and many alloys can be readily formed into complex shapes. Consequently, extrusion has developed significantly over the last two or three decades to become the second most important plastic forming operation, after rolling.

11.3.2 Deformation conditions

The overall deformation is often described by the extrusion ratio, i.e. the ratio of the billet cross-sectional area to the final cross-section area after extrusion: $R_{ext} = A_0/A_f$ and the average strain approximated by $\ln R_{ext}$. By conservation of volume, the velocity of the extruded product is the ram velocity x R_{ext}. Given that at high temperatures, the material flows at constant stress the extrusion force is related to the natural logarithm of the extrusion ratio:

$$P = k\ A_0\ \ln R_{ext} \tag{11.20}$$

where the extrusion constant k includes the effects of material flow stress, friction and inhomogeneous deformation. For the extrusion of bars, an average strain rate can be calculated using:

$$\varepsilon^{\circ} = 6VD_0^2 tg\alpha \frac{\ln R_{ext}}{(D_0^3 - D_e^3)} \tag{11.21}$$

with D_0 the diameter of the billet, D_e the diameter of the bar, V the speed of the plunger, and 2α the die angle.

Because of the relatively complex shapes formed during extrusion, the plastic strains vary substantially with position in the workpiece. They are usually close to uniaxial elongation in the centre of round bars and near plane strain compression for flat sections, but with large shear components near the surface where the metal flows past the chamber walls and the die edges. There is often a dead metal zone in the corner of the exit side of the chamber. Lubrication conditions can therefore be critical for some metals.

In fact, two metals, Cu and Mg, are particularly easy to extrude since their oxides are softer than the metal at high temperatures. The oxide layer formed during heating the billet acts as a natural lubricant.

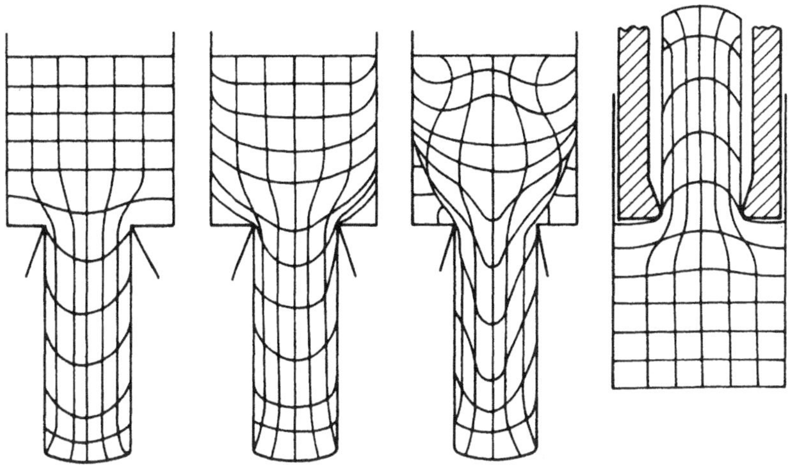

Figure 11.19. Some typical deformation patterns during direct (a–c) and indirect extrusion (d) with a square die: (a) good lubrication, (b) average lubrication, (c) poor lubrication and (d) indirect extrusion. From Schey [1968].

Figure 11.19 shows some typical deformation patterns in extrusion: (a) for the case of relatively homogeneous deformation with a well-lubricated billet or indirect extrusion as in Figure 11.19d. Figure 11.19c illustrates the deformation for a high-friction condition, where the material flow is concentrated near the centre in association with internal shearing near the surface.

The die geometry is important for facilitating material flow without excessive friction and heterogeneous deformation. Well-lubricated billets are extruded using die fronts with conical entrances of semidie angle, typically 45–60°, to reduce the extrusion pressure. However, some metals such as aluminium form a dead zone and then shear internally about it to form their own die angle; in these cases, the die entrance face is flat.

A simple method of estimating the die pressure Pd is to use the analysis of extrusion through a conical die of semidie angle α. Assuming Coulomb sliding friction (but no redundant deformation) Hoffman and Sachs [1953] gives a solution analogous to that of wire drawing through a die:

$$\mathrm{Pd} = \sigma_0 \left[\frac{1+B}{B}\right]\left[1 - R_{\text{ext}}^{B}\right] \tag{11.22}$$

where $B = \mu \cot \alpha$.

Slip line theory (without friction) yields solutions of the form:

$$\text{Pd} = \sigma_0 (a + b \ln R_{\text{ext}}) \tag{11.23}$$

with a and b taking values of order unity.

More detailed analyses of the die pressures can be found in Dieter [1988].

The ram speed has to be controlled to ensure defect-free products. Some metals such as highly alloyed aluminium and copper are susceptible to hot shortness so the ram speed is reduced to low values (<1 m/min) at avoid excessive frictional heating during the process. Hot shortness leads to periodic transverse cracking along the product, denoted fir-tree cracking. For other materials, such as the refractory metals extruded at very high temperatures, heat transfer from the billet to the tools is a problem and the extrusion has to be carried out as rapidly as possible (ram speeds of up to 30 m/min).

Extrusion develops compressive and shear forces in the stock but usually no tensile stresses so that high deformations are possible with only one extrusion operation. The tolerances can be small, particularly for cold extruded products.

11.3.3 Steels and high melting temperature alloys

Up to the invention of the Ugine–Sejournet process (Séjournet and Delcroix [1955, 1966]), the extrusion of steel was severely hampered by lubrication problems. Without a suitable lubricant the very high frictional forces developed during extrusion led to severe wear of the dies so that reproducible extrusions were very difficult to obtain. Séjournet [1955] devised a method of lubrication based on the use of molten glass in which both the chamber walls and the die surfaces are lubricated. The hot steel billet is rolled in a powder of glass which melts and forms a thin, viscous, film coating before introduction into the chamber. The front end of the billet is also placed in contact with a glass pad, located just before die orifice, whose surface gradually melts during the extrusion to provide a continuous supply of lubricant between the die and the extruded product. This process has been extended to practically all metals and alloys that are extruded in or above the steel hot working range. Figure 11.20 illustrates the strain rate distribution in a partially extruded steel billet.

Steels such as carbon steels and stainless steels are hot extruded in the typical temperature ranges 900–1200°C, i.e. in the hot working austenite range, and therefore undergo the same deformation–transformation reactions as during hot rolling. The potential for TMP is not so great since only one deformation step is performed but, even given this limitation, TMP has perhaps not yet been sufficiently developed for the extrusion process. For example, Lesuer *et al.* [1999] has shown that the properties of hypereutectoid carbon steels can be significantly improved by hot extrusion.

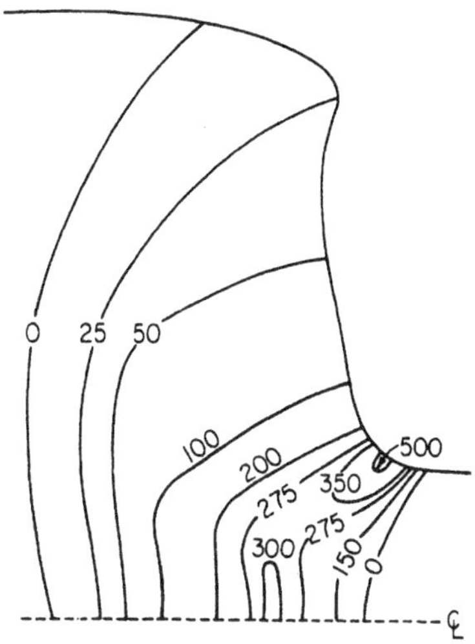

Figure 11.20. The strain rate distribution in a partially extruded steel billet ($R = 16.5$, ram speed = 12.6 m/min, T = 1440 K (Childs [1974]). Permission obtained from Maney.

Carbon steel envelopes are also used to extrude and consolidate powder metallurgy products.

There is a general tendency to develop cold extrusion of steels since the product has higher strength through work hardening, better surface finish and improved tolerances. Here again frictional effects are very important; cold extrusion of steel only became possible, thanks to the introduction of phosphate lubrication. The steel surface is coated with a spongy phosphate coat that absorbs the lubricating liquid, often a soap solution, to substantially reduce friction and wear during extrusion. The literature on cold forming, including cold extrusion, has been reviewed by Watkins [1973].

11.3.4 Aluminium alloys

Aluminium extrusions are usually carried out in the temperature range 450–500°C but some of the softer alloys can be extruded at temperatures down to 400 or even 350°C. The extruding rate depends upon the alloy and the complexity of the die shape; to first order, they can be simply classed according to the velocity of the ram used to push the billet. For example, ram velocities are in the range 0.5–2 m/min for hard 7xxx alloys, 10–80 m/min for the intermediate 6xxx alloys and 20–100 m/min for commercially pure aluminium as used for electrical conductors.

The corresponding average strain rates vary over two orders of magnitude from 0.1–50/s, i.e. the same range as hot rolling operations. The nominal strains defined as ln l/l_0 are of the order of 1–4 and sometimes higher for very soft metals. These are much higher than the strains encountered in a single rolling pass.

As noted above, the real strains and strain rates vary widely through the section of the extruded profile because of the geometrical complexity inherent to the process. Also aluminium extrusion is carried out without lubrication so the immediate surface 'sticks' to the container surface enhancing the surface shear deformations. Thus, in a relatively simple case of plane strain extrusion, the centre part undergoes a deformation close to PS compression but the outer sections deform in a mixture of compression and shear as they are forced to flow round the dead zone and into the die orifice. Even for a simple case of axisymmetric extrusion of a round bar (just involving a change of cross-section area) the off-centre particle paths can be quite complicated. Some finite element calculations of this case are given in Figure 11.21.

As for the case of hot rolling, there is also a significant increase in temperature – of the order of 20–50° – in the high deformation zone of the die during the extrusion process.

There has been a very strong effort over the last two decades to improve the technology of extrusion, essentially by improving die shapes to facilitate metal flow. Scientific studies of extrusion microstructures began with the classic work of McQueen *et al.* [1967] relating the subgrain sizes to the Zener–Hollomon parameter during extrusion of Al alloys.

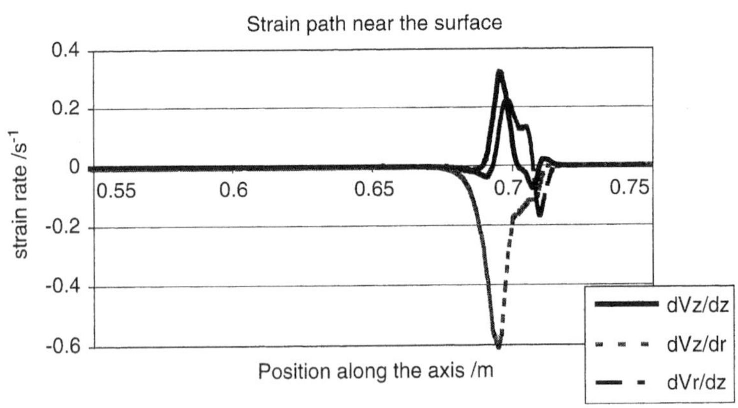

Figure 11.21. Velocity gradient components calculated for slow, axisymmetrical extrusion of a round bar of a hard 7xxx Al alloy at a point near the surface (90%). d*Vz*/d*z* is the (elongation) strain rate along the extrusion direction *z* and the other two represent the shear components. From Courtesy of D. Piot, Ecole des Mines de Saint-Etienne.

LITERATURE

Pearson C.E. and Parkins R.N. "The Extrusion of Metals" Chapman Hall, London (1961).

Storen S. and Moe P.T. 'Extrusion', in "Handbook of Metallurgical Process Design", G. Totten, K. Funatani and L. Xie (eds), Chapter 6, Marcel Dekker, New York (2004), pp. 137–204.

11.4. WIRE DRAWING

11.4.1 Introduction

In a conventional wire drawing process, the diameter of a rod or wire is reduced by pulling it through a conical die (Figure 11.22a). In industrial production lines, a large reduction is obtained by pulling the wire or rod through a series of consecutive dies. In some cases, an intermediate annealing treatment may be necessary. For most metals, drawing is carried out at ambient temperature, although the temperature inside the die can rise considerably due to heating associated with deformation and friction. Some materials (e.g. tungsten wire for incandescent lamp filaments) are drawn at high temperature.

Wire drawing is not limited to wires with a cylindrical shape; rectangular or more complex cross-sections can also be produced with appropriate dies. A possible alternative for the classical die, is the so-called 'Turk's head' (Figure 11.22b). This consists of four rolls, which can be adjusted to different positions for different products. Flat wires can easily be drawn, but the rolls can be shaped to draw other profiles.

Drawing can also be used to reduce the diameter of tubes (Figure 11.22c). The wall thickness can be reduced in the same operation, using a mandrel.

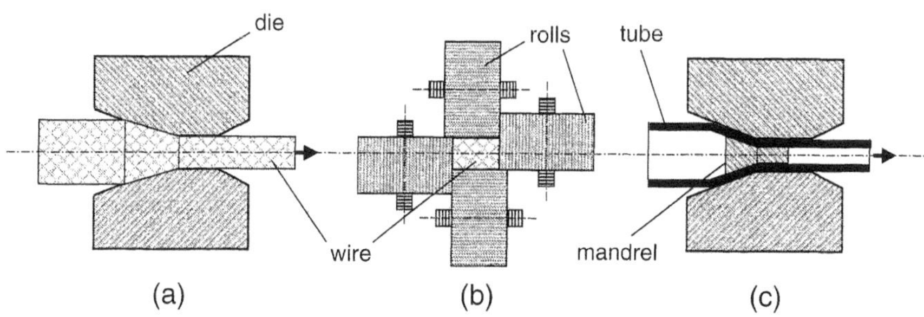

Figure 11.22. Illustration of some drawing operations: (a) conventional wire drawing with circular cross-section; (b) wire drawing with rectangular cross-section, using a so-called 'Turk's head; (c) tube drawing using a floating mandrel.

11.4.2 Wire drawing machines

Tubes and straight rods with diameters above 20 mm are usually drawn with draw benches (Figure 11.23). These machines contain a single die and the pulling force is provided by a drawing trolley. The drawing speed is relatively low (0.1–1 m/s), and the length of the tube or the rod is limited (typically less than 30 m). Longer rods and wires are drawn by a rotating drum, called a 'bull block' or 'capstan'. Many design variations are available. Single-die machines (Figure 11.23) are relatively simple and are used for breakdown or finishing operations. In most wire drawing plants, several die/capstan combinations are mounted in series to form a continuous wire drawing machine (Figure 11.24). Since the wire diameter D is reduced in each pass, the wire speed v increases after each die. In principle, the following continuity relation must be satisfied:

$$D_{n-1}^2 v_{n-1} = D_n^2 v_n = D_{n+1}^2 v_{n+1} \tag{11.24}$$

with $n-1$, n and $n+1$ successive drawing passes. In reality, because of wear in the drawing dies, the diameter of the wire after each pass is not constant in time.

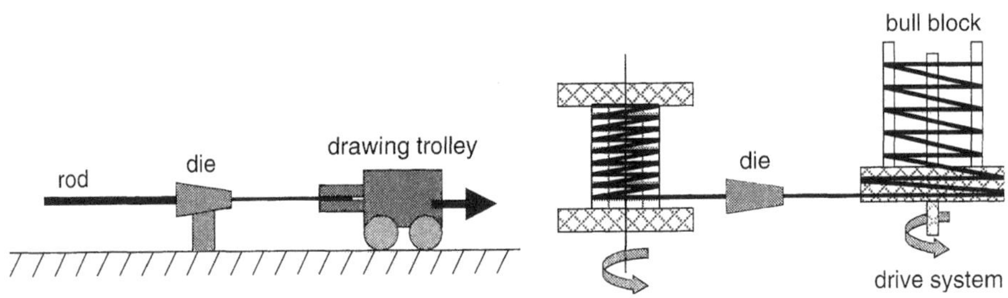

Figure 11.23. Draw bench (left) and single pass drawing equipment (right).

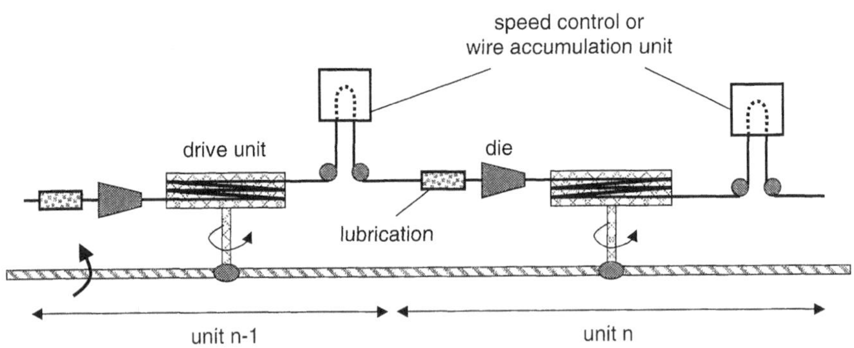

Figure 11.24. Continuous wire drawing machine of the "non-slip-type".

Suppose that the diameter after pass '*n*' (Figure 11.24) increases to a value $D'_n > D_n$. When the rotation speed of drum '*n*' is kept constant, more material per unit time leaves the die '*n*' and hence the speed v_{n-1} of the wire in unit '$n-1$' will increase. This would cause a so-called 'negative slip' on drum '$n-1$': the speed of the wire becomes higher than the circumferential speed of the drum and this generates an extra tension in the wire. This situation cannot be tolerated because of an increased risk of wire breaks. In some installations, each drum is powered by an independent dc-motor and the speed of each motor is adapted to the actual speed of the wire by a tension arm, located between the drums. In other machines, several drums are driven by one single motor and the rotation speed of drum '$n-1$' cannot be changed independently from the others. In that case an 'accumulation unit' is placed after each drum. These accumulation units can temporarily provide the required extra volume of wire. When unit '$n-1$' is exhausted, all following drums are disconnected from the central shaft and the accumulator '$n-1$' is replenished. In these 'non-slip-type' machines, a dry lubricant is used. Continuous wire drawing machines for fine wires are usually of the 'slip-type' and operate with a wet lubricant. The rotation speed of the drums is initially regulated in such a way that the circumferential speed is higher than the speed of the wire. This 'positive slip' only causes some wear on the drum and it can absorb an increase in wire speed. When the speed of the wire becomes equal to the speed of the drum, the die in the following unit must be replaced.

11.4.3 Wire drawing dies

11.4.3.1 Geometry. A schematic of a classical wire drawing die is shown in Figure 11.25. The core of the die is made from a wear-resistant material and fits into a steel frame. The entrance angle β permits the lubricant to enter the die and

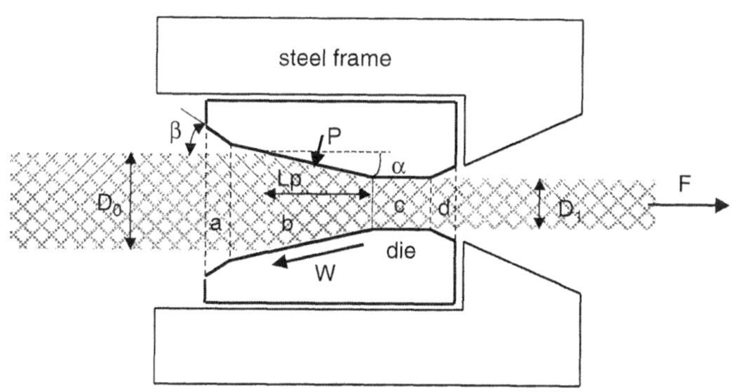

Figure 11.25. Geometry of a drawing die.

adhere to the wire. The deformation takes place in zone b. The deformation zone is characterized by two important parameters: the length Lp and the semi die angle α. In most dies, the latter takes values from 4° to 12°. A shorter die reduces the friction, but increases the redundant deformation[4] (see Section 11.4.4). In this zone, the drawing force F is balanced by the horizontal components of the friction force W and the compressive force $P = F \sin\alpha$. In zone c, called the bearing surface, the final sizing of the exit diameter is performed. Finally, zone d is the exit zone.

The reduction r and the true strain ε in wire drawing are defined as:

$$r = \frac{D_0^2 - D_1^2}{D_0^2} = 1 - \left(\frac{D_1}{D_0}\right)^2 = 1 - \left(\frac{1}{\exp(\varepsilon)}\right) \tag{11.25}$$

$$\varepsilon = \ln\left(\frac{D_0}{D_1}\right)^2 = \ln\left[\frac{1}{1-r}\right] \tag{11.26}$$

The semi die angle α and the reduction r can be combined in one parameter Δ, which will be used to express the degree of redundant deformation (Hosford and Caddell [1983]):

$$\Delta = \frac{D_g}{L_p} = \sin\alpha \frac{\left[1 + (1-r)^{0.5}\right]^2}{r} \tag{11.27}$$

with D_g the mean wire diameter in the deformation zone and α expressed in degrees.

11.4.3.2 Die materials. The inner core of a drawing die is fabricated from a wear-resistant material, usually cemented tungsten carbide or, for fine wires, from diamond. Hardened tool steel (HRC around 60) is only used for small series. Dies made from diamond are very expensive, but can outperform the cemented tungsten carbide dies by a factor of 10–200. When a die is worn out, it can be reworked and used again for thicker wires.

11.4.3.3 Lubricants. Friction between wire and die is not required in a drawing process. It causes wear of the die, possible damage to the wire surface and an increase in drawing force and temperature. Proper lubricants have to be used to minimize friction and in some cases to cool the wire.

[4] Internal distortion of the metal, not contributing to the dimensional change of the wire.

In dry drawing, the surface of the wire is coated with dry soap powder by passing the wire through a box filled with the lubricant and the wire is cooled, while it resides on the bull blocks or by cooling the die holder with water. High-strength materials are often coated with a softer material that acts as a lubricant, e.g. steel with brass.

In wet drawing, the die is completely immersed in oil or in an emulsion. In this case, the lubricant also serves as cooling medium. In general, the friction coefficient in wire drawing ranges from 0.01 to 0.1.

11.4.4 The drawing force

Although wire drawing seems to be a rather simple deformation, a precise calculation of the drawing force is not an easy task. In the simplest approach, factors such as friction, redundant deformation and work hardening are neglected. If the effective strain is considered to be a pure uniaxial elongation [$\varepsilon = \ln(D_0/D_1)^2$] and the material as ideal plastic (with constant flow stress σ_f), the drawing force F can be estimated from:

$$F = \sigma_f \ln\left(\frac{D_0}{D_1}\right)^2 \frac{\pi D_1^2}{4} \quad \text{or} \quad \sigma_F = \sigma_f \ln\left(\frac{D_0}{D_1}\right)^2 = \sigma_f \varepsilon \qquad (11.28)$$

This simple approximation predicts that for $\varepsilon > 1$ ($r > 63\%$) the tensile stress in the wire (σ_F) would exceed the flow stress of the material which would lead to fracture.

Several attempts have been done to incorporate the influence of friction and redundant deformation. Two of the more elegant formulas have been proposed by Siebel *et al.* [1947] and by Hoffman and Sachs [1953].

The formula of Siebel starts from Eq. (11.28) and simply adds a friction and a redundant deformation term:

$$\sigma_F = \sigma_f\left[\varepsilon + \left(\frac{\mu}{\alpha}\right)\varepsilon + \left(\frac{2}{3}\right)\alpha\right] \quad \text{with } \alpha \text{ in radials} \qquad (11.29)$$

The formula of Hoffman and Sachs adds a friction term, and a 'correction coefficient' Φ which, according to Hosford and Caddell [1983] can be expressed as a function of the Δ-factor (Formula 11.4):

$$\sigma_F = \Phi\sigma_f\varepsilon(1 + \mu \cot g\alpha) \quad \text{with } \Phi = \frac{\Delta}{6} + 1 \quad \text{and } \alpha \text{ in degrees} \qquad (11.30)$$

In most cases, the drawing force can be estimated with these formulas with an accuracy of about 20%. Several other approaches have been proposed in the literature. The reader is referred to overviews of Wistreich [1958], Wright [1976] and Avitzur [1983].

Both formulae illustrate the influence of the semi die angle α. A larger α (shorter die) decreases the friction, but increases the redundant work. An optimal value for α, minimizing the drawing force, can be calculated but in practice other criteria like die wear, cost and standardization may determine the choice of the angle α.

The redundant deformation is a function of α and the reduction, and hence of Δ. For a complete homogeneous deformation, Δ should in principle not exceed the value of one. But in many wire drawing operations Δ-values up to 2 or 3 are tolerated.

Equations (11.29) and (11.30) can be adapted to take into account the strain hardening. The constant value of the flow stress (σ_f) should be replaced by some appropriate hardening law, e.g. $\sigma = k\varepsilon^n$ or others (cf. Chapter 4).

In many wire drawing lines, the wire is not only subjected to a forward pulling force, but also to a 'back pull' or 'back tension', exerted by downstream capstans. The main effect of a back tension is to increase the drawing force and the risk of fracture, but with a decrease in die pressure and with possible reduction in friction (Wright [1999]).

Illustrative problem

A pearlitic wire should be drawn from 5 mm till 2.6 mm diameter, using a wire drawing machine with a maximal pulling force of 15.5 kN. The desired reduction scheme is: 5 mm → 4 mm → 3 mm → 2.6 mm. Three types of dies are available with $\alpha = 4°$, $\alpha = 8°$ and $\alpha = 16°$. The friction coefficient is always 0.05. Calculate for each pass and each die angle the pulling force. Choose for each pass the best (available) die and discuss your answer.

The pulling force is the product of the pulling stress and the wire cross-section after the die. The pulling stress can be estimated using Eqs. (11.29) or (11.30). Since pearlite shows a large work hardening during drawing (cf. next paragraph), an appropriate work hardening law should be used the estimate the actual material flow stress after each die. The results are shown in Table 11.5. In the first drawing pass, the die with a semi die angle of 4° leads to a pulling force that exceeds the capacity of the drawing machine. The dies with angles of 8° and 16° can both be used, but the die of 8° leads to a more homogeneous deformation (lower Δ-factor) and a slightly lower pulling force.

In the second pass, the pulling force is lower, but another problem occurs for the die of 4°: although the flow stress of the wire has considerably increased, the pulling stress in the wire is higher than the flow stress, which would lead to deformation and fracture of the wire after the second die. The two other dies can in principle be

Table 11.5. Calculation of the pulling force during wire drawing.

Equations	Limitations
(11.29): $\sigma_F = \sigma_f\,[\varepsilon + (\mu/\alpha)\varepsilon + (2/3)\alpha]$ with α in radials	Pulling force <15.5 kN
(11.30): $\sigma_F = \Phi\sigma_f\varepsilon(1+\mu \cot g\alpha)$ with $\Phi = \Delta/6+1$ and α in degrees	$\sigma_F < \sigma_f$
(11.27): $\Delta = \sin\alpha\,[1+(1-r)^{0.5}]^2/r$	$\Delta < 2$ to 3
Pearlite: $\sigma_f = 72 + 1304\exp(\varepsilon/4)$ (MPa)	

	5→4 mm			4→3 mm			3→2.6 mm		
α (°)	4	8	16	4	8	16	4	8	16
r (%)		36.0			43.8			24.9	
ε		0.446			0.575			0.286	
Δ	0.63	1.25	2.48	0.49	0.97	1.93	0.98	1.95	3.86
ϕ	1.10	1.21	1.41	1.08	1.16	1.32	1.16	1.32	1.64
σ_F (11.29) (MPa)	1243	1070	1090	1815	1535	1518	1011	906	985
σ_F (11.30) (MPa)	1294	1119	1133	1873	1592	1568	1073	967	1038
σ flow (MPa)		1529			1755			1880	
Pulling force (kN) (11.29)	15.6	13.4	13.7	12.8	10.8	10.7	5.4	4.8	5.2
Pulling force (kN) (11.30)	16.3	14.1	14.2	13.2	11.3	11.1	5.7	5.1	5.5

used, but the die of 16° could be preferred because it requires the lowest pulling stress, although it gives a somewhat less homogeneous deformation compared to the 8° die. In reality, most production lines operate with a pulling stress of maximum 60–70% of the flow stress of the material, so an adaptation of the proposed pass schedule is advisable. In the third pass, the pulling stresses and forces are lower, because of the smaller reduction, but the homogeneity of the deformation in the 16° die is questionable. In this step, a die of 4° seems to be the best choice.

11.4.5 Some important metallurgical factors

During wire drawing of fcc metals, classical strain hardening of the wire takes place. This hardening is related to the substructural developments, as discussed in Chapter 4. But during drawing of bcc materials, such as low-carbon steel, some unexpected hardening behaviour can be observed. After a parabolic transition, the stress increases linearly with strain (Langford and Cohen [1969]). Strange enough, this behaviour is not observed during a torsion test carried out on the same material (Gil Sevillano *et al.* [1980]).

During wire drawing, grains should elongate in the drawing direction and contract in all directions perpendicular to the wire axis. In reality this is observed in fcc materials, but in bcc materials, such as low-carbon and pearlitic steel, tungsten and niobium, a peculiar effect occurs (Peck and Thomas [1961], Hosford [1964]).

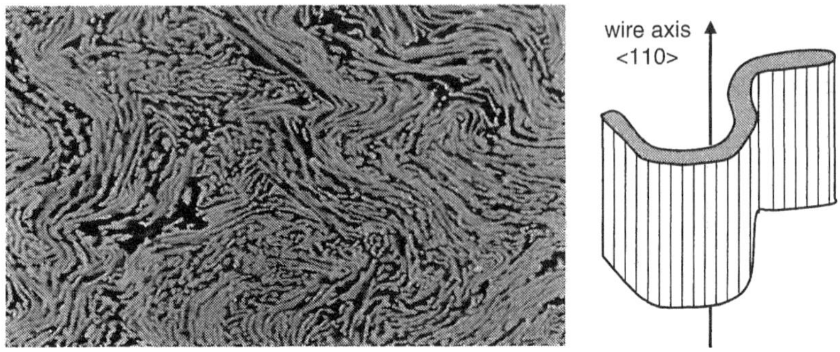

(a) (b)

Figure 11.26. Illustration of the curling effect in pearlitic steel (a) view parallel with the wire axis; (b) scheme of a cementite lamella after wire drawing.

The grains do not only elongate in the direction of the wire axis, but also get a kind of folding, visualized in Figure 11.26. This is called the *curling effect*. In order to fold the grains over each other, some extra dislocations have to be generated. These geometrically necessary dislocations (GNBs, see Chapter 4) all have the same sign and are not annihilated. As a result, the mean cell size never reaches an equilibrium as in the case of Cu or Al, but continuously decreases and the flow stress linearly increases. In a torsion test, this curling effect does not occur and hence the flow stress and cell size reach an equilibrium after a certain strain.

The curling effect is a direct consequence of the crystallographic texture of bcc metals after wire drawing. In cold-drawn wires of bcc materials, many grains have an <110> direction more or less parallel with the wire axis. This is called 'a <110> fibre texture' (cf. Chapter 8). To understand the curling effect, let us first consider a single crystal with a [011] direction parallel with the wire axis (Figure 11.27). The dominant slip directions are [111] and [−111] and also [−1−11] and [1–11]. Only the first two contribute to an elongation in the z-direction. They also provoke a contraction along the [100] axis but no length change along [0–11] (plane strain). In order to realize an axisymmetric deformation (as imposed by the drawing die), the [−1−11] and [1–11] slip directions should also be activated and generate a contraction along [0–11] and an elongation along [100]. It can however be shown (Hosford [1964]) that a plane strain deformation requires less stress than an axisymmetric deformation.

This means that in the real wire with a strong <110> texture, every grain will show a strong tendency to undergo a plane strain deformation, with the contraction axis perpendicular to the wire axis. Of course, in this way, the cohesion between the grains would be lost; and moreover, the die forces a global (macroscopic)

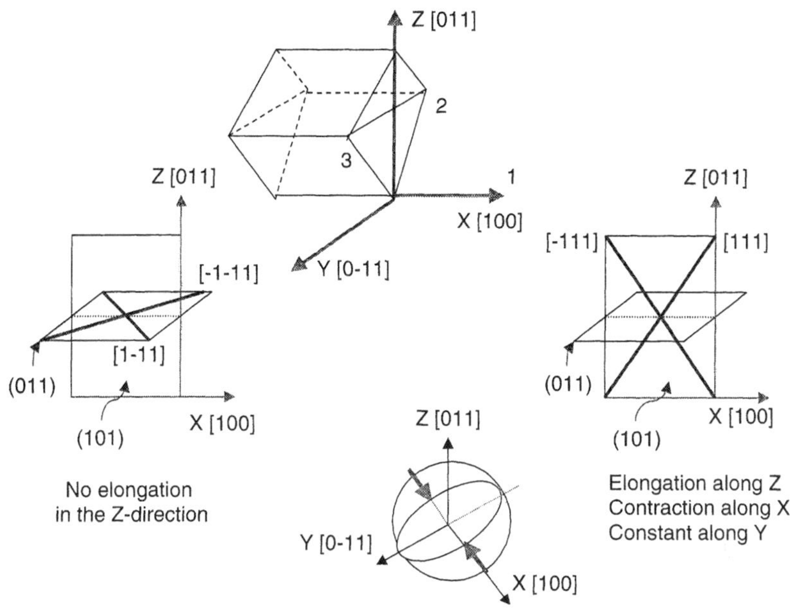

Figure 11.27. Illustration of the <111> slip directions in a bcc lattice with the [011] direction parallel to the wire axis.

axisymmetric deformation. To comply with that, the grains have to curl over each other as seen in Figure 11.26.

In pearlitic steel, but also in other lamellar or fibrous materials like Fe–Al (Wahl and Wassermann [1970]), an exponentially increasing flow curve is recorded during wire drawing. During drawing of pearlitic steel, the ferrite and cementite lamellae are gradually re-oriented and get more and more parallel with the wire axis (a 'morphological texture'). Owing to the reduction of the wire diameter, the cementite lamellae are pushed together and the distance between them diminishes (Chapter 15.5). In fine pearlite, the spacing between two lamellae is smaller than the 'natural' size of the cells in the substructure. With further deformation, the mean cell size '*d*' will artificially be reduced with increased deformation, because the spacing between the lamellae is reduced. To a first approximation, it can be assumed that the reduction in cell size is proportional to the macroscopic reduction in diameter '*D*':

$$\varepsilon = \ln\left(\frac{D_0^2}{D^2}\right) = \ln\left(\frac{d_0^2}{d^2}\right) \tag{11.31}$$

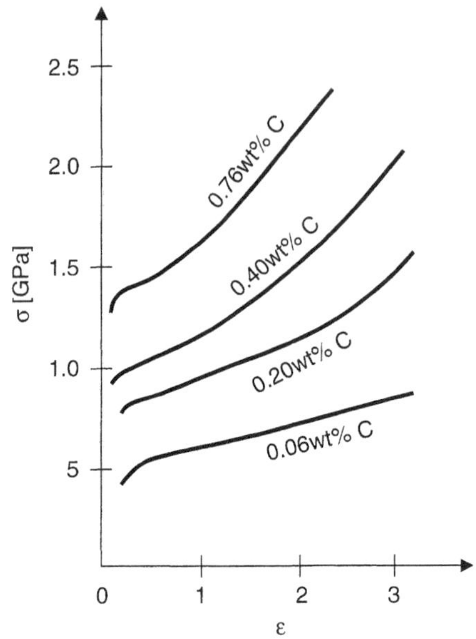

Figure 11.28. Stress strain curves during wire drawing of steel. After Gil Sevillano *et al.* [1980].

Substitution in the Hall–Petch relation $\sigma = \sigma_0 + kd^{-0.5}$ gives:

$$\sigma = \sigma_0 + k' \exp\left(\frac{\varepsilon}{4}\right) \tag{11.32}$$

This expression is in good agreement with an empirical relation (Gil Sevillano [1974]):

$$\sigma = 72 + 1304 \exp\left(\frac{\varepsilon}{4}\right) (\text{MPa}) \tag{11.33}$$

Figure 11.28 shows some stress–strain curves for steel with different carbon contents during wire drawing. The upper curve shows the exponential hardening obtained during wire drawing of lamellar pearlite. The lower curve illustrates the linear strengthening of low-carbon steel. The other curves can be described by a rule of mixture.

11.4.6 Drawing of metal fibres

Ultrafine wires ($<50\,\mu$m) and fibres have interesting applications such as filters, anti-static textiles, magnetic shielding, medical products, bonding wires in microelectronics, etc. Metal fibres can be produced in many ways such as conventional drawing, bundle drawing, shaving and melt spinning.

Most ductile metals can be drawn into ultrafine wire by pulling them through thinner and thinner dies. In most cases, drawing conditions have to be selected with care: low drawing speeds, low per pass reductions and diamond dies with low approach angles are recommended. Unfortunately, productions costs are very high and increase exponentially as the diameter decreases.

An alternative technique to produce fine fibres is the so-called 'bundle drawing'. Instead of drawing a single wire, several (in some cases up to a few thousand) are bundled, tightly packed into a tube and drawn simultaneously. The challenge is to separate the individual wires with a suitable material prior to bundling. This separating material must be easily removable after drawing, e.g. by leaching. Stainless steel fibres can be drawn to a diameter of 1 μm with this method. Most metals can be drawn up to 8–12 μm in diameter and even brittle superconductor alloys like NbTi and Nb_3Sn are processed in this way (ASM Handbook [1988]).

Another alternative is 'coil shaving'. The material is first rolled into foil, which is coiled. A sort of big razor blade cuts thin slices of material from the side of the coil. A fibre with rectangular cross-section, one dimension being the thickness of the foil (typically 25–100 μm) is obtained.

Ribbons of glassy metals, typically 20–60 μm thick, can be obtained by melt spinning. A jet of molten metal is poured onto a cool, fast rotating wheel. The solidification is very fast and amorphous or semi-crystalline materials can be obtained.

LITERATURE

Altan T., Oh S. and Gegel H.L., "Metal Forming: Fundamentals and Applications", ASM, Ohio (1983).

11.5. FORGING

11.5.1 Introduction

Forging is the oldest of the metal forming processes. Archaeological evidence of forging or simple hammering of native gold, silver and copper dates back to 8000–5000 BC (Singer *et al.* [1954], ASM [1988]). The succeeding Bronze Age reduced, albeit temporarily, the dependence on forging; as cast bronze tools and weapons dominated. The beginning of the iron age around 800 BC, however, brought the importance of forging to new heights (Schey [1970]). The blacksmith's trade became an art of strategic importance. The subject evolved slowly over centuries – cold forging of roman coins (Singer *et al.* [1954]), forging of

Damascus sword (Naujoks and Fabel [1953]), introduction of tilt hammers around 13th century (ASM [1988]) to interchangeable forging dies for rifle parts in 18th century (Schey [1970]). Today, the technology of forging has come a long way – from being a blacksmith's art to the relative maturity of being called a technology. In the subsequent sub-sections, different aspects of today's forging technology are discussed.

Table 11.6 provides a broad classification of the forging processes for bulk metallic materials. A more detailed classification can be obtained in forging handbooks (Altan *et al.* [1973], Metalforming [1982], ASM [1988]). Such detailed classifications are, however, based on minor modifications in process/equipment/die-tool/applicability. It is also to be noted that the forging of finished components often involves several steps of forming or TMP (Metalforming [1982]) as shown in Figure 11.29.

11.5.2 Forging equipment

As shown in Table 11.7, the basic forging equipment can be classified as hammers and presses. The table provides only a 'snap shot' of forging equipment, leaving

Table 11.6. Broad classification of forging processes for bulk metallic materials. The specialized forging techniques like radial forging, isothermal forging, etc. (ASM [1988]) are covered in the broad categories. Techniques like roll forging fall in the preview of both forging and rolling. The processes may involve cold or hot deformation (in special cases with isothermal condition) with intricacies in die and equipment design. The brief schematics of the processes are given in Figure 11.30.

Forging process	Brief description	Applicability
Open-die forging	Flat or curved dies, respectively attached to hammer or press and anvil	Large jobs, limited quantity, difficult to forge material
Closed-die forging	The metal is shaped between the cavities of the two dies – the finished product carries the 'impression' of die shape or cavity. Excess material is released through a 'flash'	More productive than open die forging and capable of making complicated shapes Forgability of the material and die design are important
Rotary swaging	Reduction in cross-sectional area by repeated radial blows from two or more dies	Tube manufacturing
Rotary forging	Two die process, deformation being localized to a small part of the workpiece in a continuous manner	Substitute for conventional forging, typically applied to symmetric parts

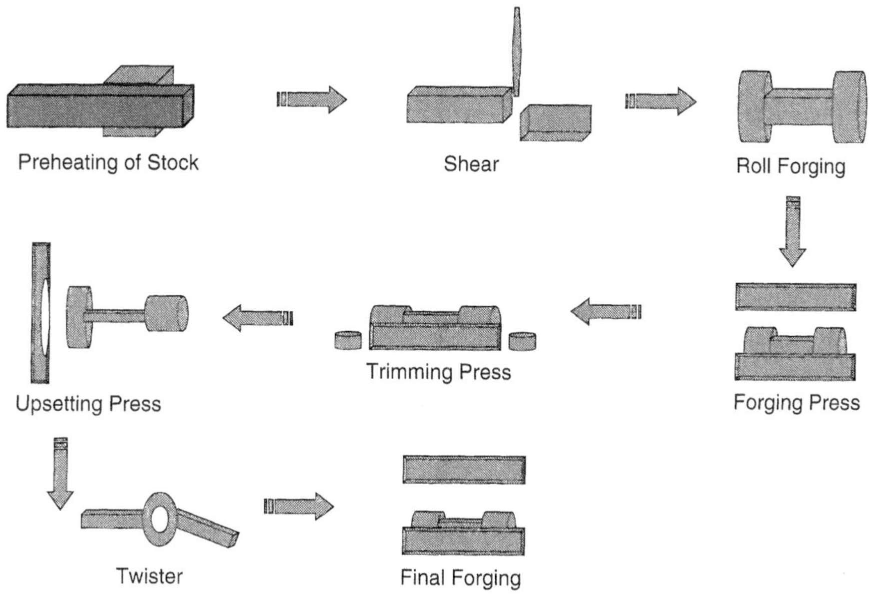

Figure 11.29. Schematic of the different stages involved in the forging of a crankshaft – intermediate annealing is common. This is a simplified picture of the actual process.

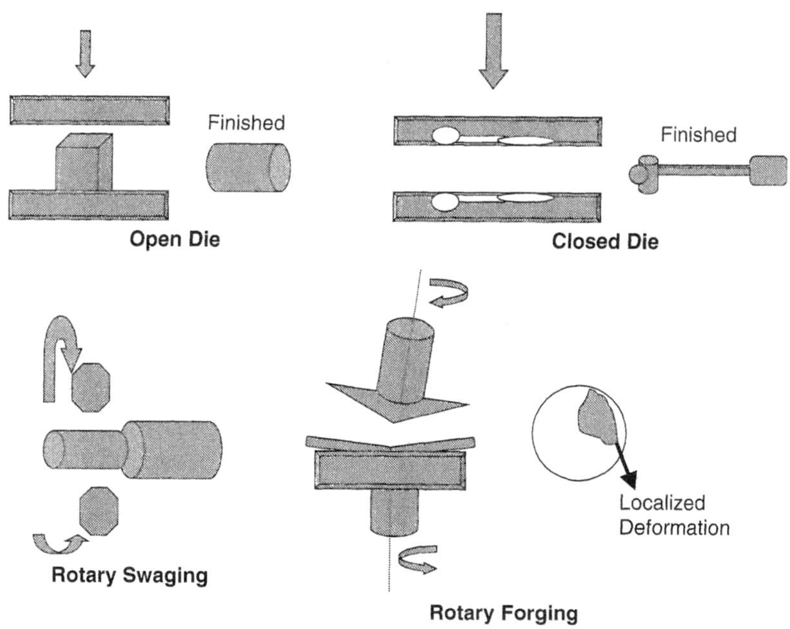

Figure 11.30. Schematics of the forging processes, as in Table 11.6.

Table 11.7. Classification of forging equipment.

Forging equipment		Description	Remarks
Hammers	Gravity drop	The lifted hammer is allowed to fall under gravity. Lift mechanisms may differ, but the energy of the blow is largely restricted by the hammer weight	'Energy restricted' Typically, available range of ram weight, blow energy, impact speed and number of blows per unit time are *lower* in gravity drop
	Power drop	The hammer blow is accelerated by air/steam/hydaulic pressure	
Presses	Mechanical	Rotational motion of eccentric shaft is translated to linear motion by appropriate mechanism(s)	'Stroke restricted' Better than hammers in all aspects except for cost and the fact that the energy cannot be varied – and hence are not suited for preliminary forging operations
	Screw	Rotational motion of flywheel is translated to linear motion by threaded screw	'Energy restricted' More energy per stroke than mechanical presses, but normally the full force is restricted near the centre of the workpiece
	Hydraulic	Direct or accumulator driven	'Force restricted' Other than high cost, these are typically of slower action. But there are serious advantages in terms of control/optimization of operation and die/tool design

aside detailed information on design, performance and applicability (Naujoks and Fabel [1953], Altan *et al.* [1973], Metalforming [1982], ASM [1988]). For example, high energy rate forging equipments are high-speed hammers with interesting design innovations, while painstaking design aspects are involved in forging presses (Altan *et al.* [1973], ASM [1988], Ishinaga [1997]).

11.5.3 Forging dies

Dies used in open die forging are restricted to flat dies, swage dies and V dies – the latter two respectively have a semi-circular and V shape cut at the centre of the dies (both upper and lower) to avoid bulging (and for faster operations with easier material flow, though die changes are essential for different sizes). Change-over from open die to closed die forging is required based on die-forging dimensions.

Thumb rules are often cited (Metalforming [1982], Ghosh and Malik [1985]) as first step towards forging die design:

- $L < 3d$ *for open die forging*
- $D < 1.5d$ and $1 < d$ *for closed die forging*

L and *l* and *D* and *d*, for open and closed die forging, are shown in Figure 11.31. It is to be noted that *L* and *l* have similar implications – respectively representing unsupported length in open and in closed die forging. Actual die design involves considerations of many more parameters, especially valid for closed die forging (dies may have single or multiple cavities and die design often calls for multistage dies), to meet both production and materials aspects. For example, closed die design is expected to provide sufficient friction so that enough pressure is built up to fill all the cavities. The gutter and flash, are often an intergral part of such design – excess material leaves the die through the gutter and gets 'collected' in the flash (see Figure 11.31b). Table 11.8 outlines some of the standard terminologies associated with die design of closed die forging; for more details, the reader may refer standard forging handbooks (Altan *et al.* [1973], Metalforming [1982], ASM [1988], Lapovok [1998], Doege and Bohnsack [2000], Groenbaek and Birker [2000]).

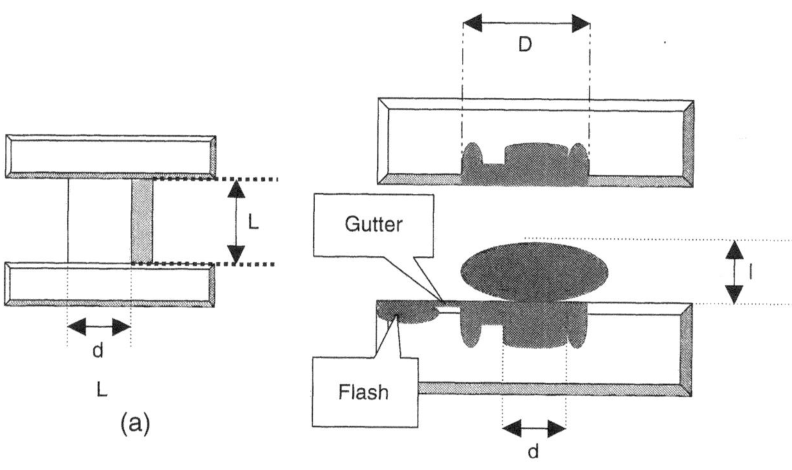

Figure 11.31. Schematics of (a) open and (b) closed die forging. Relative values of *L* & *l* and *D* & *d* provide the basic outline for die design. Closed die forging also needs considerations of several other aspects – e.g. gutter-flash, parting line, draft, etc.

Table 11.8. Standard terminologies associated with closed die forging dies and die design.

Nomenclature	Brief description
Fullers, edgers, rollers, flatteners,benders, splitters, blockers, finishers	Respectively to reduce cross-section area, to redistribute for heavier sections, to round the stock, to flatten, to bend, to make fork-type forgings, to prepare for finishers and to provide final overall shapes
Parting/flash line	Plane dividing upper and lower die
Flash land and gutter	Excess material flows through gutter to flash land
Webs and ribs	Thin sections parallel and perpendicular to parting line
Fillet and corner radii	May limit metal flow and increase stresses on the die surface
Draft	Draft or tape is used for easy removal of the forgings
Die insert	Used for economy and dimensional tolerance – plug and full types
Locks	For non-flat parting line, locations for dies to mesh/lock
Mismatch	Mismatch between dies can be a serious problem, but optimum die mismatch can also used to produce non-symmetrical parts (e.g. crank shafts)

11.5.4 Friction and lubrication in forging

Even in the simplest form of forging, pressure–temperature–velocity combinations change continuously – making formulation of any generalized velocity–stress–strain–friction distributions rather difficult. In open die upsetting, one of the simplest forms of forging, friction may change from sliding to sticking (or a combinations of the two), depending on the forging parameters and material aspects (Schey [1970]). For closed die forging, the frictional behaviour is more complicated and analytical solutions may exist only for the simplest geometries (Schey [1970], Boisse *et al.* [2003]). Though the exact historical time frame for lubricant use in forging cannot be pin-pointed (Schey [1970]); the use of appropriate lubricants – as in Table 11.9 – for frictional and other considerations – as in Table 11.10 – is common in today's forging technology.

11.5.5 Forging optimization

Forging force can be calculated as σAC, where σ, A and C are respectively stress, cross-sectional and correction factor. C typically varies betweem 4 and 9 for closed die forging, dependent on the complexity of the workpiece. It is also difficult to find a value for σ, as strain, strain rate, strain path and temperature distribution can be very inhomogeneous and only an approximate 'mean' value can be used or a point-to-point bookkeeping must be done – which, in turn, requires the use of FEM.

Table 11.9. Lubricants used in cold and hot Forging (Schey [1970], Möller [1996]).

Cold	Mineral oils, animal oils, vegetable fats, soaps, waxes, graphite in water/oil, etc.
Hot	For high-temperature forging graphite in a suitable medium is recommended (Möller [1996]). Below 400°C, molybdenum-di-sulphide has better frictional response. Also used are sawdust, mica, grease, asbestos, glasses, organic polymers, phosphates, oxides, bromides, fluorides, carbonates, etc. (Schey [1970])

Table 11.10. Factors, and their implications, determining the use of lubricants in forging.

Factors	Implications
Frictional issues	To reduce sliding friction between die and forging, to reduce the forging load and provide a better die filling (however, high friction in flash gutter is needed for die filling). Lubricants act as a parting agent by reducing sticking friction and local welding and providing better die life and good surface quality of the product
Insulating properties	To reduce heat loss from workpiece and to minimize temperature fluctuations on the die surface – better forgability
Balanced gas pressure	Primarily used for quick release of forging from die cavity, but can also be used to 'calibrate' lubricant wettablity.
Surface wettability	Uniform surface wettability – lack of lubricant may lead to sticking, while excess of lubricant may cause partial die filling
Non-abrasive	To prevent die surface wear
Residues	May accumulate and lead to forging defects and environmental issues

Typically the optimization of forging parameters, including estimation of forging loads, are obtained by:

- *Analytical solutions*: The complexities of the forging processes may vary enormously – from the simple open die upsettings to the complex precision forgings offering near net shapes (ASM [1988]). Straight analytical approach to forging optimizations are possible only for the *simplest* of the cases. For example, the maximum blow force of a open die forging hammer (p_{MAX}) can be approximated by Li *et al.* [1997] as:

$$p_{MAX} \Rightarrow \sqrt{\frac{C_1 M_1 (M_1 + M_2)}{(M_2 + 2M_1)} (v_1^2 - v_x^2)} \qquad (11.34)$$

where M_1 and M_2 are the respective masses of ram and anvil, v_1 and v_x are the blow velocity (ram velocity at the moment of 'contact') of the ram / hammer and the downward velocity of ram and anvil, respectively and C_1 is stiffness-related (given as $2K_0rE/(1-\nu^2)$; r = maximum radius of the workpiece, E and ν are the respective elastic modulus and Posson's ratio for the die material and K_0 is a constant).

- *Constitutive equations*: Use of constitutive equations in forging optimization is common (ASM [1988]), but these are typically applicable for simpler forgings.
- *Physical modelling*: Easily deformable materials like plasticine and wax (and even aluminium (ASM [1988])) are used (Vazquez and Altan [2000]) to 'approximate' material flow, friction conditions, etc. and are often used as an effective optimization tool. Such simulations are easy and inexpensive. In the case of aluminium, grids can be made to study the deformation behaviour quite effectively).
- *Computer modelling*: Use of computer modelling (both finite element and finite volume based) in forging optimization is a reality of today's technology – a list of developers of such commercial optimization/simulation codes is given in Table 11.11 (more details on semi-commercial and academic packages can be obtained elsewhere (ASM [1988], Bramley and Mynors [2000])). Such programs, though not magic boxes to solve all forging related problems (Like any other computer program, forging optimizations depend totally on the input data, material database and the physics (empirical/constitutive/analytical relationships) behind the codes.), are now used routinely.
- *Deformation modelling*: Actual forging parameters can be simulated in a deformation or forging simulator – a topic discussed in more detail in Chapter 13. The material behaviour can be optimized through forgability tests or deformation maps (the basics of deformation maps are covered in Chapter 6).

Table 11.11. List of major developers for forging simulation/optimization packages.

Primary code	Developer
Caps – Finel	CPM Gesellschaft für Computeranwendung Prozeß- und Materialtechnik mbH, Kaiserstr.100, D-52134 Herzogenrath; http://www.schraubenverband.de
Deform	Scientific forming technologies corporation, 5038 Reed Road, Columbus, Ohio 46220-2514, USA; http://www.deform.com
Forge 2/3	Transvalor S.A., Parc de Haute Technologie, Sophia Antipolis 694, av. du Dr. Maurice Donat 06255 Mougins Cedex, France; http://www.transvalor.com
MARC Superform	MSC.Software Corporation, 2 MacArthur Place, Santa Ana, CA 92707, USA; http://www.marc.com
Qform	Quantor Ltd., PO Box 39, 117049 Mocow, Russia; http://www.quantor.com

11.5.6 Forgability

Figure 11.32 shows the flow lines typical of forgings. The ease of development of such flow lines, representing plastic flow during forging, naturally depends on material and forging parameters. Together they determine the forgability of a material. The main forging parameters are the stress–strain relations, changes in strain path, temperature, strain rate, friction and other constraints. To evaluate the material under specific deformation parameters, tests are conducted which may involve simple forgability tests or more complex tests and elaborate deformation maps.

General forgability of different metals and their alloys is classed in Figure 11.33a. For closed die forging, however, the ease of die filling depends on a combination of

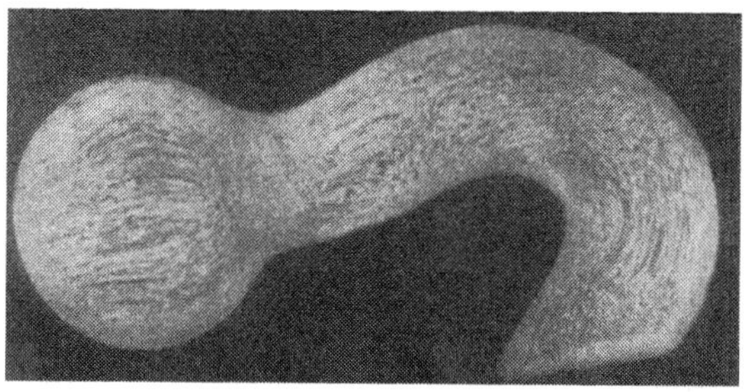

Figure 11.32. Macroetched structure of a hot forged hook – etchant hot aqueous 50% HCl. Courtesy of ASM [1988] (copyright (2007) with permission from ASM).

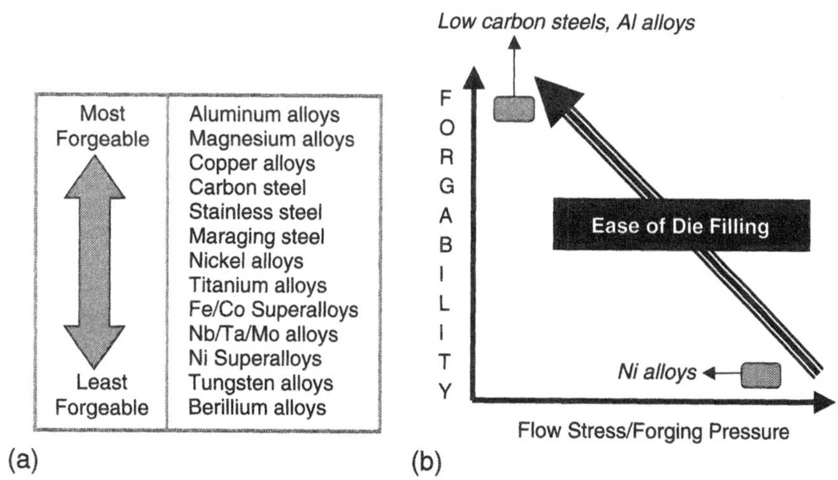

Figure 11.33. (a) Relative forgability for different metals and alloys. This information can be directly used for open die forgings. (b) Ease of die filling as a function of relative forgability and flow stress/forging pressure – applicable to closed die forging.

inherent forgability and flow stress/forging pressure (see Figure 11.33b). Table 11.12 and Figure 11.34 summarize the common forgability tests. After the tests, specimen observations (on cracking, metallography, etc.) are typically used to establish forgability. More complex tests are also designed to study flow localizations and complex forging practices.

Table 11.12. Common forgability tests. These and more complex tests are typically used to establish crack initiation through flow localization, forgability at a given forging condition and also to study structural changes. Schematics of the forgability tests/specimen geometries are given in Figure 11.34.

Forgability test	Brief description
Upsetting	Similar to a compression test. Simplest, but non-isothermal upsetting can be effective to characterize flow localizations through die chilling
Wedge forming	Wedge-shaped 'representative' sample tested between parallel/flat dies – to establish forgability at different reductions and forging conditions
Side pressing	Compressing a cylindrical bar perpendicular to the axis of the cylinder; normally performed between flat/parallel dies for un-constrained deformation – but minor modifications in die design can be used to test constrained deformation. Effective to study surface related cracking
Notched bar	Bars with notches perpendicular to loading direction – notches are used to initiate cracking through stress concentration (approaching actual forging practices). Effective in differentiating forgabilities, which otherwise may indicate marginal differences
Truncated cone	Indentation of a cylindrical specimen by a conical tool – effective in studying initiation of surface/sub-surface cracking and to optimize cold forgability against surface flaws

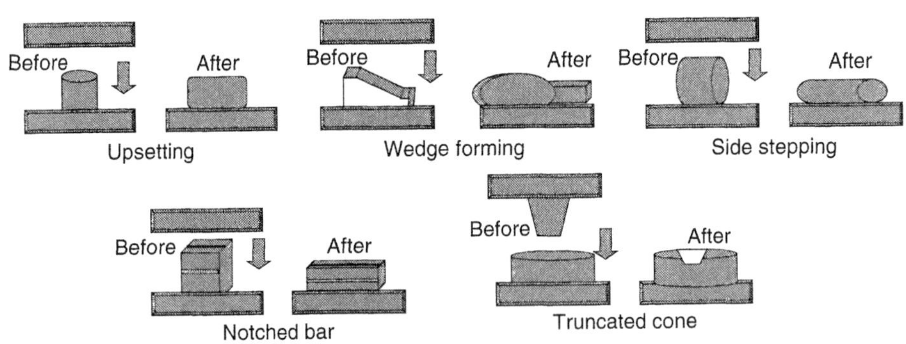

Figure 11.34. Schematic of the specimens in common forgability tests. The specimen dimensions are often standardized based on industrial standards and also on the structural heterogeneity present before the forgability tests. More details on the tests are given in Table 11.12.

LITERATURE

Altan T., Boulger F.W., Becker J.R., Akgerman N. and Henning H.J., "Forging Equipment, Materials and Practices", Metals and Ceramics Information Centre – Department of Defence Information Analysis Centre, USA (1973).

Metals Handbook "Forming and Forging", 9th Edition vol. 14, ASM, Ohio, USA (1988).

Schey J.S., "Metal Deformation Processes: Friction and Lubrication", Mercel Dekker Inc., New York (1970).

11.6. PILGERING

11.6.1 Introduction

Tubes and pipes do find wide ranging applications in the present technological society (Chapters 15.3. and 16.1. – pipe-line steel and Zr clads in nuclear reactors). Other than the welded tubes (any comparison of seamless vs. welded tube standards (e.g. DIN 2448 vs. DIN 2558) would clearly show that welded tubes are preferred for small wall thicknesses and large outside diameters), seamless tubes can also be 'formed'. Seamless 'tube-forming' typically avail several technologies – pierce and pilger rolling, plug rolling, tube extrusion, pierce and draw, assel rolling, etc. In addition to these primary processes, 'forming' the final tube may also require, both from dimensional tolerances and appropriate structure–property correlation, down-stream processes like cold drawing and cold pilgering. Intermediate operations like de-scaling (before and after hot forming, typically using water jets), pickling (before cold working), calibration (to provide a small taper before the piercing operation), elongation (1 to 2 times increase in original length along with correction for concentricicty), reeling (to loosen the hollows from the mandrel by using hyperbolically profiled rollers), etc. are also used as and if required.

This chapter does not attempt to overview all the processes of seamless tube manufacturing, but rather *concentrates on the topic of pilgering*. It all started

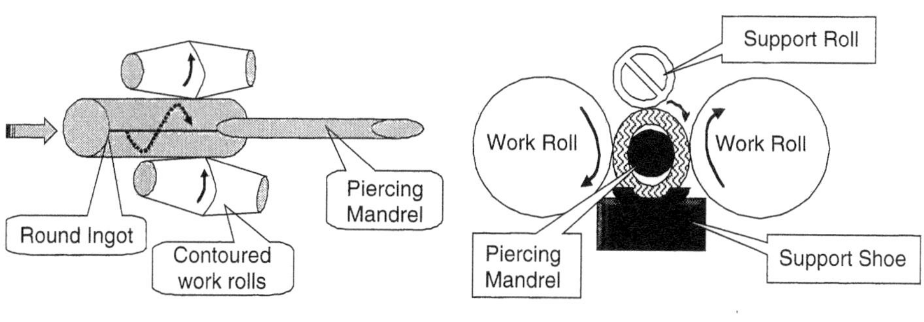

Figure 11.35. Schematics of pierce rolling at two cross-sections.

(*Source*: http://www.mannesmann-archiv.de/englisch/faq) with innovations by the Mannesmann brothers at the end of 19th century. 'Pierce rolling', was patented in 1886 (see Figure 11.35). A pierce rolling mill consists of two contoured (i.e. two halves of the rolls are at shallow angles – as shown in Figure 11.35) work rolls, driven at the same direction. These work rolls are typically placed at an angle of 3–6° around the hot billet. The roll gap is closed respectively from top and bottom by a support roll (non-driven) and support shoe. At the centre of the roll gap, the piercing point is held by the mandrel via an external thrust block. Typically, the hot billet is thrust and bitten by the tapered inlet section of the rolls. The helical movement of the rolls leads to a certain degree of 'locking' of the billet, while spiral motion of the billet over the piercing mandrel produces the thick-walled hollow shell – mother tubes or hollows.

For further processing of the mother tubes or hollows, Max Mannesmann subsequently (early 1890) developed the pilgering or pilger rolling process. In this process, discussed in more details later, a pair of grooved rolls and a moving plug or mandrel are used to reduce both the wall thickness and the diameter of the mother tubes. The repeated back and forth movement of the tube, during the rolling passes, was related to the dance (three steps forward and one back) of the 'pilger' (German for pilgrims) during Echternach dancing procession. The process was thus named as 'pilger rolling' or 'pilgering'. The Mannersmann's pilgering was a hot working process. Subsequent developments in North America by Neubert during late 1920 (Randall and Prieur [1967]), formed much of the basis of the present cold pilgering process. The reasons that pilgering is often employed over more conventional cold drawing are summarized in Table 11.13. As described in the table, typical

Table 11.13. Comparison between cold drawing and cold pilgering.

The issue	Cold drawing	Cold pilgering
The process	Hot-formed hollows are pointed, pickled and surface treated for cold drawing. The process involves drawing the hollows through reducing dies, usually supported by a plug or mandrel	Hot-formed hollows are pickled and then cold pilgered. The process involves repeated rolling through grooved conical shaped rolls and over a moving mandrel
After drawing/pilgering, finished tubes are subjected to cutting, degreasing, heat treatment (if required) and straightening operations		
The product	Close-dimensional tolerances are possible, but maximum reductions, reduction in wall thickness, are often limited	Close-dimensional tolerances, very high reductions and reductions in both wall thickness and tube diameter are possible. Superior surface finish and better metal lurgical control are possible

advantages involve reduced processing stages, superior product quality and excellent formablity (i.e. high reductions are possible without intermediate annealing).

11.6.2 Pilgering equipment and process

Figure 11.36a provides a schematic of the basic equipment of cold pilgering. A pilger stand has typically two rolls (the rolls are referred sometimes as dies – a refernce linked to the forging action associated with the pilgering process) with a tapered

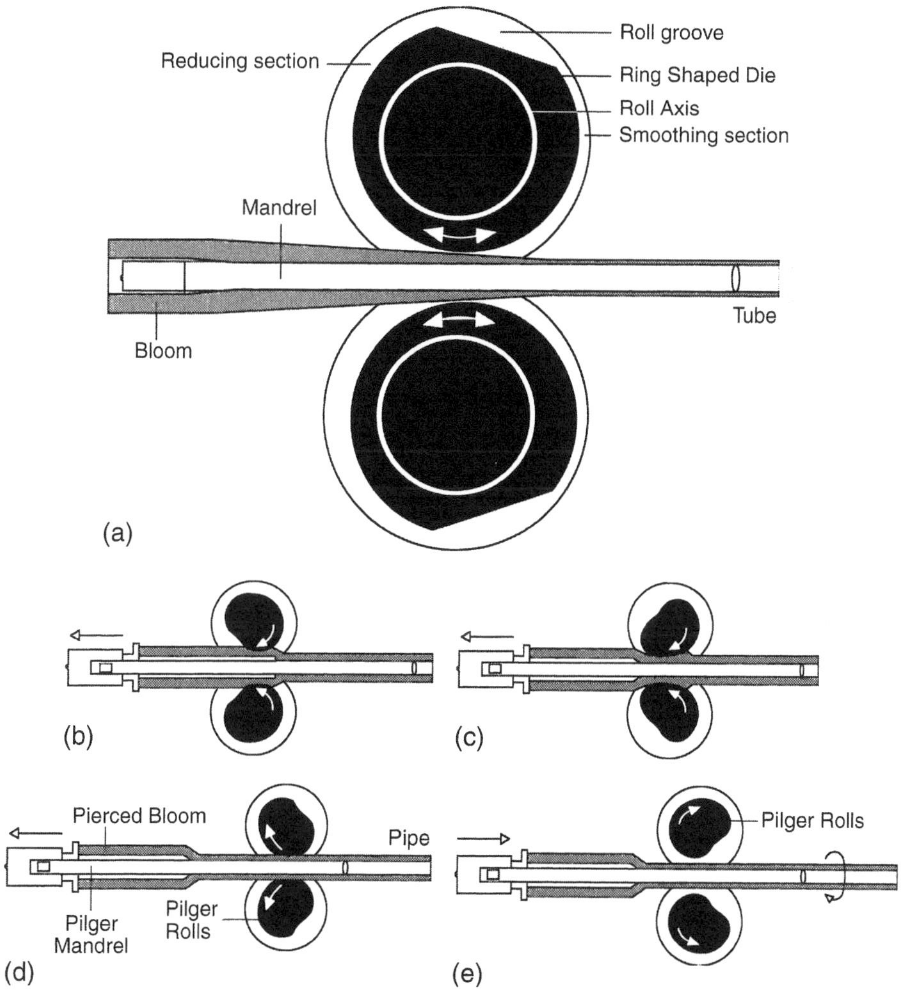

Figure 11.36. (a) Schematic of pilgering equipment. (b)–(e) different stages of pilgering. (b) start of rolling or the 'bite', (c) forging or pilgering, (d) polishing, (e) advancing or feed. After http://mannesmann-dmv.de/en/products-services/production/cold-pilgering-and-cold-drawing/

groove around their circumferences. Mother hollow or tubes are rolled *repeatedly* over an axisymmetric mandrel. Reduction of tube diameter can be obtained through a tapered mandrel, while grooved rolls enforce reduction in wall thickness. The basic pilgering operation is explained schematically in Figures 11.36b–e. Mother hollow or tube is first pushed over the lubricated mandrel and then:

- *Figure 11.36b*: Start of rolling – hollow-mandrel assembly is 'bitten' by the grooved rolls.
- *Figure 11.36c*: Forging or pilgering – the grooved rolls forge out a small wave of material to the desired wall thickness.
- *Figure 11.36d*: Polishing – the soothing section of the gooved rolls, see Figure 11.36a, reels or polishes the forged wall.
- *Figure 11.36e*: Advancing or feed – roll and mandrel movements are reversed and a fresh section of the mother hollow is 'bitten'.

After each stroke or pass, Figures 11.36b–d, the tube is rotated by 30–90°, advancing or feed – Figure 11.36e – and a new stroke, albeit in the reverse direction, starts. Appropriate advancing or feed is critical for maintaining concentricity. The synchronization of the mandrel stroke (typically 80–100 strokes are used in a cold pilgering operation) and the reversing roll movements are maintained by suitable mechanical arrangements. The usual pilgering mills are two-high single stand, though use of 4-high (or even higher) multi-stand mills do exist for specific applications.

11.6.2.1 Roll/die design. The critical aspect of roll/die design is the design of the groove (Randall and Prieur [1967], Roberts [1983], Montmitonnet *et al.* [1992]). As shown in Figure 11.37, the grooved rolls typically have 3 zones – forging/rolling (AB), polishing (BC) and idling (CA) (Roberts [1983]). Rolls bite the mother tube at the beginning of forging/rolling zone, point A of Figure 11.37, roll rotation and mandrel movement pushing the tube forward. The forging/rolling zone is followed by a polishing zone. The generalized strain patterns in the forging/pilgering and polishing zones are given in Table 11.14. After reaching the end of the polishing zone (point C in Figure 11.37) or reaching the end of a stroke, a suitable mechanical design (discussed latter in Section 11.6.2.2) reverses the motion of the rolls/mandrel bar and at the same time rotating the tube assembly. The advancing or feed also involves roll movement, albeit without any plastic deformation (see Figure 11.36e). This involves the idling zone – CA in Figure 11.37.

A successful pilgering stroke requires continuous decrease in groove diameter along the length of the contact area, AB and BC in Figure 11.37, while an increase in groove width is also necessary. The latter is called 'side relief' (the primary

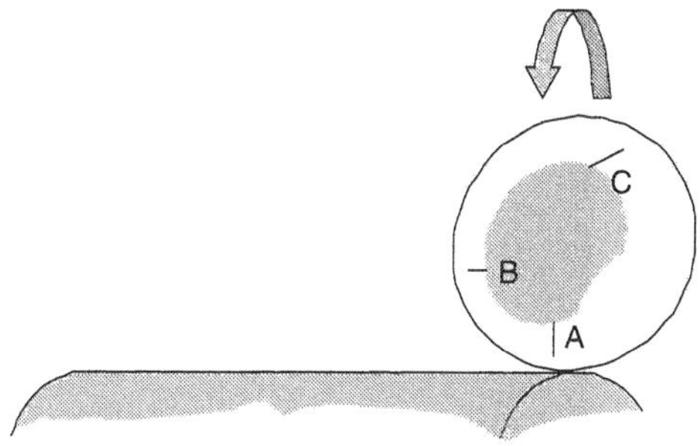

Figure 11.37. Schematic of a grooved roll. AB, BC and CA, respectively represent the die zones used for rolling/forging, for polishing and for idling.

Table 11.14. Generalized strain patterns during forging/pilgering and polishing.

Forging/pilgering	Polishing
$\varepsilon_r < 0$, $\varepsilon_\theta < 0$ and $\varepsilon_z > 0$ ε_r = radial, ε_θ = hoof and ε_z = axial strain, respectively	$\varepsilon_r > 0$, $\varepsilon_\theta < 0$ and $\varepsilon_z = 0$

basis for the 'side relief' is that the tube elongates straight forward and not down the taper – Randall and Prieur [1967]) and is a prime requisite for successful cold pilgering operation. Operating with insufficient 'side relief' may result in die pick-up, end spitting or tube shear marks, while excessive 'side relief' may cause tube splitting through excessive ovalizing, loss of dimensional tolerance and half moon tears through preferential elongation of the cross-section. For 'side relief' design, feed, elongation, die taper, die separation under load, tube springback, tube end ovalizing and groove wear are to be considered (Randall and Prieur [1967]).

11.6.2.2 Other mill details. Several other mechanisms/design aspects are equally vital to the overall pilgering operation. Some of these aspects are highlighted in the next few paragraphs.

11.6.2.2.1 Mandrel design. Mandrel is designed to keep the roll separating force within limits. Mandrel design varies from simplified straight taper mandrels to more complicated multi-tapered types, the last of the tapers being shallow enough to produce close tolerance tubes.

11.6.2.2.2 Loading and feed mechanisms. For loading a fresh tube in a high-speed pilgering mill, two alternatives are usually adopted: (I) fresh tube is fed by a pusher tube into the mandrel bar, the mandrel bar having a suitable locking–unlocking mechanism and (II) loading of the fresh tube is through the sides of the mill so that the mandrel can be inspected totally from both sides after each pilgering run (Randall and Prieur [1967]). The design of the feed mechanism is also vital (Voith [1985]) and several possibilities, mechanical as well as hydraulic, exist (Randall and Prieur [1967]).

11.6.2.2.3 Rolling drive. A large roll groove in a small machine would have a tendency to pull the tube into the dies, while the reverse would increase the tendency of excessive slippage (Randall and Prieur [1967]). Use of appropriate rolling drive and drive parameters are vital for the pilgering operations (Pavlovskii [1984]).

11.6.2.2.4 Synchronization and turning. Two of the most important design aspects of a pilger mill are 'synchronization' and 'turning'. A pilgering mill operating at 150 strokes a minute is subjected to five reversals in a second (Randall and Prieur [1967]). Each reversal requires turning the tube, by 30–90°, and reversing the roll/mandrel movement – a pretty demanding situation in any machine design. Synchronization of the mandrel, rolls and the feed mechanism can be achieved (Randall and Prieur [1967]) through a fairly complicated array of crank drive, rack and pinion, gears and rotating cam. Appropriate pneumatic clutch and breaking arrangements are also required. Turning, on the other hand, is achieved efficiently (Randall and Prieur [1967]) through a turning cam, mounted on the drive shaft and rotating at the same speed. The turning cam is 'timed' to push a sliding shaft and a worm, the latter acting as a rack in activating a worm gear – the gear executing actual turning operation. Once the turning is finished, the worm returns to its starting position.

11.6.2.3 Lubrication. Pilgering lubricants are often similar to that of drawing (Möller and Boor [1996]). Lubrication is typically applied in the inner surface once and for all by injection a suitable emulsion (e.g. stabilized animal fats, chlorinated paraffins) before the mother tube is fed, while the external surface is continuously sprinkled with the same, albeit dilute, emulsion (Montmitonnet *et al.* [1992]). However, examples of continuous lubricant application in the inner surface (lubricant being pumped through a hollow mandrel) also exist (Randall and Prieur [1967]). Cold pilgering is typically in a 'mixed' (as in Figure 11.4) lubrication mode[5] giving 'plateaux and valleys' (Montmitonnet *et al.* [1992]). High concentration of 'plateaux' does create serious risks of surface defects. The

[5] For more details on lubrication mode, the reader may refer Section 11.1.1.3.

concentration of the 'plateaux', on the other hand, can be controlled by proper lubrication – controlling the lubricant delivery and filtering.

11.6.3 Optimization in pilgering

Pilgering optimization may range from relatively simple in-plant empirical modelling for increasing productivity/efficiency (Mfnshchikov [1984], Plyatskovskii [1986]) to more complicated analytical and/or numerical modelling of the plastic deformation (Siebel *et al.* [1954], Roberts [1983], Furugen [1984], Masahiro [1985], Voith [1985], Montmitonnet *et al.* [1992], Huml [1994], Mulot [1996], Montmitonnet *et al.* [2002]). The earliest analytical model of cold pilgering (Siebel *et al.* [1954], Montmitonnet *et al.* [2002]) is based on an approach similar to the analytical solution of cold strip rolling. The model (Siebel *et al.* [1954]) estimates the vertical rolling force, $F(z)$, as:

$$F(z) = K(z)L_d(z)D(z) \tag{11.35}$$

where L_d is the projected contact length (in rolling, contact length = $L=(R\Delta h)^{1/2}$, R and Δh, respectively being the roll diameter and reduction.), D is the external diameter of the tube (L_dD being the projected contact surface with one die/roll) and K is the average normal stress (depends on yield stress and on friction). Considering two different contact lengths, material in contact with both mandrel and die/roll and material in contact with only die/roll (Siebel *et al.* [1954], Mulot [1996], Montmitonnet *et al.* [2002]), the following analytical solutions for L_d and $F(z)$ were proposed (Siebel *et al.* [1954]):

$$L_d = \left(2Rm\left(\frac{S_0}{S}\right)\right)^{1/2} \cdot \left\{(\phi-\delta)^{1/2} + (\delta)^{1/2}\right\} \tag{11.36}$$

$$F(z) = \sigma_0\left(\frac{1+\mu L_d}{2t}\right)\cdot D\left\{\frac{2Rm}{S_0/S}\right\}^{1/2}\left\{\frac{(\phi-\delta)+((\delta)^{1/2}+2t)}{D}\right\} \tag{11.37}$$

where at a section z, R is the die radius, m the feed length, S the cross-sectional area, ϕ and δ are mandrel and die/roll contact angle, t is the tube thickness and D is the tube diameter. S_0, σ_0 and m are the initial cross-sectional area, yield stress and friction coefficient, respectively.

Though simplified in approach, the 2D analytical model (Siebel *et al.* [1954]) shows serious discrepancies with experiential results (Yoshida [1975]). A 3D analytical model (Furugen [1984]) has subsequently been proposed and shows better predictive behaviour. The model has been used effectively for overall stress analysis, though shear components are somewhat neglected even in this approach, and to correct die/mandrel geometry (Furugen [1984]). A complete picture of the stress distribution, and hence a complete approach to pilgering optimization seems possible only through FEM (Mulot [1996], Montmitonnet *et al.* [2002]). It is also to be noted that the FEM approach often uses existing commercial codes (Montmitonnet *et al.* [2002]). The capability and promises for overall pilgering optimization, especially for die/roll and mandrel design, of these codes are very un-equal and relatively simpler analytical approaches are still being used effectively for optimization during the pilgering process itself – from simple load calculations (Roberts [1983]) to relatively more complicated estimations of maximum feed at a given tolerance for relative ovality (Voith [1985]).

11.6.4 Materials aspects

Because cold pilgering is more of a compression forming process, the malleability is a better measure of formability than ductility. This makes the usual high reductions a common feature for cold pilgering operations – reductions exceeding 50% are possible even in difficult to process materials like Hastelloy and Ti-6Al-4V, while reductions approaching 80% are common for more formable metals (Randall and Prieur [1967]). Cold pilgering has a large possible range of applications: from breakdown of hot-formed mother tubes to final finishing to close dimensional tolerances (less than 5% dimensional tolerance is typical), from forming of expensive material (and thus avoiding 'point-loss' usual in tube drawing operation) to forming of difficult to draw material (high strength material) and from forming bar/rod products with uniform mechanical properties across the cross-section to forming specific components (e.g. for aerospace and nuclear industry) with difficult shapes (tapering of inner/outer diameter and/or hexagonal/square shapes for inside channel).

The strains involved with the pilgering process are often generalized from the '*Q*-factor', which is the ratio of reductions in tube wall and tube diameter (Jeffrey [1991], Kiran Kumar *et al.* [2003]). Though the strain history is complex triaxial, processing with large *Q* may bring it close to plane strain. Naturally the *Q* factor has strong effects on developments of mechanical anisotropy, by controlling the deformation texture development (Jeffrey [1991], Davies *et al.* [2002], Kiran Kumar *et al.* [2003]).

Though a process with better formability, the complex history of cold pilgering may lead to defects. Table 11.15 provides a typical listing of defects in cold-pilgered Zr tubes (for reactor cladding applications). This list was formulated by considering available plant data. As shown in the table, a larger percentage of defects in cold-pilgered tubes are on the inner surface – *about 45% of the overall defects being cracks on the inner surface.* Cracks typically develop through low cycle rolling contact fatigue at locations of microstructural heterogeneity. As indicated in the table, other than quality control of the hollows or the mother tubes and eliminating defects

Table 11.15. Listing of defects in cold-pilgered Zr tubes for reactor applications.

Location	Type	Source
Defects in the outer surface (about 40–45% of total defects)	*Pilgering defects*: Cracks	Rough surface (die/mandrel), improper lubrication (both cooling and filtering), wrong calculation of feed and 'side relief', non-uniform rotation during pilgering
	Others: Cracks, handling defects (e.g. grinding and burning defects)	*Cracks*: Owing to defects present in the mother tube (in solidification, hot forming or heat treatment)
		Handling defects: Owing to improper handling or defects in the accompanying operations (grinding, sand blasting, etc.)
Defects in the inner surface (about 50–55% of total defects)	*Pilgering defects:* Cracks/ laminations and surface roughness	Use of low Q-factor, wrong mandrel size and/or feed, improper lubrication (can affect surface roughness seriously and large surface roughness, especially high plateaux concentration, may lead to other defects as well (Montmitonnet [1992])
	Others: Cracks and porosity	Owing to defects present in the mother tube or created during accompanying operations

in the accompanying fabrication/TMP steps, control (control and usage of appropriate tooling and process variable are typically the decisive issues in the overall quality control scheme) of the tooling in pilgering (roll/die and mandrel) and process control (control of feed, Q factor and lubrication) are important steps in the overall quality assurance.

LITERATURE

Montmitonnet P., Loge R., Hamery M., Chastel Y., Doudoux J.L. and Aubin J.L., J. Mater. Process Tech., 125–126 (2002), 814.

Randall S.N. and Prieur H., Iron and Steel Engineers, (August) (1967), 109.

Roberts W.L., "Hot rolling of steel", Marcel Dekker, New York (1983), 419.

11.7. SHEET METAL FORMING

11.7.1 Introduction

Large quantities of thin sheets are produced at a relatively low cost by rolling mills. They are transformed into familiar products, such as beverage cans, car bodies, metal desks, domestic appliances, aircraft fuselages, etc., by sheet metal-forming processes. Many of these processes involve a rather complex deformation path. In most cases, the latter can be considered as a superposition of some 'elementary'

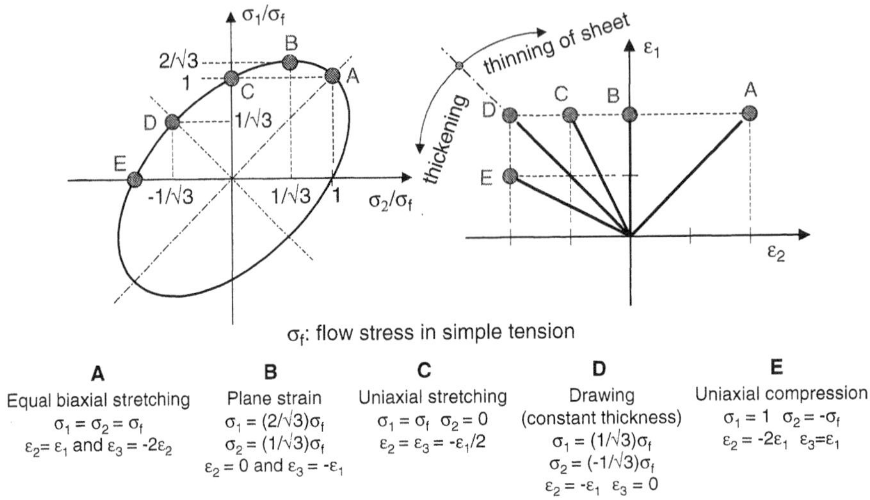

Figure 11.38. Stress and strain state of some plane stress conditions ($\sigma_3 = 0$) in an isotropic material. For more information, the reader is referred to any introductory textbooks on plasticity.

processes like bending, stretching and deep drawing, which will be discussed in this chapter.

In most sheet metal-forming processes, the stress perpendicular to the sheet surface is small compared to the stresses in the plane. It is often assumed that this 'normal' stress can be neglected. When the normal stress is zero, the stress state is called 'plane stress'. Some possible plane stress situations and their associated strain state are illustrated in Figure 11.38.

An important feature of many sheet metal-forming operations is the so-called 'plastic anisotropy' of the sheet. This will be discussed in the next paragraph. After that, a paragraph is devoted to the concept of 'forming limit diagrams' (FLDs). An FLD indicates how much deformation a sheet can withstand before failure and is an important tool for many practical press-forming operations. Subsequently, the most important sheet metal-forming operations like stretching, deep drawing, bending and others will be analysed.

11.7.2 Plastic anisotropy

The term 'anisotropy' refers to a directionality of some particular material properties. These properties can be elastic (e.g. Young's modulus), plastic (strength, formability), electrical (conductivity), magnetic (hysteresis losses), etc. In the frame of this textbook, we will mainly concentrate on 'plastic anisotropy'. It will be shown in the next paragraph that the main cause of plastic anisotropy is the crystallographic texture of the material. This is in contrast with the directionality

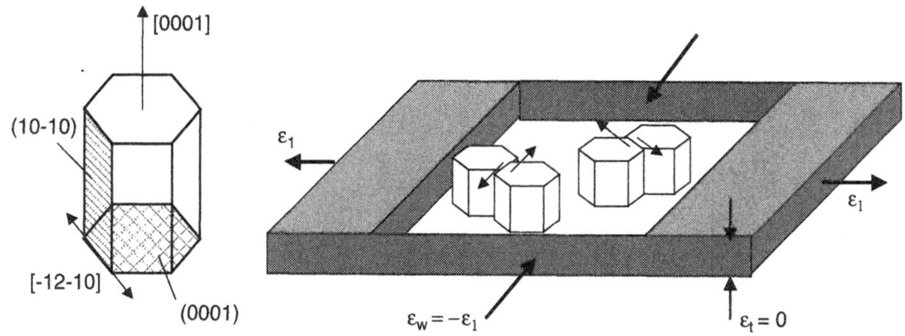

Figure 11.39. Slip systems in pure Ti. In uniaxial tension, the fibre texture prohibits thinning in the thickness direction. After Hosford and Caddell [1983].

in fracture behaviour, which results mainly from a morphological texture (elongated grains, alignment of precipitates, etc.).

11.7.2.1 Crystallographic background. As described in Chapter 10, the macroscopic deformation of a metal is the result of many microscopic shear deformations on particular crystallographic planes (for simplicity, we do not consider other deformation mechanisms like twinning, etc.). Take the example of pure Ti which has a hexagonal closed packed (hcp) structure up to 882°C. Deformation takes place by glide in the <1–210> directions in the basal (0001) plane and prismatic {10–10} planes (Figure 11.39). In none of these cases, can any deformation in the [0001] direction occur. Suppose now that a Ti sheet has an 'ideal' fibre texture and consist of grains with the basal plane (0001) parallel with the sheet surface, but with a random distribution of crystallographic orientations around the [0001] axis (see Chapter 8 for more details on crystallographic textures). A uniaxial deformation in the sheet plane will be performed by glide on the prismatic {10–10} planes. The elongation of the sheet in the tensile direction (ε_l) will be compensated by a perpendicular contraction in the width of the sheet (ε_w), but the activated slip systems will not allow any contraction in the thickness direction (ε_t) (Figure 11.39). The difference in contraction between width and thickness is due to the material anisotropy (Chapter 3). On the other hand, because of the random distribution of crystallographic directions around the [0001] axis, the tensile properties will not vary with the angle between tensile axis and rolling direction: this is called 'planar isotropy'.

11.7.2.2 The $\bar{R}$ and ΔR factors. In practical situations an 'ideal' texture as depicted in Figure 11.39 will never occur. In a real tensile test on a Ti sheet, some contraction in thickness would be observed. As briefly introduced in Chapter 3, the ratio between plastic strain in the width over plastic strain in the thickness direction in a uniaxial tensile test is called the *R*-value:

$$R(\alpha) = \frac{\varepsilon_w^{\text{pl}}}{\varepsilon_t^{\text{pl}}} \tag{11.38}$$

with α the angle between the rolling direction and the tensile axis. A large value of R means that the crystallographic orientations of the grains are such that the sheet has a large resistance against thinning. In practical circumstances, it is often difficult to measure the change in thickness during a tensile test. Because $\varepsilon_t^{\text{pl}} = -(\varepsilon_w^{\text{pl}} + \varepsilon_l^{\text{pl}})$, the *R*-value can also be calculated from width and length variations. In most cases the total (elastic + plastic) strains are measured, so a more practical definition of the *R*-value is:

$$R(\alpha) = \frac{-\varepsilon_w}{(\varepsilon_w + \varepsilon_l)} \tag{11.39}$$

To minimize the error introduced by neglecting the elastic strain, the *R*-value is usually measured between 10 and 15% length strain.

Since the *R*-value of most materials depends on the direction in the surface plane of the sheet (indicated by angle α between tensile axis and rolling direction), a mean *R*-value ($\bar{R}$) can be calculated. A rigorous definition would be:

$$\bar{R} = \frac{2}{\pi} \int_0^{\pi/2} R(\alpha)\, \mathrm{d}a \tag{11.40}$$

Because in most cubic materials a minimum or maximum of R is found at $\alpha = 0°$, 45° and 90°, the $\bar{R}$-value is often calculated based on only three sets of tensile tests with:

$$\bar{R} = \frac{(R_{0°} + 2R_{45°} + R_{90°})}{4} \tag{11.41}$$

This mean *R*-value is called 'the normal anisotropy' and is an important technical parameter because it is related to the deep drawability of the sheet, as will be discussed in Section 11.7.5 and illustrated in the case study, Chapter 15.1 on 'steel for car body applications'.

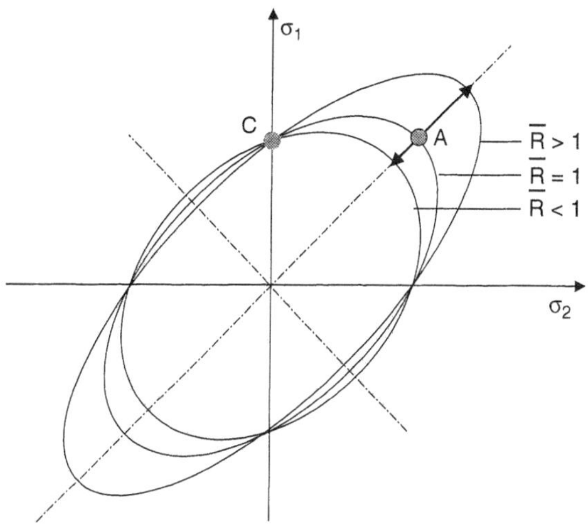

Figure 11.40. Influence of plastic anisotropy on the shape of the yield locus.

The planar anisotropy is also important from a technical point of view. The planar anisotropy reflects the variation of $R(\alpha)$ in the plane of the sheet and can be defined as:

$$\Delta R = \frac{(R_{0^\circ} - 2R_{45^\circ} + R_{90^\circ})}{2} \qquad (11.42)$$

It will be shown in Section 11.7.5 and illustrated in the case study on 'Al beverage cans' (Chapter 14.1) that ΔR can be related to the earing behaviour during deep drawing.

It is obvious that plastic anisotropy will change the shape of the yield locus. This is illustrated in Figure 11.40. It is worth noting that the uniaxial yield stress (point C) is not affected by a change in mean R-value, but that the biaxial yield stress (point A) increases with increasing $\bar{R}$.

11.7.3 Forming limit diagrams

11.7.3.1 Determination and practical use. A forming limit diagram (or 'FLD'), as schematically shown in Figure 11.41, indicates how much deformation a material can take before it fails.[6] For example, when a material is biaxially stretched along path 'a' till strain state 'A', no failure is predicted. Following strain path 'b' till point

[6] An FLD is also known as a 'Keeler–Goodwin' diagram, referring to the original work of Keeler [1965] and Goodwin [1968]. Reviews on FLDs have been given by Wagoner [1989] and Esche *et al.* [1996].

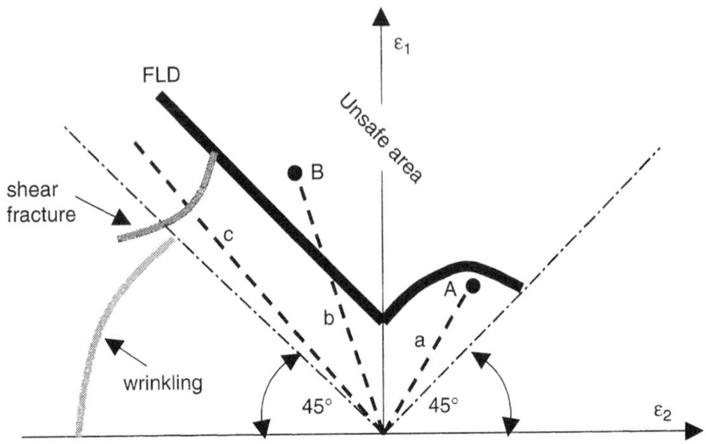

Figure 11.41. Schematic illustration of a forming limit diagram (the thick black line).

'B' will however cause failure as soon as the strain state reaches the forming limit curve. 'Failure' is considered here as the point where a local neck develops. The real 'fracture' of the material will occur immediately after the development of a local neck, at a slightly higher strain.

In the region where thickening of the sheet occurs (cf. Figure 11.38), the sheet will not fail by local necking, but wrinkling is likely to occur. This is not indicated in most FLDs. Along certain strain paths, e.g. path 'c', fracture (in this case a so-called 'shear fracture') can occur before the actual FLD is reached. Also this line is often not indicated in a simple FLD diagram.

An FLD is not only useful to predict the maximum strain one can apply in a particular stamping operation, but it can also be used to analyse a stamped part. Suppose one can measure or calculate the local strain state in every area of a stamped part. These strain states can than be plotted on an FLD and the most critical areas of the part, in the safe area but close to the FLD, can be identified and the shape of the part can be adapted if necessary. The calculation of local strain states is performed in most cases by finite element simulations. An experimental verification can be done in the following way: the sheet is covered with a grid pattern of circles using some electrochemical technique. These circles have typically a diameter of 2–5 mm. The sheet is then stamped and during deformation most circles are converted into ellipses. The large and small axes of all the ellipses are measured with a microscope or with some automatic optical system and from these dimensions the local strains ε_1^i and ε_2^i are calculated and plotted on the FLD. When certain points are close to the 'forming limit', it may be advisable to adapt in that location the shape of the part. Even when the actual prototype part has not

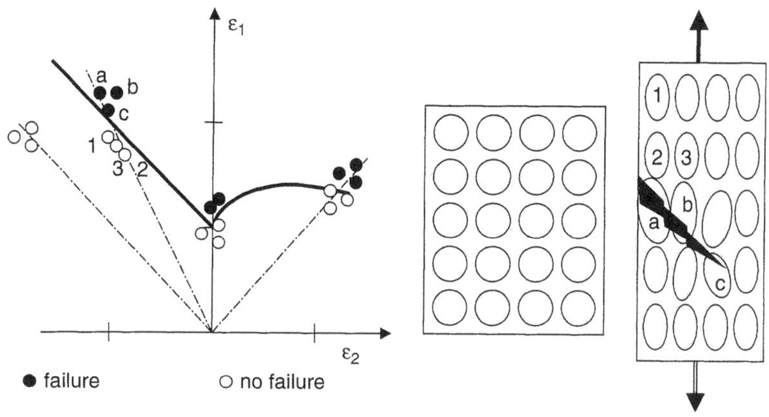

Figure 11.42. Experimental determination of an FLD diagram.

failed, a slight change of some of the production parameters (friction, deformation rate, sheet thickness, etc.) could bring the local deformation in the 'unsafe' region and several failures during a production run can be anticipated.

The experimental determination of an FLD is time-consuming and can be done in the following way. A number of test sheets are covered with a pattern of circles (or other markings). All the sheets are deformed till failure along a certain deformation path (uniaxial stretching, biaxial stretching, drawing, plane strain, etc.). The dimensions of the deformed circles are measured and the local deformation is calculated and plotted in a (ε_1, ε_2) diagram. Ellipses in the crack or in a necked region are indicated with full circles; other points are indicated with open circles (Figure 11.42). Based on this information, the forming limit curve can be determined. More information about the practical determination of FLDs can be found in ASM [1988].

11.7.3.2 Parameters that affect the FLD. Each material has its own FLD. But also the actual state of a material (cold deformed, recrystallized, etc.) will affect the position of the FLD. In the next paragraphs, the most important parameters that affect the FLD will shortly be discussed.

The strain hardening coefficient 'n'. The general effect of an increase of the strain hardening coefficient 'n' is to shift the whole FLD upwards (Figure 11.43). In fact, it can be shown (see, e.g. Marciniak *et al.* [2002]) that the left-hand side of an FLD can roughly be described by: $\varepsilon_1 + \varepsilon_2 = n$. For $\varepsilon_2 = 0$ (plane strain) $\varepsilon_1 = n$. For strongly cold-worked metals, n is close to zero. This means that the plane strain formability or uniaxial deformability will be close to zero, but that in equi-biaxial

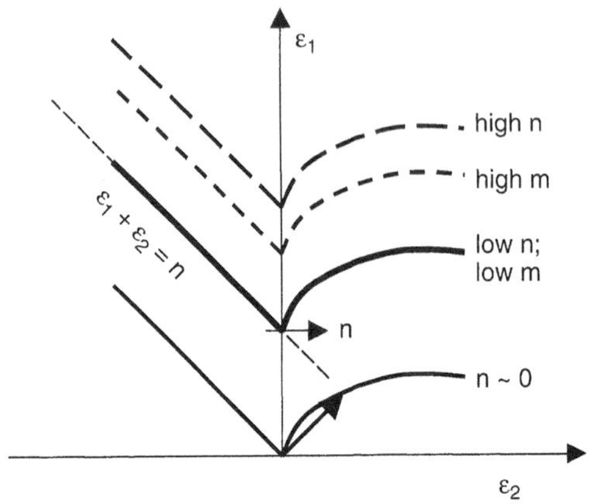

Figure 11.43. Influence of strain hardening (n) and strain rate sensitivity (m) on the forming limit diagram. For strongly cold deformed metals ($n \sim 0$) some equi-biaxial stretching is still possible.

direction some stretching is still possible, as illustrated in Figure 11.43. Especially for biaxial stretching, the strain hardening has a large effect on formability.

The strain rate sensitivity coefficient 'm'. The development of a local neck can be delayed by a high strain rate sensitivity. Increasing 'm' shifts the FLD upwards and the intersection with the ε_1 axis is higher than n.

The thickness of the sheet. The sheet thickness has more or less the same effect as the strain rate sensitivity: increasing thickness shifts the FLD upwards.

Strain path changes. Nearly all experimental FLDs available in the literature have been determined for proportional loading, which means that $\varepsilon_1/\varepsilon_2$ is constant. In many complex stampings this is not the case and changes in strain path during deformation occur. It has been shown in by Nakazima *et al.* [1968] that such strain path changes can have a drastic effect on the position of the FLD curve, as illustrated in Figure 11.44. It was shown that for a combination of equi-biaxial stretching, followed by uniaxial stretching (curve 'a') or plane strain deformation (curve 'c'), failure occurred before the 'proportional' FLD was reached. On the other hand, for a combination of uniaxial stretching followed by equi-biaxial stretching, a much larger deformation could be applied than anticipated from the conventional FLD. Since it is impossible to determine experimental FLDs for all possible strain path changes for all materials, efforts have been done to predict the influence of strain path changes with computer models (see, e.g. Hiwatashi *et al.* [1998]).

11.7.4 Stretch forming

In stretch forming, the material is clamped along the edges and stretched over a punch. A simple case of stretch forming is an equi-biaxial stretch over a

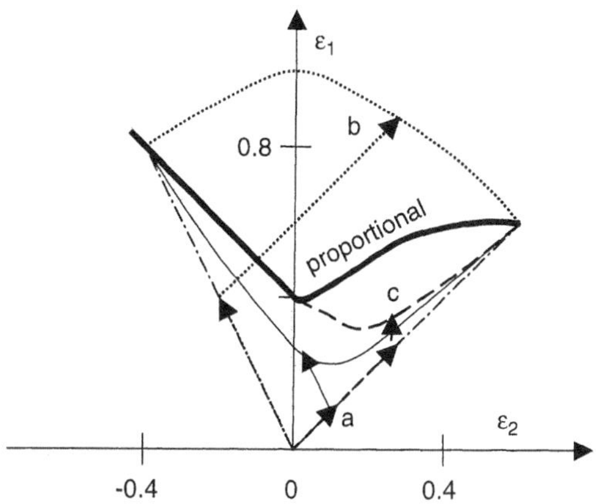

Figure 11.44. Influence of strain path changes on the position of the FLD. (a) equi-biaxial stretching followed by uniaxial tension; (b) uniaxial tension followed by equi-biaxial stretching; (c) equi-biaxial stretching followed by plane strain deformation. Adapted from Nakazima *et al.* [1968].

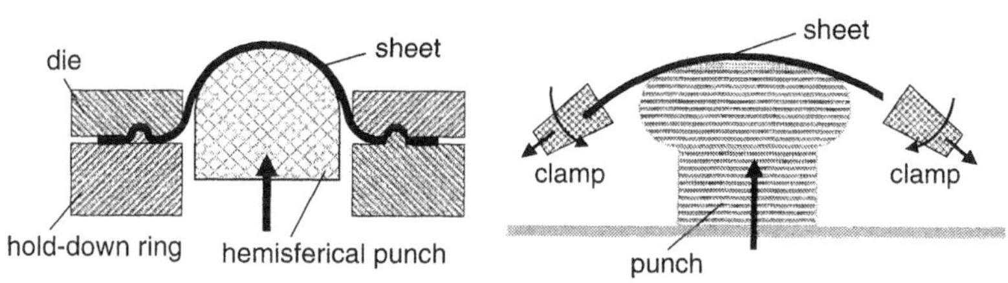

Figure 11.45. Equi-biaxial stretching using a clamped sheet and a hemispherical punch (left) and a schematic of an industrial stretch-forming operation (right).

hemispherical punch as shown in Figure 11.45 (left). The edges of the sheet are clamped by some groove in the die (the so-called 'draw-bead'). In many industrial stretch-forming operations, the situation is more complex. The sheet is often rectangular and is clamped along two sides only (Figure 11.45, right). In this case, the strain is close to plane strain. The middle section of the sheet is less stretched because of the friction between punch and sheet.

To avoid unstretched regions (without work hardening) uniaxial stretching with the clamps, before 'draping' the sheet over the punch can be a solution. In many cases, the clamps can not only exert a tensile force, but can also rotate in order to maintain the tensile force tangential to the sheet. Instead of using a simple punch, some presses make use of a male and female die (Figure 11.46). The forming operation is now very complex.

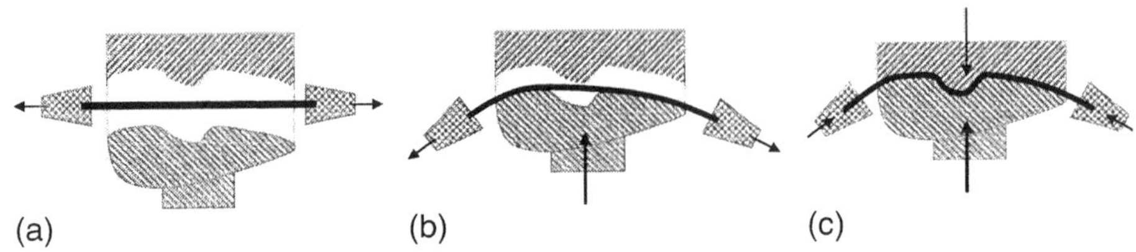

Figure 11.46. Complex stretch-forming operation using a male and a female die.

Many possible routes exist, but a standard stamping could be as follows:

- Uniaxial stretching of the sheet; because of the large width of the sheet, this will lead to a strain state close to plane strain.
- Draping of the sheet over the punch by an upward movement of the latter, the clamps maintain a tangential tensile force.
- Downwards movement of the upper die; the clamps move back inwards to feed new material in the opening between the two dies. The deformation is then a combination of stretching and deep drawing (see Section 11.7.5).
- The dies move back to their initial position, and the clamps release the pressed part.

Stretch forming is very useful to produce large parts with a large radius. It is frequently used in the aircraft and automotive industries. Another advantage is a limited springback.

11.7.5 Deep drawing

11.7.5.1 Stress and strain. During deep drawing, a sheet is pushed through an opening in a die. A blank holder prevents wrinkling of the sheet but, in contrast to stretching, the material from the flange can more or less without constraints be drawn towards and into the die opening. The most common example of deep drawing is the formation of a cylindrical cup (Figure 11.47). For a constant punch diameter D_p, the height of the cup wall will increase with increasing sheet or blank diameter (D_{bl}). There is, however, a limit to the height that can be obtained in one drawing pass, without fracturing the cup wall. This limit is called the 'limiting drawing ratio' or LDR and is defined as:

$$\mathrm{LDR} = \frac{D_{\mathrm{bl}}^{\mathrm{MAX}}}{D_{\mathrm{p}}} \tag{11.43}$$

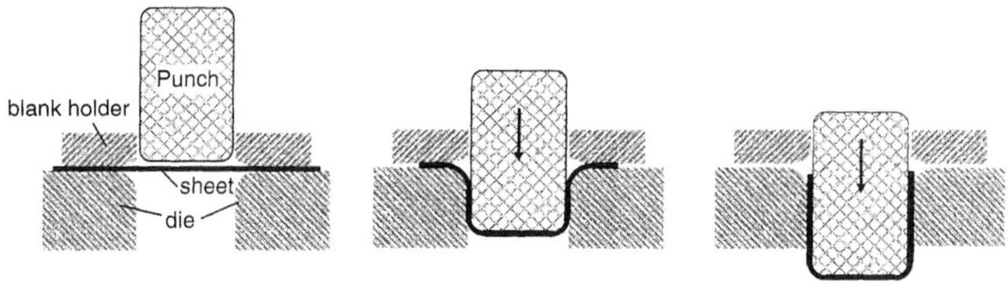

Figure 11.47. Deep drawing of a cylindrical cup from a circular blank.

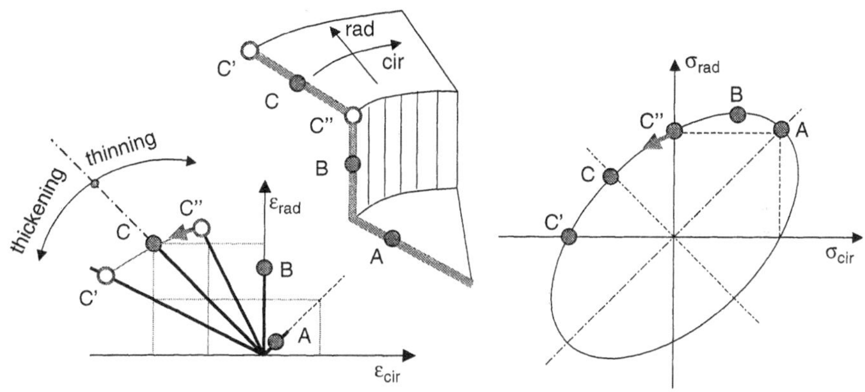

Figure 11.48. Stress and strain state in various points of the cup during deep drawing. Possible compressive stresses in flange and wall are not taken into account.

During deep drawing, four important zones can be identified: the flange, the transition between flange and cup wall, the cup wall and the bottom of the cup.

- The bottom of the cup, which is formed by the material under the punch, is equi-biaxially stretched (point 'A' in Figure 11.48). In principle, the material will thin down; but in practice, the strain in the bottom is not so large and it is often assumed that the bottom of the cup has more or less the same thickness as the original sheet.
- The situation in the flange is more complicated. The material is drawn into the die hole by a radial tensile force. An imaginary circle on the flange will shrink during drawing, so the stress state is tensile in the radial and compressive in the circumferential directions. In an ideal case, the strain can be considered as 'ideal drawing' without change in thickness (point 'C' in Figure 11.48). In reality, the stress- and strain-state change in radial direction. At the outer surface, $\sigma_{rad} = 0$ (point C′ in Figure 11.48). This means that the edge of the flange will thicken.

At the edge of the die hole, using Tresca's yield criterion (Marciniak *et al.* [2002]), the stress state can be described by

$$\sigma_{\text{rad}} = \sigma_{\text{f}} \ln\left(\frac{D_{\text{bl}}}{D_{\text{p}}}\right) \quad \text{and} \quad \sigma_{\text{cir}} = -\left(\sigma_{\text{f}} - \sigma_{\text{rad}}\right) \tag{11.44}$$

with σ_f the flow stress. Since σ_{rad} cannot exceed σ_f, the limiting drawing ratio $D_{bl}^{MAX}/D_p = e = 2.72$. In that case, $\sigma_{cir}=0$ (point C'' in Figure 11.48). In reality, most deep-drawing materials have an LDR between 2 and 2.4, which keeps $\sigma_{rad} < \sigma_f$ and $\sigma_{cir} < 0$. In any case, the strain state will be such that the material will thin down. In many cases, the blank holder – whose principle task is to prevent the flange from wrinkling – will prevent (or limit) the thickening of the blank. In that case, an additional stress σ_3 perpendicular to the sheet will be present.

- The force exerted by the punch is transmitted to the deforming flange by the material in the cup wall. Hence this material is subjected to a tensile force. Because of the geometry of the cup and punch, no contraction in the circumferential direction is possible. This results in a plane strain deformation (an elongation of the cup wall compensated by a reduction in thickness), point B in Figure 11.48. The purpose of deep drawing is to concentrate the deformation in the flange and not in the cup wall. For this reason, the thinning of the wall must be kept to a strict minimum. As will be explained in the following paragraph, this is done by choosing materials with a large $\overline{R}$-value.

  With the assumption that flange, wall and bottom of the cup retain the initial sheet thickness, the following expression for the height of the cup can be found:

$$h \sim \frac{D_{\text{p}}}{4}\left\{\left(\frac{D_{\text{bl}}}{D_{\text{p}}}\right)^2 - 1\right\} \tag{11.45}$$

  For a material with a LDR = 2.25, the maximal cup height that can be drawn in one pass is equal to the cup diameter. It must be noted that the assumption of constant wall thickness is not completely correct. In reality, the wall will thin down close to the bottom part, while the top is thickened.

- The transition between flange and cup wall is done by bending and unbending. This will add some extra component to the total force needed for deep drawing, it will increase the work hardening and it will lead to an extra reduction in thickness.

The total force P that has to be exerted on a blank to draw a cup, depends on many factors. Besides the complicated stress state described above, the friction between blank holder and flange and between cup wall and die must be considered, as well as the blank holder force and the force needed to bend the material. An approximate equation for the total punch force as function of the blank diameter D_{bl} at any stage of the process is (Dieter [1976]):

$$P = \left[\pi D_p t (1.1\sigma_f) \ln\left(\frac{D_{bl}}{D_p}\right) + \mu\left(\frac{2BD_p}{D_{bl}}\right) \cdot \exp\left(\frac{\mu\pi}{2}\right) + A \right] \tag{11.46}$$

where D_{bl} the blank diameter, D_p the punch diameter, t the thickness of the wall, σ_f the average flow stress, A the force to bend and unbend the material, B the blank holder force, μ the friction coefficient. The first part of the equation gives the ideal force to draw a cup, the second term expresses the friction force under the blank holder and the exponential factor brings the friction between wall and die into account.

11.7.5.2 Redrawing and ironing. According to formula (11.45), the cup height that can be obtained in one single drawing pass is limited. To obtain deeper cups, two techniques are frequently used: redrawing and wall ironing. Both techniques are illustrated in Figure 11.49.

In redrawing, several consecutive drawing passes are applied. After each pass, the cup radius decreases and the cup height increases. When the cup is turned inside out after each pass, the process is called 'reverse redrawing'. In wall ironing, the cup passes through a series of ring-shaped dies (Figure 11.49, right).

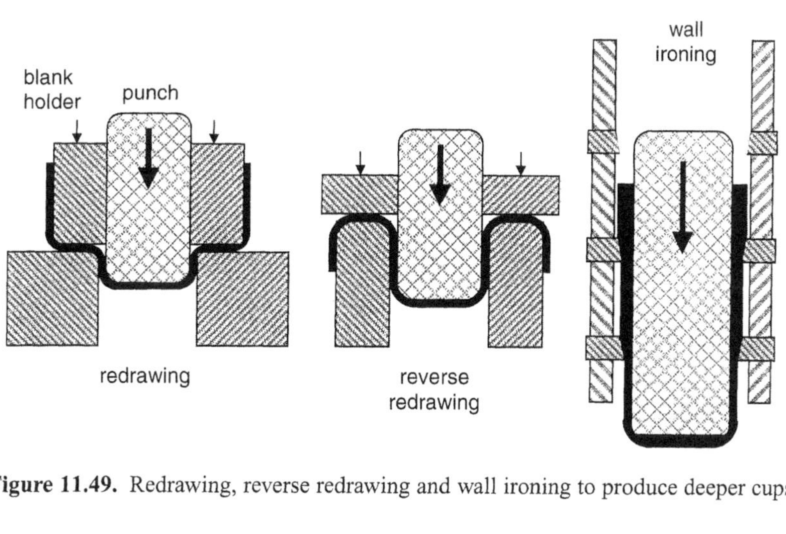

Figure 11.49. Redrawing, reverse redrawing and wall ironing to produce deeper cups.

The clearance between the punch and each die is less than the local sheet thickness, which increases the height and decreases the thickness of the cup wall.

11.7.5.3 Deep drawing and texture. In the previous paragraph, it has been shown that the deep drawability of a material can be expressed by the value of its LDR, which can be experimentally determined by drawing a series of blanks with increasing diameter till fracture occurs. Owing to the work of Held [1965] and many others, it became clear that the LDR is linearly related to the mean R-value ($\bar{R}$) of the material.

This means that deep drawability can be estimated from tensile tests. On the other hand, initial work of Whiteley [1960] and Hultgren [1968] established for low-carbon steel sheet a linear correlation between $\bar{R}$ and the texture. It is now common knowledge that in low-carbon steel, grains with a {111} plane parallel to the rolling plane will enhance the deep drawability of a sheet, while grains with a {001} plane parallel to the rolling plane are detrimental. Figure 11.50 illustrates the position of these 'good' and 'bad' crystallographic directions in Euler space. In the case study (Chapter 15.1) on 'steel for car body applications', it will be shown how a well-balanced TMP can optimize the texture and the deep drawability.

After drawing, the tops of the cups are not completely flat, but show minima and maxima (commonly called 'ears'). This earing is a consequence of the planar anisotropy of the sheets and correlates well with the ΔR value or the angular variation of R (Figure 11.51). In the directions of a low R-value, more thickening of the sheet will occur and the material is drawn faster into the die hole. Although earing manifests itself in the wall of the cup, it is a consequence of the deformation history of the flange. For $\Delta R < 0$, ears will occur at 45° with the rolling direction and for $\Delta R > 0$ at 0° and 90°. For fcc materials, it has been established that cube-oriented grains stimulate the 0°/90° ears, while orientations belonging to the β-fibre give rise to 45° ears. It will be shown in the case study on 'aluminium

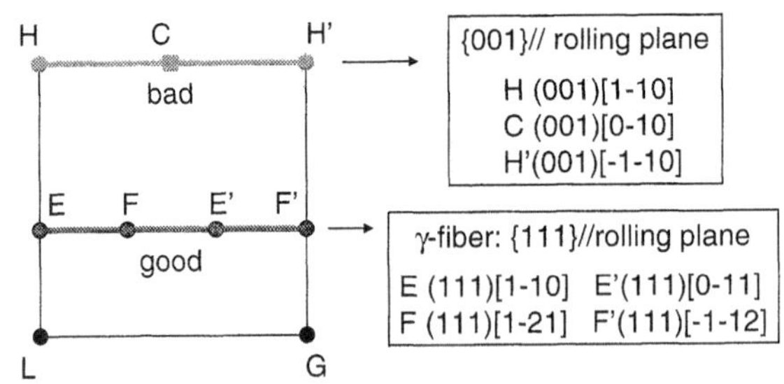

Figure 11.50. $\varphi_2 = 45°$ section of Euler space (Chapter 8), with crystallographic orientations that are 'good' and 'bad' for the deep drawability of a low-carbon steel sheet.

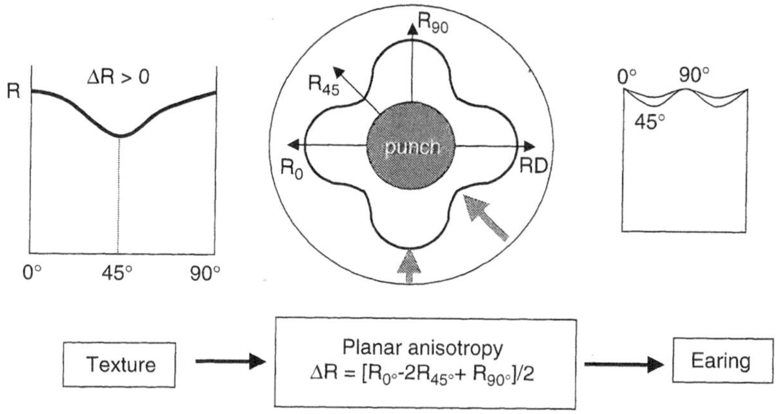

Figure 11.51. Influence of R-value directionality on material flow during cup drawing and resulting earing profile.

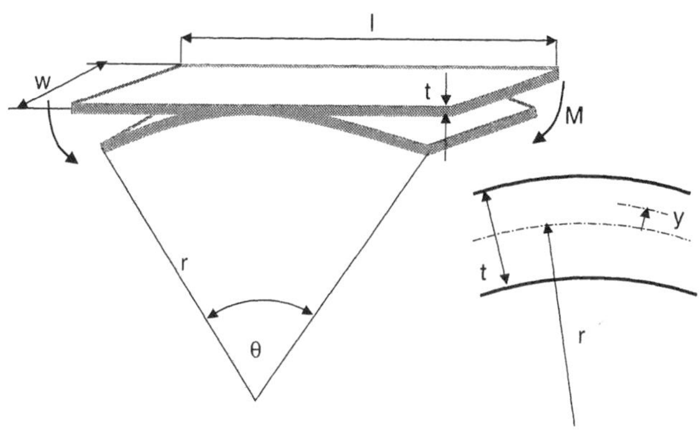

Figure 11.52. Illustration of a simple bending operation.

beverage cans', how Al sheets are processed in order to achieve a good balance between both in order to get $\Delta R \sim 0$ and to minimize earing.

11.7.6 Bending and folding

11.7.6.1 Stress and strain. Bending is a relatively simple forming operation, which can be achieved in various ways. A schematic illustration of a bended sheet is shown in Figure 11.52. During bending, the convex part of the sheet (the upper part in Figure 11.52) is in tension and the concave (lower) part in compression. The strain is zero at the neutral axis and maximum at the outer surface. The strain distribution over the thickness is linear, as illustrated in Figure 11.53. At a distance y from the neutral axis $\varepsilon = y/r$ with r the bend radius and at the surface $\varepsilon^{max} = t/2r$. The strain state during bending of a sheet (with $w >> t$) is approximatively plane strain with $\varepsilon_l = -\varepsilon_t$ and $\varepsilon_w = 0$. The stress state is biaxial with at the surface $\sigma_l = (2/\sqrt{3})\,\sigma_f$ with σ_f the flow stress in a simple tensile test, and for isotropic materials, $\sigma_w = (\sigma_l/2)$ and $\sigma_t = 0$. $S = (2/\sqrt{3})\sigma_f$ is the so-called 'plane strain flow stress'. For anisotropic material it can be calculated from:

$$S = \sigma_f \left[\frac{\bar{R}+1}{(2\bar{R}+1)^{0.5}} \right] \tag{11.47}$$

When the width of the sheet is not at least 8–10 times larger than the thickness, the stress ratio σ_w/σ_l decreases as shown in Figure 11.54.

For practical applications, it is important to know how far a sheet can be bent before cracks appear at the surface. The minimal bend radius is usually expressed as function of the sheet thickness, for example, $r_{min} = 3t$.

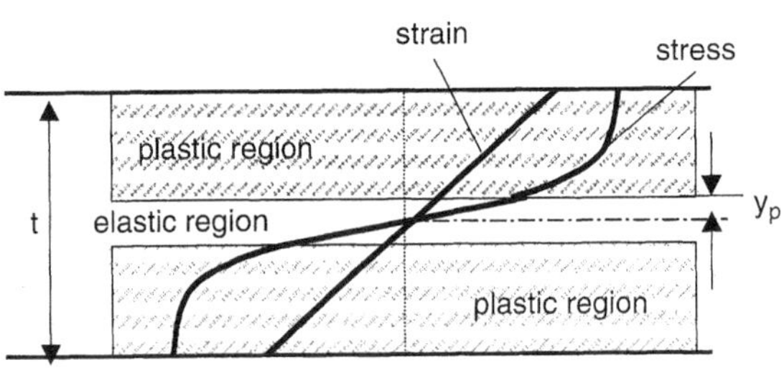

Figure 11.53. Stress and strain distribution over the sheet thickness.

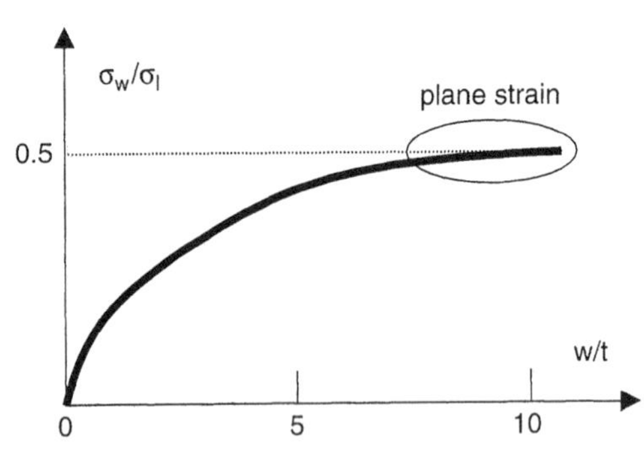

Figure 11.54. Ratio of σ_w/σ_l as function of the width over thickness ratio, the bending limit.

It is difficult to calculate an accurate bending limit. An approximate value can be estimated from the reduction in area (RA) in a uniaxial tensile test:

$$\varepsilon_{max} = \frac{t}{2r} \leq \frac{RA}{(1-RA)} \tag{11.48}$$

According to Eq. (11.48), a small bend radius can be reached for ductile materials with a large RA value; for most materials, this means a low yield strength. This rule of the thumb must however be used with some caution because microstructural effects, especially a morphological texture, can have a big influence on the bendability. Some data about bendability can be found in ASM [1985]. The bending limit predicted by (11.48) can only be reached when the length of the bended zone is sufficiently large. When the length of the bended zone ($l = r\theta$) is too small (Figure 11.55a), the maximum strain in the outer surface will not be reached because the whole length is taken by a transition zone between bended and unbended parts. Only when l increases by increasing r or increasing θ, a zone with constant maximum strain in the outer region will be present.

11.7.6.2 Spring back and residual stress. Spring back is a dimensional change that occurs after plastic forming and unloading, and is a consequence of the recovery

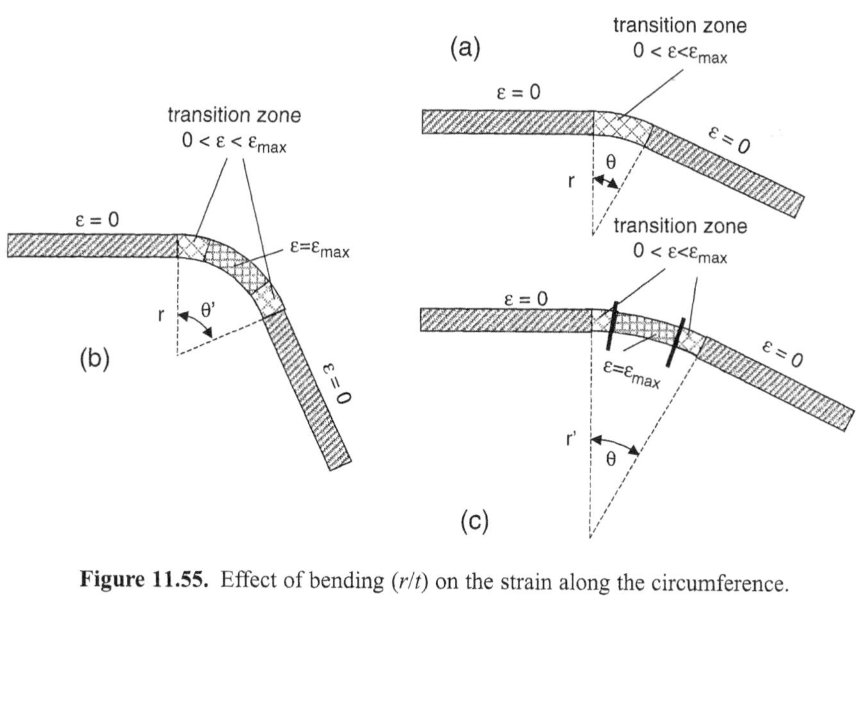

Figure 11.55. Effect of bending (r/t) on the strain along the circumference.

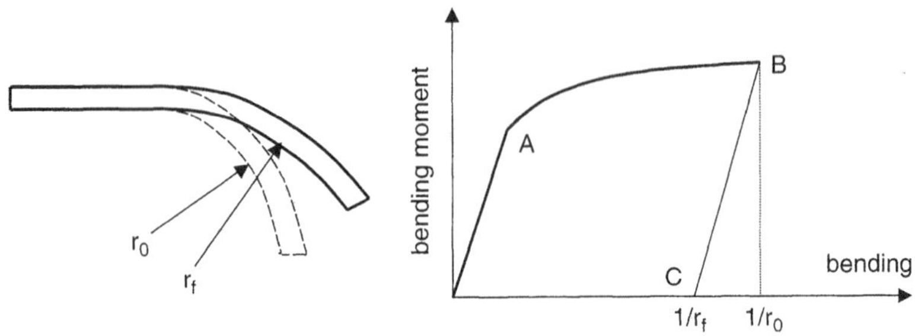

Figure 11.56. Spring back as a consequence of the recovery of the elastic part of the deformation. A: end of the elastic deformation ($\sigma = \sigma_f$); B: end of the plastic deformation; C: recovery of elastic deformation.

of the elastic part of the deformation (Figure 11.56). For an ideal plastic material it was shown by Hosford and Caddell [1983] that $\Delta\varepsilon = -3yS/tE'$, with E' the Young's modulus in plane strain [$E' = E/(1-\nu^2)$ with ν the coefficient of Poisson], t the sheet thickness and y the distance to the neutral fibre. Figure 11.56 than shows that:

$$\frac{1}{r_0} - \frac{1}{r_f} = \frac{3S}{tE'} \tag{11.49}$$

with S the plane strain flow stress given by Eq. (11.47).

In practice, it remains a difficult task to predict the spring back accurately. Formula (11.49) can be used as a first estimate and shows which parameters will have an influence on spring back. It appears that spring back will increase with increasing flow stress, increasing strain hardening coefficient, decreasing Young's modulus and decreasing sheet thickness.

After bending and subsequent relaxation of the elastic stress, an internal stress remains inside the sheet. For an elastic-ideal plastic material (Figure 11.57a), it can be shown (Hosford and Caddell [1983]) that the elastic relaxation can be estimated from $\Delta\sigma = -(3y/t)S$. Superposition of the total stress during bending and elastic relaxation leads to an internal stress profile shown in Figure 11.57c. It means that after unloading a compressive stress ($-0.5S$) is present at the outer (convex) surface and a tensile stress ($0.5S$) remains at the inner surface.

11.7.7 Other techniques

11.7.7.1 Spinning. In a conventional spinning process (Figure 11.58a), a circular blank is pushed against a mandrel and rotated. With a rigid tool (often in the form of rollers), the blank is forced against the mandrel. Spinning is a useful cold forming technique for shaping relatively simple axisymmetric parts.

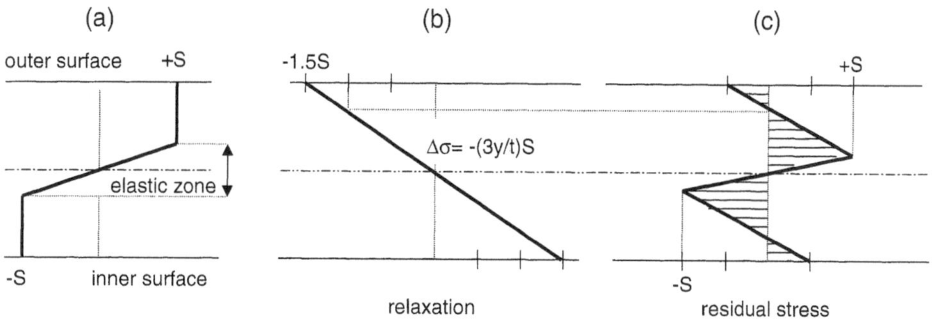

Figure 11.57. Bending stress (a), elastic relaxation (b) and residual stress (c) through the sheet thickness.

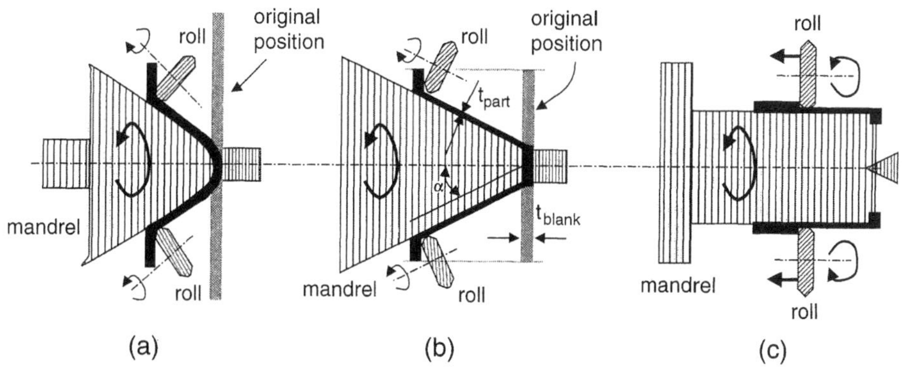

Figure 11.58. (a) Conventional spinning; (b) shear spinning; and (c) tube spinning.

Shear spinning is a variant of conventional spinning in which each element of the blank maintains its distance from the axis of rotation. A direct consequence is a reduction of the part's thickness (Figure 11.58b): $t_{part} = t_{blank} \sin \alpha$. For many materials, the maximum reduction in wall thickness is around 80%. Some materials with low ductility can be spun at elevated temperature.

In tube spinning (Figure 11.58c), the wall thickness of a cylindrical part is reduced by spinning on a cylindrical mandrel. This operation can be carried out internally or externally. The technique of spinning has recently been reviewed by Wong *et al.* [2003].

11.7.7.2 Incremental forming. In incremental forming a rigid, ball-shaped tool is used to gradually deform (emboss) a flat sheet to form a complex 3D metal part (Figure 11.59, left). Although some variants of incremental forming exist for many

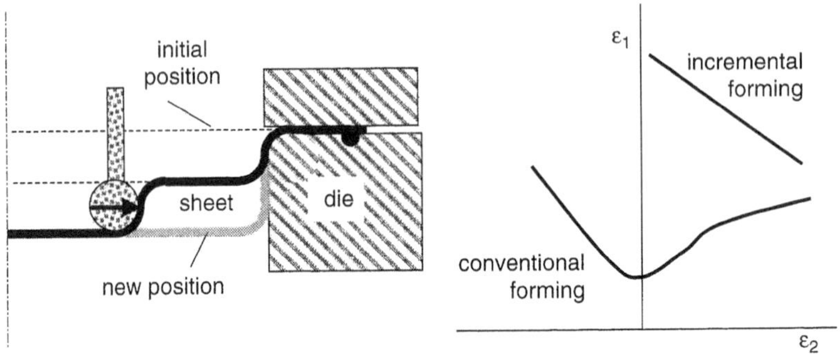

Figure 11.59. Schematic of incremental forming (left) and illustration of the increased formability (right).

years (e.g. spinning), new forms of this technique have been developed recently. This technology is well suited for small batch production of sheet metal components. The FLD in incremental forming seems to deviate considerably from that in conventional forming (Figure 11.59, right) (Kim [2002b]).

11.7.7.3 High strain rate forming. High strain rate forming of metals has been studied for many years, but up to now has not found widespread applications. The basic idea is to create a shock wave in a fluid, which drives a sheet into the die cavity. Various methods to create the shock wave have been considered, e.g. the detonation of explosives or the discharge of capacitor banks. Most metals seem to have a better ductility at very high strain rates than at conventional strain rates. This effect, which is not fully understood, is sometimes called 'hyperplasticity'.

In explosive forming, the metal sheet is clamped over a die and placed into a fluid (Figure 11.60a). The air in the die cavity is evacuated and an explosive charge is detonated. The shock wave generated in the fluid drives the sheet into the die cavity. The process is versatile and workpieces of large size can be formed.

In electrohydraulic forming, an electric arc discharge is generated between two electrodes, which are submerged in oil or water. The fluid around the electric arc is rapidly vapourized, creating a shock wave. This system is not fundamentally different from explosive forming; only the energy source is different.

In electromagnetic forming (or magnetic pulse forming), the energy stored in a capacitor bank is discharged into a coil, which is placed near or over the workpiece. An example of a possible configuration is shown in Figure 11.60b. The current pulse causes a high magnetic field around the coil. This magnetic field will induce eddy currents in the workpiece and this generates a second magnetic field, which causes a mutual repulsion between the coil and the workpiece. This repulsive force causes the deformation of the workpiece.

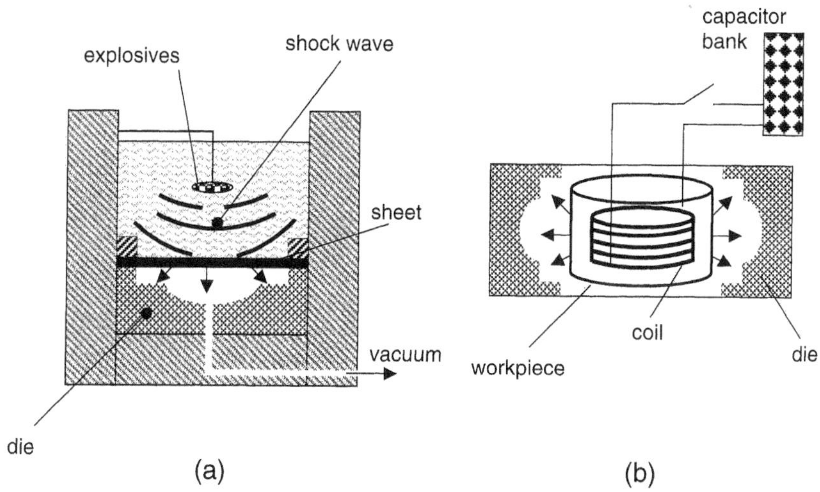

Figure 11.60. (a) Explosive forming and (b) electromagnetic forming.

11.7.7.4 Peen forming. Peen forming is a dieless process performed at room temperature. It is used to produce a curvature on sheet metals by shot peening. During shot peening, the surface of a workpiece is impacted by small, round balls from steel or cast iron, called 'shots'. Every piece or shot acts as a tiny peening hammer, producing a residual compressive stress. In many cases, shot peening is used as a surface treatment to improve corrosion resistance and/or fatigue strength of the material. When one side of a panel is shot peened, the compressive stresses tend to expand the surface layer and cause a convex curvature of the sheet on the peened side. The process is ideal for forming large panel shapes, where the bend radii are reasonably large and without abrupt changes in contour.

LITERATURE

Pearce R., "Sheet Metal Forming", Adam Hilger, Bristol (1991).

11.8. HYDROFORMING

11.8.1 Introduction

'Hydroforming' (or 'fluid forming') is a relatively new technique with an increasing number of applications in the automotive and aircraft industry. The principle is rather simple: a sheet or tube is pressed into a die cavity by fluid pressure. In most cases, the fluid is separated from the material by a flexible membrane. In some cases, the die itself can move, to assist the deformation. The production rate of hydroforming is smaller than in conventional press forming, but hydroforming has the advantage that more complex parts can be formed in one pass. Sheet hydroforming is well suited for prototyping or for low-volume production, which is often required in the aerospace industry. A closer control over the sheet thickness is possible and less springback is noticed. In the automotive industry, hydroforming is used for shaping alloys with lower formability (e.g. aluminium alloys and high-strength steels).

11.8.2 Sheet hydroforming

An illustration of sheet hydroforming is shown in Figure 11.61. The two principal deformation modes are stretch forming (the pressure of the blank holder is high and the flange material is not allowed to flow into the die cavity) and deep drawing (the flange material is allowed to flow into the die cavity). Most practical hydroform operations start with stretching. When the internal pressure increases, the tensile force on the flange increases and at a certain point the material under the blank holder is pulled inwards and the deformation becomes a combination of deep drawing and stretching, as shown in Figure 11.62. When the internal pressure is too high, two failure modes are possible. For a relatively low blank-holder force, the sealing between die and blank holder starts leaking. For high blank-holder forces, the sheet or part will burst open. Sheet hydroforming can also be used to produce hollow parts, as illustrated in Figure 11.63.

11.8.3 Tube hydroforming

Hydroforming can also be used to deform tubes. An example of a fabrication route is shown in Figure 11.64. A tube, in some cases pre-bend, is placed in a die. The

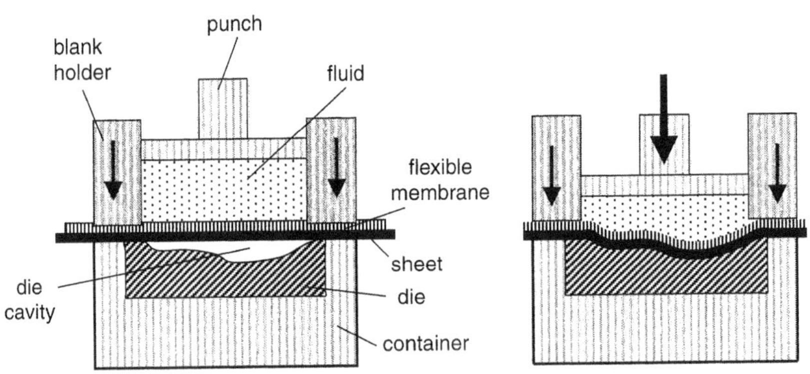

Figure 11.61. Schematic illustration of sheet hydroforming.

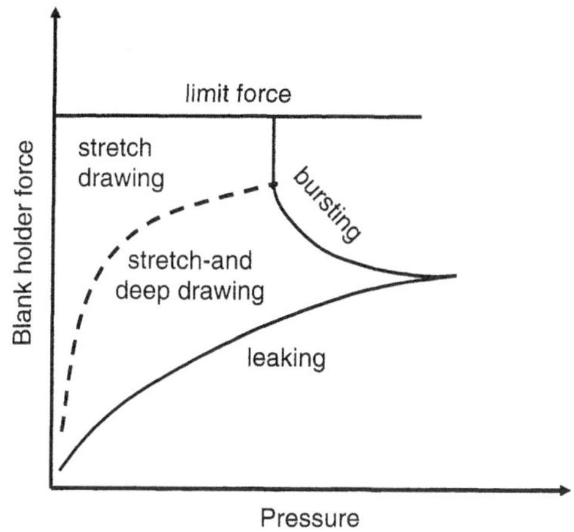

Figure 11.62. Influence of blank holder force and internal pressure on the strain mode and failure mechanism during hydroforming. After Novotny and Hein [2001].

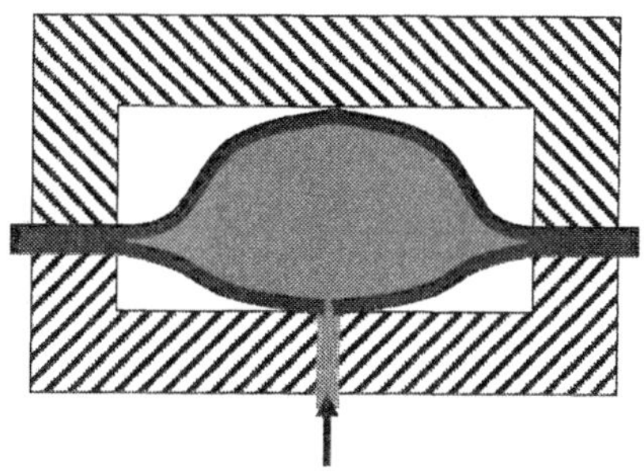

Figure 11.63. Sheet hydroforming, for the production of hollow parts.

die is sealed off on both ends and the tube is put under internal pressure. This pressure may range from 2000 to 10 000 bar, although 3000 bar will do for many applications (Koç and Altan [2001]). An axial displacement of the seals assists the material flow into the cavities of the die. A good synchronization between the pressure built-up and the axial feeding is required. During tube hydroforming, some typical defects can occur. 'Bursting' occurs when the pressure in the tube rises too fast and not enough extra material can flow into the die. It is also the typical failure mode at high blank-holder forces. A too fast 'axial feeding' of material, will however lead to buckling or wrinkling of the tube.

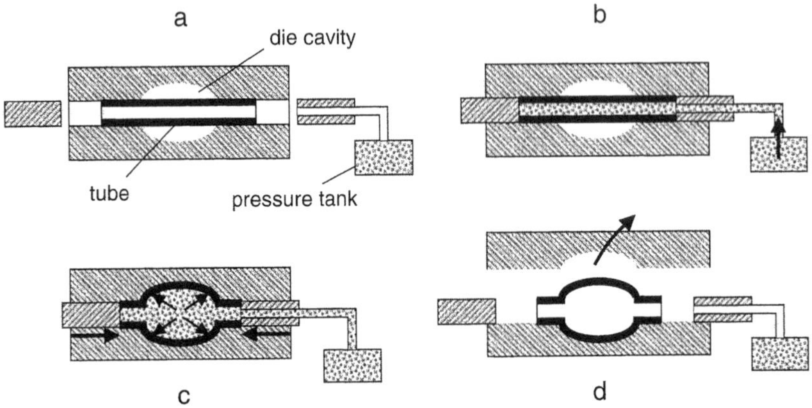

Figure 11.64. 'Tube hydroforming': (a) positioning of the tube in the die; (b) sealing off the tube and filling with fluid; (c) pressurizing combined with axial feeding; (d) opening of the die.

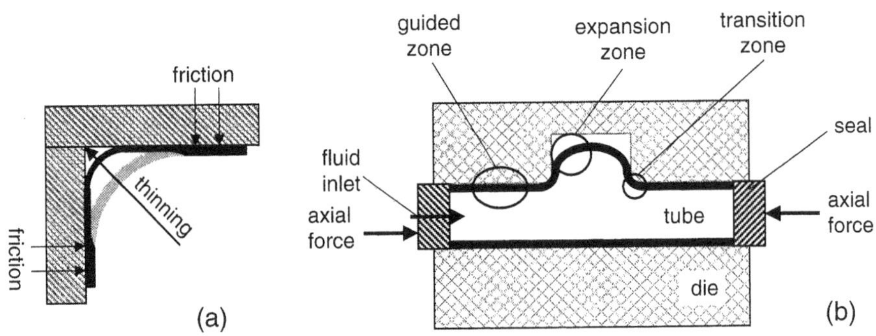

Figure 11.65. Illustration of the three friction zones in tube hydroforming. After Ahmetoglu and Altan [2000].

11.8.4 Important parameters

11.8.4.1 Friction and lubrication. Friction and lubrication are very important, especially in parts where a substantial axial feeding is required. A lubricant reduces tool wear, axial force and the risk of buckling. It prevents sticking of the material to the die as well as galling. Another critical situation is depicted in Figure 11.65a. The tube touches the wall in two contact regions. When friction between tube wall and die is high, the zone in between the contact points will thin down and eventually will burst.

In most hydroforming processes, three 'friction zones' can be found (Figure 11.65b): the 'guided' zone, the 'transition' zone and the 'expansion' zone (Ahmetoglu and Altan [2000]). Each of these zones has his own friction characteristics. In the guided zone, the material does not deform, but it is pushed towards the deformation zone by the cylinders generating the axial force. The major part of the deformation takes place in the transition zone and the so-called 'calibration' (fitting the material into the die) takes place in the expansion zone, mostly by local stretching.

Friction itself depends on many parameters like the type of the lubricant, the material of the tube and the surface conditions of the die. A suitable lubricant must be chosen in function of the internal pressure, the contact speed and the contact length. Various types of lubricants are available: solid lubricants like graphite or MoS_2, waxes, oils and emulsions.

11.8.4.2 Material parameters. The $\bar{R}$-value or normal anisotropy of a sheet has an important influence on the deformation during hydroforming. As discussed above, the deformation starts with a biaxial stretching of the sheet because the blank holder withholds the sheet to move inwards. When a certain internal pressure is reached, the axial force overcomes the pressure of the blank holder and the material is pulled inward. At that moment, a deep drawing component is added to the deformation. Since the stretchebility of a material is limited, it is important that this deep-drawing component, which feeds material into the die, occurs in an early stage of the process. It has been shown (Novotny and Hein [2001]) that for a material with large $\bar{R}$-value, the stress in the sheet raises faster with increasing internal pressure than for a material with a lower $\bar{R}$-value (Figure 11.66). This means that for higher $\bar{R}$-value, deep drawing starts at a lower internal pressure; and hence, the

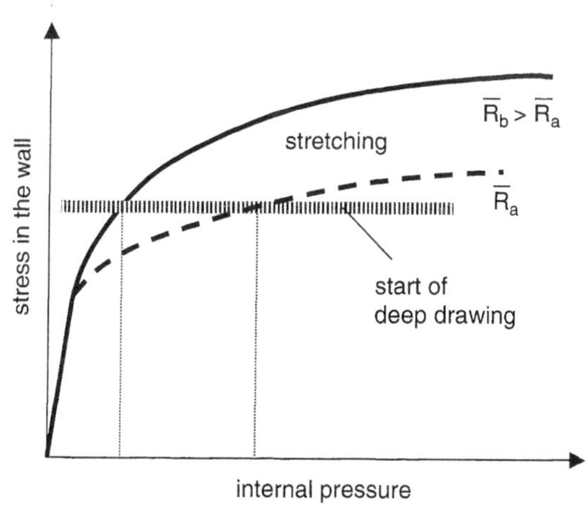

Figure 11.66. Influence of the $\bar{R}$ -value on the relation between internal pressure and the start of the deep drawing component during sheet hydroforming.

risk that the sealing of the die will leak, or that the material will burst before fresh material starts moving into the die, is lower.

The anisotropy is also an important parameter in tube hydroforming. Experiments and simulations have shown that a high $\bar{R}$-value is beneficial for the deformation (Carleer *et al.* [2000], Manabe and Amino [2002]). A high strain-hardening coefficient is also beneficial because (like in biaxial stretching), the deformation is more homogeneously distributed over the tube wall and local necking is delayed.

LITERATURE

K. Siegert (ed.), "Hydroforming of Tubes, Extrusions and Sheet Metals", vol. 1, Stuttgart University, Institut for Metal Forming Technology, Stuttgart, Germany (1999).

Ibid vol 2 (2001). Ibid vol 3 (2003).

11.9. HIPPING

11.9.1 Introduction

Hot isostatic pressing (HIP/hipping) involves the simultaneous application of heat and high pressure. A furnace is enclosed in a high-pressure vessel and the workpieces are heated while an inert gas (usually argon) applies a uniform pressure. The technique was invented at the Battelle Memorial Institute in 1955 for the diffusion bonding of dissimilar materials (a method was needed to bond zirconium to a zirconium–uranium alloy (Boyer [1992]), but other applications, usually related to the removal of pores, came gradually into focus. At this time, hipping is mainly used for the densification of high-performance castings, the consolidation of metal powders and for cladding. The temperature used in a hipping cycle is between 0.5 and 0.7 times the melting temperature and the gas pressure is between 20 and 400 MPa, and typically around 100 MPa. This is comparable with the pressure at 10 000 m below sea level. A schematic view of an HIP installation is presented in Figure 11.67. The pressure is obtained partly by pumping-in gas with a compressor, partly by heating this gas.

11.9.2 Densification mechanisms

The main purpose of hipping is the densification of a porous product (casting, powder compact, etc.). Several mechanisms can contribute to this densification. When a pressure is applied, some elastic deformation of the component will first occur, followed by plastic yielding at the contact points, e.g. between powder particles (Figure 11.68). Once these contact points can bear the internal forces

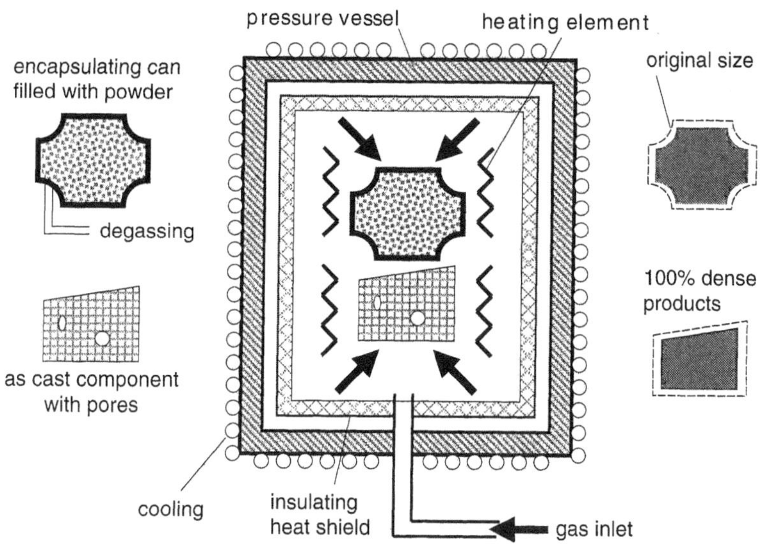

Figure 11.67. Schematic view of an HIP installation.

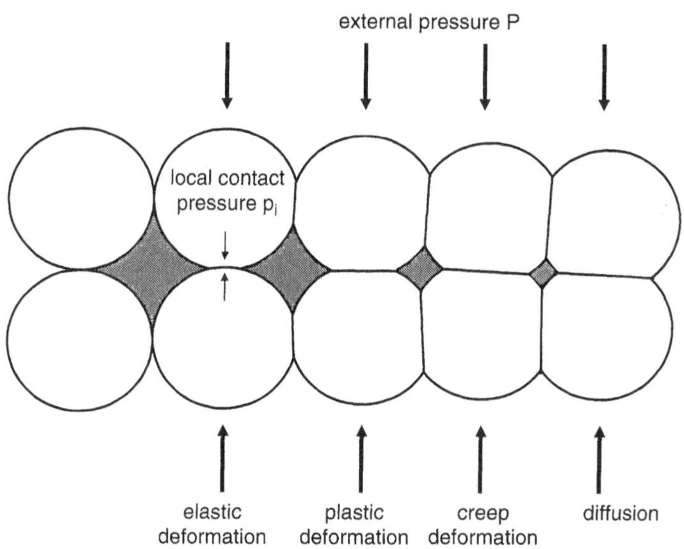

Figure 11.68. Schematic overview of some densification mechanisms occurring during hipping.

caused by the external pressure, further densification will occur by time-dependent processes such as power-law creep or by diffusion.

An elegant way of visualizing the pressure and temperature dependence of these densification processes was proposed by Arzt *et al.* [1983]. They constructed 'hot-isostatic pressing diagrams' (also called 'HIP-maps' or 'densification maps'). These maps show the dominant densification mechanism as a function of external applied pressure and temperature. A schematic representation is shown in Figure 11.69. At a given external pressure P and temperature T, the relative density gradually increases (the pore size decreases) with time. When the hipping process starts, there is first some elastic deformation (not shown in Figure 11.69), immediately followed by plastic yielding. Perhaps, it is surprising that plastic yielding occurs for $P < \sigma_y(T)$, but one has to take into account that P represents the externally applied pressure. Owing to the small contact areas at the beginning of the hipping cycle, the local contact pressures, p_i can be much higher than the yield stress of the material. During plastic deformation, the local contact areas increase and the local contact pressures decrease.

When the local contact areas are large enough to bear the local forces, yielding stops and the size of the contact areas continue to grow by power-law creep. This will lead to 100% density only at high external pressure. For low to moderate pressures, the final closure of the pores will occur by diffusion. For some materials, total densification may not be possible and in that case the HIP-map will show a region of 'no densification'.

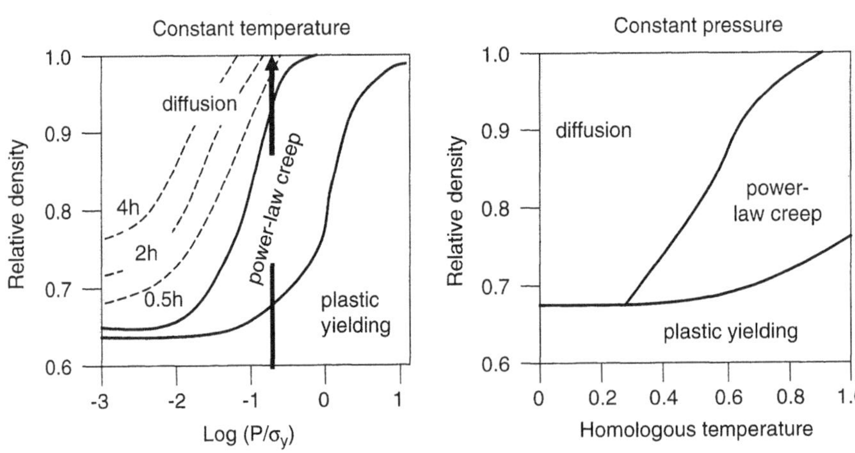

Figure 11.69. Hot-isostatic pressure maps (density/pressure and density/temperature). After Arzt *et al.* [1983].

Many HIP-maps contain lines of 'constant time'. For example, the HIP-map on the left-hand side of Figure 11.69, shows that for the external pressure indicated by the arrow, 100% density will be reached in about 1 h. This densification process will occur by plastic yielding, followed by power-law creep and the final closure of pores will occur by diffusion.[7] As an alternative for the 'pressure–density' maps, a 'temperature–density' map, as shown in the right-hand side of Figure 11.69, can be used.

11.9.3 Hipping equipment

The combination of high temperature and high pressure puts severe demands on the construction of a hipping installation. Especially, the pressure vessel must be designed with great care. The most important requirements are a high tensile strength, a high fatigue resistance and good fracture toughness. There are mainly two types of pressure vessels: forged monolithic vessels made from ductile low-alloy steel and wire wound vessels. The top and bottom closure of the vessel are critical parts of a hipping unit. Some vessels are closed with an external 'joke-type' system. The pressure in the vessel is transmitted to the joke by the end closures. Monolithic vessels have a traditional threaded closure, but with rounded threads for a better distribution of the stresses. For safety reasons, hipping units are located in concrete pits.

Most commercial hipping installations have a furnace inside the vessel, with thermal barriers to protect the vessel (Figure 11.67). The gas which is used to exert the pressure (partly by mechanical pumping, partly by heating) is usually high-purity argon. Impurities such as hydrogen, oxygen and nitrogen, even in very small concentrations, cause surface degradation in certain alloys. In nickel-base alloys, for example, nitrogen forms brittle carbonitrides and oxygen forms intergranular oxides (Eridon [1988]). Sometimes additives are required, e.g. for carburized steel, the gas should contain carbon to prevent decarburization. For encapsulated materials, like powders and for aluminium castings cheaper nitrogen gas can be used. The most common materials used to encapsulate powders are mild steel, stainless steel and nickel. Most envelopes are made by spinning, hydroforming and superplastic forming or by conventional sheet-forming techniques. For materials which have to be hipped at very high temperature (1500–2000°C), glass envelopes can be produced by glass blowing. At the hipping temperature, glass is soft enough to be deformed.

[7] It is recommended to look at the 'deformation mechanism map' shown in Figure 6.1. This map shows that for decreasing load the dominant deformation mechanism can indeed change from plastic deformation (called: dislocation glide) to creep and diffusion.

Table 11.16. Properties of an as-cast, hipped and wrought alloy steel (*Source*: Data from Atkinson [1991]).

	YS (MPa)	UTS (MPa)	Elongation to failure (%)	Toughness to fracture ($MNm^{-3/2}$)
Cast	988	1350	8.9	105
Hipped	1142	1460	13.3	122
Wrought	1155	1480	14.7	124

11.9.4 Typical applications

Hipping enables the production of materials of various shapes and sizes. About 50% of the applications are related to the densification of castings, one-third to powder metal products and about 10% to cladding. Hipped products are used in the aerospace, automotive, medical, chemical and petroleum industries.

11.9.4.1 Densification of castings. In a casting process, internal porosities may be generated because of shrinkage during solidification or because of entrapped gas. Removal of these pores will improve mechanical properties, e.g. strength, toughness, fatigue and creep resistance. In many cases, the properties of the hipped product are nearly as good as those of a wrought component. Table 11.16 shows an example for an alloy steel.

Materials used for hipping are often titanium alloys and nickel-based alloys. Typical applications are structural castings for jet engines, vanes for divers engines and turbine blades. Lower cost castings from steels and aluminium alloys are an increasing market. Many new applications for the pharmaceutical industry, food processing and vacuum applications require polished, pore-free surfaces at competitive prices.

11.9.4.2 Densification of powder metal products. With conventional sintering techniques, it is very difficult to obtain fully dense products. Extended sintering times can lead to unacceptable grain growth, which degrades the properties of the components. However, some products, e.g. wire drawing dies, require zero porosity. In such cases, hipping can transform pre-sintered components into near 100% dense products. Classical examples are the elimination of residual porosity of pre-sintered hard metals (cemented carbides) or hipping of pre-sintered ceramics like Al_2O_3, Si_3N_4 and Si–Al–O–N compounds (Atkinson and Rickinson [1991]).

Hipping can also consolidate fine powders into near 100% dense products. In that case, the powders must first be encapsulated into a container which is degassed and sealed. Typical examples of hipped powder metallurgical parts are nickel-base superalloys, high-speed tool steels and ceramic powders. Casting of these alloys is problematic, but the production and consolidation of powders offers an attractive alternative. Nickel-base superalloys, for example, undergo severe segregation during classical ingot casting. Production of powders by atomization eliminates macrosegregation and reduces microsegregation. After hipping a homogeneous product with fine grain size and excellent mechanical properties is obtained. The shrinkage of the product should in principle be isotropic because the pressure exerted by the gas is the same in all directions. However, in practice, some factors like a non-uniform filling of the container may generate slight distortions. In some cases, the shape of a rectangular bar turns into a dog bone because the corners of the container are stronger than the side walls.

11.9.4.3 Cladding. Another application of hipping is to join dissimilar materials that are difficult to join by traditional methods. Owing to the surface roughness, real contact between two surfaces that are pressed together will only occur at some local contact points. At high temperature, local plastic deformation will increase the contact area and subsequently diffusion of atoms from one material into the other will form the bond between them. Remaining porosity will be removed in the same way as illustrated in Figure 11.68 for powder compacts.

LITERATURE

Atkinson H.V. and Davies S. Metall. Mater. Trans. A, 31 (2000), 2981.

Conway J.J. and Rizzo F.J., "Hot Isostatic Pressing of Metal Powders", vol. 7, ASM-Handbook (1998).

Draney C., Froes F.H. and Hebeison J., Mat. Tech. and Adv. Perf. Mat., 19.2 (2004), p. 140

11.10. SUPERPLASTIC FORMING

11.10.1 Technology

In Chapter 6, the superplastic behaviour and the underlying deformation mechanisms (grain boundary sliding + accommodation mechanism) of certain materials have been discussed. It was shown that superplasticity occurs in fine-grained materials, deformed at high temperature and low strain rate. In the present paragraph, the technology of superplastic forming will be introduced. In most cases, superplastic forming is carried out with one of the following techniques: blow forming/vacuum forming, thermo-forming or superplastic forming combined with

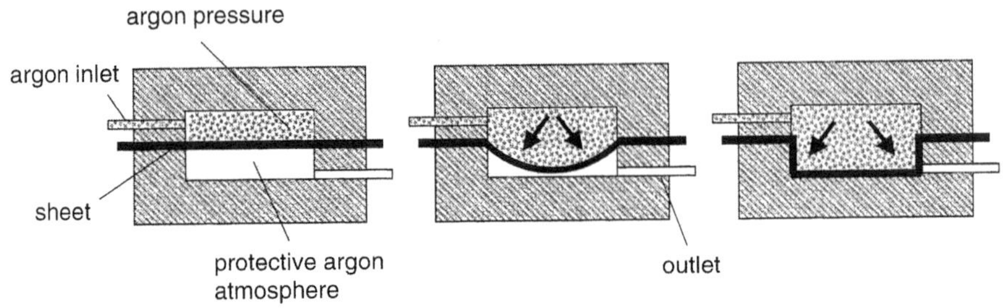

Figure 11.70. Illustration of the blow-forming technique for superplastic forming.

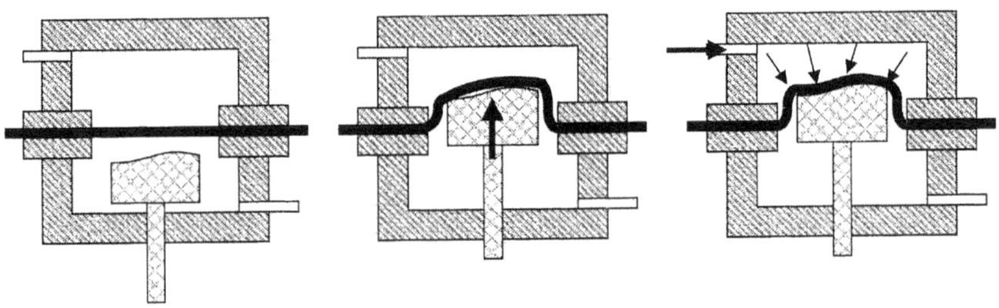

Figure 11.71. Illustration of a possible set-up for the thermo-forming technique.

diffusion bonding. Forging and extrusion can also be used to shape superplastic materials.

Blow forming and vacuum forming are related techniques. In blow forming, a gas (mostly argon) is used to create a pressure difference over the (warm) superplastic sheet (Figure 11.70). The sheet is stretched and forced to adopt the shape of the die. The outer side of the sheet is held in a fixed position and does not draw-in as would be the case in deep drawing. In some cases, a 'back pressure' is used to prevent cavitation. In vacuum forming, the pressure difference over the sheet is generated by vacuum in one of the die chambers. In this case, the pressure difference is limited to 1 atmosphere.

Thermo-forming is a typical forming technique for plastics. In this technique, a moving die is used in conjunction with a gas pressure. Several configurations are possible, e.g. a male die is used to stretch-form the superplastic sheet. After that, a gas pressure is used to force the sheet against the die (Figure 11.71). One of the most important advantages of superplastic forming is that complex shapes can be produced in one piece and most often in one pass. The absence of welds, rivets, etc. diminish the risk for fatigue damage and corrosion. An additional advantage is that little or no residual stresses are present. The low forces acting during superplastic forming reduce the tooling cost. A disadvantage of superplastic forming is

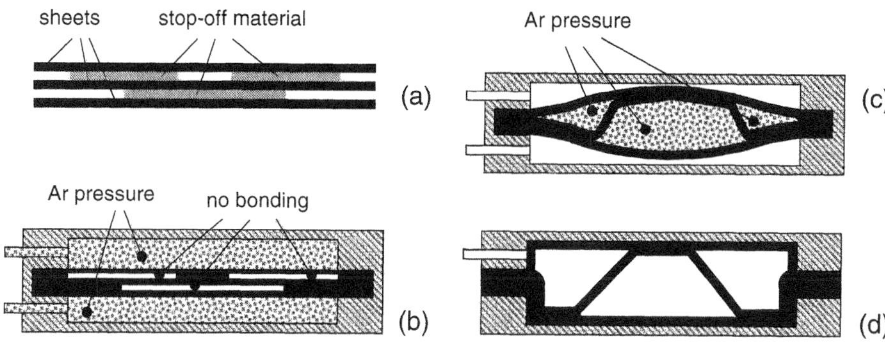

Figure 11.72. Combination of superplastic forming with diffusion bonding.

of course the low production rate, because of the limitations in maximum allowed strain rate.

The low production rate is turned into an advantage when superplastic forming is combined with *diffusion bonding* (the 'SPD/DB' process). Two or more sheets of superplastic material are stacked together but make direct contact with each other only at some well-chosen locations (Figure 11.72a). At the other contact areas, a diffusion barrier (a so-called 'stop-off' material) is used. In a first step, the sheets are heated and bonded together at the chosen locations by diffusion bonding. This bonding process is 'assisted' by an external gas pressure (Figure 11.72b). In a subsequent step, the small clefts at the stop-off material are expanded by an internal gas pressure (Figure 11.72c). The outer sheets are forced into the die cavities and the final shape is being formed (Figure 11.72d). The stop-off material depends on the material of the sheets. For titanium alloys, yttrium or boron nitride have successfully been used. The SPD/DB technique is well established for titanium alloys and has many industrial applications, for example, in the aircraft industry. For aluminium alloys, diffusion bonding is less evident because of the surface oxide film and solutions have been sought by introducing interlayers in the form of claddings, coatings or foils.

11.10.2 Thinning

Owing to the large deformation during superplastic forming, significant thinning of the sheet is unavoidable. In many cases, this thinning will not be uniform throughout the sheet. In order to deliver products with the required dimensions, this thinning behaviour must be understood and taken into account in the design and production of superplastic parts. Let us consider the case of blow forming (Figure 11.70). In the first stage of the deformation, free bulging of the sheet occurs. In this geometry there is a stress-state gradient from pole to edge, with an

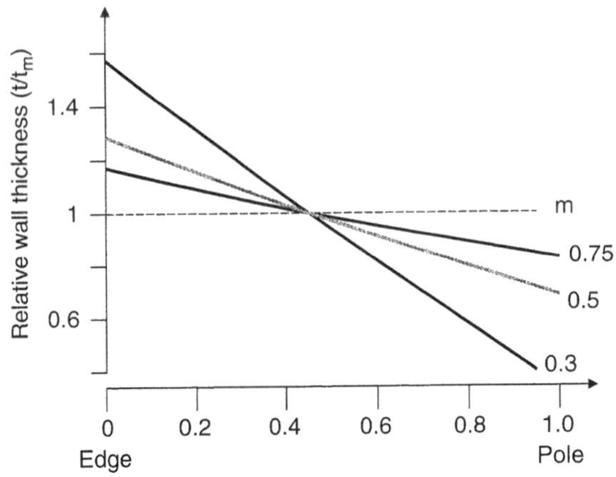

Figure 11.73. Relative wall thickness between edge and pole of a hemisphere, for different values of the strain rate sensitivity coefficient *m* (*t*: actual wall thickness; t_m: mean wall thickness). After Cornfield and Johnson [1970].

equi-biaxial stress at the pole and a plane strain stress state at the edge. It can be calculated and experimentally verified that this will lead to more thinning at the pole than at the edge. The difference in thickness will be larger when the strain rate coefficient *m* is smaller. This is illustrated in Figure 11.73.

Once the hemisphere makes contact with the die, an even more severe problem may occur. Because of friction between the sheet and the die, the corners of the die will be filled by excessive local stretching and significant local thinning of the sheet. This problem resembles a similar problem occurring during hydroforming and is illustrated in Figure 11.65.

Thinning control is often an important issue in superplastic forming. Some possible methods to reduce thickness differences are (Hamilton [1989]):

- Processing the superplastic material to obtain high *m*-values.
- Using adequate lubrication to avoid excessive local thinning in the die corners.
- Applying pressure in a controlled and profiled manner (pressure profiling) in order to maintain the strain rate at a level corresponding to a high *m*-value.
- Using microstructural gradients (e.g. grain-size gradients) (Cheong *et al.* [2003]).

11.10.3 Cavitation

The nucleation, growth and interlinkage of intergranular voids during superplastic forming has been observed in many alloys. This process, called 'cavitation', can be responsible for a degradation of the service properties of the superplastic part or for a premature fracture of the material during superplastic forming. For this reason, it is important to understand and control or inhibit cavitation.

The nucleation of cavities (Suéry [1985], Ridley [1989])
Cavities may pre-exist in many alloys due to the TMP carried out to get superplastic material. But there is also a strong suspicion that cavities nucleate continuously at grain boundaries due to insufficient accommodation of grain boundary sliding. The main nucleation points seem to be tripelpoints and hard particles at the boundaries. When the deformation rate remains below a critical value ε_c°, the formation of cavities is prevented by diffusive stress relaxation. An expression for the critical strain rate has been proposed by Stowell [1983]:

$$\varepsilon_c^{\circ} = 11.5\left[\frac{\sigma\Omega}{xd^2D}\right]\left[\frac{\delta D_{gb}}{kT}\right] \tag{11.50}$$

with σ the external stress, Ω the atomic volume, d the radius of a particle, D the grain size, x the fraction of deformation by grain boundary sliding, δ the grain boundary width, D_{gb} the grain boundary diffusion coefficient and T the temperature. This formula shows that nucleation of cavities increases when larger particles are present and when the grain size becomes larger.

The growth of cavities (Suéry [1985], Mukherjee [1992])
A cavity located on a grain boundary can grow by a diffusion-controlled growth mechanism or by plasticity-controlled growth or by a combination of both mechanisms. The first mechanism seems only relevant at very low strain rates and for small cavities. At strain rates typical for superplastic forming, the growth is controlled by plastic deformation of the surrounding region. The growth of cavities can be described by the relation Suery [1985]:

$$\frac{\mathrm{d}V}{\mathrm{d}t} = kV\varepsilon^{\circ} \tag{11.51}$$

with V the cavity volume, ε° the strain rate and k a parameter. The growth rate is proportional to the volume of the cavity and hence an exponential increase of V with strain can be anticipated. It has been shown by Cocks and Ashby [1982] that k decreases as the strain rate sensitivity m increases.

Cavity interlinkage and fracture
Individual cavities grow during deformation. At a given moment, two expanding cavities will meet and the two cavities with volume V_1 and V_2 will merge into one cavity with volume $V_3 = V_1 + V_2$. This cavity interlinkage will cause an unexpected

shift towards higher values of the cavity size distribution. Attempts to model this process of cavity coalescence have been published (Stowel [1984], Pilling [1985], Nicolaou [1999, 2000]), but the exact mechanism and precise criteria to link coalescence with premature failure seems to be lacking.

Cavition control by hydrostatic pressure

One of the most effective methods to reduce the occurrence of cavitation is to apply a hydrostatic pressure during superplastic forming. For example, during blow forming (Figure 11.70), argon pressure is applied on both sides of the sheet. Of course a pressure difference is maintained to force the sheet into the die. This hydrostatic pressure not only decreases the growth of the cavities, but also reduces the nucleation rate of new cavities. A critical stage in the formation of a new cavity is the growth of the cavity to a critical size, which is stable against shrinkage due to surface tension (Suéry [1985]). The critical radius r_c for a cavity subjected to a tensile stress σ is given by the expression $r_c = 2\gamma/\sigma$, with γ the surface tension. When a hydrostatic pressure p is superimposed, the critical radius is larger, and can be calculated from:

$$r_c(p) = \frac{2\gamma}{(\sigma - p)} \qquad (11.52)$$

This formula shows that the nucleation of stable new cavities can considerably be suppressed by hydrostatic pressure.

LITERATURE

Mukherjee A.K., 'Superplasticity in metals, ceramics and intermetallics', in "Materials Science and Technology", R.W. Cahn, P. Haasen, E.J. Kramer and H.Mughrabi (eds), vol. 6: 'Plastic Deformation and Fracture of Materials', Wiley, Weinheim, Germany (1992).

Nieh T.G., Wadsworth J. and Sherby O.D., "Superplasticity in Metals and Ceramics", Cambridge University Press, Cambridge, UK (1997).

"Superplasticity", AGARD lecture series No 168, Agard (ISBN 92-835-0525-5) (1989).

"Superplasticity", Proceedings of the International Conference Grenoble 1985, B. Baudelet and M. Suéry (eds), CNRS, Paris (1985).

Chapter 12

Defects in Thermo-Mechanical Processing

Chapter 12
Defects in Thermo-Mechanical Processing

12.1. INTRODUCTION

Discussions on fabrication defects are usually focussed on individual forming technique(s). However, in this section a more 'generic' classification is presented. As shown in Table 12.1, TMP defects may range from mostly macroscopic 'form and fracture'-related defects[1] to defects related to strain localizations,[2] and microstructure-related imperfections. More specific defect nomenclatures (e.g. forming technique related) are discussed in the subsequent sections (12.2–12.6).

The 'defects' in TMP can have two possible origins – process related and/or metallurgical. The former can be grossly related to the TMP practices, including the forming techniques and the heat treatment, while the metallurgical origin may range from the starting solidification structure to structural developments during TMP. A clear demarcation between the two 'origins' is difficult to establish – for example, the so-called roll marks (a surface defect) may include work roll marks, work roll pin holes, chatter marks or any other repetitive structures and can have both technological, or process related, and metallurgical origin. Tables 12.2 and 12.3 classify respectively the process related and metallurgical origins of TMP defects. Subsequent sections will discuss the 'generic' TMP defects, as mentioned in Table 12.1, and attempt to

Table 12.1. 'Generic' classification of defects in TMP. The nature (macroscopic or microscopic) and possible origin (process related or metallurgical) of the 'generic' defects are also indicated.

Defect classification	Remarks
Form defects: incomplete/improper macroscopic metal flow leading to improper shape/size	Mostly macroscopic – both process related and metallurgical
Surface defects: defects or flaws restricted to surface or near surface regions	Both process related and metallurgical
Fracture-related defects: total or partial fracture	Macroscopic. Both process related and metallurgical
Strain localizations: a generic name used for any plastic instability	Both process related and metallurgical
Structural defects: inappropriate structure, causing inferior properties	Mostly metallurgical in origin. Typically *microstructure related*

[1] Concepts of fracture have been treated in Chapter 3.

[2] Deformation heterogeneities or plastic instabilities; this has been treated in Chapter 4.

Table 12.2. Some process-related origins of TMP defects.

Process-related origin	Description
Machine related	Inappropriate (too small/large) forces, etc.
Process parameter related	Inappropriate strain, strain rate, working temperature, etc.
Die-tool related	Die-tool marks, stress raisers, etc.
Tribology related	Inappropriate lubrication and friction
Environment related	Scaling, internal oxidation, decarburization, etc.

Table 12.3. Some metallurgical origins of TMP defects.

Metallurgical origin	Description
Starting structure	Segregation, piping, porosity of solidification structure
Structural developments during processing	Development of inappropriate structure during processing (strain localizations, etc.)
Final structure	Inappropriate final structure property

sub-divide them following conventional TMP defect nomenclature. Present understanding on the defect-origin and possible preventive measures are also discussed.

12.2. FORM DEFECTS

Different types of form defects, defects in size and shape of the product, are reported during TMP operations (Polushkin [1956], Metals Handbook [1988a]). Some[3] of these are compiled in Table 12.4, along with their process-origin and typical corrective measures. Automation in a modern plant often incorporates machine/process-related corrective measures.[4] Freedom from major form defects is primarily assured by proper process optimization – either through numerical or physical modelling/simulation (Chapter 13). An example of the FEM-based numerical simulation in predicting form defects is shown in Figure 12.1.

12.3. SURFACE DEFECTS

The surface defects can be classified as deformation or forming process-induced, environment induced, coating related and induced by solidification

[3] For a more exhaustive listing, the reader may refer Polushkin [1956].

[4] For example, automatic roll bending, to partially compensate for buckling, is a standard feature in a modern rolling mill.

Table 12.4. Different types of form defects, their process-origin and corrective measures.

Nomenclature	Description	Origin (*process*)	Corrective measures
Bow, side and longitudinal bent	Sidewise or longitudinal bending	Inaccurate guiding or alignment in *rolling or slitting*	Use of levelling or straightening rolling
Buckling	Bending under compressive stresses	Any forming process: inaccurate stress distribution	Straightening
Camber	Curvature in rolled plates/strips	Roll bending at the middle of the rolls – *rolling*	Compensated by varying the roll diameter across the roll length
Cobbling	Accidental distortion in shape through re-rolling	Front end being caught and rolled again – *rolling*	Reduced rolling speed
Exfoliation	Disjoining during fabrication	Any forming process: pipes, unwelded blowholes, severe inclusion content	Control of solidification structure
Overfill/underfill	Inappropriate (over or under) metal flow	Extrusion/drawing: over or under feed of metal	Control of feed rate
Warping	Distortion or warpage	Any warm/hot forming: fast heating cooling	Control of thermal history

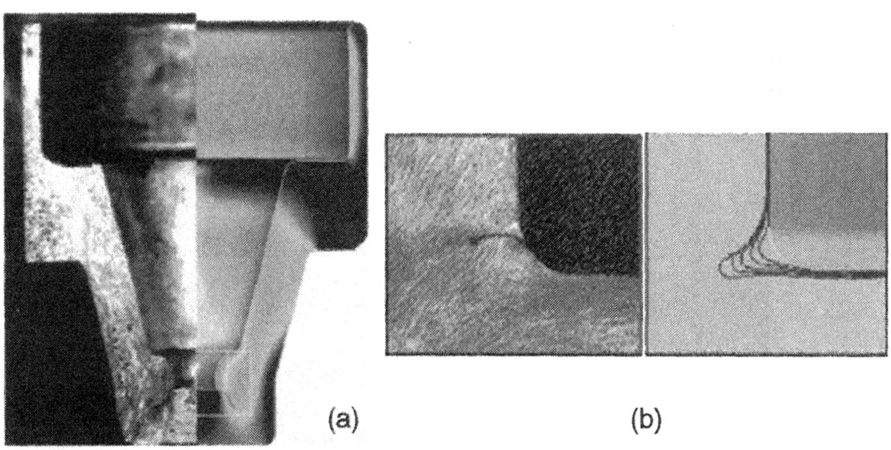

Figure 12.1. (a) Real and simulated images of 'lap' during cold extrusion. (b) Real and simulated images of a form defect in forging. Courtesy of Biba [2001] (copyright (2007), with permission from Elsevier).

structure (see Table 12.3). In this section, the first three types of surface defects are discussed.

12.3.1 Deformation or forming process-induced surface defects

Deformation or forming process-induced surface defects may include seams and slivers caused by surface inclusions (Metals Handbook [1988a]), snake skin or shark skin caused by tribological factors (Polushkin [1956]), wrinkling (Section 12.3.1.4) and a range of plastic instabilities or strain localizations. Typical examples of such surface plastic instabilities are Lüder bands, orange peel and roping. These and the wrinkling phenomena are discussed separately in this section.

12.3.1.1 Lüder bands. Lüderlines or stretcher strain marks become visible after press forming of a sheet and have a typical 'flame-like' pattern as shown in Figure 12.2c. This surface pattern is, in most applications, unacceptable for visible parts. The phenomenon is well known in most steels and in certain aluminium alloys. It can easily be detected with a simple tensile test. Materials which are prone to form Lüderlines during press forming always show a so-called 'yield point elongation' or 'Lüderstrain' in the stress–strain curves obtained in a tensile test (Figure 12.2a). The reason for these Lüderstrains is the pinning of dislocations by carbon atoms in steel or by substitutional atoms in aluminium. In steel the effect can be eliminated by giving a so-called 'skin pass' (a few per cent cold rolling reduction) or by 'roll levelling' (bending – unbending). This liberates the dislocations from their pinning points. In aluminium the effect can be avoided by grain size control (a grain size above 10–15 μm is required).

Repetitive plastic yielding is observed under certain deformation conditions – conditions with negative strain rate sensitivity (Hähner [1993]). This is also called as dynamic strain aging (DSA) – dynamic interactions between mobile dislocations and solute atoms (Hähner [2003]) or PLC (Portevin–LeChatelier effect) (Chihab [1987], McCormick [1988], Aluminum [1993], Hähner [1993], Estrin [1995], Lloyd [1997], Kovács [2000], Chmelik [2002], Hähner [2003]). This has originally been reported as 'blue brittleness' in mild steel, and the effect is typically felt as visible bands on the product surface and also as serrated flow in a typical stress–strain diagram (see Figures 12.2b, c).

The selection of exact alloy system and appropriate TMP conditions are often decided based on the severity and the nature of Lüder bands and PLC. Controversies still exist on the Lüder band and PLC nomenclature and on their mechanisms of formation. As shown in Figure 12.2a, serrations after the first yield point elongation is called Lüder band/strain or stretcher strain marks or type A bands, while serrations after Lüder bands are called PLC or type B bands. On the other hand, as shown in

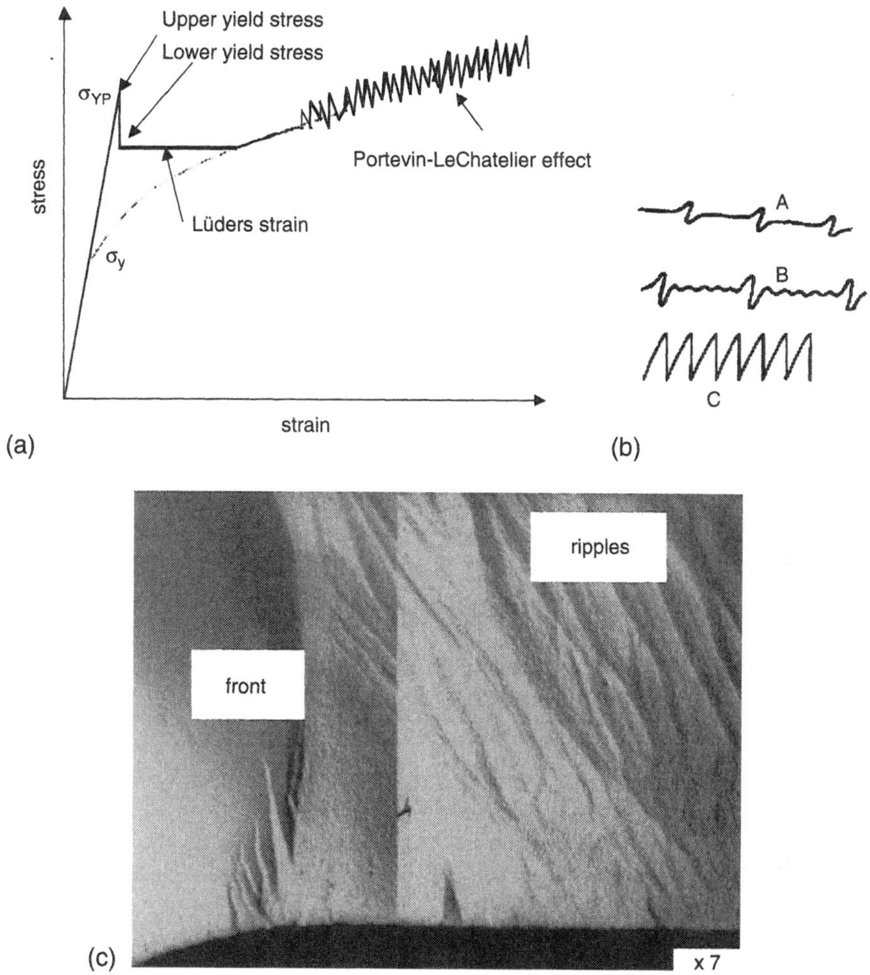

Figure 12.2. (a) Generalizing the effects of strain aging, Lüder bands and PLC (Portevin–LeChatelier effect) on a stress–strain diagram. (b) Distinct patterns of the jerky flow during PLC. (c) Lüder band progressing along the surface of a tensile specimen.

Figure 12.2b serrated or jerky flow, associated with both Lüder bands and PLC, are generalized.[5] into three categories (Hähner [1993, 2003]):

- Type A: Appears as regular and equidistant 'drops' in a stress–strain diagram. Mechanism – bands nucleating and propagating as solitary plastic waves.

[5] The differentiation is mainly phenomenological (Hähner [2003]).

- Type B: Appears as C and A superimposed on each other. Mechanism – oscillatory or intermittent propagation.
- Type C: Appears as heavily serrated flow with abrupt oscillations. Mechanism – stochastic and random nucleation without propagation.

12.3.1.2 ***Orange peel.*** Plastic deformation, for example: stretching, bending and drawing, may develop rough surface – generically termed as 'orange peel'. This has been reported in aluminium (Stachowicz [1986], Aluminum [1993], Lee [1998]) and also in steel (Rittel [1991]).

Coarse surface grains are usually considered to have less constraints[6] to plastic deformation. This, on the other hand, may cause severe non-uniform deformation between the surface grains, leading to an apparent roughening or an 'orange peel' condition. The development of 'orange peel' or surface roughness is dependent on the surface grain size – effect being negligible at finer grain sizes (Rittel [1991], Aluminum [1993]). Though cluster of small grains of the similar crystallographic orientations may act as individual coarse grain(s) and create 'orange peel' effects (Aluminum [1993]), no clear relationship between 'orange peel' and the crystallographic orientations of the surface grains has been reported (Lee [1998]).

12.3.1.3 ***Roping.*** Roping or looper lines, is also a phenomenon of surface roughening, observed during deep drawing. The name roping is linked to characteristic 'roping' marks on drawn shapes (Aluminum [1993]). The roping is also caused by non-uniform deformation, non-uniformity linked to irregularities or heterogeneities in the structure. For example, macro-segregation from the cast structure may produce striated structure or surface banding or roping marks. Alternatively, the roping may also develop from deformed grains or bands of fine grains of similar orientations – both may originate from earlier[7] coarse grains.

12.3.1.4 ***Wrinkling.*** Wrinkling (Hayashi [1984], Metals Handbook [1988a], Szacinski [1991], Hosford [1999], Livatyali [2004], Abedrabbo [2005], Loganathan [2005]) or the formation of surface roughness/wrinkles is caused by internal compressive stresses, plastic as well as elastic. It is usually observed (Abedrabbo [2005]) in the flange, but is also reported in the free-forming zone between die and tool, when the minor stress (Section 11.7) of sheet metal forming is compressive in nature (Hosford [1999]). Normally, wrinkling is more severe in metals with lower normal anisotropy ($\bar{R}$ values). Wrinkling also depends on tooling, elastic modulus and sheet thickness (Metals Handbook [1988a], Szacinski [1991], Hosford [1999],

[6] Constraints by surrounding grains.
[7] Developed during hot working or during intermediate annealing.

Livatyali [2004], Abedrabbo [2005], Loganathan [2005]) and is normally considered as a complicated phenomenon, exact initiation and growth of wrinkling limit being difficult to predict theoretically.

To avoid wrinkling, usually the blank-holder pressure is increased (Metals Handbook [1988a], Hosford [1999]), which, on the other hand, introduces additional strain in the sheet metal and hence may provide a narrow processing window between wrinkling limit and failure limit. An alternative exists by forming at elevated temperatures (Hayashi [1984], Szacinski [1991]). Thermal shrinking of sheet metal is attributed (Hayashi [1984]) to reduce compressive stresses and the resulting wrinkling, while elevated temperature forming also improves drawability (Szacinski [1991]) by reducing the yield strength. Alternate forming technologies, e.g. hydroforming and electro-magnetic forming, are often 'recommended' (Abedrabbo [2005]) as techniques with lesser 'wrinkling'.

12.3.2 Environment-induced surface defects

Environment-induced surface defects are often related to heat-treatment practices (Polushkin [1956]). These may range from simple scaling to depletion of alloying elements (e.g. decarburization in steel) and internal oxidation, and these are discussed separately.

12.3.2.1 Oxide scale. Hot working has been almost always associated oxide scaling. Except for highly reactive and expensive metallic systems, where jacketing (e.g. Zr alloys–Section 16.1) or protective coating (e.g. glass coating) is used, the oxide scale is typically a part of any TMP practice of warm/hot working. The oxide scale is often complex. For example, in hot steel the scale is reported (Sheasby [1984], Fukagawa [1994]) to consist of three layers – inner wüstite (FeO), intermediate magnetite (Fe_3O_4) and outer haematite (Fe_2O_3), with interesting surface characteristics (Sun [2003]). The scale has strong implications to die-tool and metal interactions (Li [2001], Vergne [2005]) and to heat transfer (Li [1998]); but more importantly control and removal of oxide scale is an important issue technologically. High-pressure water jets are an integral part of a modern hot rolling mill – jets employed to control the severity of oxide scale. Even then an entire 'pickling'[8] plant, and the science and technology of 'pickling' (Metals Handbook [1988b]), is usually associated with TMP technology. Improper pickling can lead to surface defects (e.g. pickling blisters, etc.) and even to embitterment.

12.3.2.2 Decarburization. Decarburization in special steels during TMP practices would lead to subsequent loss of hardenability, especially in surface/sub-surface layers (Polushkin [1956], Barlow [1984]). Preventive measures may range from

[8] Both acid pickling and electrolytic pickling are used.

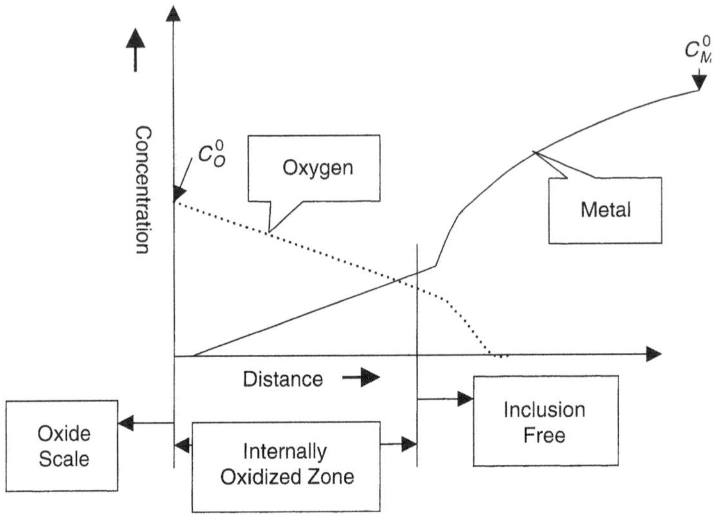

Figure 12.3. Schematic description of oxygen and metal concentration as a function of distance. The figure outlines existence of oxide scale, internally oxidized zone and inclusion-free interior. After Van Vlack [1977].

reduced time/temperature of heat treatment to use of protective atmosphere and coating the stock with suitable (which may reduce oxygen potential at the surface) refractory compound.

12.3.2.3 Internal oxidation. Though internal oxidation (Van Vlack [1977]), formation of relatively fine sub-surface oxide inclusions, is mainly talked about in steel, similar phenomena have also been reported in Ag–Al, Cu–Al and Ag–In alloys. Similar mechanism has also been attributed to the formation of nitrogen, sulphur, selenium and tellurium bearing internal inclusions. The mechanism and kinetics of internal oxidation is fairly well charted (Van Vlack [1977]). As indicated in Figure 12.3 and described further in Eq. (12.1), the stability of internally oxidized zone can best be described in terms of relative fluxes of oxygen and metal and η – the number of oxygen atoms per metal atom of the oxide compound.

$$D_{\mathrm{O}} \frac{\partial C_{\mathrm{O}}}{\partial X} > -\eta D_{\mathrm{M}} \frac{\partial C_{\mathrm{M}}}{\partial X} \tag{12.1}$$

where D_{O} and D_{M} are the respective diffusivities for oxygen and metal. Equation (12.1) has two interesting[9] corollaries – enhanced internal oxidation in oxidizing

[9] Both important to TMP of metallic systems.

environments and also in lean-alloys. In simplified conditions, the thickness of internally oxidized zone (κ) can be approximated as

$$\kappa = \sqrt{\left(\frac{2C_{\mathrm{O}}^{0}D_{\mathrm{O}}}{\eta C_{\mathrm{M}}^{0}}\right)t} \tag{12.2}$$

where t is the time and C_{M}^{0} and C_{O}^{0} the respective compositions of metal and oxygen inside the metal and at the free surface, respectively.

Usually the internal oxidation is harmful and can be classified as a surface/subsurface defect strongly affecting the property. For example, in non-grain-oriented electrical steel (Section 15.4), a substantial decrease in magnetic properties may result from internal oxidation of laminations (Kozmanov [1968]). Similarly, internal oxidation in low-carbon steel for packaging/decorative applications can have disastrous effects (Bradford [1964]). Though there has also been discussions on employing internal oxidation in synthesizing oxide-dispersed steels for high-temperature applications (Van Vlack [1977]) – a clear technology remains yet to be formulated.

12.3.3 Surface defects related to coating

Coatings (Metals Handbook [1988b] in metallic systems are common. These may range from organic and inorganic coatings used for specialized[10] applications to Zn and Zn–Al coatings (e.g. galvanizing[11]). The latter is of large techno-commercial interest. The coatings may relate to surface defects – defects during coating/handling and defects originating from subsequent (i.e. after coating) forming. The former may include scratches, fingerprints and crevice/pitting corrosion, while the latter can often be associated with 'galling'. As shown in Figure 12.4, forming

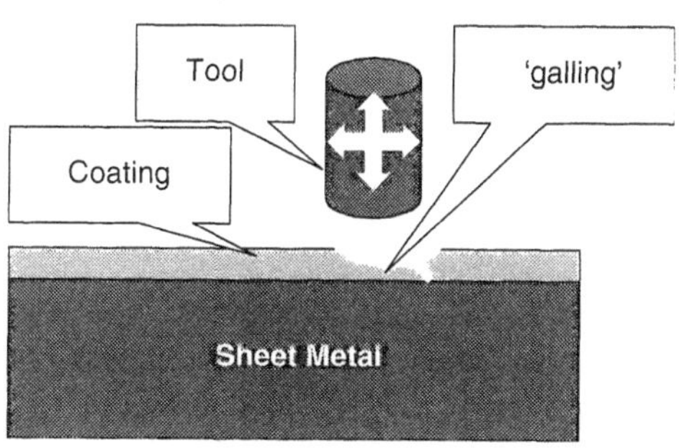

Figure 12.4. Schematic describing 'galling' through tool movement.

[10] For example, electrical steel laminations are coated for reducing the air gap and also to provide insulation.

[11] Hot-dip galvanized and electrogalvanized coatings.

operations subsequent to coating may remove the coating at some locations – the origin of this defect or 'galling' (Metals Handbook [1988a, b]) is due to adherence of coating to the die-tool surface. This is primarily a tribological problem and usual remedies involve appropriate lubrication and/or application of suitable low-friction wear-resistant coating(s) to the tools (Carlson [2005]).

12.4. FRACTURE-RELATED DEFECTS

Though the science of fracture mechanics has evolved considerably over the past few decades, it is still difficult to relate macroscopic fracture with exact microstructural origin. The fracture-related defects, on the other hand, usually mark the limit for any forming – classical example is the so-called forming limit diagram[12] (Section 11.7). Classifying all fracture-related TMP defects will be a very tall order and the present section only attempts to narrate four common fracture-related defects.

12.4.1 Edge cracking

Edge cracking (Dodd [1980], Thomson [1980], Boehlert [2000]) is a common problem in 'compressive' bulk-forming processes, such as rolling and forging. As shown in Figure 12.5, though the severity of edge cracking may not cause complete fracture, but the phenomena may turn the TMP product useless or may require expensive material removal. Typically, edge cracking involves appearance or initiation of small surface cracks at the edge and subsequent propagation of these cracks, often along the transverse direction, into the bulk of the material (Thomson [1980]). Table 12.5 summarizes the causes of edge cracking.

12.4.2 Alligatoring

As shown by the schematics of Figure 12.6a, alligatoring (Polushkin [1956], Metals Handbook [1988a], Turczyn [1996], Ko [2000]) initiates as a crack along the centre plane of the rolled material and can vary in severity – from partial separation of the upper and lower halves of the rolled material to a complete separation and even tangling on the rolls (Turczyn [1996]). Origin of alligatoring is inhomogeneous deformation – severe forms are reported (Metals Handbook [1988a]) for a combination of inhomogeneous deformation and non-uniform recrystallization during primary rolling in Al–Mg alloys, Zn- and Cu-based alloys.

[12] Different yield criteria, isotropic as well as anisotropic, has been reasonably successful (Cao [2000]) in predicting macroscopic necking and fracture behaviour in sheet metal forming. Relating such microstructural features with exact microstructural causes is, however, far more difficult – and can be considered at best as an evolving science.

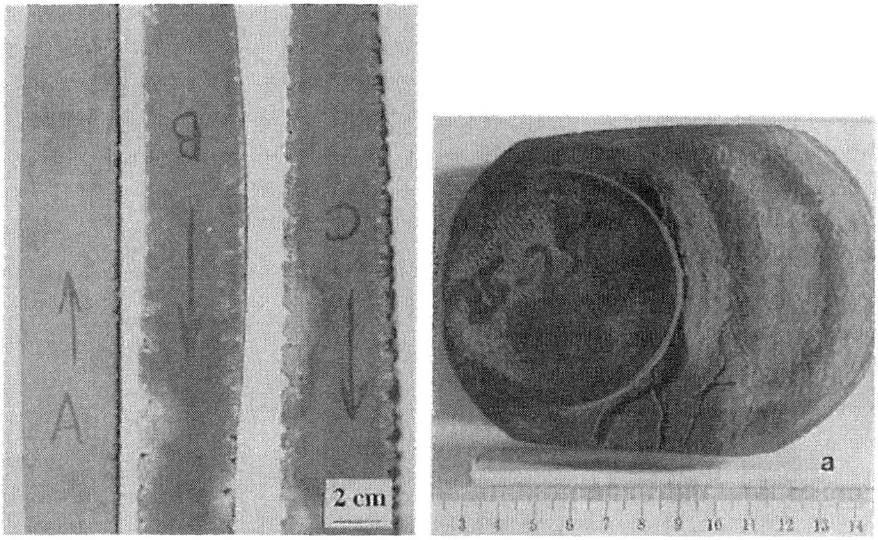

(a) (b)

Figure 12.5. Edge cracking in Ti–Nb–Al alloy for (a) hot rolling and (b) hot forging/upsetting. Courtesy of Boehlert [2000] (copyright (2007), with permission from Elsevier). The severity of edge cracking did depend on the prior bcc grain size (Boehlert [2000]).

Table 12.5. Summarizing the causes of edge cracking.

Causes of edge cracking	Description
Metallurgical	
Limited ductility or workability	The limited ductility, inherent to the material or caused by some microstructural feature – e.g. large prior transformation grain size (Boehlert [2000]), presence of embrittling phases or elements and inclusions (Thomson [1980]), etc.
Process related – uneven deformation and corresponding variations in stresses	
Camber	Heavily cambered rolls in wide-strip rolling can create edge cracking (Dodd [1980])
Spread	Deformation along the roll-axis or transverse direction is called spread. It depends on workability and width–thickness ratio of rolling/forging and frictional conditions. Excessive spread can cause edge cracking (Dodd [1980])
Edge shape	Inappropriate prior-deformation edge shape (Dodd [1980]) can also lead to edge cracking
Friction and pass sequence	Inappropriate friction and pass sequence leading severe inhomogeneity in deformation may also create edge cracking

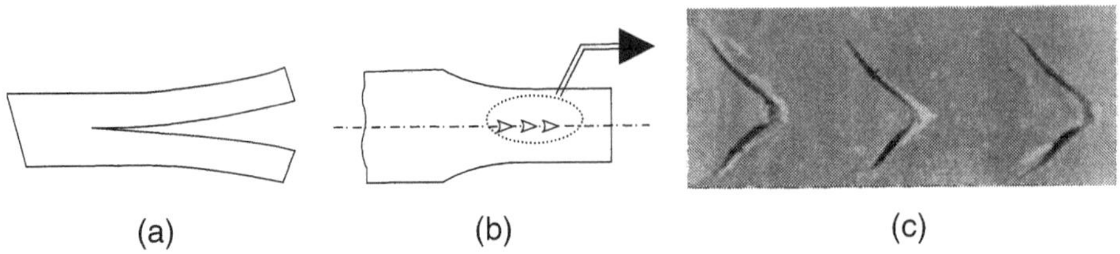

Figure 12.6. Schematics showing (a) alligatoring in rolling and (b) central burst in extrusion. (c) Image of an actual central burst in extrusion. Courtesy of Ko [2000] (copyright (2007), with permission from Elsevier).

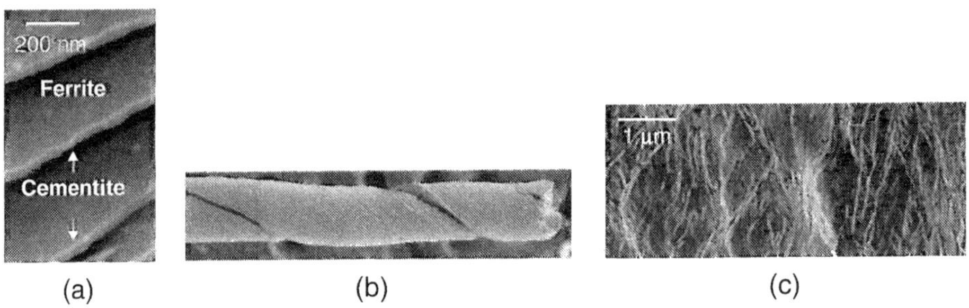

Figure 12.7. Split in wire-drawn ('patented') pearlitic steel. (a) Microstructure of the patented wire. (b) Delaminated or split filament. (c) Image showing localization of plastic flow. Courtesy of Zelin [2002] (copyright (2007), with permission from Elsevier).

12.4.3 Central burst

Central burst (Polushkin [1956], Avtizur [1968], Metals handbook [1988a], Turczyn [1996], Ko [2000]) or chevroning, as in Figures 12.6b and c, is the formation of internal voids at the centre. This has been reported in rolling, extrusion and wire drawing.[13] Poor die design and structural heterogeneity are reported (Avtizur [1968]) to be the primary causes. The same alloy with slight difference in composition and a correspondingly higher strain hardening may avoid central burst (Metals Handbook [1988a]). Small reductions and large die angle also reduce this phenomenon (Avtizur [1968], Ko [2000]).

12.4.4 Wire-drawing split

Split, see Figure 12.7, during high-speed wire drawing can be disastrous to productivity. The origin of 'split' can be both process related and metallurgical[14] – contributions from different metallurgical and process parameters being possible.

[13] A similar phenomenon has also been reported (Metals Handbook [1988a]) in forging.

[14] Remedial measures through changing process parameters, such as drawing speed, are possible. The split can often be subjective to a particular batch.

Examples, the list is not an exhaustive one, of apparent causes for split in wire-drawn metals are

- Split in patented wire through localized shear bands (Zelin [2002] – see Figure 12.7).
- Split in Cu wire through a combination of back-tension and relative inclusion presence (Cho [2002]).
- Split in ferrite–pearlite–martensitic steel wire through crack nucleation at the ferrite–martensite interface (Bae [1994]).
- Split in tungsten wire through possible crack nucleation at the grain boundaries along transverse direction (Briant [1991]).

12.5. STRAIN LOCALIZATIONS

Strain localizations, as in Figure 12.7, can be both microscopic as well as macroscopic. These can be identified through clear patterns of localized plastic flow and/or as distinct dislocation substructure – and are often responsible for the initiation of micro-cracks, ultimately leading to fracture. The present section does not attempt to describe or to discuss the microstructural aspects of strain localizations, an aspect covered in Chapter 4. The basis of their formation, an aspect touched briefly in Chapter 3, however, needs to be discussed in further details.

Considered as deformation heterogeneities or plastic instabilities, instability criteria (see Eq. (3.22)) can be used to describe the formation of strain localizations. For example, a suitable instability criterion is also used in identifying the safe process regimes in a processing map (Chapter 6). The criterion can be formulated either from localized deformation (as in Eq. (3.22)) or from softening. A much cited (Dillamore *et al.* [1979], Gil Sevillano [1981], Wagner [1995], Samajdar *et al.* [1998c]) example of the latter is the so-called Dillamore's criteria

$$\frac{1}{\sigma}\frac{\mathrm{d}\sigma}{\mathrm{d}\varepsilon} = \frac{n}{\varepsilon} + \frac{m}{\dot{\varepsilon}}\frac{\mathrm{d}\dot{\varepsilon}}{\mathrm{d}\varepsilon} + \frac{1+n+m}{M}\frac{\mathrm{d}M}{\mathrm{d}\varepsilon} - \frac{m}{\rho}\frac{d\rho}{d\varepsilon} \leq 0 \tag{12.3}$$

where σ and ε are the macroscopic stress/strain, n and m the strain-hardening exponent and strain rate sensitivity, $\dot{\varepsilon}$ the strain rate, M the Taylor factor[15] and ρ the mobile dislocation density. Condition(s) making Eq. (12.3) negative would favour formation of strain localizations. For example, a negative textural softening ($\mathrm{d}M/\mathrm{d}\varepsilon$ – a grain rotating to a plastically softer orientation) or an increase in ρ with strain is expected to enhance the formation of strain localizations (Dillamore *et al.*

[15] $\sigma_y = M\tau$, where σ_y is yield stress and τ the critical resolved shear stress.

[1979], Gil Sevillano [1981], Wagner [1995], Samajdar *et al.* [1998c]). Though stunningly simple in its approach (Dillamore *et al.* [1979]), the criteria have been successfully used in describing the macroscopic angles of shear bands and also in describing their preferred appearance at certain orientation(s).

12.6. STRUCTURAL DEFECTS

The objective of a TMP is two-fold: (I) to achieve defect-free desired size and shape, and (II) to achieve a desired property. Defects relevant to (I) have already been discussed in the earlier sections, while (II) would typically relate to the structural defects – more specifically microstructure-related defects. In other words, the inability to achieve a desired microstructure and hence required microstructure-related property can also be considered as a TMP defect. The basis of such 'defects' is, however, covered in the Part I (*Science*) of this book.

LITERATURE

Metals Handbook – "Forming and Forging", 9th Edition, vol. 14, ASM, Ohio, USA (1988).

Metals Handbook – "Surface cleaning, finishing & coating", 9th Edition, vol. 5, ASM, Ohio, USA (1988).

Polushkin E.P., "Defects and Failures of Metals", Elsevier, UK (1956).

Chapter 13
Physical Simulation of Properties

Chapter 13
Physical Simulation of Properties

13.1. INTRODUCTION

Thermo-mechanical processing (TMP) is carried out with the basic aim of improving the physical properties of the material. These properties are characterized by a wide variety of tests, such as tensile, fatigue, rupture and corrosion tests and which have usually been standardized for different types of metals. The tests for TMP, used to characterize material behaviour during processing, are, however, relatively less established and often difficult to perform satisfactorily. This chapter aims at describing different possibilities for characterizing the responses of metallic materials to TMP.

One could ask why not use standard industrial equipment such as extrusion presses and rolling mills to characterize TMP responses in real operating conditions? The basic answer is the cost; these equipments operate at such high productivities that the time spent developing process and/or material modifications (which may or may not lead to improvements) translates into very high costs – and sometimes cause damage to the machines. TMP tests are therefore usually carried out either in scaled down semi-industrial equipment (laboratory rolls and presses) or, increasingly by 'simple' laboratory tests, e.g. tension or compression, designed to simulate the processing conditions.

The downsized simulation of an industrial process on a laboratory scale has the advantage that the mode of deformation remains approximately the same and that the influence of specific process parameters such as lubrication or die geometry can be studied. On the other hand, it is difficult to obtain exactly the same conditions since in most cases the cooling conditions are different and for many processes the high industrial strain rates are not attainable on a lab-scale. Also, with this method it is difficult to separate out the respective roles of material behaviour and processing geometry (i.e. the local thermo-mechanical history of the material as influenced by friction and tool geometry).

For these reasons, it is becoming increasingly common to evaluate the material behaviour during TMP by 'simple' laboratory tests before going onto scaled down industrial equipment and finally the production process. Clearly, the latter implies that the influence of typical industrial conditions be known, i.e. temperature, deformation, strain rate and interpass or annealing times. These conditions are typically determined by FE (finite element) simulation for standard industrial processes and the laboratory tests carried out under the same, or similar, conditions. The usual tests are performed in tension, compression or shear but, particularly at high temperatures, should attain similar strains and strain rates as in industrial processing. In order to compare results of different hot deformation tests, it is useful to express the results in terms of true stress and true strain (also called equivalent stress and equivalent strain).

One of the major practical problems is the sample-cooling rate after deformation processing since microstructural evolution can take place rapidly in a few seconds. Either one attempts to simulate the industrial cooling rates using special equipment or, on materials which do not undergo allotropic transformations on cooling, one can quench as rapidly as possible to retain the TMP microstructures.

13.2. TENSILE TESTING

Standard tensile testing machines can usually be equipped with a high-temperature furnace (resistance, induction or radiation). The usefulness of such a set up for hot deformability testing is, however, fairly limited because the strain that can be obtained is too low, usually limited by necking to maximum values in the range $\varepsilon \sim 0.2$–0.4. Also, in standard screw-driven tensile machines the maximum strain rates are too low (about 0.1/s) and constant strain rate tests are very difficult, if not impossible once necking starts. There is one notable exception to these limitations and concerns tests for superplastic behaviour (Chapter 6). In this case, the slow strain rates and large elongations by reduced, or negligible, localized necking allow extensive use of hot tensile tests. A practical difficulty is maintaining a constant temperature over the length of a superplastic sample as it can extend 1000%.

For more general typical metallic alloys which undergo necking some special 'fast' tensile test machines have been built using servohydraulic rams controlled by computers. An example is the 'Gleeble' machine which can rapidly heat and cool the samples in a controlled way. The computer control keeps the true strain rate constant both during homogeneous deformation and in the zone of necking. Multi-pass tests are possible in principle, but due to the limited deformation the number of passes is restricted. These fast tensile machines are in most cases also suitable for compression tests.

The hot tension test does not, however, suffer from friction problems and inhomogeneous strain distributions (before the onset of necking), so it is often used to calibrate the other test results in the low-to-moderate strain regimes.

13.3. HOT TORSION TESTS

Much of the first scientific work on hot deformation processing was carried out by hot torsion tests (see, e.g. Rossard and Blain [1960], Hardwick and McG Tegart [1961], McQueen and Jonas [1971], Sellars and McG Tegart [1972]). The basic

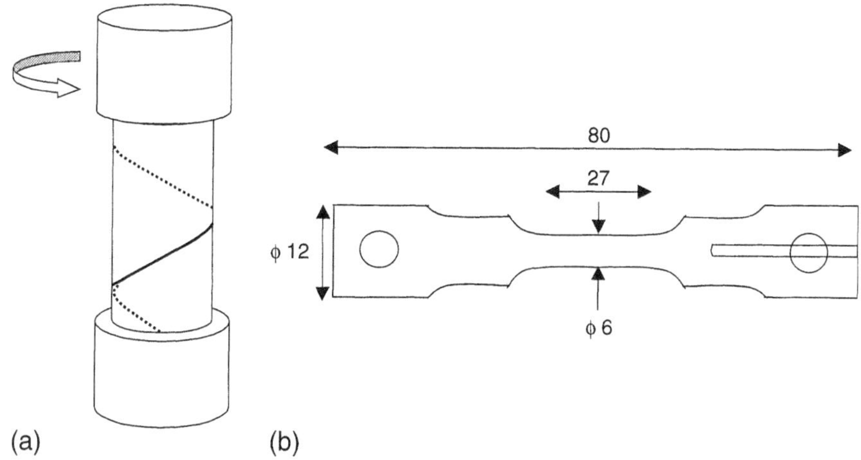

Figure 13.1. (a) Schematic of torsion test; during torsion an initially vertical line winds round the sample gauge as indicated (as do the grains), (b) typical sample dimensions.

advantages are the high strains and strain rates that can be achieved in the near-surface zones of the sample.

A torsion test is performed on cylindrical samples that are heated with an induction coil or a radiation furnace (Figure 13.1). The sample dimensions are usually optimized to minimize the temperature gradients that can develop during deformation. The torque M is recorded as a function of the angular displacement ($\theta = 2\pi N$, where N is the number of turns). In most set-ups, the length of the sample is kept constant during the test. In principle, a simple shear is applied during torsion, but due to texture development, some axial stress can be developed during deformation. The interpretation of the results is not immediate since the applied strain and strain rate vary linearly from zero in the centre of the specimen towards a maximum at the surface. For a sample of length L and radius R, the true strain and strain rate at the surface are given by geometry as

$$\varepsilon = \frac{R\theta}{L\sqrt{3}}; \qquad \frac{d\varepsilon}{dt} = \frac{R(d\theta/dt)}{L\sqrt{3}} \tag{13.1}$$

and the true stress can be derived from torsion tests with the formula, first derived by Fields and Backofen [1957]:

$$\sigma = \frac{M\sqrt{3}(3+m+n)}{2\pi R^3}; \quad \text{with} \quad m = \frac{\partial M}{\partial \varepsilon^{\circ}|\varepsilon} \quad \text{and} \quad n = \frac{\partial M}{\partial \varepsilon|\varepsilon^{\circ}} \tag{13.2}$$

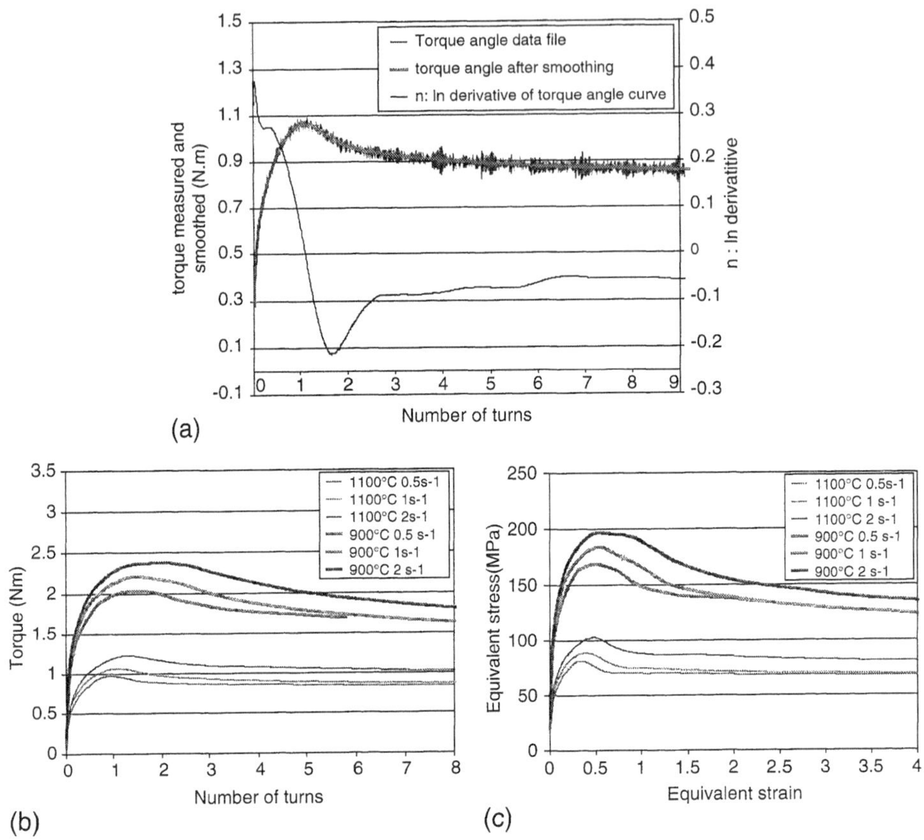

Figure 13.2. Example of deriving high-temperature stress–strain plots from torque–twist curves of 0.25%C–1%Cr–Mo steel at different strain rates; (a) gives an original torque–turn plot (both with noise and after smoothing to obtain the n values), (b) displays a set of smoothed torque–turn curves at two temperatures and three strain rates and (c) the resulting equivalent stress–strain curves. Courtesy of C. Desrayaud – Ecole des Mines St Etienne.

The n coefficient is in many cases taken as zero, although this is strictly only valid at the peak maximum and in the steady state. In order to determine the value of the strain rate sensitivity coefficient m, torsion tests at different strain rates must be carried out. Figure 13.2 gives an example of translating experimental high-temperature torque (twist) plots at different strain rates into equivalent stress–strain curves for steel. It also illustrates the practical problems of evaluating the n coefficient by deriving the torque–turn curves after appropriate smoothing. When an axial force is recorded, the flow stress can be calculated from

$$\sigma_T = (\sigma^2 + \sigma_A^2)^{0.5} \qquad (13.3)$$

with σ from Eq. (13.2) and σ_A the axial stress. The flow curves that are obtained with this procedure are representative for the material at the outer surface of the sample, which should be used for microstructural investigations. This is the major disadvantage of the test; only the outer thin layer of the sample can be considered to have undergone the strain history given by the above values of ε and σ. Torsion tests do, however, enable very large strains up to about 100 to be obtained on ductile materials and are also suitable for multi-pass tests. Also, since the sample geometry remains constant so does the strain rate. It is not too difficult to quench the sample surface immediately after deformation by either spraying with coolant or pushing the sample into a bath.

13.4. COMPRESSION TESTS

Compression tests can be sub-divided into axisymmetric and plane strain tests. In *axisymmetric* tests a cylindrical sample is compressed between two parallel platens. The axial compression is therefore imposed by the displacements of the platens, with the sample widening freely in the two other orthogonal directions. This test is often used to simulate forging deformation. In *plane strain compression* the sample, usually of rectangular shape, is compressed but the orthogonal deformation restricted to only one of the other directions (metal flow taking place in one plane). This approximately simulates rolling deformation.

In both cases the use of modern, computer-controlled, servohydraulic machines enables high and constant strain rates to be achieved on relatively large samples. For both types of tests, attention should be paid to controlling friction, between die and sample surfaces, and in maintaining temperature homogeneity during the tests – unless, of course, the influence of temperature gradients is of particular interest.

13.4.1 Uniaxial compression

The upper platen is displaced at a velocity, v, to compress the sample placed on the fixed lower platen. A large strain compression test at constant v implies increasing strain rate as the sample height, H, decreases, in accordance with the definition of strain rate $(\mathrm{d}H/\mathrm{d}t)/H$. Since the strain rate can have a significant effect on the flow stress (and microstructure), one should maintain a constant plastic strain rate by progressively decreasing the platen velocity. This used to be done by a cam plastomer but now is often performed by computer control using measured values of the instantaneous height. A standard closed-loop servo-controlled tension/compression machine is required. (These strain rate corrections are necessary for all high temperature, large strain compression tests.)

If required, quenching is carried out by pushing the deformed sample out from the platens into a coolant; this is not too difficult if the sample is well lubricated.

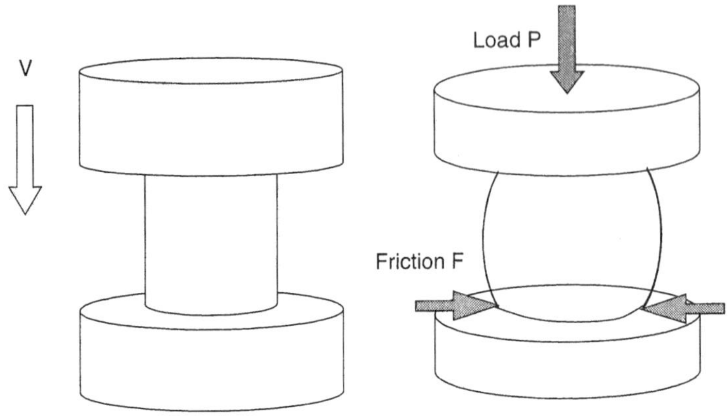

Figure 13.3. Schematic compression of cylindrical sample between two flat dies.

A frequently encountered problem in compression tests is 'barrelling' (see Figure 13.3). Friction, and hence 'restricted' deformation, in the sample-die contact area often results in the sample widening more near the centre than at the contact area between sample and dies.

The effects of friction can be reduced by proper lubrication. Optimum lubricants for compression tests up to 600°C have been described by Lovato and Stout [1992]. They include PTFE (polytetrafluoroethylene), MoS_2, graphite and boron nitride. For higher temperatures graphite or molten glass, as in the Sejournet extrusion process for steels (Section 11.3), can be used. It is also common practice to machine concentric grooves into the two sample surfaces in contact with the platens; the grooves retain some of the lubricant which is then continually squeezed out into the contact zones as the grooves are compressed with the sample. It is also possible to use conical dies. By these techniques one can obtain a homogeneous strain up to 0.7 without barrelling, or 1.2 with slight barrelling. Figure 13.4 illustrates the strain distributions calculated for a typical metallic sample deformed in compression with a friction coefficient of 0.1. Higher than average strains occur first in the centre and then, after further straining, at the edges.

The equivalent stress (σ) can be calculated from the experimental pressure:

$$p = \sigma\left(\frac{1+\mu D}{3H}\right) \tag{13.4}$$

i.e. the friction coefficient μ as well as the sample diameter and height as a function of the deformation must be known.

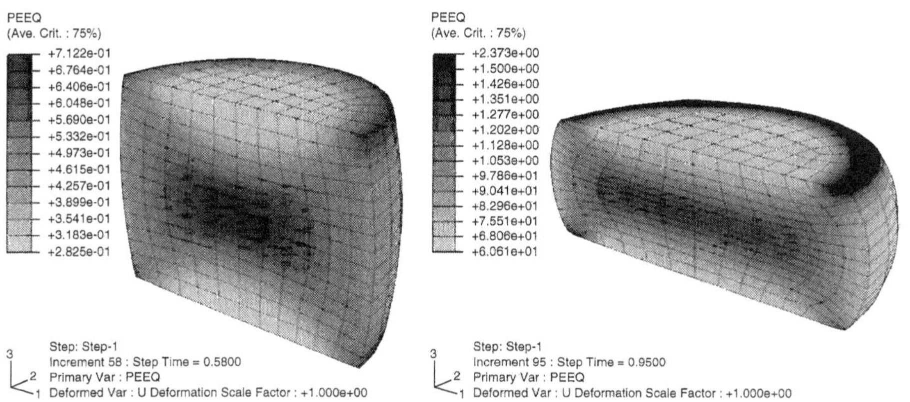

Figure 13.4. FE elasto-plastic simulation of equivalent strain distributions in compressed low C steel sample (strains of 0.5 and 1; friction, μ, and work hardening coefficients, n, are both taken = 0.1, from C. Desrayaud, Ecole des Mines de Saint-Etienne).

Friction coefficients for the compression test are often determined by means of the 'ring test' (Male and Cockroft [1964]). Using the same conditions of lubrication, temperature and strain rate as the real test, samples of the work piece in the shape of rings are compressed to different reductions, and the new inside ring diameters measured. The percentage change of the minimum inner diameters is then plotted as a function of the height reduction and compared with calibration curves for the same geometry and different friction coefficients (e.g. Schey [1983], Dieter [1988]). A common geometry is a ring with outer diameter:inner diameter:thickness ratios of 6:3:1. The ring test is described in more detail in Section 11.1 of Chapter 11.

13.4.2 Plane strain compression

In plane strain compression the aim is to compress along one direction (say Z) and elongate by the same quantity along a perpendicular direction (say X) so that by volume conservation the deformation along the third axis Y is zero. In practice, this is performed by either pressing a rectangular die into a strip of sheet or slab sample or by compressing a near-cube sample in a channel die. Here the two methods are denoted respectively strip PSC (plane strain compression) or channel-die compression.

For high-temperature PSC the strip method has been extensively used (Figure 13.5 and 13.6).

The heated sample is placed between the two heated dies and compressed so that the strip of metal between the dies elongates along the X direction (equivalent to the rolling direction). Ideally, there would be no widening along Y (or TD in rolling). This is roughly true when there is a significant amount of friction between the dies

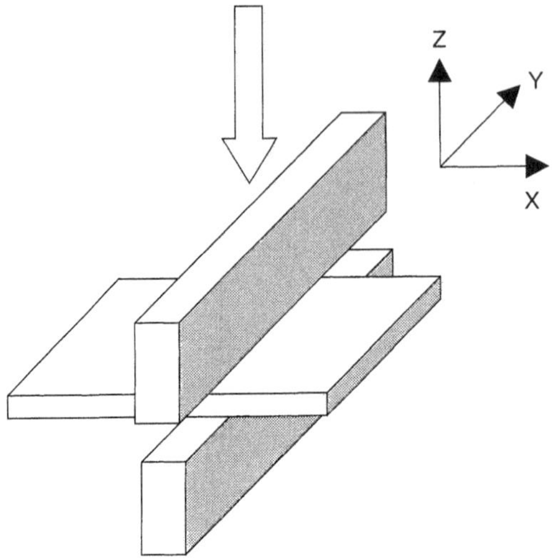

Figure 13.5. Schematic strip plane strain compression arrangement.

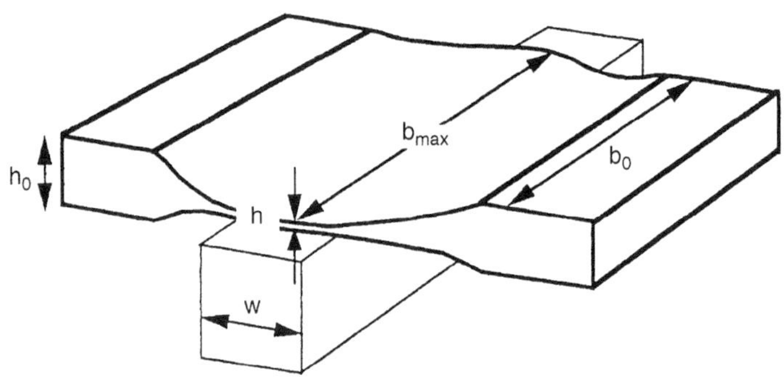

Figure 13.6. As deformed strip plane strain compression sample. Courtesy of J.-M. Feppon, Alcan Voreppe.

and the sample so that flow along *Y* is then restrained by friction. However, this surface friction induces inhomogeneous deformation in the sample thickness and a form of barrelling. If, as is usual, the sample–die surfaces are lubricated then some deformation along *Y* occurs and so $\varepsilon_{xx} \neq -\varepsilon_{zz}$. This then requires additional corrections to the stress–strain curves. The general problems and optimum conditions for hot strip PSC have been reviewed by Shi *et al.* [1997b] and the necessary corrections for non-zero ε_{yy} deformations by Kowalski *et al.* [2003].

After deformation, and if necessary, the entire sample can be quenched into an appropriate coolant. Unfortunately, the cooling rates cannot be very high since the deformed volume of interest is also in contact with a large mass of undeformed metal.

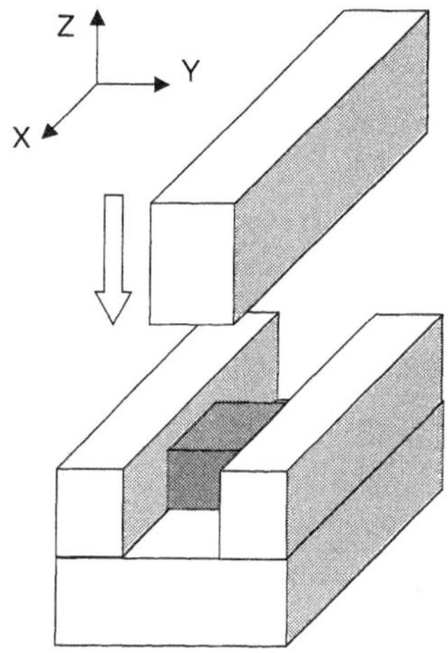

Figure 13.7. Schematic channel-die plane strain compression set-up.

Despite these limitations, strip PSC test has been widely used because of its simplicity and the possible range of temperatures (up to 1200°C for steels). Some applications to Al alloys are described by Marshall [1998]. A nominal deformation of $\varepsilon = 2.3$ (90% reduction) is possible, and multi-pass tests are quite feasible using servohydraulic machines. Ignoring the lateral widening problem, and therefore to first order, the stress (σ') as a function of the applied pressure p is expressed by

$$p = \sigma'\left(\frac{1 + \mu B}{2H}\right) \tag{13.5}$$

with B the width of the platens. The equivalent stress σ and equivalent strain ε are calculated from the plane strain stress σ' and strain ε':

$$\sigma = 0.5(\sigma'\sqrt{3}); \qquad \varepsilon = \frac{2\varepsilon'}{\sqrt{3}} \tag{13.6}$$

The *channel-die compression* test, Figure 13.7, has been mostly used for room and low-temperature plane strain compression but good high-temperature tests are feasible up to temperatures of 500°C using the equipment developed by Maurice and Driver [1993] and Maurice *et al.* [2005].

In this arrangement, the sample is placed in the corridor of a die block of the same width as the sample to effectively limit the transverse strain (along Y) to zero so that true plane strain compression is obtained. Given that four surfaces are in contact with the dies lubrication is critical to ensure a relatively homogeneous

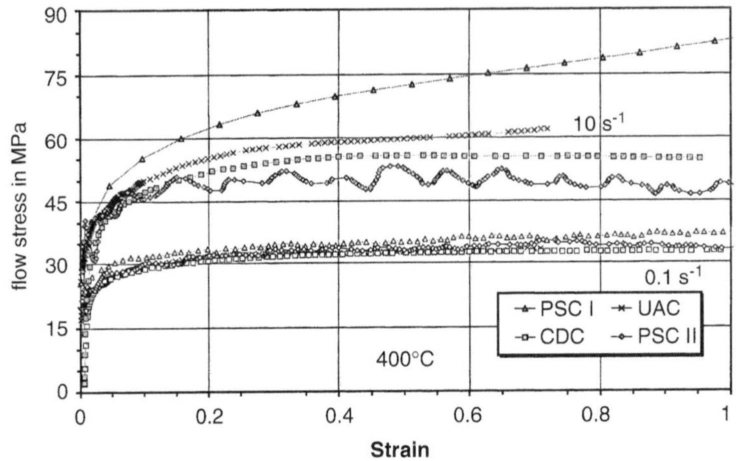

Figure 13.8. Equivalent stress–strain curves of an AA 3103 alloy deformed by compression at 400°C and two strain rates 0.1 and 10/s; UAC = uniaxial compression, PSC = strip plane strain compression (I and II refer to two laboratories using different machines) and CDC = channel-die compression (Maurice *et al.* [2005] and the VIRFAB European project). With kind permission of Springer Science and Business Media.

deformation. This can be achieved using graphite plus PTFE (Teflon) up to the temperature at which PTFE does not decompose too quickly (about 500°C). PTFE has a negligible friction coefficient of less than about 0.015 up to these temperatures. At higher temperatures, where more standard lubricants have to be employed the friction coefficients with values of the order of 0.1 imply inhomogeneous deformation. High heating and cooling rates are possible using preheated dies and the mobile die wall technique (Maurice *et al.* [2005]).

The maximum homogeneous strains that can be obtained by channel-die compression using PTFE are of the order of 1.4 (equivalent strain of 1.6 or a reduction of 75%).

Figure 13.8 compares some equivalent stress–strain curves for a hot-deformed AA3103 (Al–1%Mn) alloy as measured by the three compression techniques (uniaxial compression, strip PSC and channel-die compression). It is seen that at the lower strain rate all methods give reasonable results. At the higher strain rate of 10/s channel-die compression and uniaxial compression give similar results but, in this particular case, there are problems with the strip PSC method. The wavy curve of PSC II is due to non-constant strain rates during the test – a machine problem that illustrates the difficulties of hot deformation testing in general.

In Figure 13.9, the saturation flow stresses of an Al–Mg alloy are compared for hot channel-die compression and strip PSC as a function of the Zener–Hollomon parameter under conditions where both tests have been performed carefully. The agreement is very good.

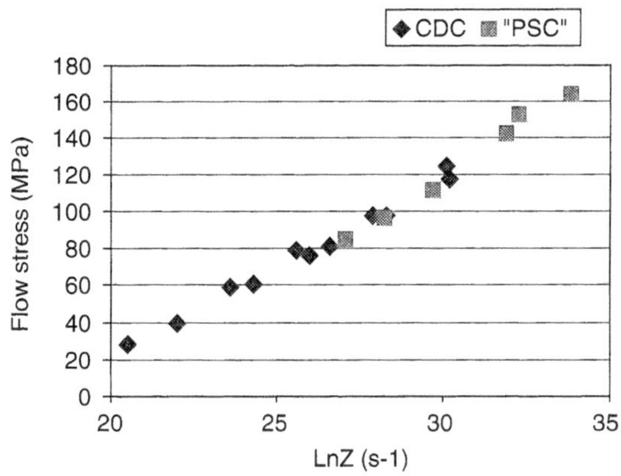

Figure 13.9. High temperature saturation flow stresses at a strain of unity of an AA 5052 (Al–2.5%Mg) alloy as a function of the Zener–Hollomon parameter (temperature compensated strain rate with an activation energy Q of 156 kJ/mol) as measured by strip PSC and channel-die compression at temperatures between 300 and 450°C (Maurice *et al.* [2005]). With kind permission of Springer Science and Business Media.

13.5. MIXED STRAIN PATH TESTS

13.5.1 Lab-scale tests

The above tests can give detailed information about plastic flow behaviour and microstructure evolution for a given strain path at high temperatures. During industrial TMP the strain path can vary both through the sample thickness (as in rolling and extrusion) and also, for a given relative thickness, through the process (e.g. the near surface regions of hot-rolled metals which undergo alternating shears of opposite sign).

Some recent developments in machine design have therefore been carried out to enable a variety of strain paths in particular those with different amounts of compression and shear (Figure 13.10). They can now be computer controlled to provide virtually any strain path combination of interest. Examples are the so-called deformation simulators, Gleeble, etc.

13.5.2 Downgrading of industrial processes

This basically involves carrying out similar deformation schedules as in the industrial process using laboratory rolling mills, small extrusion presses, etc. In principle, this is very attractive but in practice often suffers from poor temperature control (e.g. rolling of hot samples between much cooler rolls), variable lubrication and unreproducible quench, etc. Some of the problems of trying this have been vividly described by McQueen and Mecking [1987]: 'Many difficulties were encountered in rolling Al because it always stuck to one of the rolls; as a result it

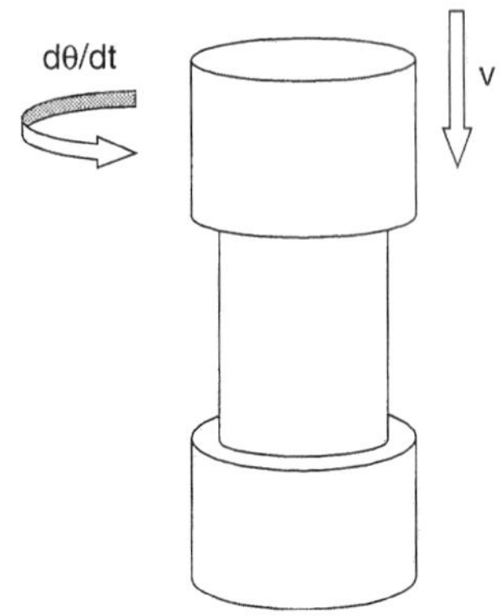

Figure 13.10. Schematic mixed torsion–compression test.

was not water quenched but rapidly cooled by conduction to the roll. Subsequently it had to be peeled off the roll and flattened'.

More successful laboratory-scale experiments have been described by Van der Winden and Sellars [1996] but require a large number of assistants to manipulate the metal during several passes.

When carrying out laboratory rolling it is recalled that the rolling reduction $[\rho = (h_0 - h)/h_0]$ can be expressed in equivalent strain with the formula

$$\varepsilon = \frac{-2\ln(1-\rho)}{\sqrt{3}} \tag{13.7}$$

and the equivalent strain rate (ε°) can be obtained from

$$\varepsilon^\circ = V \ln\left(\frac{h_0}{h_f}\right)[R(h_0 - h_f)]^{-0.5} \tag{13.8}$$

with V the surface speed of the roll, R the roll radius, h_0 and h_f the initial and final thickness of the sheet.

13.6. TYPICAL SHEET FORMABILITY TESTS

The general problem of sheet metal forming is described in Section 11.7 together with the mechanical principles involved in standard formability tests such as forming limit diagrams (FLD) and deep-drawing tests. This section describes some of the basic procedures for some other common tests. More details can be found in Taylor [1985].

13.6.1 Bending

A wide variety of bending tests are employed for different practical applications. The simplest involves clamping a sheet metal sample and a bending die in a vise and then bending the sample over the die. If it bends more than 180° without failure the experiment is repeated with dies of increasingly smaller radius until failure giving what is known as the limiting die radius. More systematic measurements have been devised, notably by the automobile industry, and an example of a bend test for assembling sheet metal panels is illustrated in Figure 13.11. Metal panels are often joined by simply folding one sheet over the other (termed hemming) and the local strain attains values near unity in this deformation mode.

A *stretch bending test* involves clamping the ends of a rectangular strip of sheet metal as shown schematically in Figure 13.12 and then punching out to the onset

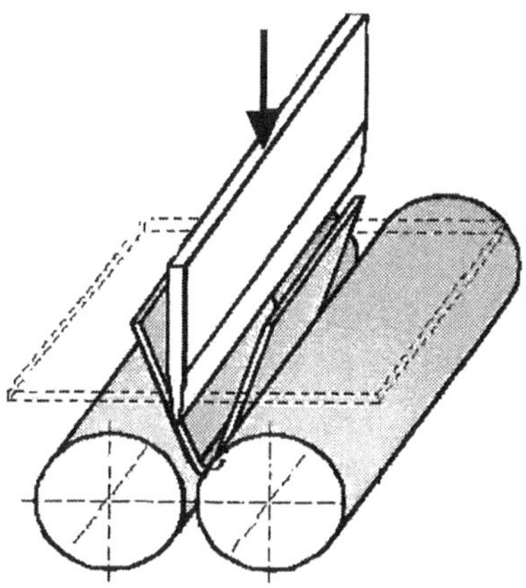

Figure 13.11. A hemming – or high strain bending test for sheet metal joining.

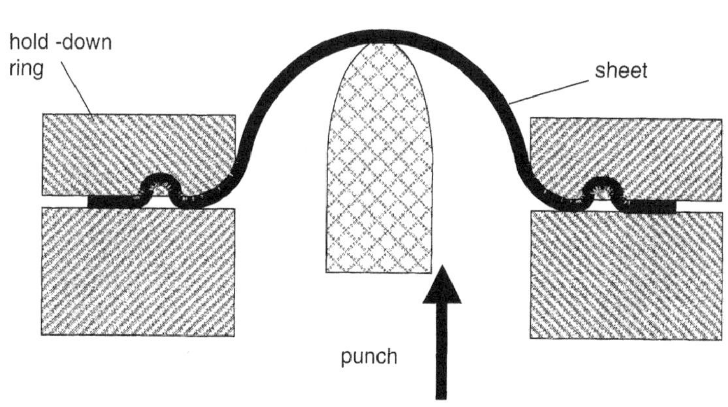

Figure 13.12. Schematic of stretch bending test.

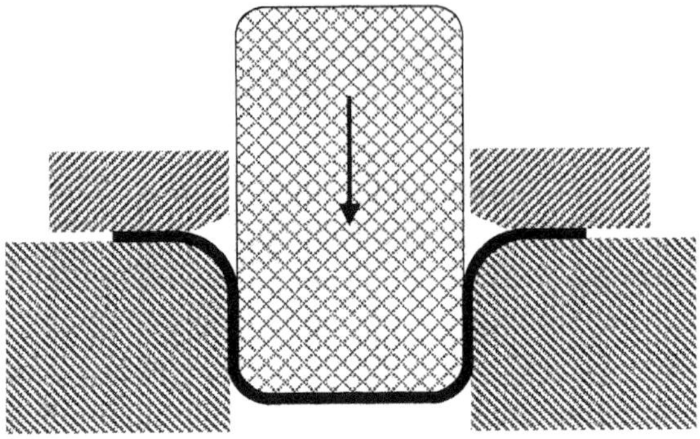

Figure 13.13. Schematic of the Swift cup deep drawing test to determine the ratio of the largest blank diameter to punch diameter.

of failure – which should occur in the punch zone. Two types of punches are used, hemispherical or wedge shaped. In both cases, the sheet metal is simultaneously stretched and bent as occurs in many practical forming operations. The angular or wedge test produces plain strain deformation, whereas the hemispherical punch test involves a range of strain states.

The principles of deep-drawing tests have been described in Section 11.7. The practical procedures are usually based on the Swift cup test. According to the schematic of Figure 13.13, it involves clamping a circular blank in a die ring and then deep drawing with a flat-bottomed cylindrical punch. A set of tests is carried out with blanks of different diameters and the largest blank diameter D that can be successfully drawn to the punch diameter d is used to define the sheet drawability as characterized by the limiting draw ratio: LDR $= D/d$.

An alternative method for determining the LDR is based on measurements of the fracture load for different clamping forces. Blanks of a single diameter (below the critical D of the above standard test) are drawn to the maximum load, usually well before the draw failure. The clamping force is then increased to restrict drawing of the flange and the punching load increased to fracture. The LDR is then obtained from the limiting blank diameter [Fracture load × (blank diameter − die diameter)]/Maximum drawing load] + Die diameter by: LDR = Limiting blank diameter/Punch diameter.

PART III: CASE STUDIES

Chapter 14
Thermo-Mechanical Processing of Aluminium Alloys

Chapter 14

Thermo-Mechanical Processing of Aluminium Alloys

14.1. ALUMINIUM BEVERAGE CANS

14.1.1 Introduction

In 1810, an English citizen, Peter Durand, received from King George III a license for the production of tin-coated iron containers for the storage and conservation of food. These containers were 'handmade' and the best skilled craftsmen could produce about 60 containers a day. As a comparison, today, a modern production line produces about 1 million cans a day! With time the process was gradually improved and automated. The basic concept was a so-called 'three-piece can' (see Figure 14.1a): the wall is cut from a sheet, folded and the edges are welded together. Both ends of the cylinder are flanged to facilitate the connection with the base and the lid. In the first containers the lid was soldered by hand after the food was introduced in the can.

In 1930, the technology was well enough developed to start with the production of beverage cans. The first beverage containers had a cone-shaped top closed by a cork. In 1935, the first flat-top can was introduced in Virginia. In 1963, Alcoa introduced the 'aluminium easy-open end'. A 'score' (a well-dimensioned groove) is machined in the lid and with the help of a small ring-pull tab, part of the lid can easily be torn off. This invention was a milestone in the development of beverage cans because it improved the convenience for the consumer. Until then the consumer needed a special steel opener to make holes in the lid.

Because the formation of a reliable score was not possible in steel, an aluminium lid was used in all beverage cans. However, a mixed steel and aluminium can is a major problem for recycling since both metals are undesirable alloying additions for each other, at least in more than trace quantities. Aluminium is readily oxidized away from steel, but iron is very difficult to remove from aluminium. The first complete aluminium beverage can was introduced in 1964. These first aluminium cans were produced by impact extrusion, but very soon this process was abandoned and replaced by the 'drawn and ironed' can. Such a can consists of only two pieces and is fabricated by deep drawing and wall ironing (Figure 14.1b). These processes have been discussed in Section 11.7. In 1990, the original 'easy-open end' system was improved by the introduction of the so-called 'eco-tab' or 'retained ring-pull ends'. Part of the lid is torn off and folded inwards but remains partially connected with the rest of the lid.

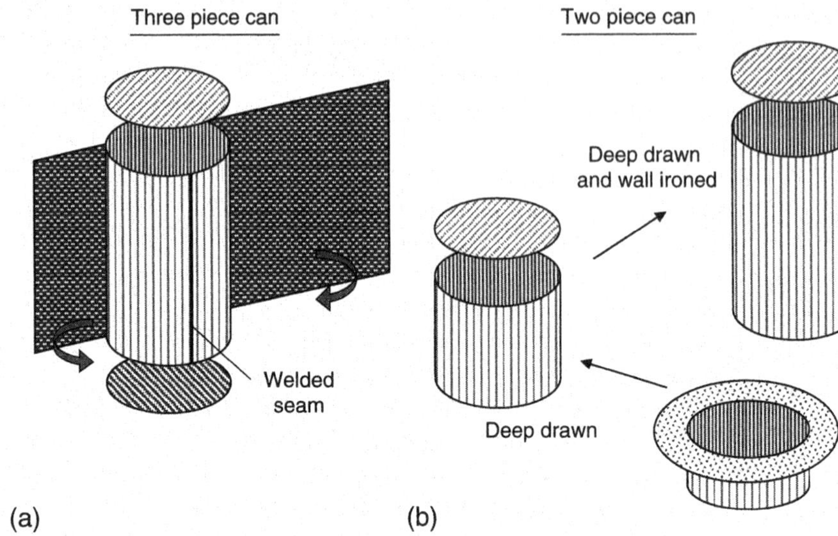

Figure 14.1. (a) A 'three-piece' container with welded seam; (b) 'two-piece' container, deep drawn or deep drawn and wall ironed.

In 2000, about 180 billion aluminium beverage cans were produced. Although the weight of a standard aluminium can is now only 13 g (about half its weight of 30 years ago), can manufacturing is a very important market for the aluminium industry. For an annual production of 100 billion aluminium cans in the US, more than 1 million tons of aluminium are needed and this represents about 10% of the total aluminium production in the US. In contrast with the US, where almost all the metal beverage cans are made from aluminium, the situation is a little different in Europe where in 2004 about 17 billion beverage cans were produced from steel and about 23 billion from aluminium (*Source*: http://www.aluminium.org/).

A two-piece aluminium beverage can is in most cases made from three different aluminium alloys. The drawn and wall-ironed cup is made of an AA3xxx alloy, in most cases an AA3104 (~1 wt% Mg and ~1 wt% Mn). The lid is usually made from an AA5xxx alloy, e.g. AA5182 (4–5 wt% Mg and 0.3–0.4 wt% Mn), and the pull ring is fabricated from another high-Mg AA5xxx alloy. The material used for the cup is called 'can body stock' or 'can body sheet' or simply 'can stock', and the material for the lid is called 'end stock'.

Although the scrap value of aluminium is fairly high, the 'value' of an empty beverage can is only about 2 Euro/US dollar-cent. Nevertheless, the US have succeeded in setting up an efficient recollection and recycling system. About 65% of the cans are recycled. In Europe, serious efforts for recycling have been made in the last decade and the recycling rate of aluminium cans was about 45% in 2001 and increases every year.

In the following paragraphs the general production scheme of aluminium beverage cans is illustrated. Then the thermo-mechanical processing (TMP) of can body sheet will be discussed, explaining the importance of a strict texture control and illustrating how this can be achieved by a good understanding of the deformation and recrystallization processes. Finally, some aspects of recycling will be addressed.

14.1.2 The production of a beverage can

Beverage cans (both steel and aluminium) can in principle be produced in three ways:

A three-piece can (Figure 14.1a): The wall is cut from a sheet, folded and the edges are welded together; both ends of the cylinder are flanged to facilitate the connection with the base and the lid. This method is still in use for the production of food cans. Most of these food cans are still made of steel.

Drawn and redrawn: A cup is obtained by one or more deep-drawing operations; this method is often used for shallow cans (with the height < diameter).

Drawn and wall ironing: A cup is obtained by one or two deep-drawing operations; the wall of this cup is then elongated (and consequently made thinner) by a wall-ironing process. Most beverage cans (both steel and aluminium) are produced by this method.

Beverage cans are used in extremely large quantities and have fairly well-accepted standard dimensions (33 cl and 50 cl; diameter ~ 64 mm). This product is thus suitable for mass production on fast and sophisticated machines. A beverage can for pressurized drinks, should comply with the following requirements:

- suitable for a very fast production rate;
- be able to withstand an internal pressure of 6.5 kg/cm^2 and a longitudinal force of 120 kg (for stacking purposes);
- compatible with an 'easy-open end' that can be mechanically connected with the cup after filling;
- the bottom of a can must fit into the top of another can (to improve the stacking of the cans);
- it must be possible to apply a coating in the inner side of the can (to avoid any interaction between the liquid and the metal);
- it must be possible to paint (decorate) the outside wall (for commercial reasons);
- be as light as possible (for both cost and transport reasons); and
- be as cheap as possible (competition from polymer containers).

14.1.2.1 The 'cupping' press. The manufacturing of aluminium beverage cans starts from coils of rolled aluminium sheet. The size of these coils depends on the details of the line, but typical dimensions are: 150–180 cm wide, 5–10 km long, a sheet thickness of about 0.3 mm and a weight of 5–12 tons. These coils are loaded in an 'uncoiler' which unrolls the sheet, feeds it into a lubrication unit and subsequently into a 'cupping press'. This 'double-action' press first cuts 10–14 circular discs from the sheet and converts these blanks into cups by deep drawing. These days, most aluminium companies are able to deliver 'pre-lubricated' sheets, which eliminate the need for a lubrication unit before the cupping press. The holed sheets from which the blanks have been stamped are sold back to the aluminium producer as primary scrap.

During drawing, the thickness of the sheet is changed as illustrated in Figure 14.2. The bottom of the cup keeps more or less the initial thickness of the sheet (~0.3 mm). The variation in wall thickness, as given in Figure 14.2, can be understood from the fact that during deep drawing, the circular blank is exposed to a circumferential stress. An imaginary circle on the blank moves inwards during drawing, reduces its diameter and increases the thickness of the blank. In the cup wall, it is exposed to a tensile stress and becomes thinner. These two effects lead to a steady decrease in thickness from the top to the bottom of the drawn cup. In many cupping presses, the radial clearance between the punch and the drawing die is restricted to about 1.05 times the initial sheet thickness in order to keep the wall thickness more uniform. This implies that some wall ironing is carried out during deep drawing. Modern cupping presses have an impressive production rate. Depending on the coil width, each stroke provides 10–14 cups at a rate of up to

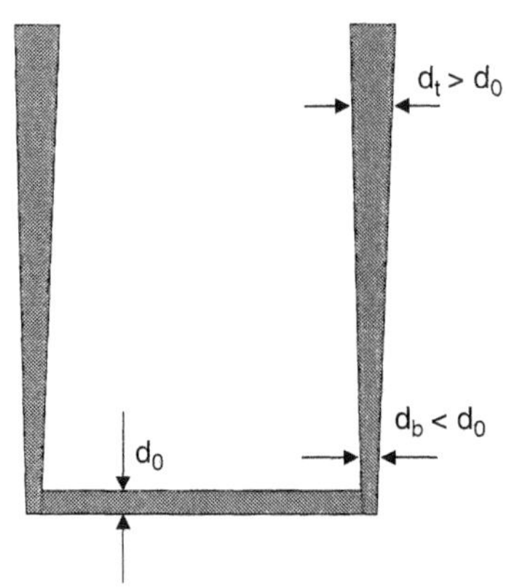

Figure 14.2. Schematic showing the variation in wall thickness along the height of the cup. The thickness of the bottom (d_0) is closed to the original sheet thickness.

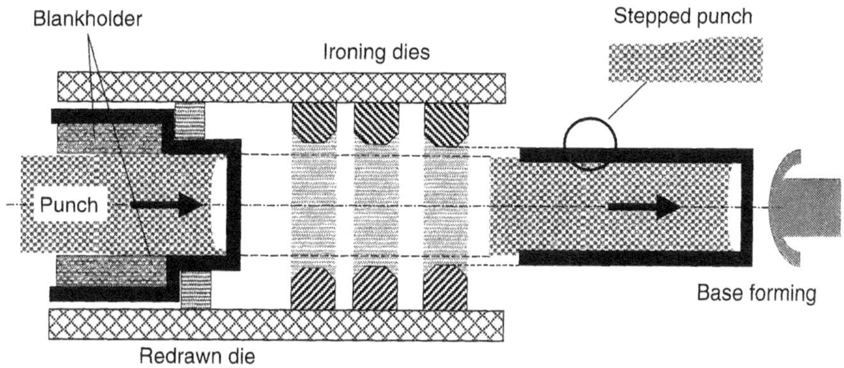

Figure 14.3. Scheme of the second deep-drawing pass, the three-wall ironing passes and the formation of a dome-shaped bottom in the 'bodymaker' press.

180 strokes a minute. This gives a production of about 2500 cups a minute, 1 million a day or 250 million cups a year.

14.1.2.2 The 'bodymaker' press. The cups produced by the cupping press are transported to a 'bodymaker' press. This is a long-stroke press: in one single stroke of the punch, the cup is first redrawn and subsequently wall ironed by three ironing rings and finally at the end of the punch stroke the bottom is formed into a dome shape that strengthens the bottom of the can (up to this point the bottom of the cup has hardly been deformed). A scheme of these subsequent forming steps is shown in Figure 14.3.

During the three ironing steps, the thickness of the cup wall is reduced from about 0.3 to 0.11 mm. The top of the wall remains thicker (~0.16 mm) because in one of the next production steps a 'necking' operation will be carried out. This local increase in thickness is obtained with a so-called 'stepped punch' (Figure 14.3). In the bodymaker press a large amount of lubricant is needed. This lubricant reduces the frictional heat and serves as a cooling agent. It must be compatible with the lubricant used in the cupping press. The production rate of a modern bodymaker press is about 320 cans/min, which is about 5 strokes/s! For each cupping press about 6–7 bodymakers are needed.

When the cans leave the bodymaker, the top is trimmed mechanically to obtain a uniform height and thereby removing the ears that were formed during deep drawing (cf. Sections 11.7 and 14.1.3.1).

14.1.2.3 Cleaning and decoration. The cans formed in the bodymaker are conveyed to a 'can washer'. The cans are degreased and rinsed so that all remaining

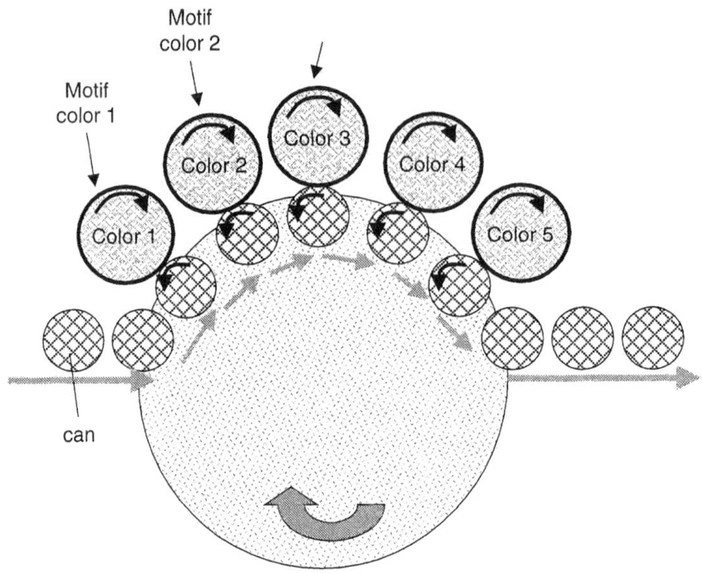

Figure 14.4. Schematic illustration of the decoration of the outer can wall.

traces of lubricants are removed. A special chemical pre-treatment is applied in order to facilitate the coating and decoration of the cans. Next they are transferred to a 'decorator'.

With the help of special 'rollers' an external base coat is applied which is used as an undercoating for the decorative print. This decoration is carried out with a dry offset technique with 4–6 colours. For each individual colour a printing plate is fixed to a separate roller, with all the rollers arranged around a central drum (Figure 14.4).

The rotation speeds of drum, rollers and cans are synchronized and the different colours are transferred to the can wall in sequence. Modern decorators run at a speed of 1200 cans/min. After this decoration, a protective coating is applied and the cans move through an oven to dry the external printing. Then an internal coating is sprayed into the can and cured in a second oven (Figure 14.5). This curing treatment is carried out at temperatures between 150 and 200°C and takes about 20 min. Since the material has been heavily cold deformed, some recovery takes place and the strength of the can wall is reduced by 15–30 MPa. The amount of softening with recovery can be reduced by having increased amounts of Mg, Si or Cu in solution before cold rolling. Such extra solute also gives a higher strain-hardening response, giving can body stock that is stronger than AA3104 while maintaining its drawability (Doherty and McBride [1993b], Doherty *et al.* [1994]).

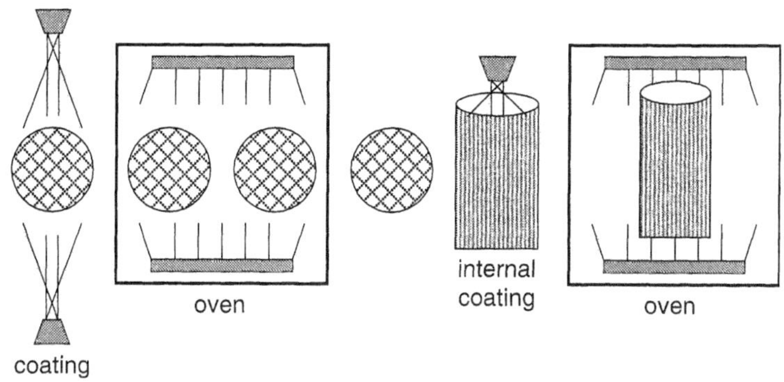

Figure 14.5. Application of an external and internal coating, both followed by curing in an oven.

Sadly that discovery has not been utilized by the industry since extra strength is not needed as the can wall thickness is now limited not by its yield strength but by its elastic modulus. Cans with thinner walls would buckle at the base of a stack of filled cans – the longitudinal load limit noted previously.

14.1.2.4 Mechanical finishing. First a tapered neck and a flange are formed. This is done in order to facilitate the connection with the can lid, after filling the can. The tapered neck was developed to reduce the amount of the more expensive can lid alloy AA5182. After these mechanical finishing operations, the cans pass through a light tester which rejects the damaged cans. The cans are palletized and dispatched to the filling plant, that is ideally situated very close to the can-making plant, usually as part of the beverage company operations. The price of such a can is about 10 US¢.

14.1.2.5 Formation of the can lid. The can lids are made of an AA5182 (Al–4.5 wt% Mg) alloy that combines a good formability with a higher strength than AA3104. This higher strength is needed due to the flat nature of the can lid, since resistance to outward bulging is more difficult to achieve in a flat lid than with the appropriately curved can base. In a first step, the plain can lid is formed in a simple press and the edge is curled in such a way that it fits onto the flange of the can body. The thickness of a lid is about 0.3 mm. The pre-formed lids are transferred to a multi-tool press to form a score line and to attach the ring-pull tab on the lid by means of an integral rivet (Figure 14.6). The score line has the form of a truncated V to a depth leaving about 0.1 mm of residual metal. The depth of the score line is very critical because on one hand, the score line must be strong enough to withstand the internal pressure and normal handling of the can, but on the other, it must not require too much force to tear it open.

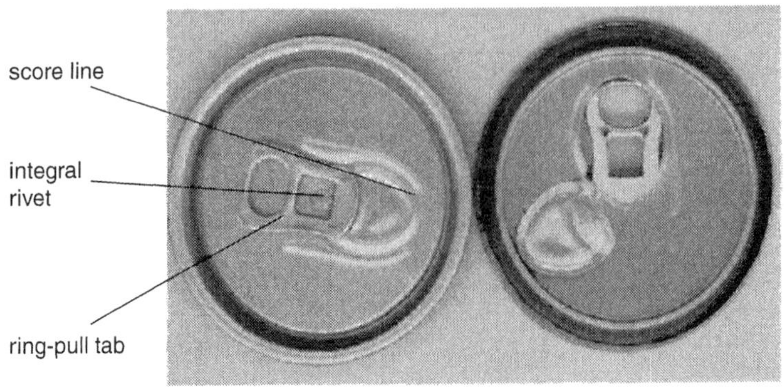

Figure 14.6. Can lid with an 'eco-tab' or a 'stay-on tab' (Talat [1999]). (Courtesy of the European Aluminium Association).

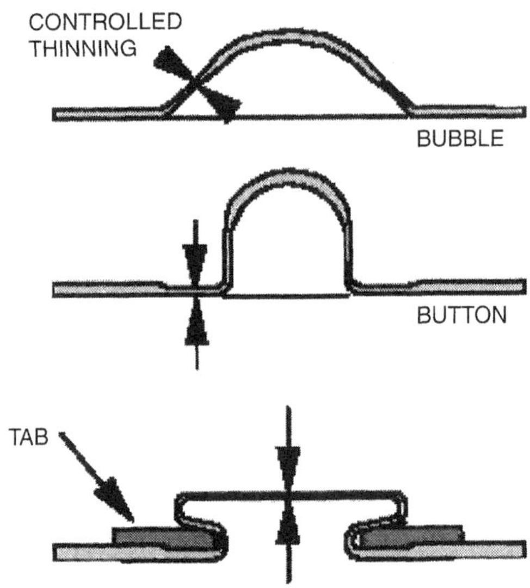

Figure 14.7. Steps in forming an integral rivet (Talat [1999]). (Courtesy of the European Aluminium Association).

The subsequent steps of the formation of the integral rivet are shown in Figure 14.7. First a bubble is formed. The flank of the bubble is thinned and the bubble is re-formed into a button (or rivet). When the tab has been placed over the button, the actual riveting takes place.

14.1.2.6 Filling and closing the can. Cans and lids are dispatched to the beverage company and carefully cleaned. The cans are filled and pressurized with CO_2.

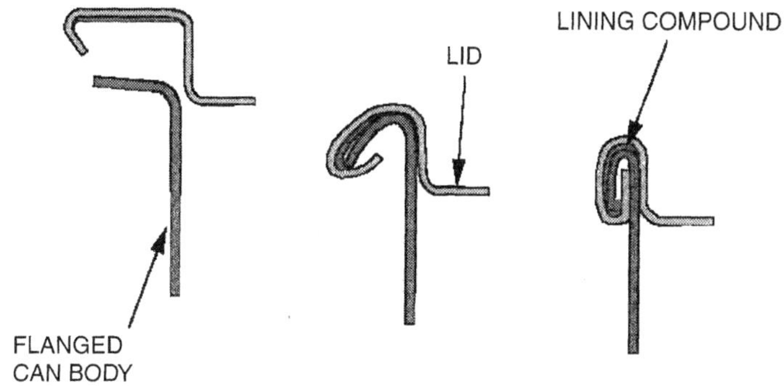

Figure 14.8. Formation of a double-lock seam (Talat [1999]). (Courtesy of the European Aluminium Association).

Table 14.1. Composition of can stock alloys.

Wt%	Mg	Mn	Si	Fe	Cu	Zn
AA3004	0.8–1.3	1–1.5	0.3 max	0.7 max	0.25 max	0.25 max
AA3104	0.8–1.3	0.8–1.4	0.6 max	0.8 max	0.05–0.25	0.25 max

The lid is attached to the can with a so-called double-lock seam (Figure 14.8). After a final screening and printing of an identification number, the cans are dispatched to the distribution points.

14.1.3 The production of can body sheet

14.1.3.1 Materials for can body and can lid. The most convenient aluminium alloys for the fabrication of beverage cans are the AA3xxx alloys. The AA3104 variant, which is nearly exclusively used as can stock alloy (Table 14.1), is a modified version of the AA3004 alloy that was used several years ago. The Mn content is required to give good corrosion resistance by its combination with the iron constituents to make the corrosion potential of the resulting α-$Al_{12}(Mn,Fe)Si_3$ constituents match that of the aluminium matrix. The Mg is needed for the strength it gives arising from the much increased strain-hardening response compared to an Mg-free alloy. Although the $\bar{R}$-value of that alloy is rather low, compared with steel (~0.8 against ~1.8), the AA3104 can easily been deep drawn into a shallow cup. In order to obtain a sufficient strength, the initial sheet is delivered in the cold-rolled condition (in most cases with 88% reduction). A good surface quality after deep drawing and ironing is obtained by a careful control of the 2nd phase particles, dispersoids and constituents (cf. Section 14.1.3.2). Special attention is given to the earing behaviour of the can stock sheet. In fact, it is mainly the amount of earing that will serve as a criterion to accept or reject an alloy. Large ears will not only produce more scrap, but will also increase the risk of blocking the can in the bodymaker.

Control of earing properties by a suitable TMP essentially relies on a careful control of the texture. This is done by a careful control of all stages of the processing. Since the can bodies are made from heavily cold-rolled sheets, the typical fcc cold-rolling texture components, S, brass and Cu will be present. As shown in Table 14.2, these components lead to the formation of 45° ears. In principle, these 45° ears can be compensated by the right amount of 90° ears, as illustrated in Figure 14.9. These 90° ears are generated by some particular texture components, e.g. the cube component. This means that the correct amount of cube must be present in the hot-rolled sheet in such a way that the texture change during cold rolling leads to the desired mixture of cube (provoking 90° ears) and β-fibre components (45° ears). This is illustrated in Figure 14.10.

The material requirements for the lid are different. Deep drawability and earing are less important, but strength and formability must be excellent. These requirements are fulfilled by the AA5182 alloy (Table 14.3). The much higher Mg content of AA5182 promotes the even stronger strain hardening compared to AA3104.

Table 14.2. Influence of some texture components on earing.

Texture component	Name	Earing
{100}<001>	Cube	4 ears; 0/90/180/270°
{110}<001>	Goss	2 ears; 0/180°
{110}<112>	Brass	4 ears; 45/135/225/315°
{112}<111>	Copper	4 ears; 45/135/225/315°
{123}<412>	R or S	4 ears; 45/135/225/315°

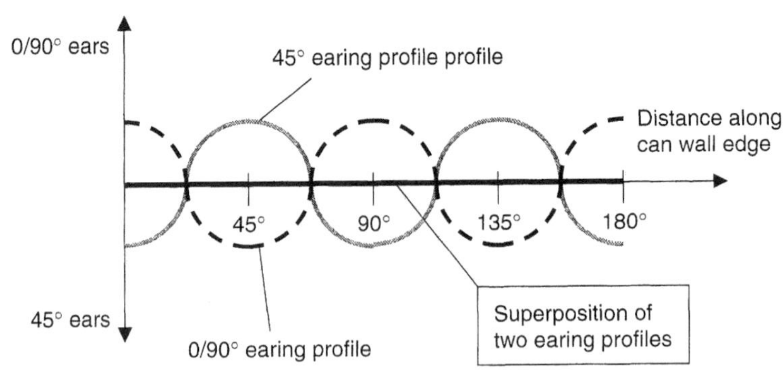

Figure 14.9. Illustration of how 45° and 0/90° ears can compensate each other.

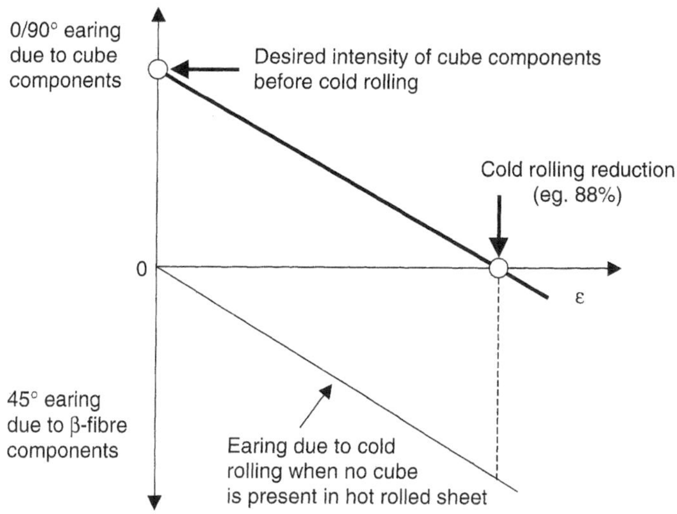

Figure 14.10. The 0/90° ears are gradually compensated by the 45° ears as a function of increasing cold rolling.

Table 14.3. Composition of an AA5182 alloy.

Wt%	Mg	Mn	Fe	Si	Cu	Zn
AA5182	4–5	0.2–0.5	0.20	0.20	0.15	0.25

The strength of the can body alloy can be increased greatly by the use of higher Cu and Si, but with high Mg contents the alloying additions can only be kept in solution by high-temperature annealing followed by rapid cooling. This is possible in modern continuous annealing lines. Without this technology, however, the only alloy addition that can be used for promoting extra strain hardening in alloys that are not continuously annealed is the highly soluble Mg.

14.1.3.2 Texture control. A minimal earing in can stock alloys can only be obtained by a well-balanced texture. Control of the texture development during the processing of these alloys is not simple and was achieved by a well-chosen composition and extensive and largely empirical development of the homogenization and hot-rolling parameters.

A key concern is to obtain a sufficient amount of cube-oriented grains after hot rolling and recrystallization. It is known (and has been discussed in some previous chapters) that a strong cube texture is formed during recrystallization of cold-rolled pure fcc metals. In can stock, the situation is however more complex. First of all, we are dealing with hot rolling and recrystallization events will not only occur after hot rolling, but can also take place on the run-out table in between two hot rolling passes. This implies that a grain that is formed during recrystallization after pass 'n', will be deformed again during subsequent rolling passes. We also deal with a complex commercial alloy containing dispersoids and large constituent particles, which can have a strong influence on recrystallization. In general, after hot rolling and recrystallization of a can stock alloy, a mixed texture is obtained, consisting of cube-oriented grains (mainly formed during recrystallization events), with weaker retained deformation components (Cu, S and brass) and 'random' grains.

A lot has been written about the origin of the cube grains after both hot and cold rolling and still no absolute agreement is found in the literature. Nevertheless, most authors seem to accept the following picture. The cube-oriented grains nucleate during recrystallization, at least after hot rolling, at the so-called 'persistent cube bands'. The presence of 'persistent cube bands' after hot rolling of 3104 was demonstrated by Daaland and Nes [1996]. The behaviour of cube grains during hot rolling has been the subject of much discussion but most authors, e.g. Ricks [1999], now accept that the cube grains are stabilized by high-temperature non-octahedral slip on $\{110\}<011>$ as proposed by Maurice and Driver [1993] following detailed single crystal studies. At lower temperatures, or high Z, where classical octahedral slip dominates, the cube grains are unstable and tend to break up and progressively rotate away from cube. The subgrains in these persistent cube bands are relatively large and so have a rather low stored energy. In other words, they are excellent nuclei and can easily grow and consume part of their surroundings, before other nuclei become active. For more details, the reader is referred to Maurice and Driver [1993], Doherty *et al.* [1995, 1998] and Vatne *et al.* [1996a, b]. These studies reflect an almost universal result that the strength of cube texture in rolled fcc metals after recrystallization always increases strongly with the rolling reduction. The much weaker cube texture on recrystallization after cold rolling commercial purity aluminium appears to arise predominantly by the decreased as recrystallized grain sizes after equivalent reduction by cold rolling (Doherty *et al.* [1998]).

The deformation components are, just as in cold rolling, the result of a natural texture development during deformation (rotation of grains towards these stable-end orientations). These β-fibre grains are being formed during deformation and partially consumed by other grains during recrystallization.

As a result of this analysis, it can be concluded that the rolling schema has to be designed in such a way that it leads to a completely recrystallized material (non-recrystallized grains will mainly have the undesired β-fibre orientations), with a

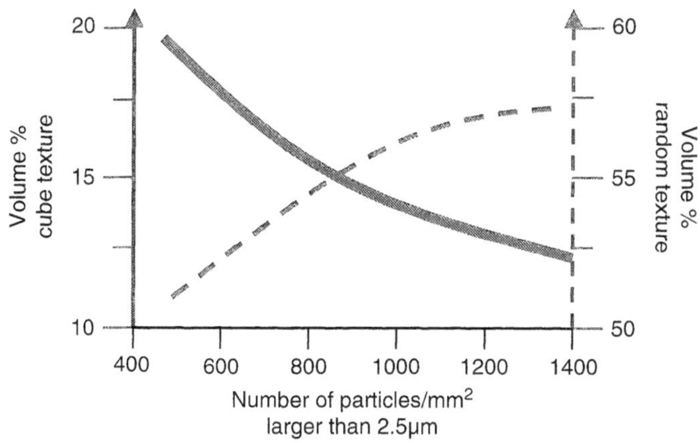

Figure 14.11. Influence of large particles (> 2.5 μm) on the texture after hot rolling and recrystallization. After Marshall [1996].

maximum amount of cube grains. As will be described below, this can only be achieved by a careful control of the processing.

Since *large particles* (Ø > 2.5–3 μm) will promote particle-stimulated nucleation (PSN) and, by decreasing the recrystallized grain size, will generate a weaker cube texture, their presence should in principle be avoided, for example by limiting the amount of Mn, Si and particularly Fe. The influence of large particles on the texture after hot rolling is illustrated in Figure 14.11. In addition, very large iron-rich constituents can promote fracture during the can-forming operations.

Unfortunately, two specific types of particles, including the constituents, are needed for other reasons. During wall ironing the ironing rings can accumulate small amounts of material from the can wall. This material will stick for sometime onto the ring and then it gets free and is squeezed into the can wall. This results in a serious surface defect that cannot be tolerated. This phenomenon is called 'galling'. Experience has shown that a minimum amount of hard α-$Al_{12}(Fe,Mn)Si_3$ particles, combined with a quantity of finer Mn-rich 'dispersoids', also α but here $Al_{12}MnSi_3$, can eliminate galling. The iron 'constituents' are formed during eutectic solidification since the iron content of the liquid alloy, usually about 0.4 wt%, is much greater than the maximum solubility of iron in solid aluminium, which is at a maximum of less than 0.04 wt%. The finer 'dispersoid' particles are formed by solid-state precipitation of Mn that is kept in supersaturated solid solution during cooling due to its very slow diffusion rates in solid aluminium. The precipitation of these dispersoids takes place during the slow heating of the large ingots, the 'homogenization' annealing, carried out before hot rolling. It is thought that these hard α-particles perform a sort of continuous cleaning of the die rings.

Hence, the composition of the can stock alloy and the homogenization and hot-rolling parameters must be chosen in such a way that an optimal amount of α-$Al_{12}(Fe,Mn)Si_3$ particles of suitable size ranges is present and that other larger particles are avoided as much as possible. The scale and uniformity of the distribution of the dispersoids is also important in controlling the recrystallization kinetics – too high a density of these fine particles can inhibit the necessary recrystallization prior to the final cold rolling.

As a result of these interacting requirements, a great deal of empirical development was required to optimize the properties of the can alloys. The specific details of the homogenization and hot-rolling techniques used by different companies are kept secret and they are probably different depending on the particular hot-rolling equipment used in each plant. However, the resulting microstructures, in terms of particle distributions, and recrystallization texture are all rather similar. In all cases, there are coarse constituents, fine scale dispersoids, a fine typically 20–30 μm grain size with a rather strong cube texture, with about 20% of the grains within less than 15–20° from the exact cube orientation.

14.1.3.3 Hot rolling schedule. Since for most companies the production parameters of can body sheet is confidential, it is difficult to find details of the production scheme in the open literature. The main lines of a typical production scheme are shown in Figure 14.12. The details can vary from one producer to another and the indicated figures are tentative but quite realistic.

The ingots are produced with a DC (*direct casting*) or EMC (*electro-magnetic casting*) method. Those ingots could have a thickness of 75 cm and a width of

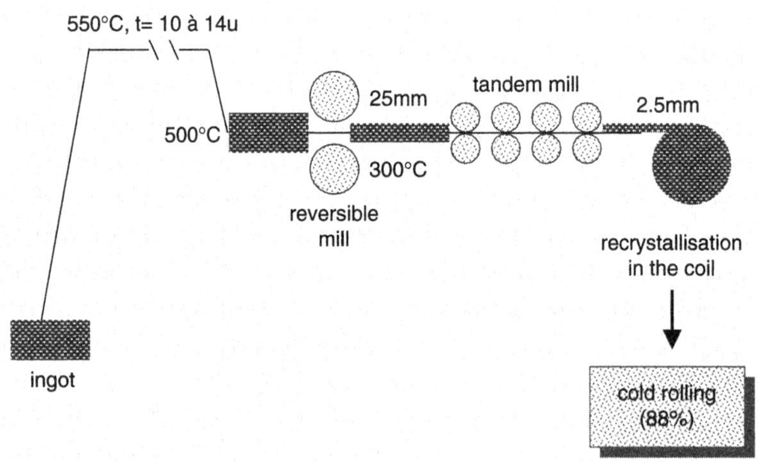

Figure 14.12. Typical production scheme for can body sheet.

about 150–180 cm. Mainly three types of eutectic constituents can be found in these ingots: β-Al_6(Fe,Mn), Mg_2Si and some α-Al_{12}(Fe,Mn)Si_3. The as-cast structure is supersaturated with Mn (up to 0.9 wt%) and there is some microsegregation (coring) of Mn, Mg, Si and Cu over a cell size of 100–150 μm.

After scalping some 2–5 mm from the surface (always needed after DC), the ingots are slowly reheated and *annealed* for several hours at a temperature around 550°C. During this homogenization treatment the microsegregation of the fast diffusing elements Mg, Cu and Si is eliminated. The non-equilibrium Mg_2Si eutectic particles are dissolved and the β-Al_6(Fe,Mn) constituents partially transform into α-Al_{12}(Fe,Mn)Si_3. This transformation is promoted by the Si atoms that are released by the dissolving Mg_2Si. Some β-constituents remain present into the material. The Mn atoms, which are supersaturated after casting, will partially precipitate as fine dispersoids during slow heating, re-dissolve partially at higher temperature and the surviving particles will coarsen. This three-step process is very important because it provides the possibility to obtain a rather uniform dispersion of particles of intermediate size and avoid very fine particles, which would hamper the recrystallization after cold rolling. If the Mn is precipitated isothermally, as was attempted in some unpublished studies, the distribution of the dispersoids is found to be highly variable, with regions of fine and regions of coarser precipitates, reflecting the microsegregation of the slow diffusing element, Mn. The precipitation, resolution and coarsening mechanisms active during the homogenization anneal, lead to a much more uniform distribution of the α-$Al_{12}MnSi_3$ dispersoids with a size of about 0.2–0.3 μm.

After this homogenization treatment, the ingots are taken out of the furnace and transported to a reversible mill to be hot rolled in several passes into plates of about 25 mm thickness. After the reversible mill, the plates are transported to a tandem mill for further reduction into sheets of about 2.5 mm thickness. This material is then coiled and will recrystallize in the coil. When the exit temperature at the tandem rolling mill cannot be kept high enough, the material will not fully recrystallize in the coil and a separate recrystallization treatment is necessary. However, in well-run plants the extra cost of this extra anneal is avoided by controlling the exit temperatures to ensure that the 'self-anneal' in the coil does lead to full recrystallization.

During hot rolling and recrystallization, it is of prime importance to avoid as much as possible the formation of fine precipitates. This is done by limiting the Mn content of the alloy (compare the composition of AA3104 with the older AA3004), although enough Mn must be present to form the desired large α-particles. The Fe content is also restricted because a low Fe content restricts the solubility of Mn at high temperature and hence the amount of precipitates on cooling.

In a later stage, the hot-rolled material will be *cold rolled* into sheets of about 0.3 mm (88% reduction). The cold-rolled material has typically an ultimate tensile strength (UTS) of 300 MPa and a yield strength (YS) of 285 MPa. The fracture strain is rather low (about 2%). During cold rolling the large particles crumble and a deformation texture is generated. The cube grains, which have been formed during recrystallization will partially re-orient, but are as a whole relatively stable and remain present as an important texture component.

14.1.4 Recycling

14.1.4.1 Background. The energy that is required to remelt aluminium scrap is about 5% of the energy needed to extract the same amount of material from the ores. It is obvious that under such conditions as much aluminium as possible must be recycled. The recycling of beverage cans is possible, but some specific problems need to be addressed.

Although the intrinsic scrap value of aluminium is much higher than that of steel, the value of an empty beverage can is still only about 0.02 Euro/US dollar. This price, though small, can be compared to the total production cost per can of only about 10 US¢. The price paid for used beverage cans in the US, in $/ton, remains at about 75% of the price charged for the finished coils of can body stock delivered to the can makers. The price varies proportionally and in phase with the cost of primary aluminium. The high value of the scrap reflects the high cost of aluminium, currently about $1700 a ton, that is largely determined by the energy cost of extracting the metal from its very stable oxide. Provided a viable recollection system can be set up, as it has been in the US, then recycling of this 'secondary' or post-consumer scrap is clearly economic. Recycling of 'primary' or industrially generated aluminium scrap, such as the can sheet from which the circular blanks have been cut and the machining chips from thick aerospace plate (Section 14.4) has always been a feature of aluminium technology. The value of the beverage can scrap has an important role to play in contributing to general materials recycling with its clear environmental benefit. Used beverage cans are economically valuable in their own right and help to cover the cost of recycling other materials such as steel and polymers whose scrap value is very much less than that of well-characterized aluminium cans. It should be remembered however that if different aluminium alloys are co-mingled, the scrap then has much less value. This arises since only Mg of the alloying additions in aluminium can be readily oxidized away from molten aluminium. In contrast with steel, it is not possible to remove all sorts of impurities during melting and a careful control and selection of the scrap is necessary.

In the United States, an efficient recollection system has been operational for several decades (Figure 14.13). In 1965, when the first aluminium beverage cans were successfully introduced, the steel lobby feared to lose an important market and launched the famous 'ban the can' campaign: 'when you throw away a steel can, it will rust and disappear; if you throw away an aluminium can, it will pollute

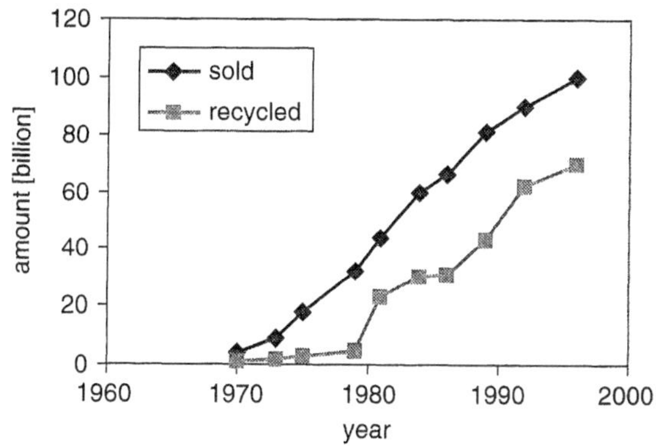

Figure 14.13. Evolution of the amount of recycled and sold beverage cans in the US. After van Linden van den Heuvel [1996].

Table 14.4. Typical composition (Alcoa standard) of the aluminium alloys used in beverage cans.

	Alloy	Si	Fe	Cu	Mn	Mg
Wall	AA3004	0.18	0.45	0.13	1.1	1.1
	AA3104	0.18	0.45	0.15	0.9	1.25
Lid	AA5182	0.10	0.24	0.03	0.35	4.5
Handle	AA5042	0.10	0.24	0.03	0.30	3.5
	AA5082	0.10	0.17	0.03	0.05	4.5
Complete can		0.16	0.38	0.11	0.88	1.94

nature for many decades'. The reaction of the aluminium industry was to set up a recollection system for used cans. During the energy crises in the 1970s, the usefulness of such a recollection system was appreciated and the system was improved. Actually in the US about 65% of the beverage cans are recycled. This means that on average an aluminium beverage can consists for 55% of recycled aluminium (some scrap is lost during processing). The time span between production of a can and its recycling is extremely short (about 60 days). An aluminium atom that is always recycled, can in one year be part of six beverage cans! In Europe the situation is less good, but the recycling rate is improving quickly and about 45% of the cans were recycled in 2001.

14.1.4.2 Contamination of the scrap. The recollected beverage cans are contaminated in several ways: contamination of the can itself (coating, ink, varnish, . . .), presence of other metals (e.g. steel cans) and presence of different aluminium alloys in one can. Table 14.4 shows some typical compositions of alloys used in beverage cans. The last row gives the 'average' composition of a can, which is in principle equal to the average composition of the scrap.

This average composition has too much Si and Fe to be remelted as an AA5182 alloy. At present the recycled metal is used to make the can body alloy, with the can lid alloy made from relatively pure aluminium direct from the smelter. The Mg content of the scrap is a little too high, but this is in most plants no problem because during the remelt some Mg is lost due to oxidation.

In some recycling units it is possible to separate the AA5xxx from the AA3xxx alloys using the lower melting temperature of the AA5182 alloy. Some attempts have been made to introduce lids of an AA3xxx alloy, but this has not been a commercial success until now.

14.1.4.3 Main steps in recycling

- Recollection of the scrap and separation of other garbage fractions
- Crushing and shredding of the cans
- Separation of steel parts (magnetic) and other metals (sink-and-float baths)
- Burning of the interior lacquer coating (around 500°C) and the outside printing inks
- Optional: separation of AA3xxx and AA5xxx
- Remelting and correction of composition

14.1.4.4 Weight savings. Over the last 30 years, a lot of progress has been made in reducing the weight of aluminium beverage cans. As shown in Figure 14.14, both the weight of the can wall and the can lid have been reduced by a factor 2. This has been achieved by a better can design, improvements in can manufacturing and TMP of can stock and by a better balanced composition of the alloy. This is very beneficial for the environment but at the same time it gives an important

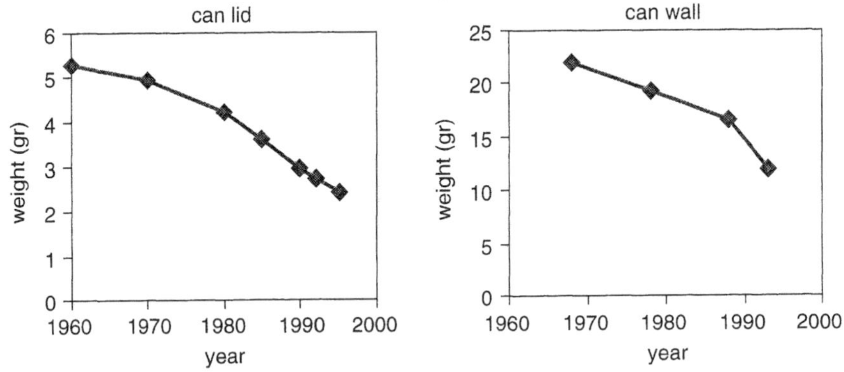

Figure 14.14. Evolution of the weight of aluminium beverage cans over the last 40 years.

Table 14.5. Typical composition of tinplate.

Element	C	Mn	P	S	Cu	Sn
Wt%	0.120	0.25–0.60	0.015	0.050	0.200	0.050

economical benefit. A weight reduction of about 1% means a worldwide saving of approximately 26 000 tons of aluminium or about 50 million Euros.

14.1.5 An alternative material: steel

In Europe about 45% of beverage cans are made of steel. The manufacturing of a steel can is approximately the same as for an aluminium beverage can. A typical composition of the steel used for beverage cans is given in Table 14.5. This 'tinplate' is delivered in cold-rolled condition, with a thickness of 0.25 mm (0.3 mm for aluminium) and with a tin coating. The final wall thickness of a steel can is about 0.09–0.10 mm (0.12 for aluminium) and its weight is about 25 g (13 for aluminium). Nearly all steel cans have a lid from aluminium (AA5182), because it is difficult to make a reliable score line in steel. Some years ago Corus introduced a steel lid in the market, but it is not widely used.

For more information about beverage cans out of steel, the reader is referred to E. Morgan [1985].

From the previous paragraphs it can be concluded that beverage cans are a 'high-tech' product, both from the point of view of production and from the point of view of material aspects. It is difficult to compare the benefits of steel cans with those of aluminium cans, but at the present moment both seems to be more or less equivalent.

14.2. ALUMINIUM SHEETS FOR CAPACITOR FOIL

14.2.1 Introduction

A capacitor, or condenser, is a device for storing electricity, which can be repeatedly charged and discharged. Capacitors are widely used in the electrical industry at scales ranging from a few farads in a high-speed train down to microcapacitors of picofarads; there are over 200 capacitors in a typical magnetic recorder. Basically, they are composed of two metal plates (anode and cathode) separated by an insulator known as the dielectric. There are a wide variety of capacitors with different characteristics but one of the most common types is made of two aluminium foils separated by a thin aluminium oxide layer and an electrolyte – the aluminium electrolytic capacitor. They are used for high capacity requirements (typically a few hundred and more microfarads).

As described below, one of the parameters that controls the efficiency of the device is the amount of reactive surface of the metallic foil. In the case of Al capacitors, this surface can be significantly increased by chemically etching the foil surface to create microscopic 'tunnelling' pits. The pitting angle is strongly dependent on the orientation of the grain and is a maximum of 90° for the cube orientation. It follows therefore that maximum capacity is obtained by etching a perfectly cube-textured sheet.

A remarkable method was discovered in 1970 by Francis Boutin of the Pechiney research centre (Boutin [1970]) for producing near-perfect cube-textured sheet. The method, described below, is based on a subtle control of texture evolution during rolling and recrystallization of high-purity aluminium. It illustrates some of the principles of texture control described in Chapter 8 together with some aspects of recrystallization and work hardening (Chapters 4 and 5). The application of this special technique of texture control to the production of high-performance capacitor foil is standard over the world.

14.2.2 Capacitor requirements

The capacity C of a capacitor is a function of the capacities of the anode C_A and the cathode C_C according to

$$\frac{1}{C} = \frac{1}{C_A} + \frac{1}{C_C} \tag{14.1}$$

where $C_A = \varepsilon^\circ \varepsilon_r S_A / t_A$ and $C_C = \varepsilon^\circ \varepsilon_r S_C / t_C$, t_A and t_C, respectively the thicknesses of the anode and cathode, ε° and ε_r the absolute (As/ Vm) and relative permittivities and S_A and S_C the surface areas of the anode and cathode.

To avoid electric discharges by arcing across the plates, t_A is proportional to the maximum applied voltage of the capacitor which usually is required to be as high as possible. On the other hand, t_C can be small to increase C. For this reason, $C_A << C_C$ (although $S_A > S_C$) and so $C \approx C_A$. From the above two equations it is then seen that the capacity is roughly proportional to the effective surface area of the anode S_A.

To increase S_A the sheet can be chemically etched to form long etch pits or micro canals which extend well into the surface grains. As noted above, etch pitting is orientation dependant; in fcc metals the etch pits grow along the <100> directions (Figure 14.15). If the grain is non-cube, then the canals are oriented at angles below 90° to the foil surface and the effective surface area is given by their *projected areas* onto the sheet surface. This area is obviously a maximum for the cube grains, which have the growth direction perpendicular to the sheet plane. It is therefore of some importance to develop a sheet with a strong cube texture which can be etched to a high specific surface area. A schematic section of this type of capacitor is given in Figure 14.16.

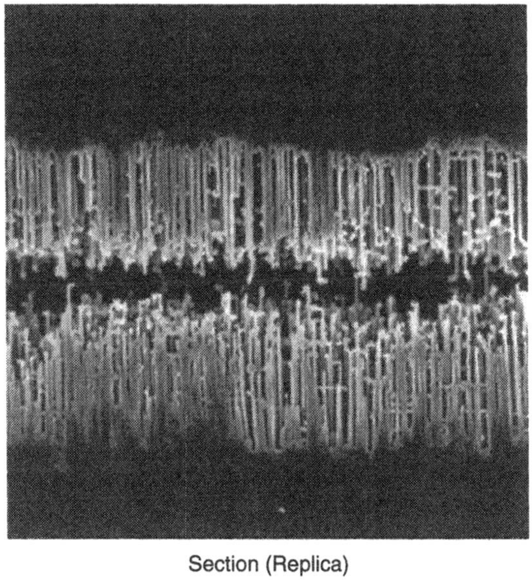

Figure 14.15. Tunnelling pits in cube-textured Al foil for anodes (cross-section).

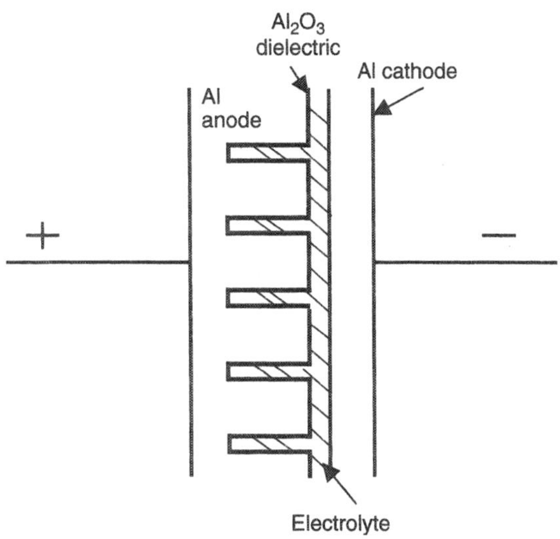

Figure 14.16. Schematic of capacitor using etched Al anode.

In a real Al electrolytic capacitor the dielectric is the thin oxide layer produced by anodizing the Al anode (relative permittivity $\varepsilon_r = 9.5$). The cathode is a conductive liquid or electrolyte (since this can penetrate the tunnelling pits) plus some paper. The so-called Al cathode foil serves just as a large surface contact area for passing the current to the operating electrolyte.

14.2.3 The process

The metal is a high-purity aluminium denoted AA1199 of minimum purity 99.99% Al (by weight). A high-purity aluminium is required first for the quality of the alumina Al_2O_3 dielectric. This is formed by anodizing the Al foils and suffers from electrical leakage if it contains too many impurities. It so happens that it is this high-purity aluminium which is amenable, by thermo-mechanically processing, to the development of a strong cube texture by recrystallization without the extra problems of say PSN.

4N aluminium is therefore required for capacitors and is produced by one of the two special techniques: three-layers refining or the more recent natural segregation method. Three-layers refining is a variant of the electrolysis of a fluoride salt solution to produce aluminium with a soluble anode of standard liquid commercial purity aluminium. The metals less electropositive than aluminium are trapped at the anode and only aluminium metal is transferred to the cathode. The bath composition is improved and modified to operate at lower temperatures of 700–800°C. The other method by segregation involves cooling the liquid metal so that solidified crystals which are purer than the initial solution form in the liquid phase, where the impurities are concentrated. By separating the crystals from the liquid, a highly purified fraction of the initial solution is obtained. 4N aluminium typically contains 10 ppm Si, 10 ppm Fe, 10 ppm Mg and 40 ppm Cu as impurities.

After casting the metal is homogenized, mostly to reduce the solute segregation of Fe that occurs during solidification. It is then hot rolled in a reversible mill down to 25 mm thickness, then transferred to a tandem mill for further hot rolling down to 3.5 mm. The recrystallized sheet is then cold rolled down to about 120 μm thickness (Figure 14.17). At this stage it is heavily cold worked. The following is the more original part. The 120 μm-thick sheet is lightly annealed at 250°C to *partial* recrystallization. It is then cold rolled about 15–20% before the final complete recrystallization at 500°C.

The partial recrystallization is the one of the keys to cube texture control. As described in Chapters 5 and 8, standard recrystallization of cold-rolled high-purity aluminium will automatically create a cube texture with typically 50% cube (and therefore 50% of other orientations). The Pechiney process – and its variants – will develop 95% or more cube grains.

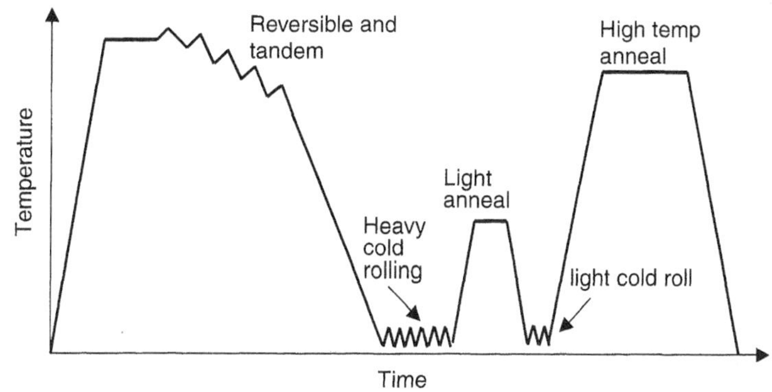

Figure 14.17. Processing route to form cube-textured Al foil.

14.2.4 Cube texture control mechanisms

In the heavily cold-rolled state, after a strain about 5, the Al sheet is made up of strongly dislocated and fragmented grains which possess the standard β-fibre orientations, i.e. Bs {110}<112>, S {123}<412> and Cu {112}<111>. However, it also contains some small fragments (but <1%) of cube-oriented zones which are left over from the recrystallized state before cold rolling. Many studies have shown that during cold rolling cube grains break up by deformation banding into strongly misoriented bands (misorientations >45°), separated by transition bands (Dillamore and Katoh [1974], Akef and Driver [1991], Liu and Hansen [1998a], Liu *et al.* [1998b]). The transition bands, as their name suggests, accommodate, over distances of a micron to a few microns, the misorientations between the deformation bands. They often include segments of the original cube orientations. The latter can recover very quickly so that they become recrystallization nuclei during a light anneal. This is what happens during the first short anneal at 250°C. Although many cube nuclei develop into small recrystallized grains during this stage, so do the other nucleation sites, e.g. SIBM and other transition bands. These other sites create recrystallization nuclei or small recrystallized grains with other orientations. At this stage of partial recrystallization, there is a recovered cold-rolled microstructure with some small, recrystallized grains of which roughly half are in cube form and the other half non-cube.

The subsequent light cold rolling to a strain of order 0.16 deforms this microstructure mix in different ways. The recovered grains with the β-fibre orientations are deformed as usual and retain their stable orientations with increased dislocation density. The non-cube recrystallized grains are also plastically deformed and rapidly develop significant dislocation densities in Stage III. The recrystallized cube grains, on the other hand, while being plastically deformed to the same amount, harden less than the other orientations, and therefore accumulate fewer dislocations for this particular strain range. As shown by Basson and Driver [2000], Al cube crystals deformed at room temperature in plane strain compression scarcely harden in the homogeneous strain range of 0.1–0.2; they harden significantly more when they begin to break up into deformation and transition bands at strains >0.3. It is usually considered that this absence of hardening is due to the operation of a particular slip system configuration with two sets of orthogonal Burgers vectors which interact very weakly (Ridha and Hutchinsen [1982]). It has also been shown that after hot deformation cube grains possess lower stored energies than the other orientations (Samajdar *et al.* [2001]).

When this material is annealed at 500°C complete recrystallization takes place, but the first grains to begin recrystallizing are the moderately deformed cube grains. They recover quickly and grow rapidly into the surrounding environment, including the previously recrystallized non-cube grains. The latter cannot grow as before since their high dislocation densities and low recovery rates inhibit new recrystallization growth, driven by stored energy differences. In a sense they become the prey of the growing cube grains which rapidly dominate the entire microstructure. The result is a structure of 100% near-cube grains developed by this rather special process in which the non-cube grains are 'killed off' by the light cold rolling and annealing. The final structure obviously contains low angle sub-boundaries where the near-cube grains meet. The typical orientation spread around cube is a few degrees.

The cube-textured sheets of about 100 μm thickness are finally etched to develop the very high specific surface areas required for capacitor anodes. In this case the surface area and therefore the capacity are increased by two orders of magnitude.

14.3. ALUMINIUM MATRIX COMPOSITES

14.3.1 Introduction

Aluminium matrix composites or AMCs have been investigated since the 1920s, but their potential has emerged rather recently. AMCs aim to combine the inherent ductility of the aluminium matrix and the strength of the reinforcing particles or fibres. Table 14.6 highlights a few examples of the present usage of AMCs – a complete list would indeed be exhaustive.

The particle and short fibre-reinforced AMCs, which are of interest to the present case study, typically use 5–30 vol% of reinforcing 2nd phase (SiC, Al_2O_3, etc.) for structural and engineering applications. The only exception is the electronic substrate applications, see Table 14.6, where AMCs with more than 30 vol% particles, using multi-modal particle size blends, are common.

Table 14.6. Selected examples of the use of AMCs in the present technological age.

Usage	Remarks
Transportation	MMC cylinder liners for Honda Prelude (http://mmc-assess.tuwien.ac.at/) Push rods for racing car engines – 50% weight saving with more strength and stiffness and high vibration damping (http://www.3m.com/market/industrial/mmc/) Brake rotors for German high-speed train – Al–Si–Mg alloys with SiC particles, developed by Knorr Bremse AG with AMC supplied by Duralcan – a weight saving of 37% (http://mmc-assess.tuwien.ac.at/)
Defence/ aerospace	Helicopter rotor sleeve to replace titanium (http://www.amc-mmc.co.uk/), Hubble Space Telescope Antenna Waveguide Mast (http://mmc-assess.tuwien.ac.at/)
Electronic packaging	Material: A356 Al with 37–63 vol% SiC particles. Advantage: Isotropic-controlled thermal expansion, high thermal conductivity and high strength/stiffness. Applications: Microprocessor lids/heat sinks, microwave housing, optroelectronic housing/base, power substrates, etc. (http://www.alsic.com/page3.html)
Power supply	The 3M^{TM} 'Aluminum Matrix Conductor Composite Reinforced' is a non-homogeneous conductor consisting of high-temperature aluminium–zirconium strands covering a stranded core of fibre-reinforced composite wires. Higher strength and superior power transmission (http://www.3m.com/market/industrial/mmc/)
Sporting goods	Cycle frames for golfing equipment (http://www.amc-mmc.co.uk/contents.htm)

14.3.2 Processing

Perhaps the most important issue in AMC production is cost-effective processing. Table 14.7 summarizes the major processing routes of metal–matrix composites. Of all the processing routes, stir casting and powder metallurgy are the most attractive ones, both in terms of cost-effectiveness and in terms of mass production, for both particle reinforced and short fibre-reinforced AMCs.

14.3.3 Hot extrusion

The biggest handicap in bulk AMC processing (by stir casting or by the powder metallurgy route) is a non-uniform distribution of the 2nd phase and also possible casting and sintering defects. Hot extrusion is the most common secondary processing route to obtain a relatively uniform 2nd phase distribution and elimination of earlier defects. Typically, cast or sintered AMCs are canned in soft aluminium foil to minimize die and tool damage and are hot extruded. Figure 14.18 is a schematic showing the 'issues' involved in the secondary processing or hot extrusion. The subsequent sections elaborate on these 'issues'.

Table 14.7. AMC processing at a glance.

Processing type	Processing details	Remarks
Liquid phase processing	Stir casting: stirring the melt with solid ceramic particles	Relative particle wetting, segregation and particle pushing by advancing solidification front may lead to non-uniform 2nd phase distribution
	Squeeze infiltration: liquid metal injected into the interstices of an assembly of fibres (pre-form)	Often used as finished product in as-cast form
	Spray deposition: atomized stream of droplets and injected ceramic powder deposited on substrate	Example: commercial OspreyTM process Finished product in as-cast form
	Reactive processing: chemical reaction produces the 2nd phase	Typical examples are Fe–TiC (by combustion casting), Al–TiB2 (patented Dimox process)
Solid phase processing	Powder metallurgy: powder blending, cold compaction, canning, evacuation and degassing	Typically requires a high-temperature consolidation stage, such as hot extrusion
	Diffusion bonding: foil–fibre–foil bonding	Arrays of fibres between metallic foils, require secondary processing
Vapour phase processing	Physical vapour deposition: evaported matrix (metal) is deposited on fibre	Typically used for coating of fibres. Assembled coated fibres require secondary processing

14.3.3.1 Engineering issues. The typical high extrusion ratios require extensive hot workability. The billets are homogenized (above 500°C) and dies are heated (above 300°C). The latter is often a production problem. In a typical production facility, the dies are heated (in conventional convection furnaces) round the clock in order to maintain 'planned' extrusion schedules. But this leaves little scope for real time changes in die tooling or in extrusion schedules, as alternative dies or heating involve pre-planning. Faster heating (e.g. fast infrared heating) technologies may emerge as an alternate technology.

The most important engineering aspect in hot extrusion of AMCs is the die design. The usual die design is aimed at minimizing the extrusion load and stresses, at tool–workpiece interface using upper bound (Reddy *et al.* [1995]) or finite element analysis (Balaji *et al.* [1991]).

The die design can and does affect the strain and strain rate, which in turn affects the structure/property of the extruded product (see Section 14.3.3.2). The particle redistribution, a crucial issue in the performance of AMCs, at a given strain depends on the strain rate – relative uniformity of strain rate is crucial for obtaining uniformity

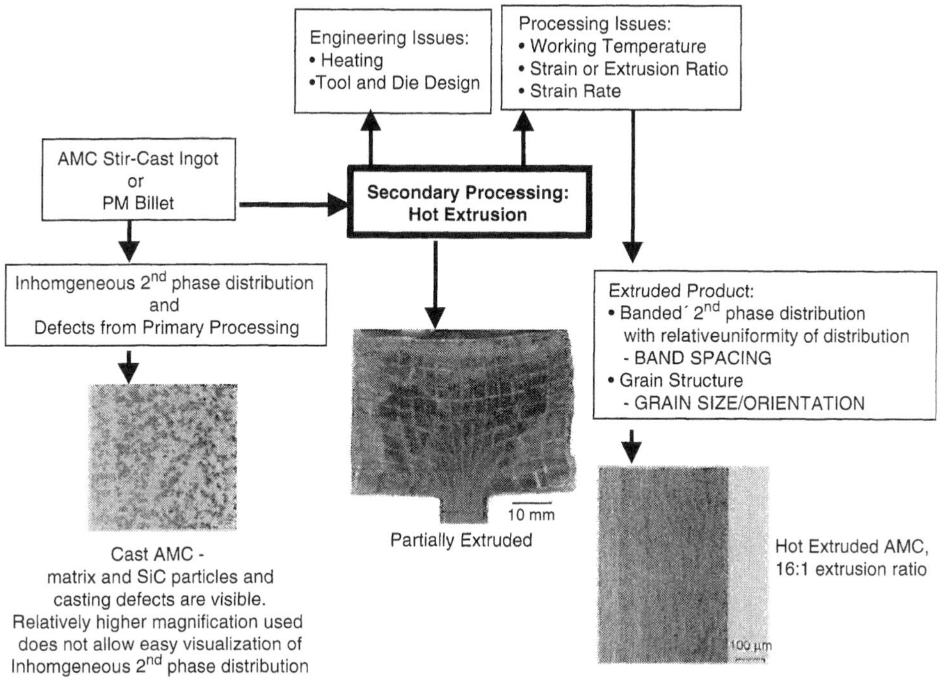

Figure 14.18. Issues involved in AMC hot extrusion. The partially extruded structure (with grid lines) shows the changes in strain in a regular die (Dutta *et al.* [1998]). The engineering issues are vital in ensuring defect-free processing, while the processing issues determine the structure and the property of the extruded product. In addition to the direct microstructural developments, hot extrusion may also account for improved bonding between particles and matrix.

in particle redistribution. This has been one of the rationale for constant strain rate (CSR) die design. CSR die design either involves a simple analytical approach or complex (but more effective) iterative methods (Kim *et al.* [2001]).

As in Figure 14.19, the extrusion dies for AMC can be generalized as regular, conical and CSR. In a regular die (Figure 14.19a and also in the inset of 14.18), both strain and strain rate vary with large strain concentrations near the die edge (Dutta *et al.* [1998]). The conical die reduces such sharp strain concentrations, while an appropriate CSR design may yield a uniform strain rate distribution. The critical issue in the design of a CSR dies is to optimize the $r(z)$ (die radius as a function of axial distance) profile. A simple analytical treatment (Kim *et al.* [2001]) shows

$$r(z) = \left\{r_0^2/\left[(r_0/r-1)z/L+1\right]\right\}^{1/2} \tag{14.2}$$

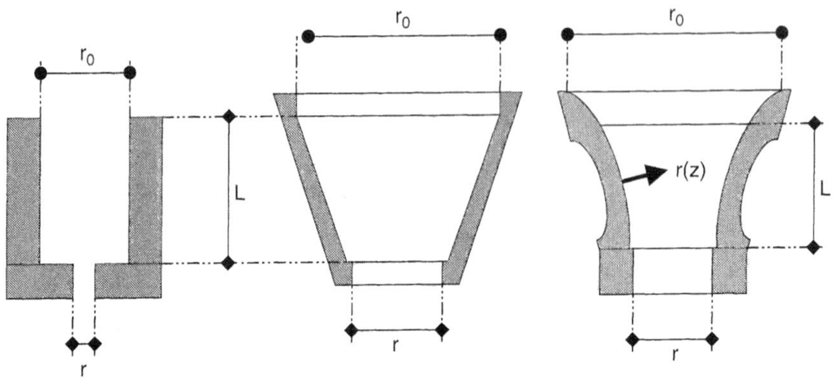

Figure 14.19. Schematics of three generalized types of dies – (a) regular, (b) conical and (c) constant strain rate (CSR). The extrusion ratio is r_0/r and L is the die length. The die radius as a function of axial position, $r(z)$ is a critical design parameter in CSR dies.

A more effective die profiling (especially for higher extrusion ratios), can be calculated following a more rigorous numerical optimization using the so-called Beizier curves (Anand [1996], Kim *et al.* [2001]).

14.3.3.2 Microstructures of extruded AMC. Before getting into the hot extrusion parameters, including die design, and their effects on AMC structure, it is important to describe typical structures of extruded AMCs (see Figure 14.20). Important features of AMC microstructures, which determine the AMC property, are: particle band spacing (see Figure 14.18), alignment and distribution of particles, effectiveness of particle–matrix bonding, residual stress developments in matrix and in 2nd phase (Sun *et al.* [1992]), thickness/orientation/substructure of the deformed grains, elimination of porosity and absence of TMP defects. Naturally the extrusion parameters, namely extrusion temperature, extrusion ratio, strain rate and die design, determine the extruded microstructure – an example of this is shown in Figure 14.21. Appropriate extrusion parameters and die design are also important for avoiding TMP defects (Chapter 12).

14.3.3.3 Optimization in AMC extrusion: optimized processing parameters and die design. AMC extrusion is often viewed as 'difficult' (Dutta *et al.* [1998], Seo and Kang [1999], Kang *et al.* [2000], Kim *et al.* [2001]) – difficult from both optimized process parameters and from optimized die design. One needs to view this 'difficulty' from the 'desired' objectives of AMC extrusion: breaking of particle agglomerates, avoiding particle rupture, distribution of 2nd phase, preferential axial alignment of 2nd phase, improved bonding between particle and matrix,

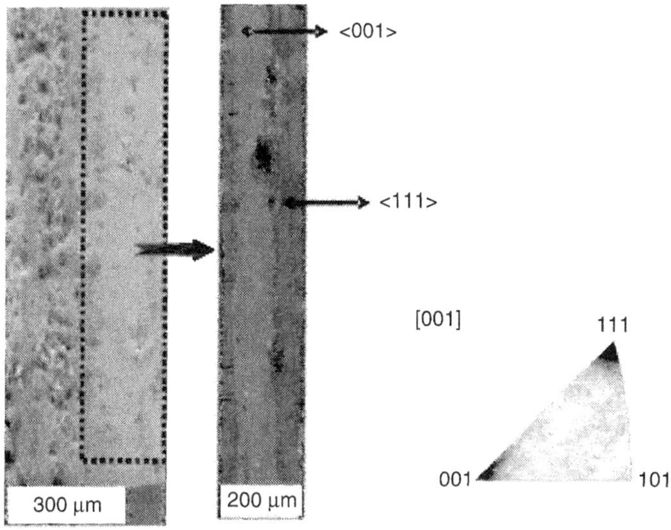

Figure 14.20. The microstructure (EBSD images) of an extruded AMC – an AA8090 alloy with 8 vol% 40 μm SiC particles (Bauri *et al.* [2005]). Microstructure typically consists of bands of particles and deformed grains. The particle bands spacing and the particle distributions are important indices of the AMC microstructure. Thickness, orientation and substructure of the deformed grains are also important structural issues.

elimination of casting defects, sufficient workability, freedom from extrusion defects, control of residual stress, lower extrusion load and extended tool life. Often such objectives are in conflict with each other – this is the root cause of the 'difficulty' in AMC extrusion. Examples of conflicts between the stated objectives of AMC extrusion are listed in Table 14.8. Optimized processing parameters and die design are needed to manage such conflicts and a discussion on the rationale behind such optimization is given below:

Working temperature: A 'minimum' working temperature is needed to ensure hot-workability and also to operate under acceptable values of extrusion pressure and tool life. Higher working temperature, on the other hand, may cause TMP defects and insufficient bonding between the matrix and 2nd phase. An 'optimum' working temperature would depend on matrix composition, but more on the 2nd phase volume fraction and size distribution. For example, optimum working temperatures of AA8090 (Al–Li alloy) and the same alloy reinforced with 8 vol% 40 μm SiC particles are 475 and 530°C, respectively (Bauri *et al.* [2005]). Increasing the size or volume fraction of the 2nd phase would increase the working temperature. It is, however, fair to admit that no physical relationship has been developed so far between material

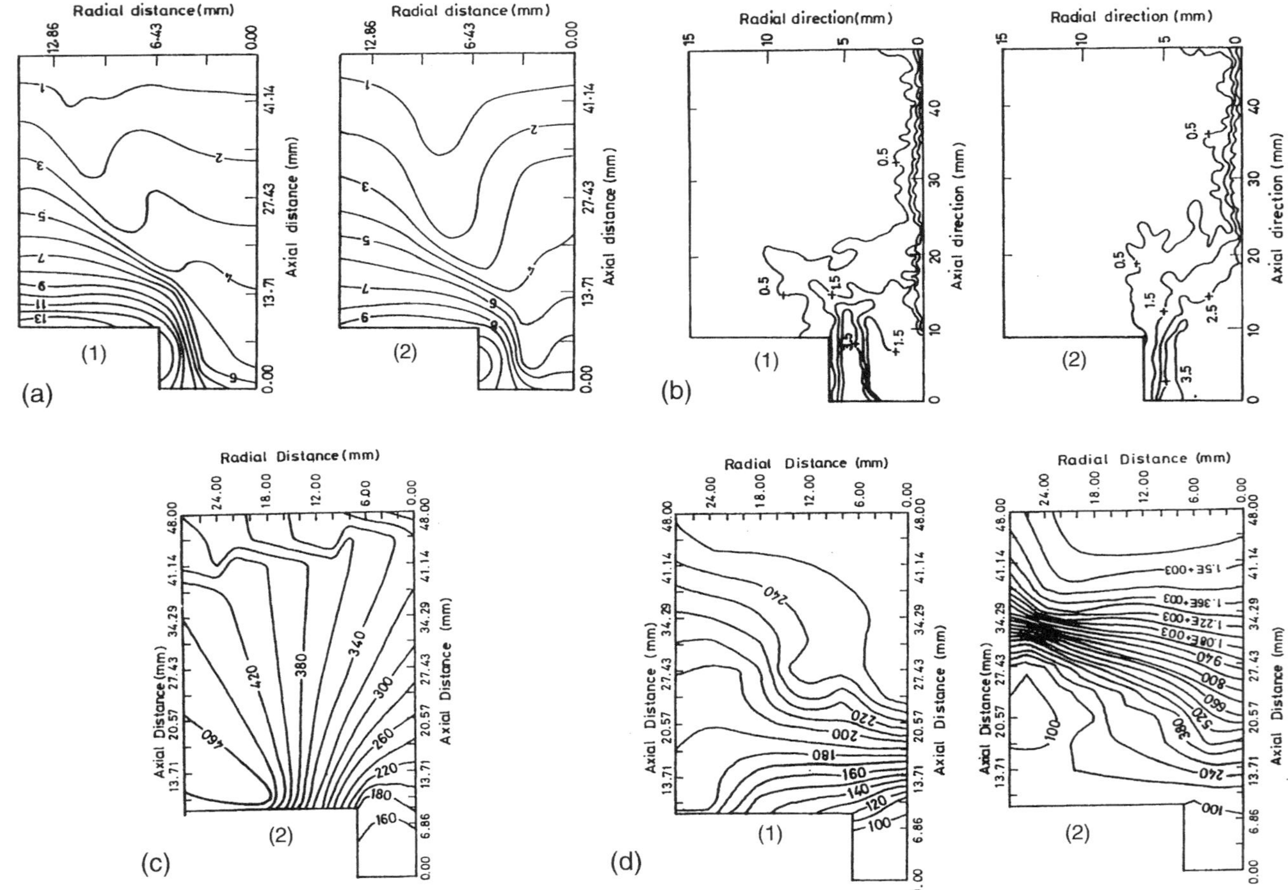

Figure 14.21. Distributions of (a) effective strain (%), (b) effective strain rate (%/s, as estimated from (a)), (c) particle band spacing (μm) and (d) deformed grain size (μm) in partially extruded (1) aluminium 3.2 wt% Cu alloy and (2) the same alloy with 7 vol% 40 μm SiC particles. Regular extrusion die, 18:1 extrusion ratio and 320°C extrusion temperature was used for the partial extrusions (Dutta *et al.* [1998]).

Table 14.8. Examples of conflicts between stated objectives of AMC extrusion.

Objective I	Objective II	Conflict
Free from macroscopic extrusion defects	Breaking of particle agglomerates	Objective I can be achieved by avoiding non-uniformity in material flow, while the same (in terms of localized high strain rates (Dutta *et al.* [1998])) can be used to satisfy objective II
Elimination of porosity	Lower residual stresses	Hydrostatic stresses satisfy objective I, but counteract objective II
Breaking of particle agglomerates	Avoid particle rupture	Parameters (high extrusion ratio, extrusion speed, etc.) supporting objective I would oppose objective II
Improved bonding between 2nd phase and matrix	Lower extrusion load	Parameters enhancing objective II are normally detrimental to objective I

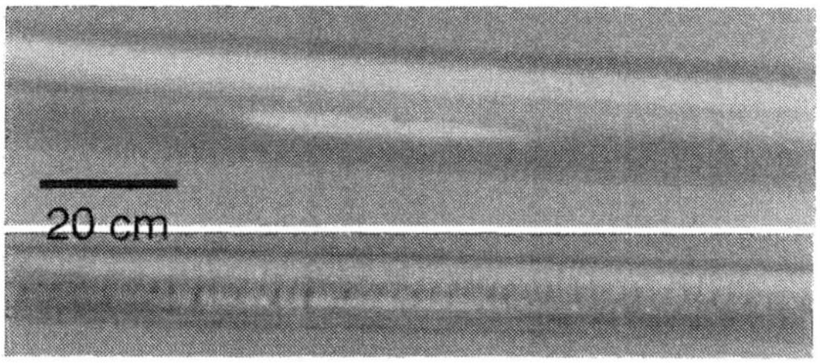

Figure 14.22. Severe TMP defects (surface cracks) visible at 25:1 extrusion ratio. No such defects were visible to 20:1 extrusion. Aluminium 3.2 wt% Cu alloy with 15 vol% 40 μm SiC particles.

variables, press conditions and optimum working temperature – typically the latter is identified from an empirical database or from pure trial and error.

Extrusion ratio: Normally high extrusion ratios are used. This enables elimination of porosity and appropriate 2nd phase distribution. High extrusion ratios are associated with large extrusion pressures and can also create particle rupture and may lead to TMP defects (see Figure 14.22). Though critical extrusion ratio has been related (Kang *et al.* [2000]) to the maximum strength of extruded AMC, the usual practice is to use extrusion ratios on the higher side – but not high enough to cause visible TMP defects or demand too large an extrusion pressure.

Strain rate: Strain rates are related to extrusion speed. High strain rates are effective in breaking particle agglomerates, but also can promote visible TMP defects.

Naturally, a compromise is sought mainly by trial and error and a 'minimum' strain rate, capable of effective redistribution and alignment of 2nd phase, is used.

Die design: This area of AMC extrusion has obtained considerable scientific and applied focus (Seo and Kang [1999], Kang *et al.* [2000], Kim *et al.* [2001]). Most of these studies highlight the design and optimization of CSR dies – an area which has seen a truly multi-disciplinary mix of finite element analysis and advanced computer graphics (Kim *et al.* [2001]). Literature, however, fails to answer a simple question unambiguously – 'what is the best die shape for AMC extrusion'. Uniformity in exit velocity, avoiding dead metal zone and large variations in local strain rates – all point to CSR dies. But this may not always provide an 'optimum' die shape for all types of AMC extrusion. Large local shear, an avoidable commodity in normal extrusion, or local variations in strain/strain rate has been reported (Dutta *et al.* [1998]) to improve the relative plasticity of the AMCs (especially in regions containing 2nd phase clusters) and facilitate the formation of the typical banded structure. This, on the other hand, can be achieved through the so-called regular dies more effectively. An optimum die shape or design would depend on the starting structure, especially the particle agglomerates, and on extrusion parameters. It appears that the state-of-art in AMC extrusion is also shifting towards CSR dies, but that may not be the best solution for all types of AMC extrusion.

14.4. THICK PLATES FOR AEROSPACE APPLICATIONS

14.4.1 Introduction

The development of modern air travel would have been practically impossible without the use of aluminium. The first all-aluminium plane, the DC3 'Dakota' that came into service in 1932, was the first to carry passengers profitably – previously, in the USA, air transport was used to carry only mail. The bodies and wings of modern civil aircraft are basically made of high-strength aluminium alloys. For example, the Boeing 747 and the Airbus 380 'jumbo jets' contain 60–70% by weight of aluminium alloys. Al therefore remains the primary aircraft material and is likely to continue in this role for some years.

Al base alloys combine good corrosion resistance, high formability and high strength with a low density, a particularly important property for aircraft. Also the material and manufacturing costs are relatively low compared with new, competing, composite materials so that Al alloys are still the most versatile materials for aircraft construction. However, the improving manufacturing methods for the much higher specific strength ratios (modulus/density and yield strength/density) of the carbon fibre-reinforced polymers present a growing challenge to aluminium in this application.

The alloys used most widely for these applications are the heat-treatable alloys of the Al–Cu system (2xxx series), where damage tolerance and fatigue resistance are of the highest importance and, for components requiring highest strength, the Al–Zn–Mg system (7xxx series). For many years, two standard alloys, 2024 and 7075, dominated in aircraft applications. Over the last 30 years there have been significant improvements in these type of alloys, particularly by limiting the Fe and Si contents and by optimizing the compositions and TMP, so that stronger, more damage-tolerant variants of these alloys have been established (see, e.g. Staley [1989], Rendigs [1997], Saintfort *et al.* [1997], Liu and Kulak [2000]). As described below, they have been particularly optimized for use as integral structures machined from thick plate.

The specific alloy (and its thermo-mechanical treatment) is now selected according to the loading of the final aircraft component for which it is designed. Thus the fuselage is subject to a variety of loadings induced by the pressurization cycle, taking off and landing cycles and the weight of the plane. Due to pressurization there are mainly tensile loads in the fuselage which consists of a skin with longitudinal stiffeners (stringers) and regularly spaced circumferential frames. However, the lower part of the fuselage is also under compression and different alloys are therefore used for the upper and lower sections of the fuselage.

Wings consist of upper and lower wing covers mechanically fastened to an internal structure comprising spars (longitudinal members) and ribs. The covers are reinforced with longitudinal stringers, which are generally either integral (more frequent for smaller aircraft) or riveted to the skins. Upper wing components are compression-dominated structures since the wing tip bends upwards during flight (by almost 1 m in the Airbus 380 super jumbo). They are made of 7xxx (Al–Zn–Cu–Mg) alloys. Conversely, the lower wing structures, where periodic tensile stresses can induce fatigue damage and crack propagation, are designed for high damage tolerance by the use of 2024 alloy or its modern variants. These usually have lower Fe and Si impurities and limits to the Cu and Mg strengthening additions to ensure that almost all of the Cu and Mg can be put into solution. Intermediate spar and rib components must achieve the best compromise between the requirements of upper and lower wing sections. Thick longitudinal spars and transverse ribs are also used to stiffen the wing box. Loads on the horizontal stabilizers are typically the opposite of those on a wing, requiring compression-resistant solutions for the lower parts and damage-tolerant alloys for the upper structures. As far as the vertical stabilizers are concerned, complex loads involving overall bending and local gust loads may occur and here higher stiffness composites seem to have already established a dominant role. Boeing have announced that their new 'Dreamliner' plane will use a very much large fraction of composite materials for major structural parts, so competition with the aluminium plates will now be tested for commercial air travel.

Figure 14.23. Example of Al thick plate machined to an aeronautics component. Courtesy of Alcan CRV.

14.4.2 Integral structures

Thick plates are now used in the manufacture of commercial aircraft in the form of integral or monolithic structures to substitute for assemblies of sheet or light gauge plate and stiffeners. A conventional assembly usually consists of formed sheets and other parts (extruded, forged and/or machined) which are riveted together. An integral structure, on the other hand, integrates the function of all these individual parts into one large structure with the double advantage of weight reduction due to the reduced number of joints and production cost savings due to the lower number of components. Manufacturing of the integrated structure requires a lot of machining from the original, fully heat treated, thick plate but is cost efficient due to the high-speed milling machines (up to 25,000 rpm) that are now available (Figure 14.23).

14.4.3 Metallurgical improvements through TMP

To replace the conventional assemblies, the major challenge for the material suppliers is to ensure that the property combination of the thick plate material should not limit the performance of the integral structure as compared to a classical assembly. This required a number of metallurgical improvements to conventionally manufactured thick Al plates, e.g. AA7075 which had not been designed for these applications. We recall that these plates are produced from ingots by homogenization and hot rolling before the heat treatment of solutionizing, quenching and ageing to harden the alloy. They are also stretched a few per cent, after quenching, to straighten the plates and to reduce the internal stresses (Chapter 9). In some alloys but not the 7xxx series, stretching provides increased strength both by strain hardening (2xxx series) and by providing more dislocations as in-grain nucleation sites for the strengthening precipitates (Al–Li–Cu–Mg alloys). Thick plate processing was therefore modified to allow for the following requirements:

The maximum thickness of conventional Al plates was increased in order to allow machining of very heavy sections (up to 150 mm or more). This meant bigger ingots and lower amounts of working by hot rolling with the attendant problems of increased material inhomogeneity. Simultaneously the residual stress level of the plate material had to be reduced to a very low level to guarantee distortion-free machinability even with the large dimensions and the high aspect ratios involved.

The alloy had to be less sensitive to quench rates since the latter obviously decrease with increasing thickness. Quench sensitivity, the reduction of properties such as strength and fracture toughness due to less than ideal quenching, is caused by precipitation of solute elements on microstructural heterogeneities during quenching from the solutionizing treatment, with a correspondingly reduced solute content available to participate in the final age hardening. Three principal types of microstructural feature can act as nucleation sites: grain boundaries, sub-grain boundaries and dispersoids. The key modification from the earlier generation of plate alloys (7075, 7049, 7178) was the replacement of Cr by Zr, as exemplified by AA7050. The coherent Al_3Zr dispersoids that are formed in this alloy are significantly less favourable nucleation sites than the $Al_{18}Mg_3Cr_2$ dispersoids found in the previous generations of alloys. The correspondingly reduced quench sensitivity observed for Zr-containing alloys enables them to maintain good static properties up to thicknesses of order 200–300 mm. The grain boundary precipitates formed on slow quenching also reduce fracture toughness (Staley *et al.* [1993]). This loss of toughness is due to grain boundary ductile fracture that is also promoted by the grain boundary precipitation during low-temperature precipitation hardening (Vasudevan and Doherty [1987]). More recently, the development of new alloys such as 7010, 7040 and 7085 with lower quench sensitivity related to higher Zn/(Cu+Mg) ratios led to improved through-thickness properties of thick plates and improved toughness.

Additionally, several important properties of the plate material such as corrosion and fatigue resistance, toughness and ductility had to be improved in order to compete with properties of other product forms. Stress corrosion cracks (SCC) propagate along grain boundaries. The sensitivity of 7xxx alloys to SCC correlates well with the degree of ageing, with overaged conditions being more resistant. The ageing treatment is therefore carried out in two or three stages, usually with a slight over-aging beyond the maximum strength to improve the stress corrosion resistance while maintaining nearly the maximum possible strength. It is not clear exactly what is achieved in these empirically developed ageing processes, but it clearly affects the geometry and the solute content of the PFZs.

A key feature of combining high strength with good fracture toughness is the use of largely unrecrystallized grain structures which are retained after hot rolling during solutionizing by the use of strong anti-recrystallizer agents such as Zr. All structural aluminium alloys contain small amounts of transition metals, Mn, Cr, Zr, that are largely held in solution during casting but are precipitated as 'dispersoids' during the homogenization heat treatment aimed at minimizing the heterogeneity of the strengthening solutes, Zn, Mg, Cu, etc. The transition metals have much smaller diffusion coefficients than the strengthening solutes. Iron, because of its negligible solubility in solid aluminium can only be used in this role for alloys processed by very rapid solidification. The use of Zr in the recent alloys is due to its homogeneous nucleation, giving many small precipitates of the coherent, cubic Al_3Zr phase.

In general, fracture toughness of metallic alloys depends on yield stress, chemistry and on the distribution of 2nd phase particles (Santner [1978], Vasudevan and Doherty [1987], Dorward and Beernsten [1995], Deshpande *et al.* [1998]). As mentioned above, the amount of intermetallic particles has been drastically reduced by lowering the Fe+Si constituent levels. The latest alloy AA7085, registered in 2002, has the lowest level of Fe and Si (the values for these two elements in Table 14.9 are the maximum allowed). In the case of 7xxx alloys, fracture toughness is also significantly improved by retention of the unrecrystallized structure with its many grain boundaries parallel to the plate surface. Figure 14.24 shows the microstructure of a 7010 alloy after hot plane strain compression (to simulate hot rolling) and a solution treatment, displaying a fine recovered subgrain structure

Table 14.9. AA7xxx alloy compositions.

Wt%	Zn	Mg	Cu	Fe	Si	Zr or Cr	Year
7075	5.1–6.1	2.1–2.9	1.5–2.0	0.5	0.4	0.18–0.28 Cr	1954
7050	5.7–6.7	1.9–2.6	2.0–2.6	0.15	0.12	0.08–0.15 Zr	1971
7010	5.7–6.7	2.1–2.6	1.5–2.0	0.15	0.12	0.1–0.16 Zr	1975
7055	5.9–6.9	2.0–2.7	1.9–2.5	0.15	0.12	0.08–0.15 Zr	1991
7449	7.5–8.7	1.8–2.7	1.4–2.1	0.15	0.12	0.25 (Zr+Ti)	1994
7040	5.7–6.7	1.7–2.4	1.5–2.3	0.13	0.10	0.05–0.12 Zr	1996
7085	7.0–8.0	1.2–1.8	1.3–2.0	0.08	0.06	0.08–0.15 Zr	2002

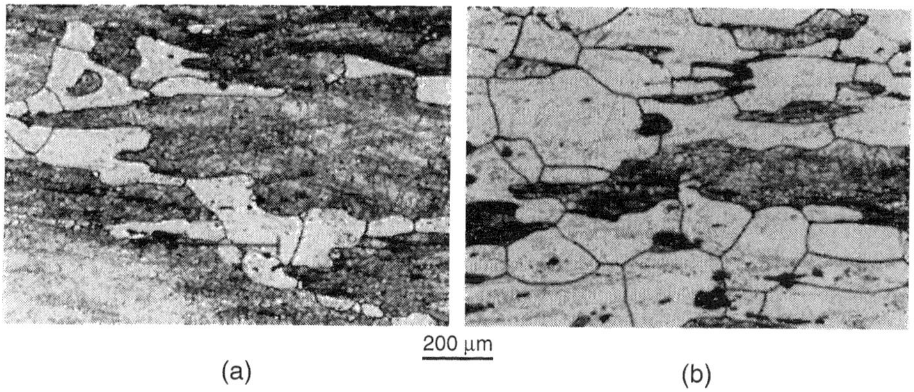

Figure 14.24. The influence of Al_3Zr dispersoid size on recrystallization of a model 7010 alloy (0.082% Zr). Optical micrographs (orthophosphoric etch) on longitudinal (ND/RD) section after PSC to a true strain of 1 at 440°C $1s^{-1}$ and solutionization 6 h at 470°C. Recrystallized grains are light while the non-recrystallized (dark) areas contain subgrains decorated by precipitates. Figure (a) is after slow heating (dispersoids 15–20 nm) and Figure (b) after rapid homogenization heating (dispersoids 30–35 nm) (Morere *et al.* [2001]). With kind permission of Springer Science and Business Media.

plus some recrystallized grains. In order to suppress recrystallization, elements such as Cr but better Zr are added but the size and distributions of these fine dispersoids are critical to the amount of recrystallization. The Al_3Zr dispersoids are formed during the initial heating up of the cast alloy to the homogenization temperature and their size depends upon the heating rate. Figure 14.24 shows that the finer dispersoids lead to less recrystallization in accordance with the increased Zener drag on the moving grain boundaries (e.g. Chapter 10). Since recrystallization is also controlled by the stored energy after rolling, the hot rolling temperature is also controlled to minimize the stored energy. The origin of the improved fracture toughness of this layered, unrecrystallized structure is not fully established but is believed to arise from the *embrittling* of the grain boundaries that are mainly parallel to the plate and thus the loading direction. These weakened grain boundaries can then deflect cracks that are normal to the loading axis (Doherty [2005]).

Using these combinations of alloy development and hot rolling practice, plate of up to 250 mm thickness can be produced for integral structures. They are currently used in the most recent generation of airplanes such as the Boeing 777 or the new Airbus A380.

Further developments are under way in the form of both higher strength alloys and lighter alloys. Higher strength, and more recently introduced alloys such as AA 7055, 7449 and 7085 (Table 14.9) are now available up to thicknesses of 100 mm. The higher Zn/(Cu + Mg) ratios are designed to improve, that is reduce, the quench sensitivity. Lighter alloys are based on Al–Cu–Mg–Li systems such as AA 2196 and 2090. These, though more expensive are likely to become more important under competition from composite materials through the influence of rising fuel oil costs. A previous generation of these alloys was developed in the early 1980s when oil reached $35 a barrel; the alloys were dropped when oil prices fell in the mid-1980s, but are now likely to be needed again. The addition of Li not only decreases the alloy density, but it surprisingly (for a low melting temperature element like Li) raises the alloy's elastic modulus. A further new development in aerospace technology is component assembly by novel welding techniques (friction stir welding and laser welding) as opposed to conventional riveting to reduce productions costs.

ACKNOWLEDGEMENTS

The author acknowledges Tim Warner of the Alcan research centre Voreppe for pertinent comments on the aerospace applications.

LITERATURE

"Aluminium Alloys – Contemporary Research and Applications", A.K. Vasudevan and R.D. Doherty (eds), Treatise on Materials Science and Technology, vol. 31, Academic Press, San Diego, CA 92101 (1989).

"Aluminium Packaging Alloys", JOM, 48 (1996), 17–42.

"Fundamentals of MMCs", Suresh S., Mortensen A. and Needleman A. (eds), Butterworth-Heinemann, Boston, MA (1993).

Chapter 15

Thermo-Mechanical Processing of Steel

Chapter 15
Thermo-Mechanical Processing of Steel

15.1. STEEL FOR CAR BODY APPLICATIONS

15.1.1 Introduction

During many decades, steel has been the most widely used material for the fabrication of car bodies. Although lighter materials like aluminium, polymers and composites are gradually taking an important share at the expense of steel, for the time being, the latter remains the principal material in cars (see Figure 15.1).

A steel sheet for car body applications must comply with a number of requirements, such as a very good formability, sufficient strength, good weldability, corrosion resistance or suitable for corrosion protection, good surface quality, etc. These requirements can only be met by low-carbon steel, since medium or high-carbon steels are not ductile enough and are difficult to weld. Within the family of low-carbon steels, several variants like Al-killed steel, interstitial-free (IF) steel, etc., have been developed and will be discussed in the present chapter.

The formability of sheets has already been treated in Section 11.7. It cannot be expressed with one single parameter, but important quantities are: the Lankford value ($\bar{R}$) (deep drawability), the planar anisotropy ΔR (earing) and the strain-hardening coefficient n (resistance against strain localization). Although a high n-value is very important for a car body sheet, the present case study will mainly focus on the processing parameters that will lead to a good deep drawability. Since the deep drawability is closely related to the crystallographic texture of the material, the present study is a good example of 'texture control' during thermo-mechanical processing (TMP).

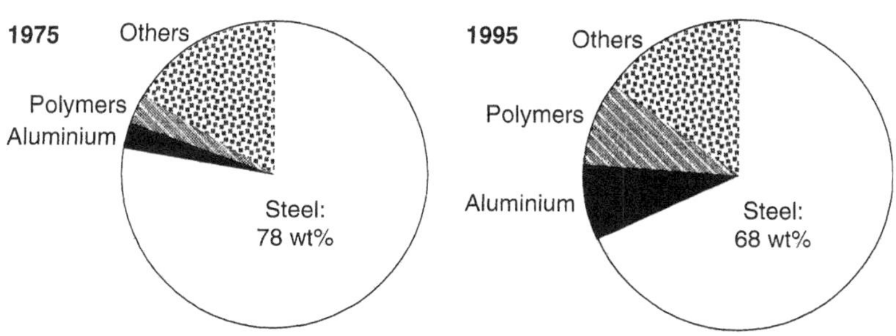

Figure 15.1. Weight fraction of different materials in a standard passenger car.

In the last decade, a clear interest for the development of high-strength steels has emerged. With stronger steels, thinner car parts can be fabricated and weight savings can be obtained. Some examples of this new development will be illustrated at the end of this chapter and two cases will be treated in detail in Section 15.2.

15.1.2 Batch annealed Al-killed low-carbon steel

A traditional Al-killed low-carbon steel has a typical composition of about 0.05 wt% C, 0.3 wt% Mn, 0.05 wt% Al and 0.006 wt% N. It is processed in a rather conventional way, involving the following production steps: basic steelmaking, continuous casting in slabs, soaking above 1200°C, hot rolling with a finishing temperature around 900°C, coiling below 600°C, cold rolling (70–90% reduction) and finally batch annealing. The latter is a fairly slow process carried out around 650–700°C for several hours. The heating rate towards the annealing temperature varies between 20 and 200°C/h. During annealing the cold-deformed sheet will recrystallize, but this process interacts with the precipitation of fine AlN particles. This interaction turns out to be the key to a successful texture control!

Although the recrystallization mechanism of low-carbon steel is not yet completely unravelled (cf. Chapter 5), it is commonly observed that cold-deformed grains with an ND//<111> orientation, have a much stronger tendency to recrystallize than grains of other orientations and that those recrystallized grains have again a ND//<111> orientation, which is a favourable orientation for deep drawing (cf. Section 11.7.5.3). The basic idea of texture control in Al-killed low-carbon steels is to enhance the balance between 'favourable' and 'detrimental' grains by an interaction of fine AlN precipitates with recrystallization. The nucleation and growth of ND//<111> type grains is somewhat hindered by the presence of fine AlN, but remains possible, due to the high driving force present in these grains. The appearance of nuclei in the other grains, however, is seriously hampered (Dillamore *et al.* [1967]). This 'filtering' of recrystallization nuclei increases the amount of 'favourable' grains in the recrystallized sheet.[1] An important prerequisite for this processing is that the AlN precipitates are small enough when the material starts recrystallizing. For this, they have to be formed during the slow heating towards the annealing temperature, just before the recrystallization starts.

[1] The recrystallized grains do not only show a strong ND//<111> crystallographic texture, but also have a characteristic elongated 'pancake' shape. It is not clear to which extend this morphological texture contributes to the good deep-drawing properties.

From these requirements, some general *processing conditions* for batch annealed Al-killed low-carbon steel can be deduced:

- The soaking temperature must be high enough to bring all Al and N in solid solution after casting. Leslie developed a relation between the Al and N content and the minimum solution temperature (T in Kelvin) (Leslie *et al.* [1954]):

$$\log[\text{wt\% Al} \times \text{wt\% N}] = -(6770/T) + 1.033 \qquad (15.1)$$

  where wt% Al and wt% N refer to atoms in solid solution. Most N atoms are dissolved, but some Al may combine with O atoms to form Al_2O_3. According to Abe [1992], the wt% Al in solid solution can be estimated from

$$\text{wt\% Al} = \text{wt\% Al}_{\text{total}} - 1.125\,\text{wt\% O}_{\text{total}} \qquad (15.2)$$

  For example, an alloy with 0.006 wt% N, 0.05 wt% Al and 0.005 wt% O, has a minimum solution temperature of about 1200°C.
- Precipitation of AlN during and after hot rolling must be avoided because in that case the precipitates would already be too coarse and no longer effective during batch annealing. The schematic CCT curve in Figure 15.2 illustrates that after hot rolling, a fast cooling till about 600°C is necessary. From that point, the material can be coiled since further slow cooling will not induce any AlN precipitation.

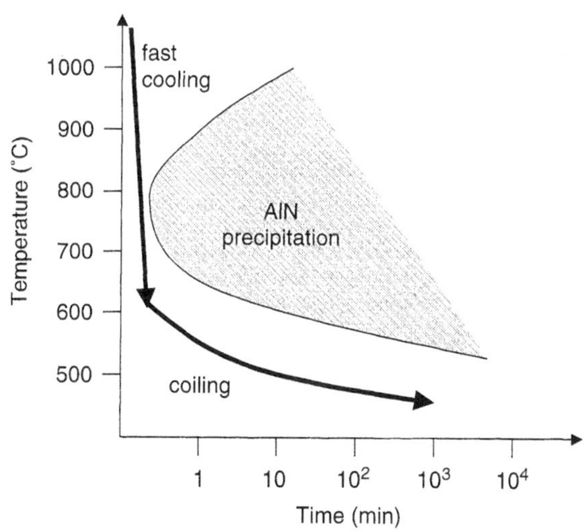

Figure 15.2. Schematic CCT curve for AlN precipitation in an Al-killed low-carbon steel. After Leslie *et al.* [1954].

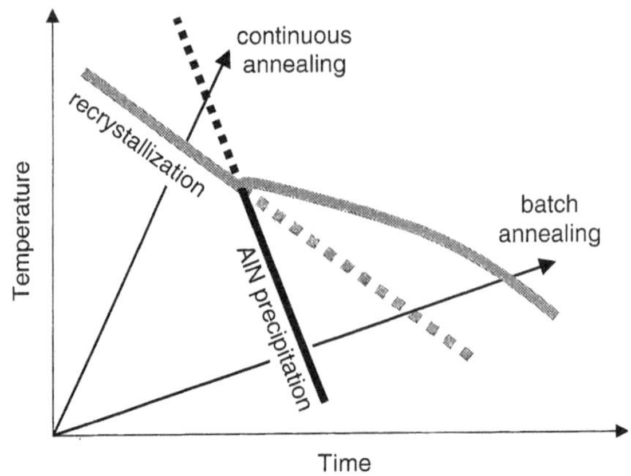

Figure 15.3. Start of recrystallization and of AlN precipitation during reheating of cold-rolled Al-killed low-carbon steel. After Hutchinson [1994]. (Courtesy Trans Tech Publications and professor B. Hutchinson).

- The reheating rate towards the recrystallization temperature may not exceed a critical value. As illustrated in Figure 15.3, precipitation and recrystallization have a different time–temperature trajectory. When the material is heated too fast, the sheet will recrystallize before the fine AlN have been formed, and hence the 'filtering' of nuclei cannot take place. A formula to calculate the optimal heating rate ΔT^{op} has been proposed by Takahashi and Okamoto [1974]

$$\Delta T^{op} = 18.3 + 2.7 \log[\text{wt\% Al}][\text{wt\% Mn}][\text{wt\% N}]/\%\text{CR} \qquad (15.3)$$

with %CR the cold-rolling reduction.

Other factors that have an important influence on the texture development during recrystallization are the carbon content, the grain size after hot rolling, the degree of cold reduction and the recrystallized grain size.

The negative influence of *carbon* on the recrystallization texture of low-carbon steel has been shown convincingly by Fukuda [1967] and discussed in a number of review papers, e.g. Hutchinson [1984]. Part of the carbon atoms will be present as carbides. These '2nd phase particles' will stimulate the recrystallization (cf. the particle stimulated nucleation (PSN) effect in Chapter 5). The recrystallized grains around these carbides will however show a randomized orientation and will hence compete with the desired ND//<111> grains. Some carbon atoms will be dissolved in the ferrite matrix and in turn have a deleterious effect on texture formation. A well known, but perhaps not generally accepted hypothesis was put forward by Abe [1984], who suggested a mechanism of C–Mn dipole formation which hinders the formation of ND//<111> grains.

A fine *grain size after hot rolling* seems to be beneficial for the deep drawability of the cold-rolled sheet. The current understanding of this experimental fact is that ND//<111> grains nucleate preferentially in the vicinity of the boundaries of the deformed ferrite grains, while {011}<100> oriented grains nucleate at heterogeneous deformations such as shear bands (Abe [1992]). Since a material with finer initial grains deforms more homogeneously (less shear bands) and has a larger total grain boundary area, a favourable texture after cold rolling and recrystallization is obtained.

An increase in deep drawability with increasing *cold-rolling reduction* can be understood from the same argument: an increase in total grain boundary area. For unknown reasons, undesirable texture components like {411}<148> and {411}<011> develop above a certain critical cold-rolling reduction. The optimum cold-rolling reduction seems to be composition dependent.

Finally, a *post-recrystallization anneal* sharpens the ND//<111> texture because after recrystallization the {111} grains are on average somewhat larger than grains with other orientations and hence they grow preferentially because of their size advantage.

From these observations, some additional processing conditions can be deduced:

- The finishing rolling temperature must be as low as possible in the austenite region, just above the A_3 line. This leads to a fine grain size before cold rolling.
- An optimal cold-rolling reduction must be applied, in accordance with the steel chemistry. In most cases this optimum is found between 70 and 90% reduction.
- Some post-recrystallization grain growth must be allowed.

15.1.3 Continuous annealed low-carbon steel

In a continuous annealing line, the steel sheets are heated at a speed of about 10°C/s. Annealing is carried out at a relatively high temperature but for shorter times, typically 1–5 min. Due to this high heating rate, recrystallization of the cold-deformed sheet will occur before the precipitation of AlN particles (Figure 15.3) and hence 'filtering' of recrystallization nuclei, as applied during batch annealing, is no longer possible. Under these circumstances, an alternative approach has to be followed. It is known that very pure iron has a natural tendency to form a strong ND//<111> recrystallization texture. Hence, a possible strategy could consist of a limitation of the carbon content (in most cases with wt% $C < 200$ ppm). During hot rolling, these carbon atoms are in solute solution, but during cooling most carbon atoms will precipitate out to form carbides. Of course, during reheating towards the annealing temperature (after cold rolling), part of these carbon atoms will re-dissolve and this will hamper the ND//<111> texture development as discussed by several authors, such as Abe [1992] and Hutchinson

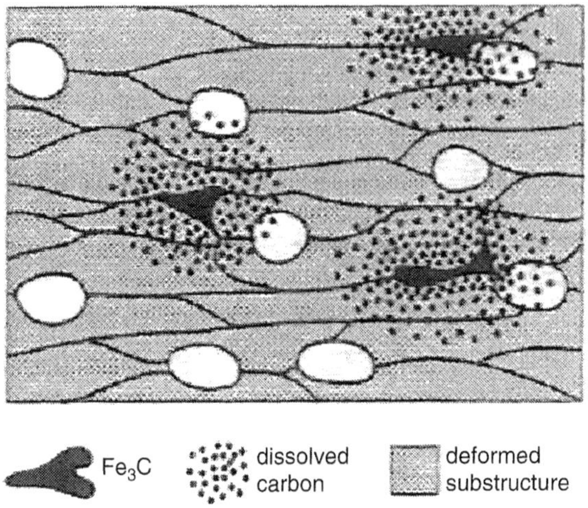

Figure 15.4. Illustration of the competition between recrystallization and the distribution of re-dissolving carbon atoms into the ferrite matrix (Hutchinson [1994]). (Courtesy Trans Tech Publications and professor B. Hutchinson).

[1984, 1994]. This leads to a competition between the dissolving carbides, 'infecting' the ferrite matrix with carbon atoms and the nucleation of recrystallized grains, which should occur as much as possible in the 'carbon-poor' regions. This competition is illustrated in Figure 15.4. Ushioda *et al.* [1986] have been able to model these processes and have shown that in respect to ND//<111> texture formation, a few coarse carbides are less harmful than several small ones and that a high heating rate decreases the average carbon content in the ferrite, as illustrated in Figure 15.5. Such large carbides can be obtained using a high coiling temperature (>730°C) after hot rolling. During the slow cooling of the coil, the carbides have time to coarsen.

A problem with continuous annealing is the short time available for post-recrystallization grain growth. To enhance grain growth, the annealing temperature can be raised, but should be kept below 850°C to avoid significant ferrite-to-austenite transformation that would ruin the favourable ND//<111> texture.

Based on these observations and models, the following *processing conditions* for continuous annealing of low-carbon steel can be formulated:

- The soaking temperature (before hot rolling) has to be low (~1100°C) in order to avoid significant dissolution of MnS and AlN. These particles can then further grow during hot rolling, without severely affecting the annealing behaviour.
- The coiling temperature is kept high (>730°C) in order to form large carbides during slow cooling of the coil.

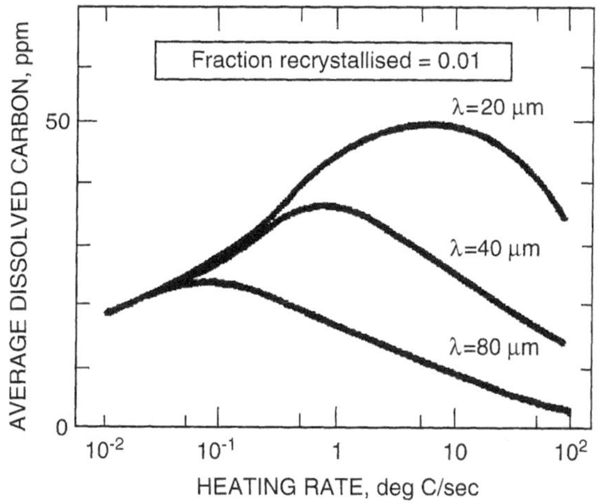

Figure 15.5. Calculated carbon fraction for different sizes of cementite particles as a function of the heating rate (Hutchinson [1994]). (Courtesy Trans Tech Publications and professor B. Hutchinson).

- The heating rate during final annealing is high (~10°C/s) to avoid excessive diffusion of carbon atoms into the matrix.
- In most cases, a high annealing temperature (up to 850°C) is used in order to benefit from some grain growth after annealing.

15.1.4 Interstitial-free steels

In the previous paragraphs it has been explained that the deep-drawing properties of low-carbon steel become better with decreasing carbon content. In IF steel, the content of interstitial elements (mainly carbon and nitrogen) is reduced as much as possible. A classical IF steel contains 20–30 ppm carbon and 30–40 ppm nitrogen. The remaining interstitials are then captured with Ti and Nb. Figure 15.6, compiled by Hutchinson [1994] from different sources, illustrates the tremendous increase in $\bar{R}$-value when the Ti and Nb content reach the stoichiometric value. IF steels typically have $\bar{R}$-values between 1.8 and 2.2.

A problem with these pure steels is to obtain a fine grain size after hot rolling. The rolling schedule must be designed in such a way that during hot rolling some carbo-nitrates can precipitate and can hinder an excessive austenite grain growth. The last hot-rolling passes are typically carried out with large reduction at relatively low rolling temperature. This also leads to a refinement of ferrite. The processing conditions for IF steel can be summarized as

- The steel chemistry must be well controlled (low C and N content; well-balanced Ti and Nb content).

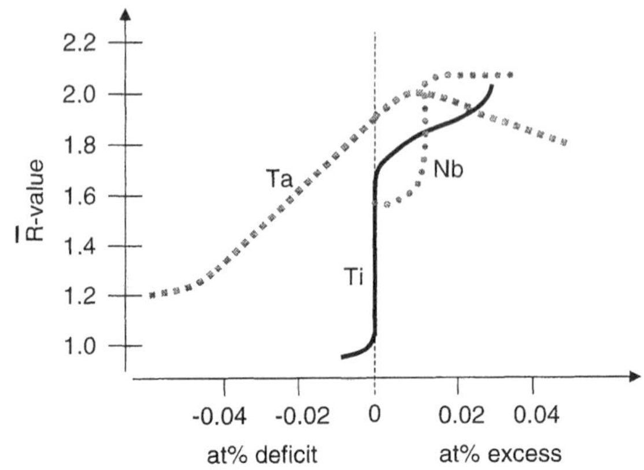

Figure 15.6. Influence of Ti, Ta and Nb content on the $\bar{R}$-value. The graph shows the deficit/excess in at% of the alloying elements relative to the C+N content. After Hutchinson [1994]. (Courtesy Trans Tech Publications and professor B. Hutchinson).

- Grain size control during hot rolling is needed and can be done by precipitation of carbo-nitrides.
- The finishing rolling temperature should be low (around A_3).
- Annealing can be carried out at high temperature (up to 850°C) because of the low carbon content. In general, the annealing conditions of IF steel are not very critical. A high annealing temperature is beneficial, but a lower temperature reduces the $\bar{R}$-value only slightly. This makes IF steels very suitable for a hot dip galvanizing treatment.

15.1.5 Trend towards higher strength steels

The need to reduce the fuel consumption and the emissions of passenger cars has driven the automobile manufacturers towards an increasing use of high-strength steels. This allows to down-gauge the body panels and to reduce the weight of the car. Unfortunately, an increase in strength goes in general hand in hand with a decrease in formability. This is illustrated in Figure 15.7, where the deep drawability (expressed by the $\bar{R}$-factor) is plotted as a function of the tensile strength for a number of steel grades. Another problem is that thinner sheets and higher strength also increase the spring back (cf. Section 11.7).

It is a continuous challenge for the steel industry to develop steel grades that combine a high strength with an acceptable ductility. In the following paragraphs the most important higher strength grades used in car body applications will be discussed.

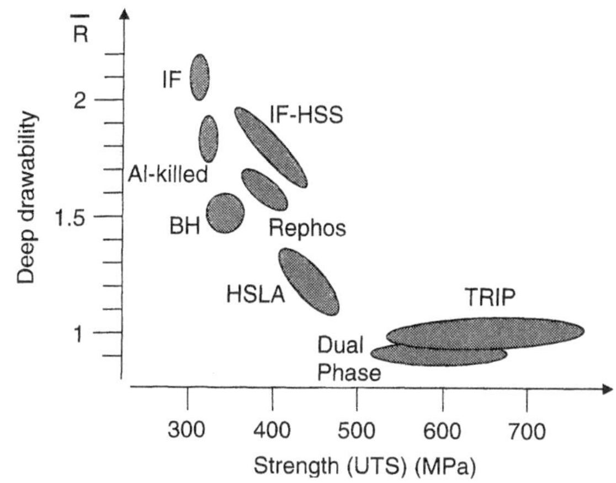

Figure 15.7. Deep drawability as a function of the ultimate tensile strength (UTS) for several steel grades used in car body applications.

15.1.5.1 Rephosphorized steel. A first possibility to increase the strength of an alloy is to increase the content of atoms in solid solution. Elements such as phosphorus, silicon and manganese will strengthen the steel, but the last two will reduce the formability to an unacceptable level. Only small amounts of phosphorus (between 0.04 and 0.08 wt%) can be tolerated and will increase the strength roughly by 40–80 MPa and decrease the $\bar{R}$-value to a value around 1.6.

Limited amounts of phosphorus can also be added to IF steel.[2] These grades are called IF-HSS (high-strength steels). The ultimate tensile strength (UTS) of 360–400 MPa and $\bar{R}$-values of 1.8–2 are possible.

15.1.5.2 Micro-alloyed or high-strength low-alloy (HSLA) steels. A second possibility to strengthen a material is to take benefit of precipitation hardening. This is achieved through the addition of small amounts of carbide or carbo-nitride-forming elements like Nb and Ti. During hot rolling, undissolved fine particles restrict the grain growth of austenite grains. In the last hot-rolling passes, austenite will not recrystallize anymore and the grains will be flattened out. During subsequent cooling many ferrite nuclei will be activated and will provide a fine ferrite grain size, which is a first important contribution to strengthening. Additional precipitates are also formed which can further strengthen the ferrite. A typical HSLA steel can have a yield strength (YS) of about 320 MPa and a UTS of 440 MPa, but

[2] Boron may also be used in IF steels.

the $\overline{R}$-value is reduced to 1.3, which is too low for car body panels. HSLA steels can however be used for structural applications, e.g. beams.

15.1.5.3 Bake hardening (BH) steels. A sheet for car body parts must be relatively soft and highly deformable. After forming, the car part should however be as strong as possible. Nature helps us to solve this dilemma by strengthening the material during pressing. However, after pressing and assembling, the body in white must be painted and this is done in a so-called 'paint baking cycle', carried out at about 170°C for 20–30 min. The idea of a bake hardening steel is to further increase the strength of the car body during this 'heat treatment' by a kind of strain ageing reaction.

In most continuous annealing lines the cooling rate after recrystallization anneal is rather fast. The carbon content after cooling will be higher than its equilibrium value. This situation, where carbon atoms will diffuse during room temperature storage towards dislocations and lock them, is called 'strain ageing' (see Chapter 12 on Lüderbands). In normal low-carbon steel grades, this over-saturation of carbon atoms is avoided as much as possible, for example by applying a two-step annealing treatment: the sheet is recrystallized in the first annealing step (for a few minutes above 710°C), rapidly cooled to 400°C, equilibrated during a few minutes in a second annealing step and finally cooled slowly to room temperature. For bake hardening steels, however, a sort of compromise is sought. The parameters of the annealing cycle are adapted to leave a small amount of over-saturated carbon atoms at room temperature. During press forming of the car part, a large number of new dislocations are created and during the paint baking cycle, these dislocations are locked by the carbon atoms. This increases the YS of the part by roughly 50 MPa.

In IF steels the interstitial atoms are combined with elements like Ti, Nb, Al, etc. and hence no paint baking effect can be anticipated. Nevertheless, bake hardenable IF steels have been developed. The amount of alloying elements is carefully controlled to leave some free carbon atoms after the recrystallization treatment. As shown in Figure 15.7, this will deteriorate the $\overline{R}$-value, but it will create some bake hardening potential. Another possibility consists of using a high recrystallization temperature (~850°C) to retake some carbon back into solution and to apply a fast cooling rate to keep some carbon in solid solution.

15.1.5.4 Dual phase (DP) and TRIP steels. Dual phase steels are low-carbon steel grades with carbon content around 0.1 wt%. They consist of soft ferrite and 10–20% hard martensite. This microstructure is obtained by soaking in the intercritical ($\alpha+\beta$) range (~800°C), followed by rapid cooling. During intercritical annealing, part of the austenite transforms into ferrite. The transformed ferrite grains reject carbon atoms into the remaining austenite. During rapid cooling, this carbon-enriched austenite transforms into martensite.

Dual phase steels combine a high strength with good ductility and high work hardening, but have a rather poor $\bar{R}$-value. For example, a steel with 0.1 wt% C, 0.7 wt% Mn and 0.2 wt% Si, continuously annealed, has a UTS of 600 MPa, a YS of 350 MPa and a fracture strain of 27%, but a $\bar{R}$-value of 0.9 (Bleck [1996]).

TRIP steels (the acronym 'TRIP' stands for 'TRansformation-Induced Plasticity') consist of ferrite, bainite and retained austenite. As for dual phase steels, the heat treatment starts by soaking in the intercritical ($\alpha+\beta$) region. In a second soaking step (usually around 350–400°C), the remaining austenite is transformed into bainite, but the last fraction of austenite (5–10%) is maintained. The carbon content in this residual austenite is so high that it remains (meta)stable during cooling to room temperature. During subsequent cold forming, this metastable austenite will transform into martensite and this transformation assists the deformation.

TRIP steels combine a high strength (600–1000 MPa) with good ductility (total elongations around 30%), but with a low $\bar{R}$-value (1–1.2).

The processing of dual phase and TRIP steels will be treated in more detail in Section 15.2.

15.2. DUAL PHASE AND TRIP STEELS

15.2.1 Introduction

In the previous section, some important developments concerning the use of higher strength steels in car body applications have been briefly discussed. An important emerging family of steel grades is the 'multi-phase' or 'transformation-strengthened' steels. These grades contain a significant proportion of a hard phase (e.g. bainite and/or martensite) mixed with soft ferrite. The 'dual phase steels' have a typical structure of ferrite and martensite. Further developments have led to 'TRIP' steels, which consist of ferrite, bainite and retained austenite. The TMP, needed to obtain these new steels, will be discussed in the following paragraphs.

15.2.2 Dual phase steel

Dual phase steels are low-carbon steels with some Mn and Si, and in most cases alloyed with other elements. An example of a possible composition (in wt%) is given in Table 15.1. Dual phase steels are produced by soaking in the intercritical ($\alpha+\gamma$) region, mostly around 800°C and by subsequent rapid cooling. In order to avoid the transformation from austenite into pearlite or bainite, the cooling rate $\Delta T/s$ must be faster than

$$\mathrm{Log}(\Delta T/s) \geq 5.36 - 2.36\mathrm{Mn} - 1.06\mathrm{Si} - 2.71\mathrm{Cr} - 4.72\mathrm{P} \qquad (15.4)$$

Table 15.1. Composition and properties of a typical dual phase steel (Bleck [1996]).

Composition in wt%	0.1 C	0.2 Si	0.7 Mn	0.05 P	0.04 Al	0.005 N
Properties	UTS: 600 MPa	YS: 350 MPa		$\bar{R}$: 0.9	El: 27%	

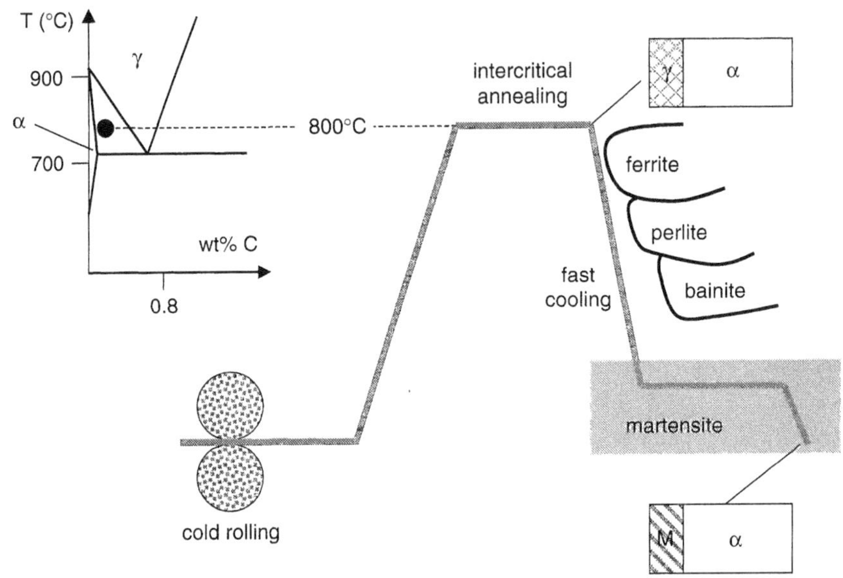

Figure 15.8. Processing of dual phase (DP) steel by cold rolling and a two-step annealing treatment.

with compositions in wt% (Pickering [1992]). In principle, both hot- and cold-rolled grades can be produced, but cold rolling followed by continuous annealing seems to be the most flexible route and the processing is schematically shown in Figure 15.8.

After conventional hot and cold rolling, the sheet is reheated into the $(\alpha+\gamma)$ region, for example around 800°C. At this temperature the structure consists of ferrite and 10–20% austenite, depending on the composition. After annealing, mostly carried out on a continuous annealing line with fast cooling possibilities, the sheet is rapidly cooled and austenite transforms into martensite. In some grades some retained austenite may still be present at room temperature. The UTS of such steel grades can vary roughly between 500 and 1000 MPa according to the exact steel composition and seems mainly to be proportional to the amount of martensite.

Dual phase steels combine a high strength with an excellent work hardening (high UTS/YS ratio), but with a low $\bar{R}$-value. Most grades show a bake hardening effect. An additional advantage is the absence of yield point elongation (Lüderlines). A sufficient amount of dislocations is generated in the ferrite near the martensite plates at the onset of plastic deformation. These dislocations, formed by the induced stresses, are not locked by interstitials and hence no yield point elongation occurs.

These steel grades can, for example, be used for energy absorbing elements in car body parts.

Dual phase steels can in principle also be produced in the hot-rolled condition. In this case, the steel can develop a suitable ferrite–austenite structure immediately after hot rolling and must be cooled fast enough to avoid formation of bainite or pearlite. A major problem is, however, that coiling must be done above martensite start temperature (M_s) but at a relatively low temperature (e.g. 250°C), because when slow cooling in the coil at higher temperatures, bainite would be formed (cf. Figure 15.8). In most production lines it is not possible to use such a low coiling temperature because the steel would be too strong to be coiled. Another possibility is to shift the start of pearlite and bainite formation towards longer times by suitable alloying (more Mn and Si and addition of Mo). A high Si content leads however to a poor surface quality of the hot-rolled product. In practice, most dual phase steels are cold rolled and annealed grades.

15.2.3 TRIP steel

TRIP steels can be considered as a further development of dual phase grades. The starting structure is again a mixture of ferrite and austenite, formed during a first annealing treatment in the $(\alpha + \gamma)$ region. A second annealing step is carried out, usually between 350 and 400°C, so that the austenite is partially transformed into a mixture of bainite and retained austenite. The presence of this retained austenite is of utmost importance because during further cold processing it will gradually transform into martensite and this transformation assists the deformation. This is called the 'TRIP effect'. The reason that retained austenite does not transforms into martensite during cooling is due to its stabilization by carbon atoms (cf. below). For some TRIP steel grades, the M_s can, however, remain above room temperature and in that case, with the martensite finish temperature (M_f) below room temperature, the final structure is a mixture of four phases: ferrite, bainite, martensite and some retained austenite.

15.2.3.1 Composition of TRIP steels. Since only a few TRIP steels are commercially available now, it is difficult to give a 'typical' composition. However, based on the published literature of the last decade, the main ingredients of TRIP steel grades can be identified.

- *Carbon*: As will be discussed below, a minimal amount of carbon is necessary to stabilize the austenite during processing; too much carbon would however impair the weldability of the steel, so most TRIP steels have a carbon content between 0.1 and 0.2 wt%, although variants with higher carbon content (up to 0.4 wt%) have been studied.
- *Silicon*: The addition of Si to steel retards the formation of carbides; this is important because the carbon atoms, rejected by the transforming austenite should be absorbed by the retained austenite and may not precipitate as carbides. Si helps to delay this precipitation reaction; a second benefit is that it gives some solid solution hardening; most TRIP steels have a Si content between 0.7 and 1.5 wt%; larger amounts give rise to hot shortness problems and to the formation of oxide scales during hot rolling.
- *Manganese*: The addition of Mn increases the relative stability of austenite towards martensite and helps to retain a sufficient amount of austenite at room temperature; Mn also shifts the start of the pearlite nose towards longer times and helps to avoid pearlite formation during cooling.
- *Aluminium*: Al has been proposed as a substitute for Si.

Some compositions of experimental TRIP grades are shown in Table 15.2.

15.2.3.2 Thermo-mechanical processing. The main target in the processing of TRIP steels is to obtain at room temperature a mixture of ferrite, bainite and a sufficiently large amount of retained austenite (~10% in low-carbon TRIP steels). In most cases these steels are hot and cold rolled as for a normal low-carbon steel and then annealed in two steps, as illustrated in Figure 15.9. During reheating towards the first annealing temperature, the ferrite in the cold-deformed sheet will recrystallize. During subsequent soaking in the intercritical region, pearlite transforms into austenite. The soaking temperature is kept just above Ac1 because this gives the highest carbon content in the austenite and limits the austenite grain growth.

Table 15.2. Composition (in wt%) of some experimental TRIP steels.

Steel	C	Si	Al	Mn	P	S	N	Reference
1.5 Si	0.11	1.50	0.04	1.53	0.008	0.006	0.0035	Girault *et al.* [2001]
0.8 Si	0.12	0.78	0.04	1.51	0.010	0.006	0.0035	Girault *et al.* [2001]
Si–Al	0.115	0.49	0.38	1.51	0.003	0.009	0.030	Jacques *et al.* [2001]
1.5 Al	0.110	0.06	1.50	1.55	0.012	0.007	0.017	Jacques *et al.* [2001]

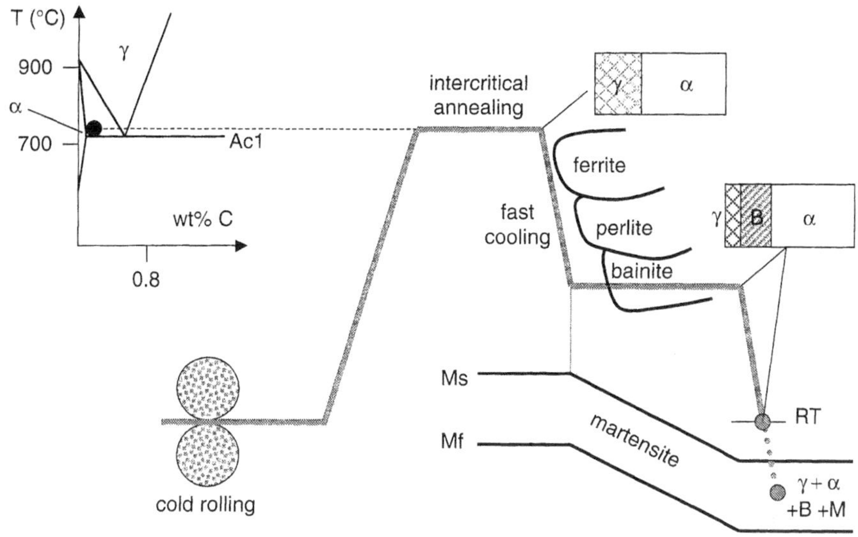

Figure 15.9. Processing scheme and structural evolution during the two-step heat treatment of TRIP steel.

For a low-carbon high Si steel, a typical soaking temperature is 750°C. The soaking time must be long enough to dissolve all carbides, but not too long to avoid grain growth. A typical soaking time at 750°C is 4 min.

After the first soaking treatment, the material is rapidly cooled to avoid the formation of additional ferrite and/or the formation of pearlite (see Figure 15.9). A second heat treatment is carried out in the upper bainite range, usually between 350 and 400°C. During this second treatment some important transformations take place, which are schematically illustrated in the Figure 15.10. The residual austenite[3] inherited from the intercritical annealing treatment, will transform into bainite. Since bainite has a very low solid solubility for carbon, the carbon atoms from the transforming austenite are rejected from the transformation product into the residual austenite. Due to this increase in carbon content, the residual austenite becomes more and more stable against martensite and a decrease in the M_s temperature can be noticed (Figure 15.9). After some soaking the residual austenite is so loaded with carbon that the transformation into bainite ceases. Nevertheless, the remaining austenite is in metastable condition and after prolonged annealing it will decompose in carbides and ferrite (called 'secondary' ferrite). Silicon is known to delay this decomposition.

[3] Residual austenite: austenite that is present in the microstructure at a certain stage during the heat treatment.
Retained austenite: austenite that remains untransformed in the final product at room temperature.

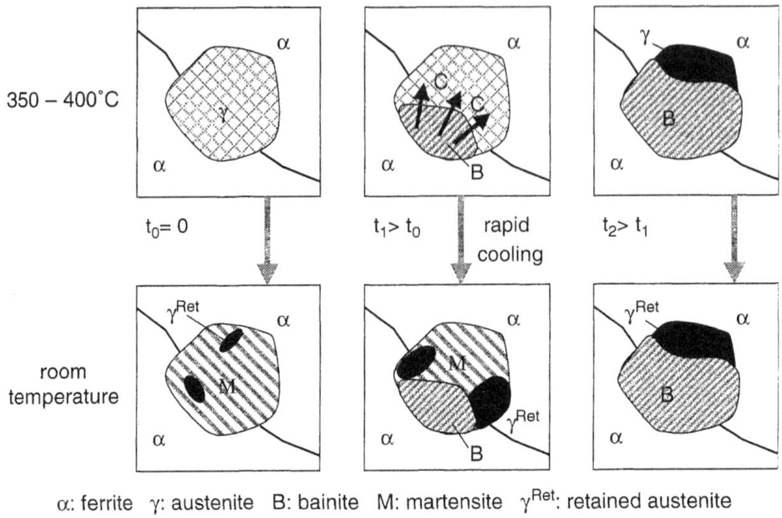

Figure 15.10. Illustration of the structural changes in TRIP steel during the second isothermal treatment (above) and transformation products after rapid cooling (below).

After this second heat treatment, the material is rapidly cooled to room temperature. The final structure obtained at room temperature, depends on the previous soaking time. If no second annealing is applied ($t = 0$ in Figure 15.10) austenite transforms into martensite, although some retained austenite will be present because for most TRIP steels the M_f temperature is below room temperature.

After a short time (t_1), the residual austenite still transforms into martensite, but larger parts of austenite do not transform. After a time t_2 the residual austenite present at the soaking temperature is completely retained during cooling.

During further cold deformation of these alloys, the retained austenite will gradually transform into martensite by a strain-induced transformation. An illustration of this TRIP effect for some of the alloys mentioned in Table 15.2 is shown in Figure 15.11.

15.2.3.3 The hot-rolling route. TRIP steels can also be produced in a hot-rolled condition. The TMP scheme follows the same principles as the one from the cold-rolling route. The material is hot rolled in the austenite region or in the intercritical ($\alpha + \gamma$) region and must be adequately cooled to a 'quench finish' temperature, mostly around 400°C. During cooling, ferrite can be formed, but the formation of pearlite must be suppressed. During annealing in the bainite region the residual austenite will partially transform into bainite, and the rejected carbon

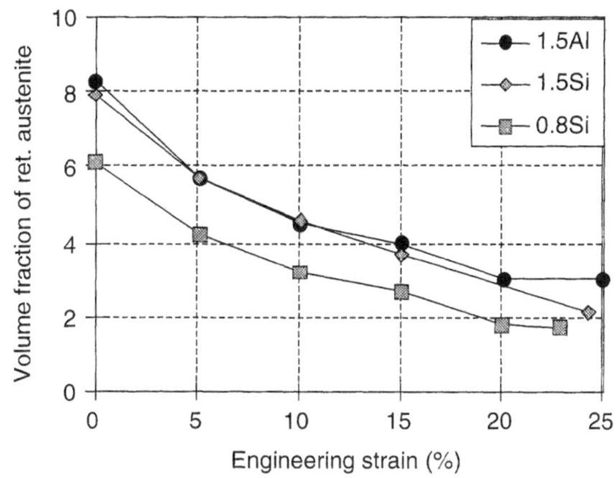

Figure 15.11. 'Consumption' of retained austenite during a uniaxial tensile test (alloy composition in Table 15.2) (Girault *et al.* [2001]).

atoms stabilize the remaining austenite. After further cooling some austenite will be retained at room temperature.

A very important aspect of this processing is the fact that austenite can be deformed (hot rolled) below the recrystallization temperature. This leads to an elongation of the austenite grains and the formation of shear bands in austenite. During subsequent cooling more nucleation centres for ferrite are present, leading to more, finer ferrite and to smaller bainite plates. It has been reported (see, e.g. Basuki and Aernoudt [1999], Godet *et al.* [2002]) that the stability of retained austenite, at room temperature, increases with prior austenite deformation. During further cold deformation this retained austenite transforms more gradually into martensite and better mechanical properties are observed: higher strength, increased work hardening and higher fracture strain.

A disadvantage of the hot-rolling route is that the processing conditions are somewhat difficult to control and the low coiling temperature is often problematic.

15.2.3.4 Properties and applications. A selection of mechanical properties of some typical TRIP steels is shown in Table 15.3.

Most TRIP steels have a YS between 350 and 600 MPa and a tensile strength between 600 and 1000 MPa. TRIP steels are suitable for bake hardening, for example, after 2% tensile deformation and a typical paint baking treatment, an increase in strength up to 75 MPa can be obtained. Work hardening is better than for most dual phase steels and certainly better than for HSLA steels. Consequently, TRIP steels have a fairly high uniform elongation. The $\overline{R}$-value is on the other hand

Table 15.3. Compilation of mechanical properties of some selected TRIP steels.

Type	C (wt%)	Si (wt%)	Mn (wt%)	YS (MPa)	UTS (MPa)	ε_{total} (%)	Lüders strain	n (5–15%)	$\overline{R}$	LDR	UTS × ε_{total}	Reference
IF	0.006		0.12	133	298	51	0	0.24	2.1	2.34	1.5×10^4	
Dual phase	0.10	0.2	0.7	350	600	27	0	0.18	0.9		1.6×10^4	Bleck [1996]
TRIP 0.8 Si	0.12	0.78	1.51	374	635	29	1.1	0.25	1.14	2.20	1.8×10^4	Girault [1999]
TRIP 1.2 Si	0.11	1.18	1.55	339	614	35	Yes	0.244	0.86		2.2×10^4	Hiwatashi *et al.* [1993]
TRIP 1.5 Si	0.11	1.50	1.534	452	698	31	1.5	0.24	1.0	2.24	2.2×10^4	Girault [1999]
TRIP 1.9 Si	0.14	1.94	1.66	530	890	32	Yes				2.6×10^4	Itami *et al.* [1995]

rather low (around 1). Most TRIP steels suffer from Lüderbands, but this can be reduced when some martensite is present in the structure at room temperature.

Because of their high strength and fair elongation, TRIP steels seem particularly suitable for medium-strength structural parts responsible for crash energy absorption, but requiring good formability.

15.3. CONTROLLED ROLLING OF HSLA STEELS: PIPELINE APPLICATIONS

15.3.1 Introduction

The problem of transportation of fluids over large distances emerged as soon as people began to live in large communities, creating the need to divert water from its natural course. The first aquaducts were constructed from clay, wood or lead. None of these materials was really satisfactory, and a first significant improvement was made in the 16th and 17th century with the introduction of cast iron pipes. One of the first full-scale cast iron pipe systems was built in 1664 to carry water from Marly-on Seine to the castle of Versailles, France over a distance of about 25 km (Cast Iron Soil Pipe and Fittings Handbook [1994]).

Drilling of the first commercial oil wells in the middle of the 19th century, created new demands. Transport in old whiskey barrels,[4] wooden pipes and rail was soon replaced by transport through long pipelines made of wrought iron and later of steel. The first specification for pipeline steel, issued by the American Petroleum Institute (API) in 1948, prescribed a weldable low-carbon high-manganese steel, grade X42, which means that it has a YS of 42 ksi or 290 MPa. Since that time higher strength steels have been developed up to grade X80 (80 ksi or 552 MPa) (Llewellyn [1994]). These high strengths are reached by grain refinement obtained by a special TMP scheme, called 'controlled rolling' and by precipitation hardening.

In this chapter, the principles of controlled rolling will be discussed and the main characteristics of steels for pileline applications will be illustrated.

15.3.2 Controlled rolling

15.3.2.1 Limitations. Steels, commonly used in a wide variety of engineering applications such as general building applications, pressure vessels, ships, pipelines, bridges, etc., are called 'structural steels'. These steels have a pearlite–ferrite structure and carbon and manganese as basic elements. In some grades small amounts of Nb, Al, V and Ti are added to increase the strength. These grades are called HSLA steels.

[4] These whiskey barrels were made a standard size; that's why volumes of oil are measured in 'barrels' [www.pipeline101.com].

Besides a minimal strength, structural steels need an excellent weldability and a high toughness (low ductile to brittle transition temperature). The general weldability of a steel can be expressed by its carbon equivalent (CE): the lower the CE, the better the weldability. Several equations can be found in literature to calculate the CE. Traditionally, Eq. (15.5) has been used (Llewellyn [1994]), but some authors prefer 'adapted' equations such as (15.6) (Kalwa and Kaup [1993]).

$$\text{CE (wt\%)} = \text{wt\%C} + \text{wt\%Mn}/6 + \text{wt\%(Cr+Mo+V)}/5 + \text{wt\%(Ni+Cu)}/15 \quad (15.5)$$

$$\text{CE (wt\%)} = \text{wt\%C} + \text{wt\%Si}/25 + \text{wt\%(Mn+Cu)}/20 + \text{wt\%(Cr+V)}/10 + \text{wt\%Ni}/40 + \text{wt\%Mo}/15 \quad (15.6)$$

In order to obtain a good weldability, the carbon content of structural steels should be kept as low as possible and alloying elements can only be tolerated in small concentrations. To fulfil the other two requirements (minimal strength and good fracture toughness), a fine ferrite grain size is of prime importance, as expressed by Eqs. (15.7) and (15.8). In these equations, A, B and k are constants and σ_0 the friction stress. The YS increases and the 'fracture appearance transition temperature' (FATT) shifts to lower temperature with smaller ferrite grain size D.

$$\text{YS} = \sigma_0 + kD^{-0.5} \quad (15.7)$$

$$\text{FATT} = A - BD^{-0.5} \quad (15.8)$$

Before World War II a fine ferrite grain was obtained using a normalized Al-killed steel. Stimulated by the bad experience with the break-up of welded Liberty ships and guided by the work of Hall and Petch in the 1950s, new processing routes for fine-grained material were investigated. In a number of European countries (Sweden, Holland and Belgium), the concept of 'controlled rolling' was established (see next paragraph). In the 1970s, stronger steel grades were developed, using small amounts of Nb, Al, Ti and V. Most of these HSLA steels were hot rolled using the concept of controlled rolling. A last improvement in structural steels was obtained in Japan with the introduction of accelerated cooling (Kozasu [1992]).

15.3.2.2 Principles of conventional controlled rolling. Traditional hot rolling of steel takes place in the austenite region. As long as the temperature stays above a critical value, called 'the temperature of no recrystallization' (T_{NR}), the austenite grains remain more or less equiaxed, due to subsequent waves of recrystallization on the run-out table in between the rolling passes. After hot rolling the austenite grains transform into ferrite and some pearlite during cooling of the hot-rolled plate.

Since most ferrite grains nucleate on the former austenite grain boundaries, a finer austenite grain size results in finer ferrite. In conventional hot rolling the reheating temperature before entering the roughing mill is usually around 1200–1250°C. In this temperature range it is difficult to prevent the growth of austenite. Finishing rolling is carried out above T_{NR} to avoid high rolling forces as a result of accumulation of strain hardening due to the absence of recrystallization.

In *controlled rolling* the reheating temperature is limited in order to avoid excessive austenite grain growth (e.g. 1100°C). The final rolling passes are carried out below T_{NR}. This results in pancaked austenite with an increased grain boundary fraction/volume and hence in more ferrite nuclei and smaller ferrite grains. The consequence is of course an accumulation of strain hardening in the last rolling passes, leading to high rolling forces. A common practice is to apply more reduction in the first passes (roughing mill) and to reduce the strain in the final passes.

Accelerated cooling starts above A_3 and is maintained till about 500°C. It activates additional ferrite nuclei inside the austenite grains and hence contributes to additional ferrite grain refinement. A faster cooling can however also lead to the formation of some bainite. This contributes to the strength of the material but it impairs the toughness. Accelerated cooling should be limited to 10–15°C/s.

In Table 15.4, the main processing parameters for hot rolling of a typical structural steel are shown and the influence on strength and fracture toughness is

Table 15.4. Main rolling parameters for conventional rolling and controlled rolling of a structural steel. After Kozasu [1992].

Hot rolling of a slab from 220 mm thick into a plate of 20 mm (0.12 wt% C, 1.4 wt% Mn, 0.025 wt% Nb)		
	Conventional	Controlled rolling
Reheating temperature	1200°C	1100°C
Number of passes in roughing mill	9	11
Temperature window for roughing	1100–1000°C	1050–950°C
Thickness after roughing	100 mm	67 mm
Time in between roughing and finishing mill	25 s	250 s
Start finishing rolling	1020–1000°C	800°C
Number of passes in the finishing mill	6	9
Exit temperature	1000–950°C	770°C
Cooling speed	0.8°C/s	8°C/s
Fracture appearance transition temperature (FATT)	−15°C	−85°C
Change in flow stress	0 (basis)	+50 MPa

Note: Typical reduction per pass: roughing: 5–13%; 'finishing': conventional 25–35% and controlled rolling 10–18%.

illustrated. The main differences between conventional and controlled rolling are the lower reheating temperature of the latter and the larger reduction during roughing. The time between roughing and finishing mill is much higher in controlled rolling in order to cool down the thick plate. This is considered as one of the drawbacks of controlled rolling because it decreases the productivity. In spite of the lower plate thickness, more rolling passes are needed in the finishing mill during controlled rolling, because the reduction per pass is limited due to the high rolling forces required at this temperature. After controlled rolling, an accelerated cooling of 8°C/s is applied. The combined effect of controlled rolling and accelerated cooling is an increase of 50 MPa of the YS and a decrease by 70°C of the fracture appearance temperature. The reader is referred to Kozasu [1992] for a more detailed discussion.

Some rolling mills may not be strong enough to exert the high rolling forces required in the last passes of conventional controlled rolling. In such situations, *recrystallization controlled rolling* can be appropriate. Rolling is then completely carried out above T_{NR} and austenite grain refinement is obtained by static recrystallization in between the rolling passes. Coarsening of these recrystallized grains is prohibited by fine TiN dispersoids. More information about recrystallization controlled rolling can be found in Siwecki [1997].

15.3.3 HSLA steel for pipelines

Until 1950, pipelines for oil and gas transport were made from conventionally hot-rolled C–Mn steels, as in rolled or normalized condition. They had a YS up to 360 MPa. New developments including a reduction of the carbon content and alloying with Nb and/or V increased the YS to 420 MPa by grain refining and precipitation hardening. Thanks to controlled rolling and further adjustments of micro-alloying elements, the YS was raised to 552 MPa (grade X80) (Kalwa and Kaup [1993]). New grades (X100 with YS ~700 MPa) are currently being developed (Kalwa [2002]).

The chemical composition of HSLA steels must be well balanced. Equations (15.5) and (15.6) show that alloying elements cannot be added in large quantities because this would impair the weldability. The carbon content is kept low (usually below 0.1 wt%) and only Mn is added in larger quantities (usually 1.4–1.9 wt%) because it has been shown to improve notch toughness.

Micro-alloying elements such as Nb, V, Ti and Al form carbides, nitrides and carbo-nitrides. They play a three-fold role:

- Fine precipitates prevent the excessive growth of austenite during reheating towards the starting rolling temperature. Especially, TiN remains stable up to temperatures above 1200°C.
- Nb in solid solution (the same holds to a lesser extend for Ti) retards the recrystallization. In the case of recrystallization controlled rolling the use of Nb will thus be avoided or restricted, but in case of conventional controlled rolling, it

Table 15.5. Chemical composition of some pipeline steels (Llewellyn [1994]).

Grade	C	Si	Mn	P	S	Ni	Mo	Nb	V	Ti	B
X65	0.02	0.14	1.59	0.018	0.003			0.04		0.017	0.001
X65	0.03	0.16	1.61	0.016	0.003	0.17		0.05		0.016	0.001
X80	0.07		1.65		0.002		0.22	0.05	0.075		0.001
X80	0.02	0.26	1.95	0.022	0.003	0.38	0.31	0.04		0.019	

will increase T_{NR} and this allows to carry out the last passes at higher temperature and hence with lower rolling force. Vanadium is reported to show only a weak retardation effect.

- During the last low-temperature rolling passes and during cooling, fine precipitates will form and will contribute to the strength of ferrite. Vanadium is especially effective because it is completely in solid solution above 1000°C. It re-precipitates on cooling and contributes significantly to the strengthening of ferrite.

Standards for pipeline steels do not specify in detail the compositions of the different steel grades. They put some upper limits concerning the levels of carbon, manganese, sulphur and phosphorus and leave it to the customers to narrow down the specifications and to the suppliers to choose an appropriate processing route and an adapted composition. Table 15.5 gives some examples of chemical compositions. In general, the lower grades (up to X60 or YS ~415 MPa) can be produced by controlled rolling of ferrite–pearlite steel with some Nb. For X70 grades, some additional precipitation strengthening (e.g. with V) is needed and the highest grades (X80) elements such as Mo and/or Ni are often added (Llewellyn [1994]).

15.4. ELECTRICAL STEELS

15.4.1 Introduction

The use of steel in 'electrical' applications such as laminations for motors, dynamos and transformers, dates from the mid-19th century. The use of silicon as an alloying element to minimize core loss (see Section 15.4.2) was however discovered in the late 19th century and such soft magnetic steels remain the mainstay of the present technological age (Takashi and Harasse [1996]). A generalized classification of electrical steels is given in Table 15.6, while Figure 15.12 provides a typical picture of the world market for soft magnetic materials.

The grain-oriented steels typically contain large grains (millimetres to several centimetres, depending on the processing route) with a typical orientation, {110}<001>, called the 'Goss' orientation. Relative deviations from the exact Goss orientation are a primary issue in the core loss (Takashi and Harasse [1996]).

Table 15.6. A generalized classification of electrical steels. The reported core loss values are in watts/kg (for 0.27 mm thick sheet), estimated at 1.5 Tesla, 50 Hz (Lyudkovsky *et al.* [1986], Matsuo [1989], Takashi and Harasse [1996]). The grain-oriented steels (with 3% or more Si) are classified according to the processing route.

Types		Details	Remarks
Non-oriented (CRNO)	Ordinary lamination	Normally does not contain Si, but may contain Mn and P	Typical core losses are higher than Si-bearing grades, in excess of 5.5
	Si-bearing	Contains Si, but may also contain Al, P and Mn. Sold as fully processed and semi-processed grades	Grades with watt loss as low as 2.35 are available. For inferior grades, highest watt loss may even approach 8
Grain oriented (CRGO)	Armco	Other than Si, it contains Mn, S or Se	Watt loss 0.793
	Nippon steel	HI–B. Other than Si, it contains Al, N, Mn, S and Sn	Watt loss 0.696
	Kawasaki	RG-H. Other than Si, it contains Mn, Se or S and Sb	Watt loss is comparable or even lower than Nippon steel

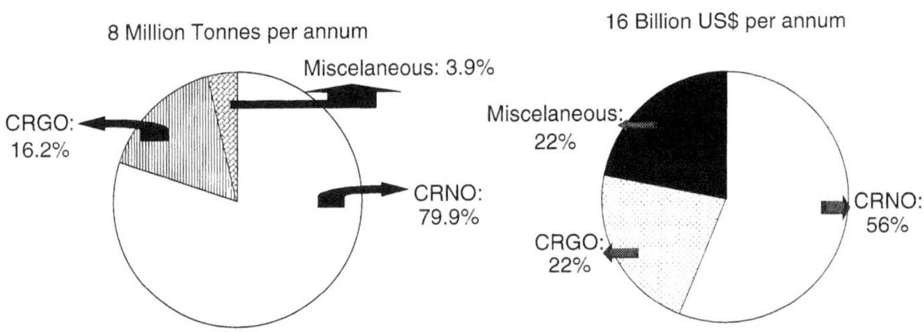

Figure 15.12. World market for soft magnetic materials. CRNO and CRGO, respectively, represent non-oriented and grain-oriented grades, while miscellaneous stands for a range of other soft magnetic materials – ferrites, amorphous alloys, nano-crystalline alloys, Ni–Fe alloys, etc.

Grain-oriented steels are used in applications involving unidirectional flux paths with no air gaps, like transformer cores. Non-oriented electrical steels, on the other hand, are used in places where the flux direction may change or rotate as in electrical motors (Lyudkovsky *et al.* [1986]).

15.4.2 A few relevant basics on magnetism

Interaction between electricity and magnetism is the basis of all electrical equipment. To use this interaction effectively, ferromagnetic materials, with high magnetic induction, are required. The transition elements, Fe, Ni and Co, exhibit ferromagnetism at room temperature. Of these, Fe has the highest saturation magnetization of 2.15 Tesla, and this property is the basis of electrical steel. Other than saturation magnetization, the two issues of vital interest are low core loss and high permeability.[5]

When a ferromagnetic material is subjected to an alternating magnetic flux, part of the energy is lost through generation of heat. This is the core loss or iron loss. It has three components – eddy current loss, hysteresis loss and anomalous loss (Lyudkovsky *et al.* [1986], see Figure 15.13). The classical *eddy current loss* is due to the eddy currents generated in the steel, creating magnetic fields opposed to the field induced by the magnetization. Modifying Golding's classical treatment (Golding [1963]), Lyudkovsky *et al.* [1986] found:

$$\text{Classical Eddy Current Loss} = \frac{10^{-13}\pi^2 t^2 B^2 f^2}{6\sigma\rho} \qquad (15.9)$$

where t is the thickness in cm, B the induction in Gauss, f the frequency in Hz, σ the specific density in g/cm^3, ρ the resistivity in ohm cm and the core loss is estimated in watts/kg.

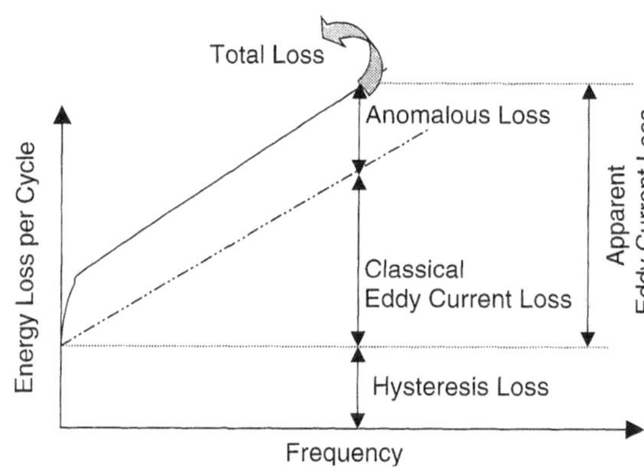

Figure 15.13. Different components of the core loss in electrical steels. This is shown following the classical theories. After Lyudkovsky *et al.* [1986].[6]

[5] For the basics on permeability, watt loss and saturation magnetization, the reader may refer to Figure 2.17 in Chapter 2.

[6] The modern magnetization theory (Benford [1984], Stephenson [1985]) considers such separation as 'artificial', rejects anomalous loss and separates the losses as synchronous and asynchronous. The classical theory, however, still finds wide applications by manufacturers and users of electrical steels and is hence used in the present case study.

The easiest way to reduce the classical eddy current loss is through increased resistivity or ρ. Fe has a high eddy current loss due to its low ρ. A natural solution for this is to increase ρ in electrical steel through 'suitable' alloying additions (see Section 15.4.3).

Hysteresis loss is caused by the migration of magnetic domains and contrary to classical eddy current loss, hysteresis loss is structure sensitive and can be reduced significantly through texture and grain size control.

$$\text{Hysteresis Loss} = KfBtH \tag{15.10}$$

where K is a constant, B the magnetic induction and H the magnetizing field. The processing of electrical steels, both non-oriented and grain-oriented types, is aimed at minimizing the hysteresis component of the total loss, while the composition is important for both eddy current loss (as it may affect ρ) and also for hysteresis loss.

The other important factor, permeability, is a measure of the ease with which a material can be magnetized. At a specified induction (typically 1.5–1.7 T), magnetic permeability (μ) is estimated as B/H. Permeability is sensitive to the same factors as hysteresis loss. The relevance of permeability, however, depends more on the applications. For motors operating at high flux densities, a high-permeability material is of utmost importance.

A natural question emerges – 'to what extent do these magnetic properties affect the present society or technology'? The answer is apparent and can be viewed from any major electrical steel manufacturers advertisement brochure. The core loss or iron loss during electricity generation is about 2.4% and during transmission is about 1% – a yearly loss of at least 400 billion kWh across the world.[7] Continuing developments in electrical steels are not only making more efficient and downsized machineries, but have tremendous implications in both sustainable technology and economics.

15.4.3 Role of chemistry

The first breakthrough in electrical steel development was in the late 19th century (Barret *et al.* [1900]) with the discovery of Si additions as a means to increase the resistivity (and hence decrease core loss and increase permeability) without affecting the saturation magnetization significantly. As shown in Figure 15.14, the negative effects of Si on saturation induction and Curie temperature are offset by the positive effects of increased resistivity (Lyudkovsky *et al.* [1986]). The Si

[7] The 1990 statistics from the Institute of Energy Economics, Japan. This statistics does not consider losses during usage of electrical machineries.

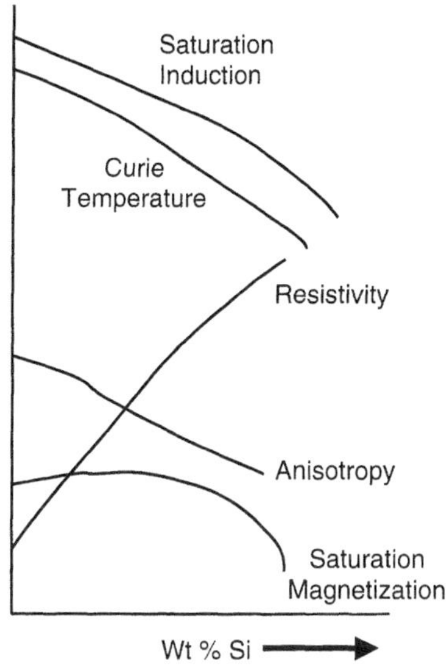

Figure 15.14. Electrical resistivity and intrinsic magnetic properties in Fe–Si alloys. After Lyudkovsky *et al.* [1986].

percentage, however, does not exceed 3.5 wt% in grain-oriented electrical steel and is even lower in non-oriented steel. This is due to reduced workability of high Si alloys during conventional processing (also the fact that in CRNO an increase in Si is typically associated with a drop in permeability). Several emerging technologies may offer possibilities of increasing the Si percentage. These are: new rolling technologies (Masuda *et al.* [1988], Shoen [1990]), chemical (Nakaoka *et al.* [1986], Takada *et al.* [1988], Krutenat *et al.* [1991]) and physical (Aldareguia *et al.* [1999]) vapour deposition of Si on Fe–3 wt% Si strips, rapid quenching (Narita and Yashimiro [1980]), strip casting and spray forming (Moses [1990], Shin *et al.* [2001]). It has to be noted that only chemical vapour deposition is used commercially at present (NKK Corporation, Japan), but the costs remain prohibitively expensive.

The use of other alloying elements depends on the type of electrical steel – non-oriented or grain-oriented. In grain-oriented steel, abnormal grain growth or secondary recrystallization is the tool for 'complete' texture control. The so-called growth inhibitors, fine 2nd phase particles responsible for 'selective' pinning of the grain boundaries, play a crucial role. In the Nippon steel process, AlN[8] is used as a growth inhibitor, while in the Armco and Kawasaki steel process, MnS and

[8] Al and N being added in the melting stage.

MnSe particles act as inhibitors. Sb, in the Kawasaki steel grade, is expected to have 'selective' grain boundary segregation. The exact compositions of the grades, especially the possible use of the trace elements, remain a closely guarded industrial secret.

In non-oriented electrical steel, the alloying additions are usually aimed to improve the following factors:

- Resistivity increase: In general, substitutional elements improve the resistivity. P is the best, even better than Si. P in excess of 0.14 wt% is however detrimental for workability. Al causes similar improvements to resistivity as Si and is often used. An excess of Al is avoided, as it is detrimental to workability and also causes problems in continuous castings, especially with thin slab casters. Mn also offers resistivity improvements.
- Grain growth and texture control: Grain growth, albeit normal and not abnormal, is also one of the tools for texture control in non-oriented grades. An 'optimum' grain size (Shimanaka *et al.* [1982], PremKumar *et al.* [2003]) is required for the best magnetic properties. This combination calls for suitable growth inhibitors, as in the case of grain-oriented grades.[9] Mn is one of the examples of typical alloying additions to generate growth inhibitors.
- Eliminating interstitials that, in general, are harmful. Their presence (especially C, O, N and S) seriously affects the magnetic properties. Modern steel-making technologies may offer low interstitials, such as less than 10 ppm S content, and offer non-oriented steels for specialized applications (Oda *et al.* [2002]), but often a cheaper and more convenient option is to remove 'harmful' interstitials through 2nd phase precipitation (Lyudkovsky *et al.* [1986]). The classical examples are Mn additions to arrest S and limited B doping (within 30 ppm) to form BN in preference to AlN.

It also needs to be noted that 'clean' steel, or steel free from major inclusions is a necessary requirement for all electrical steels.

15.4.4 Role of crystallographic texture, stress and grain size

The influence of crystallographic orientation(s) on magnetization is known from the early 20th century (Honda and Kaya [1926]). As shown in Figure 15.15, depending on crystallographic direction, the magnetization may be hard, medium or easy. More recent models have established a similar approach to saturation – cube fibre (ND//<001>) being the easiest and γ fibre (ND//<111>) being the

[9] Though strong inhibitors like AlN are avoided.

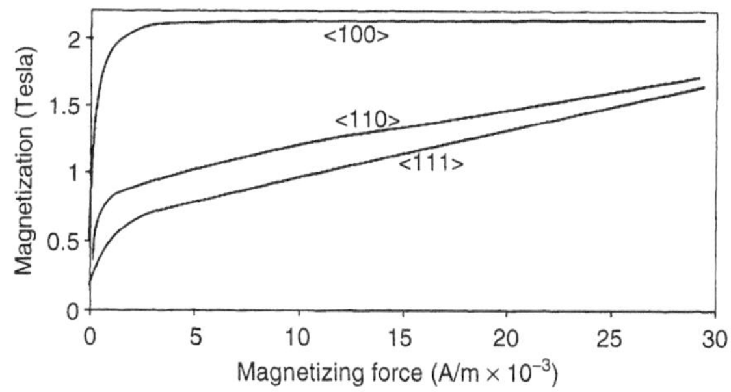

Figure 15.15. Magnetization curves in single crystal bcc Fe as a function of three distinct crystallographic directions. After Lyudkovsky *et al.* [1986].

hardest (Rollett *et al.* [2001]). In grain-oriented grades, following the Goss patent (Goss [1934]), *only* {110}<001> Goss grains are present. Decreasing the spread around exact Goss seems to be the road to better magnetic properties (Takashi and Harasse [1996]). In non-oriented grades, the relative role of texture is often viewed from different perspectives and it needs to be 'delinked' from the effects of grain size and chemistry (see Section 15.4.5).

Another important aspect to a manufacturer of electrical equipments is the role of stresses. Cutting, punching and stacking operations, used in both CRNO and CRGO, grades can affect the properties severely. However, it needs to be pointed out that unlike the role of crystallographic texture, our understanding of the role of stresses on magnetic properties remains qualitative at best. For example, plastic deformation can degrade magnetic properties severely – degradation related to interactions between dislocations and magnetic domain walls. But quantification of such effects remains at the levels of individual case studies, see Figure 15.16, and not as a comprehensive model. Residual stresses can also affect magnetic domain movements and such effects are usually detrimental. Interestingly, no comprehensive study has been undertaken, at least in the domain of published literature, to bring out the relative role of residual stresses on magnetic properties in CRNO/CRGO.

In CRGO the grain size is determined by the process – millimetres in the Armco process and centimetres in Nippon/Kawasaki process (see Section 15.4.6). Grain sizes affect magnetic properties critically in both CRGO and CRNO – an optimum grain size is often considered desirable (see Figure 15.17). The existence of an optimum grain size can be explained from domain theory – below the optimum grain size hysteresis loss due to domain wall interactions dominates, while above the optimum grain size losses are linked to domain wall movement. In CRGO,

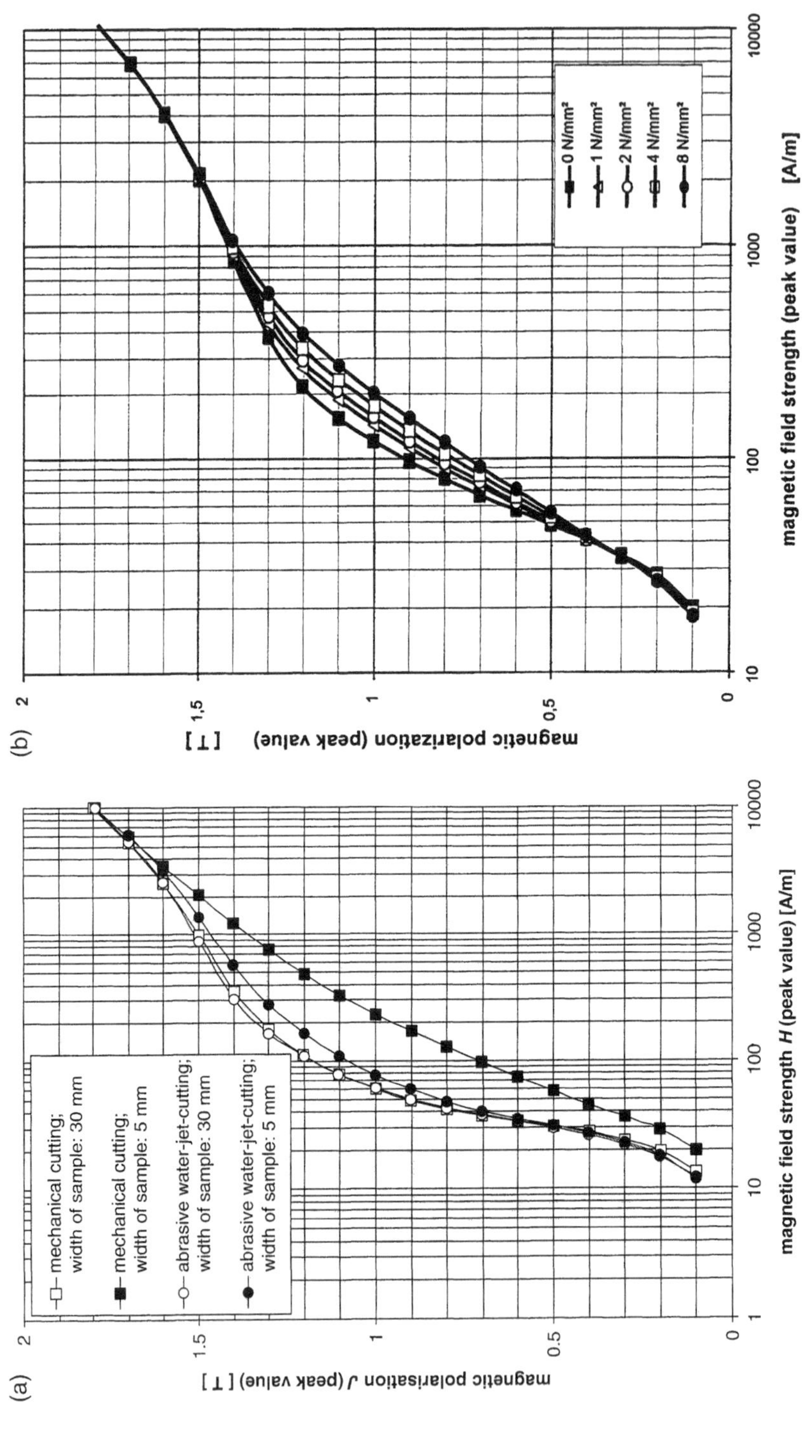

Figure 15.16. Role of (a) cutting and (b) pressing on the magnetic properties of electrical steel. (a) (Schoppa [2003]); (b) (Schoppa [2000]). (Copyright (2007), with permission from Elsevier).

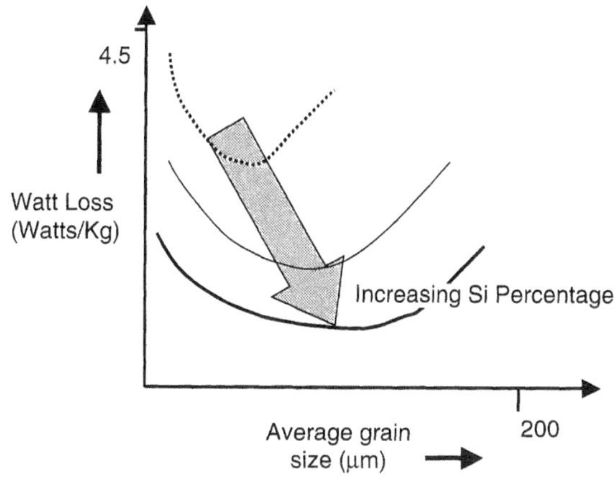

Figure 15.17. Effects of grain size on core loss in electrical steels of different Si concentration. After Shimanaka *et al.* [1982].

mechanical or laser scribing is used to defeat the effects of very large grain size, while in CRNO an optimum grain size is obtained through controlled TMP.

15.4.5 Non-oriented electrical steels (CRNO)

The non-oriented grades are produced in significantly larger tonnages (see Figure 15.12) than their grain-oriented counterparts. CRNOs include ordinary lamination material, but also more controlled grades (Oda *et al.* [2002]) for high-performance applications. The typical processing stages, and the main issues in such stages, are given in Figure 15.18. Though control of deformation can be an effective tool in controlling the recrystallized texture and grain size in typical CRNO grades, multi-stage reductions with intermediate annealing are quite uncommon because of cost economics.[10]

The control of texture and grain size is usually done through the final decarburization grain growth annealing, where reducing the carbon content is also a critical issue. In both annealing stages (as in Figure 15.18), prevention (Lyudkovsky *et al.* [1986]) of oxide scale (Jenko *et al.* [2000], Mandrino and Jenko [2001]) is crucial. Inorganic insulating coatings are used only for the highest grades.

An important issue in CRNO is an appropriate index for crystallographic texture. Unlike CRGO, a specific orientation cannot be considered, as the directions of the applied electric field vary continuously. A usual practice (Shimanaka *et al.* [1982], Lyudkovsky *et al.* [1986], PremKumar *et al.* [2003]) is to consider extreme fibres,

[10] Grain-oriented grades being significantly more expensive, intermediate annealing is used in both the Armco and Kawasaki routes.

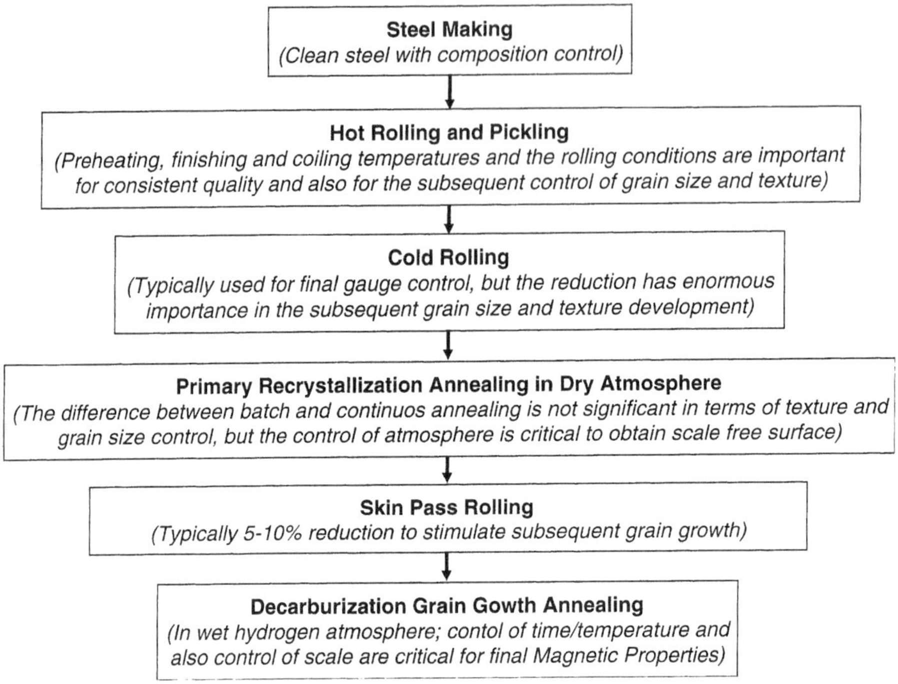

Figure 15.18. Typical processing stages of non-oriented electrical steel. The main issues involved in such steps are indicated.

(001)<*uvw*> and (111)<*uvw*>, respectively as good and bad. (001) Planes compensate the <111> directions, hard directions of magnetization. Similarly, (111)<*uvw*> is the worst texture for magnetic properties.

Another important technological issue for CRNO is the link between magnetic properties and the metallurgical factors. In other words, what a motor manufacturer understands, i.e. permeability and watt loss, and what a controlled TMP can offer, grain size and texture, are different. Naturally, individual manufacturers of electrical steel have their own models, often empirical, linking magnetic properties with metallurgical factors – but such models are rarely published. Table 15.7 offers a glimpse of such inter-relationships. More rigorous linking appears to be far more complicated and can be attempted only through artificial neural network (ANN) type brute curve-fitting (Chaudhury [2005]).

15.4.6 Grain-oriented electrical steels

Grain-oriented electrical steels are the *only* class of material with 'complete' texture control through TMP. It is interesting to note that even the original inventor did

Table 15.7. Linear statistical correlation coefficients (r) between magnetic properties (x) and metallurgical factors (y) in an ordinary lamination (Chaudhury [2005]). Where

$$r = \frac{n\sum xy - \sum x \sum y}{\sqrt{\left[n\sum x^2 - (\sum x)^2\right]\left[n\sum y^2 - (\sum y)^2\right]}} \tag{15.11}$$

Magnetic properties	Percentage silicon	Grain size	Texture factor (001)<*uvw*>/(111)<*uvw*>
Permeability	−0.29	0.35	0.28
Watt loss	−0.11	−0.64	−0.25

not visualize[11] the complete scientific implications of his invention – 'the product is characterized by a high magnetic moment approaching that of a single crystal, and the grains of the material being substantially oriented at random throughout the structure as shown by an X-ray pattern' (Goss [1934], Matsuo [1989]). This puzzle was 'cleared' subsequently and the characteristic crystallographic texture resulting from the TMP proposed by Goss, was named as 'Goss Texture' or {110}<001> preferred orientations (Matsuo [1989]). In addition to Goss's original process (the Armco route), alternative routes have emerged – the Nippon and Kawasaki routes (Takashi and Harasse [1996]).

The details of the processes are, however, shrouded in industrial secrecy and there remain large unanswered issues in our present scientific understanding. Figure 15.19 provides schematics of the processing stages, indicating important differences between the three processing routes. The steel making is important, especially in terms of compositional control. Other than the high preheating temperature, little has been published on the hot-rolling schedule. The hot band microstructure has a bimodal grain size distribution and significant presence of 2nd phase particles in the bands of finer grains (Cicalè *et al.* [2002]). The bands of finer grains seem to be generated during austenite-to-ferrite transformations (Matsuo [1989]). The hot band texture is reported to contain a strong α fibre (<110>//RD) and a weak γ (<111>//ND), remarkably different from the so-called randomized hot band texture in a forming grade steel (Cicalè *et al.* [2002]). This difference is due to the presence of silicon that suppresses the ferrite–austenite–ferrite transformations, responsible for texture randomization in forming grade steel. In grain-oriented grades, such transformations are expected to happen only in zones where carbon is segregated, resulting in bands of fine grains. The significant 2nd phase

[11] This does not belittle the inventor or the invention. On the contrary, it was and is a breakthrough for any age.

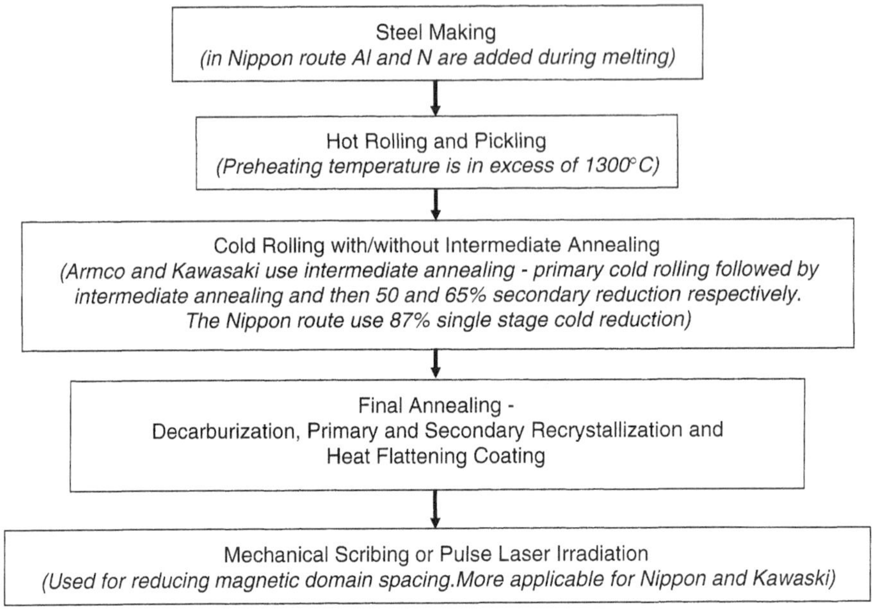

Figure 15.19. Processing stages of grain-oriented electrical steel. Important differences between the three processing routes (Armco, Nippon and Kawasaki) are indicated (Takashi and Harasse [1996]).

concentration in such bands is linked to the higher solubility of interstitials in the austenite phase (Matsuo [1989]).

Cold rolling leads to the formation of a deformation texture and also to the developments of deformation heterogeneities (Cicalè *et al.* [2002]), which are a major source for recrystallized grains, including recrystallized Goss grains (Samajdar *et al.* [1998a]). The cold-rolling texture development is reported (Cicalè *et al.* [2002]) to follow a pattern, predictable through Taylor-type deformation texture modelling, while the relative presence of deformation heterogeneities are 'selective' in orientations or grains with negative textural softening (Cicalè *et al.* [2002]). A typical deformation microstructure is given in Figure 15.20.

The selection of intermediate annealing, as used in both the Armco and Kawasaki routes, is mainly to 'age' the 2nd phase, but may also have strong implications on the subsequent recrystallization microstructure. During laboratory 'inert' atmosphere annealing two 'types' of Goss grains are observed (Ushioda *et al.* [1989] and Samajdar *et al.* [1998a]), inclined respectively at about 37° and 20° with RD (rolling direction). These Goss grains originate from deformation heterogeneities of similar spatial orientation. It remains unclear if such inert atmosphere results can be related to the actual production environment of decarburization annealing, with both

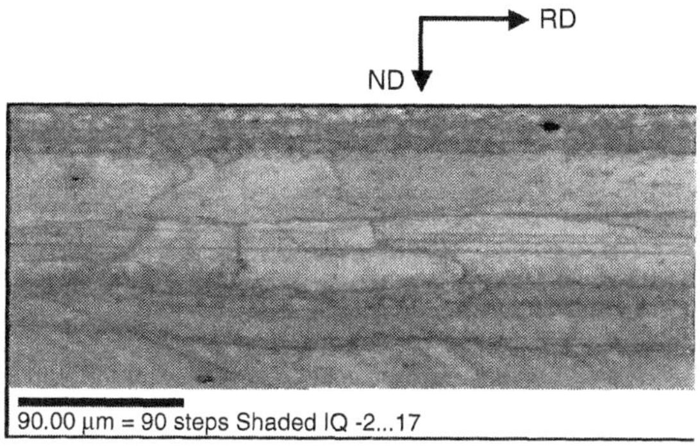

Figure 15.20. Typical deformation microstructure of cold-rolled grain-oriented steel after 25% reduction in thickness (Cicalè *et al.* [2002]). The rolling direction (RD) and normal (ND) direction are marked. The 'selective' appearance of the grain interior strain localizations and even the pre-deformation (hot band) typical bimodal grain size distributions are apparent – contrast being image quality in electron backscattered diffraction (EBSD).

primary and secondary recrystallization taking place simultaneously with MgO coating.

The final annealing, typically in hydrogen-containing atmosphere,[12] is crucial. It involves decarburization, primary and secondary recrystallization and heat-flattening coating or coating with MgO. The outcome of this processing stage is massive grains (from millimetres in the Armco route to centimetres in the others) with a near-Goss orientation (within a few degrees of exact Goss, depending on the route (Takashi and Harasse [1996])). Naturally, the cause(s) for this 'complete' texture control or precise selection of Goss in secondary recrystallization has been an object of intense scientific interest. Interestingly, three different theories have emerged – theories which are in conflict.

- Size advantage: The 'selection' of Goss has been linked to the size advantage of Goss grains (Inokuti and Maeda [1984], Etter *et al.* [2002]).
- Advantage of low CSL (coincidence site lattice) boundaries: The 'selection' has been linked to higher probabilities of Goss in having low CSL or special (low energy) boundaries with respect to other matrix grains or orientations (Abbruzzese *et al.* [1988], Harase *et al.* [1991]). A special boundary is expected to offer less Zener drag (see Chapter 5) and thus may provide a growth advantage for Goss.
- Advantage of high-energy boundaries: The 'selection' has also been linked to higher probabilities of having random boundaries (Hayakawa and Szpunar

[12] Cracked ammonia is used for Nippon steel route (Takashi and Harasse [1996]) and there seems to be an 'optimum' heating rate (Dzubinsky and Kovac [2001]).

[1997]). It is argued that such random or high-energy boundaries may help in the effective dissolution of the pinning particles during secondary recrystallization and thus account for the growth advantage.

Direct experimental evidence is the only means to resolve this issue. But it is not as simple as it sounds. A simple calculation shows that there is a probability of the order of 10^{-8} to find the 'origin' of a final Goss grain – considering average recrystallized grain size of 10 μm, final grain size of 10 mm and about 2% primary recrystallized Goss grains. Any direct experimental validation remains statistically questionable, even given the enormous improvements in local orientation measurement techniques like electron backscatter diffraction (EBSD).

Possibly the most interesting aspect in electrical steels, especially in grain-oriented grades, are the continuing innovations. Any 'recent' patent search will reveal such innovations in areas ranging from product and process developments to inorganic coatings and domain refining. One may hope that the serious commercial interests will act as an incentive for better scientific understanding in the future and not as a 'blanket' for continuing industrial secrecy.

15.5. PATENTED STEEL WIRES – FROM BRIDGES TO RADIAL TYRES

15.5.1 Introduction

Steel wires are used extensively throughout the world for many critical engineering applications; high-strength cables for bridges (Figure 15.21), cable and ski lifts, general haulage, e.g. ship moorings and, on a large scale, for reinforcing radial tyres. They are also widely used for more cultural activities such as piano and violin strings. In all cases, their properties of very high strength and toughness are just about unique. In applications where high corrosion resistance is necessary, special stainless steel wires are employed but otherwise for high-strength requirements, as in vehicle tyres, patented steel wires are essential. As detailed below, patenting consists of heating carbon steel rods into the austenite phase region, cooling to pearlite and then wire drawing down to thin wire. The resulting work hardened pearlite is extremely strong – probably the strongest material known which possess some ductility and therefore toughness. It is this strength that enables architects to design spectacular suspension bridges and engineers to make reliable tyres for vehicles and aircraft, such as the jumbo jets.

15.5.2 The patenting process

The steel composition is basically a fully eutectoid 0.8%C steel containing some Mn and Si (more exactly the carbon content is between 0.7 and 0.9%C) designed

Figure 15.21. (a) Millau bridge, France and (b) fitting the cable stays into position.

to develop the maximum amount of pearlite, i.e. about 100% ferrite+cementite Fe_3C. To produce fine wire it starts as coiled rod of diameter around 5 mm. In this condition, the strength is about 1100–1200 MPa and the aim of the patenting process is to multiply this by a factor of about 3.

The rod is first given a preliminary drawing reduction down to 1–2 mm diameter without lubrication. It is then passed into a furnace and heated in the range 950–1000°C, i.e. in the single-phase austenite domain for transformation into homogeneous austenite with all the carbon in solid solution. It is then very carefully cooled to transform into fine pearlite. This can be done isothermally by cooling in liquid lead or in a fluidized bath but can also be carried out by controlled air cooling. The cooling rate is one of the critical parameters in the process. Figure 15.22

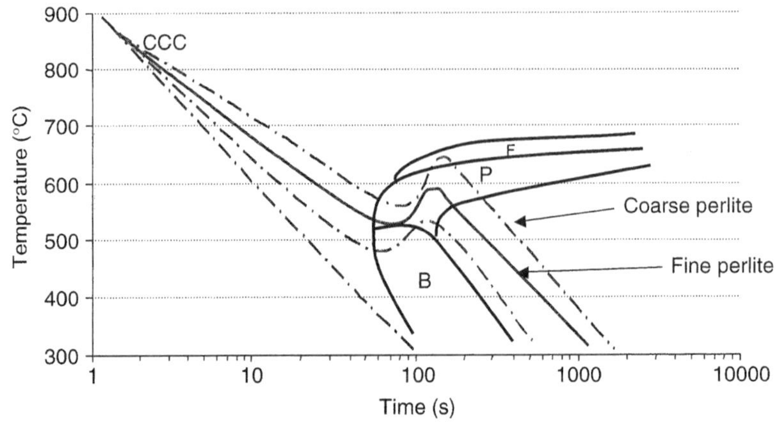

Figure 15.22. Continuous cooling transformation diagram of eutectoid steel for wire production. Courtesy of E. Depraetere, Michelin.

shows that according to the cooling rate one can form pearlite, bainite or even some ferrite mixed with pearlite. Also during cooling the austenite to pearlite, the exothermic transformation gives off sufficient heat, as indicated, to significantly increase the temperature in the range of transformation. The cooling rate allows for the heat of transformation to follow a finely judged path into the pearlite domain that leads to the formation of a very fine pearlite, i.e. in the range of 500–600°C (and without any bainite). The microstructure is then basically 100% pearlite with a spacing of about 0.25 μm (Figure 15.23).

Before wire drawing the rod is coated with a fine layer of brass which acts as a lubricant. Wire drawing is described in Section 11.4 together with some comments on drawing pearlitic steels. The important feature is that the material should work harden sufficiently during each drawing pass so that the extra hardening compensates for the reduced section to avoid rupture during pulling through the dies. This two-phase material does in fact work harden extensively (see Chapter 4). In practice, this means reductions of 15–25% per pass. To produce wire for tyre reinforcements, the rod is drawn down to thicknesses of about 0.3 mm or less (i.e. strains of 3 to over 4).

Figure 15.24 shows the concomitant lamellar alignment and pearlite spacing reduction which ultimately decreases to values of order 20–50 nm. This is in fact a composite nanomaterial used on a large scale before the word came into fashion! Unlike most other nanomaterials it is also not very expensive.

As pointed out in Section 11.4, the interlamellar spacing is eventually much smaller than the typical dislocation cell size giving unusual hardening effects at very high strains. The wires can fail during the drawing operation if they contain defects such as inclusions. The inclusion content must therefore be reduced to very

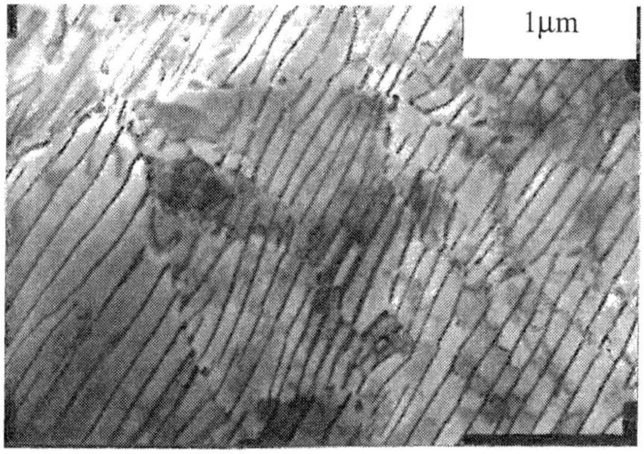

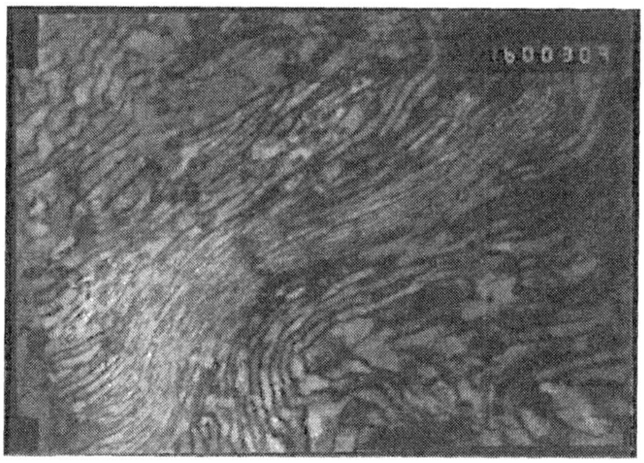

Figure 15.23. (a) TEM microstructure (cross-section) of fine 0.7%C pearlitic steel before wire drawing and (b) after cold drawing to a strain of 2.9. Courtesy of E. Depraetere, Michelin.

low levels in the initial steel casts. These very fine wires are then usually twisted together into strands, which are used in practice (Figure 15.25). The twisting operation is also quite critical since if the material has insufficient ductility it can fail by localized shear, often initiated at surface defects.

15.5.3 The mechanical properties

Figure 15.26 shows the work hardening that occurs in the drawn pearlitic wires as a function of strain. The material can harden from about 1200 to 4000 MPa. As pointed out in Section 11.4, the particular work hardening behaviour of this class of materials can be described by an exponential function up to strains of about 4. This fits in with the empirical law of Embury and Fisher [1966]: $\sigma_f = \sigma_y + k \exp(\varepsilon/4)$ with $\varepsilon = 2 \ln(D/D_0)$.

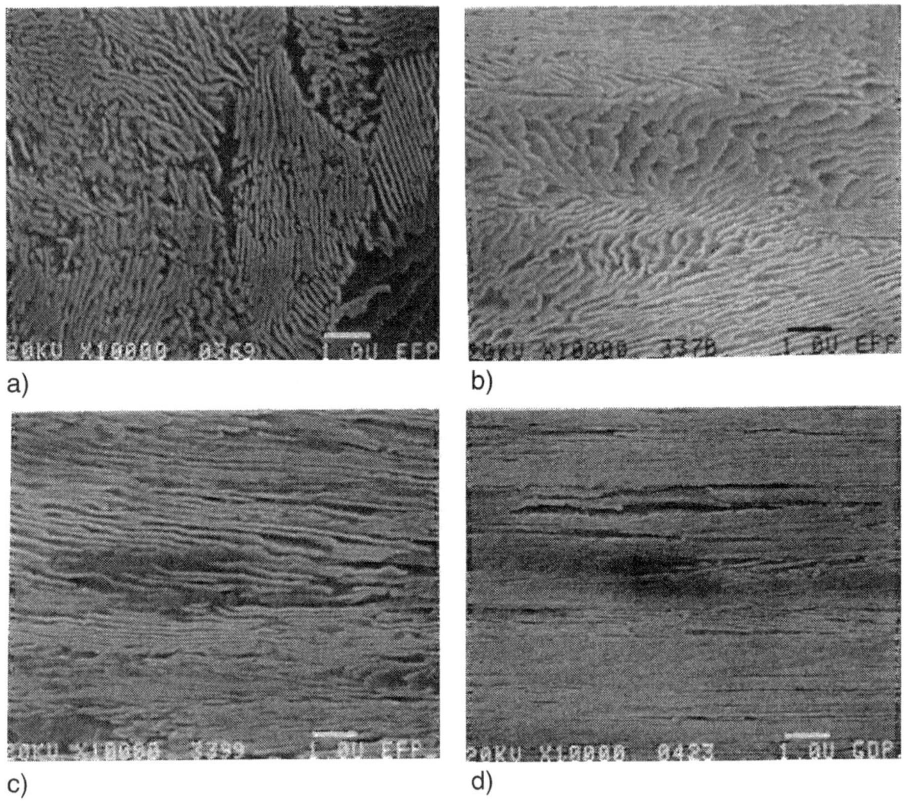

Figure 15.24. SEM microstructure (longitudinal section) of fine 0.7%C pearlitic steel after cold drawing strains of 0, 1, 2 and 3.5. Courtesy of E. Depraetere, Michelin.

The strength is also inversely proportional to the square root of the lamellar spacing in accordance with the Hall–Petch law (Embury and Fisher [1966], Langford [1977]). In fact, the material work hardens in the Stage IV regime over almost the entire stress–strain curve as a consequence of the accumulation of geometrically necessary dislocations at the ferrite–cementite interfaces and because of the curling effect visible in Figure 15.23b. It is also worth pointing out that the cementite phase, unlike most carbides, must possess extensive ductility during the wire drawing to maintain some compatibility. It appears to deform plastically by dislocation slip under the high hydrostatic pressures of the drawing operation, as for the bcc iron.

Patented steel wires are usually drawn up to strengths of about 3000 MPa for use in tyre reinforcements. However, there is a trend to develop even higher strength

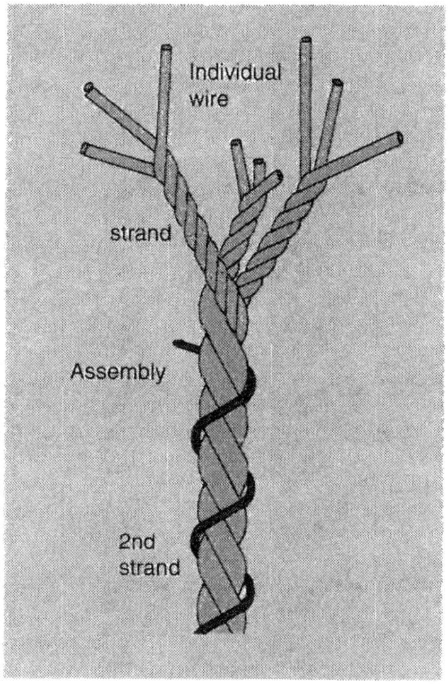

Figure 15.25. Successive wire-twisting operations into strands.

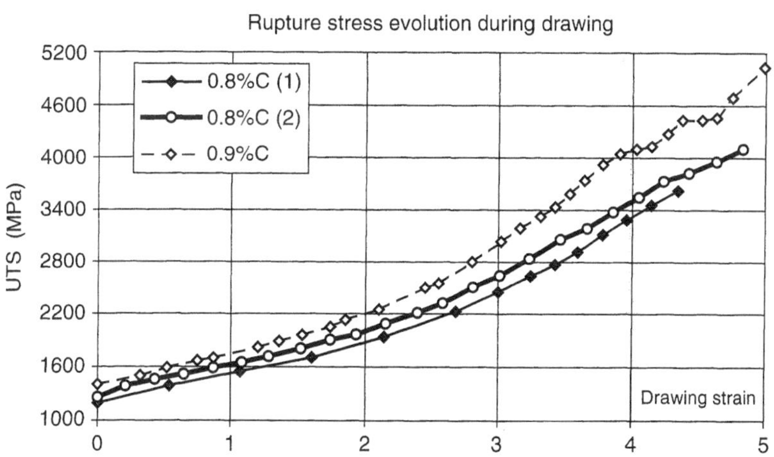

Figure 15.26. The UTS of drawn pearlitic wire as a function of drawing strain. The two plots for the 0.8%C steel refer to different initial lamellar spacings. Note the higher strength of the 0.9%C steel. Courtesy of E. Depraetere, Michelin.

wires by increasing the drawing strain and, in some cases, the carbon content (Figures 15.5 and 15.6). These wires, aimed to attain 4000 MPa, could be used in a new generation of tyres for heavy duty trucks and personal vehicles. At these levels of strain, and even before, the cementite layers break up and, surprisingly begin to dissolve locally so that excess carbon goes into solution in the ferrite (Languillaume *et al.* [1997], Read *et al.* [1997]). This behaviour is the object of current research.

ACKNOWLEDGEMENTS

The author acknowledges Eric Depraetère of the Michelin research centre Clermont Ferrand for figures and pertinent comments concerning Section 15.5.

Chapter 16

Thermo-Mechanical Processing of Hexagonal Alloys

Chapter 16
Thermo-Mechanical Processing of Hexagonal Alloys

16.1. ZIRCONIUM ALLOYS FOR NUCLEAR INDUSTRY

16.1.1 Introduction

All fission reactors, irrespective of their configuration and moderator/coolant combination, operate on the same basic principle – heat produced by a nuclear chain reaction is used to turn a turbine and generate power. The fuel in a typical thermal nuclear reactor is fed through fuel channels (Lamarsh [1983], Mckay [1984], Sundaram [1986], Ganguly [2002]) – a complex assembly of fuel bundles, pressure and calandria tubes.[1] To show this in a more descriptive way, Figure 16.1 makes use of the schematic of a Canadian deuterium uranium CANDU-type pressurized heavy water reactor (PHWR) and the cross-sectional schematic of a typical PHWR fuel channel. As shown in the Figure 16.1a, the reactor containment vessel, houses the calandria or the actual reactor vessel, containing the moderator (deuterium

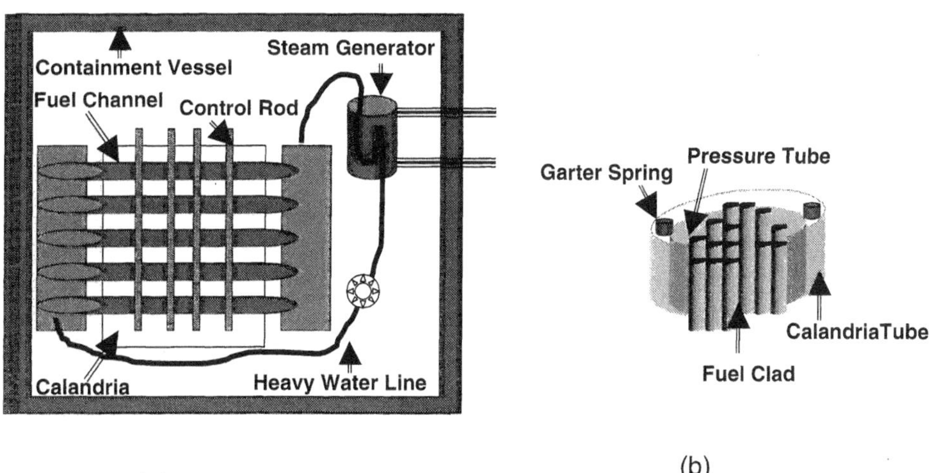

Figure 16.1. (a) Schematic of a CANDU-type PHWR and (b) schematic of the cross-section of a PHWR fuel channel.

[1] Also included are smaller components like garter springs, square channels, bellows, etc.

oxide or 'heavy water'), fuel channels and control rods. The heat of fission is extracted through circulating heavy water, by using a suitable steam generator. The most critical issue in the entire reactor design is the fuel channel. As shown in Figure 16.1b, fuel (UO_2) pellets are covered in fuel tubes or fuel clads, which typically constitute a fuel bundle. The fuel clads allow efficient heat transfer to the coolant and also prevent highly radioactive fission products from coming into direct contact with the coolant. The fuel bundles are housed in a pressure tube, through which coolant, heavy water, flows. The primary purpose of the pressure tube is to hold the fuel bundles and to allow the coolant to flow around the bundles at a specified pressure and temperature – coolant or heavy water inlet and outlet temperatures are of the order of 260 and 290°C, respectively. Calandria tubes envelop the pressure tubes and are separated from them by an annulus, which provides the thermal insulation – CO_2 is passed through the annulus for efficient heat transfer. Calandria tubes also act as containment for the contents of coolant channels in the unlikely event of a pressure tube rupture. The pressure tubes are supported by the garter springs located between the pressure tubes and calandria tubes at periodic intervals along the length of the calandria tubes. The garter springs are expected to prevent sagging during service. A typical 235 MW CANDU-type PHWR has an inventory of about 5 tonnes of calandria tubes, 15 tonnes of pressure tubes and 5 tonnes of fuel tubes.

The design of the fuel channel, which also involve feeders, end shield, end fitting, bearings, shield plug, etc., is 'fascinating' from an engineering point of view. An equally fascinating aspect is the material design aspects of the fuel channel components – components exposed to one of the harshest environment, where zero failure is not a 'goal' but an absolute requirement. For example, in a typical CANDU-type PHWR, 208–400 pressure tubes are present. The pressure tubes of 31 CANDU-type reactors in six different countries are approximately subjected to 150 000 years of full-power operation, but only two pressure tube ruptures have been reported.

The selection of the material for a fuel channel is decided by three factors: low neutron absorption cross-section, excellent mechanical properties and good corrosion properties at reactor working temperatures. As shown in Table 16.1, in terms of neutron economy, only magnesium (Mg) and beryllium (Be) are superior to zirconium (Zr), while additional strength consideration (relative neutron absorption for design strength) for the structural components do clearly bring out Zr as the ideal fuel channel material. Zirconium is usually[2] the material of choice for today's thermal nuclear reactors.

[2] The exceptions are: fast breeder reactors (where stainless steel is used), gas-cooled Magnox reactors (magnesium alloy for cladding) and in low-power water-cooled research/training/materials reactors (1XXX series aluminum as cladding).

Table 16.1. Approximate neutron economy and ultimate tensile strength of different base metals. The combination of these two factors, makes Zr the ideal structural material for fuel channel in thermal nuclear reactor.

Base metal	Neutron economy: Macroscopic thermal neutron cross-section (cm^{-1})	Ultimate tensile strength (MPa) at 300°C
Ni	0.31	1100
Ti	0.26	1000
Fe	0.17	1100
Al	0.014	90
Mg	0.005	90
Be	0.001	350
Zr	0.01	900

For individual components of the fuel channel, naturally, specific properties are important and different types of Zr alloys are used. Other specialized applications of Zr such as chemical applications in corrosive environment, bio-medical applications such as (medical implants and surgical tools) and metallurgical applications (alloying additions in steel, super-conducting magnets in conjecture with Nb) are over-shadowed by its usage in nuclear industry.

16.1.2 Zirconium and its alloys

The cost and the production of Zr has evolved since the beginning of the nuclear age. For example, in 1943 the production of Zr was in kgs and the cost approximated 1500 US$/kg, while in 1950s the production was already approaching 1 million ton and the cost approximated 10–18 US$/kg. The industry has also fast adapted to the property requirements – an issue discussed in the next section. In this section, three aspects of zirconium metallurgy are covered – processing of Zr from its ore, Zr alloys in service and phase transformations in Zr alloys.

16.1.2.1 Processing of Zr ingots/billets from its ore. Earth's crust contains approximately 0.02 wt% Zr, mostly as Zircon ($ZrSiO_4$) and also as Baddeleyit (ZrO_2). The processing of Zr (Zr-sponge) from its ore (Northwood [1985], Sundaram [1986], Ganguly [2002]) is shown schematically in Figure 16.2. The most important step in the Zr process metallurgy is the Kroll process: established first by Dr. W.J. Kroll at the Albany Research Center of the US Bureau of Mines in the late 1940s and it continues to be employed for producing reactor-grade Zr. Kroll process, also known as 'magnesio-thermic reduction process', involves the reduction of zirconium chloride vapour with molten magnesium. Other than the Kroll process, the ore is also subjected to a sequence involving chlorination and chloride

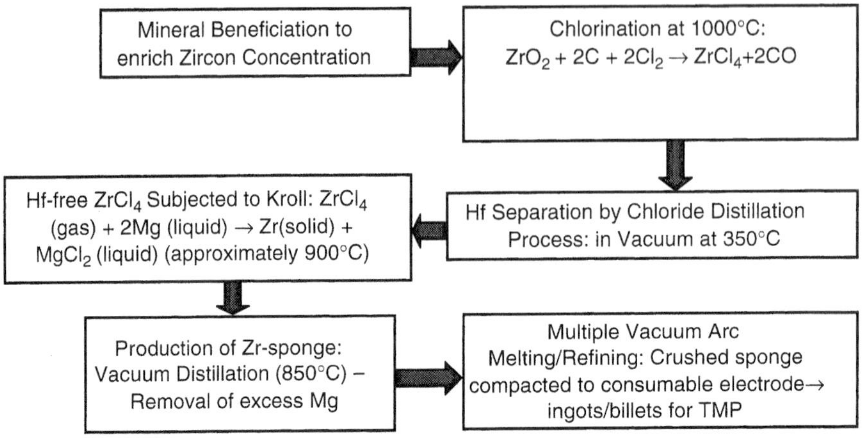

Figure 16.2. Flow sheet showing preparation of Zr ingots/billets from its ore.

distillation, mainly to remove hafnium (Hf) – a step crucial for improvement in neutron economy, and vacuum distillation, to produce Zr-sponge. Subsequently, the sponge is crushed to size, compacted to consumable electrodes, after incorporating alloy additions, and then vacuum arc melted in water-cooled copper moulds. Typically the vacuum arc melting/refining is conducted twice, to ensure compositional homogeneity and absence of undesirable tramps. The arc-melted ingots constitute the starting billets for final TMP/fabrication stages for the production of a variety of structural components.

16.1.2.2 Zr alloys in service. It is important to point out that the attractive neutron economy of the Zr base metal (see Table 16.1), needs to be supplemented with dependable corrosion resistance and/or mechanical properties. To this purpose, a range of Zr alloys have been developed (Cheadle [1977], Sundaram [1986], Bhardwaj [2002], Ganguly [2002]). Part of this development, both in single and two-phase alloys, is given below:

- *Single phase*: Zircaloy-1, Zr with 2.5 wt% Sn (Sn being a stabilizer of hcp α), was developed for increased corrosion resistance over Kroll Zr. The long-term corrosion rate of Zircaloy-1 is, however, almost comparable to pure Zr. Development of Zircaloy-2 (with 1.5 wt% Sn, 0.12 wt% Fe, 0.10 wt% Cr and 0.05 wt% Ni) provided an alloy with far superior corrosion properties than the earlier Zircaloy-1. Zircaloy-4, where Ni was eliminated and Fe content was increased to 0.24 wt%, does not reduce the corrosion rates of Zircaloy-2, but does show a considerable drop in hydrogen absorption rates.

- *Two phase*: A two-phase structure (hcp α & bcc β) may enhance mechanical properties, and can improve corrosion properties (e.g. stable oxide layer – especially at high-fuel rod burn-ups) and hydrogen/deuterium absorption. Zr-2.5 wt% Nb (10–15% bcc β, rest hcp α) is a material of choice for PHWR pressure tubes. Zr-1 wt% Nb (3–5% β), the so-called 'M5 alloy' of French/Russian origin, is often used for fuel clads in boiling water reactor (BWR). Zr-2.5 wt% Nb with 0.5 wt% Cu, an alloy of improved resilience, has been developed for garter springs.

16.1.2.3 Phase transformation in Zr alloys. In terms of diversity of phase transformations, zirconium alloys can easily match and even out-class steel. The phase transformations of zirconium alloys, relevant to TMP and/or to properties, can be classified in the following categories:

- *α (hcp) to β (bcc) Transformation*: In pure Zr, this allotropic transformation happens at 862°C. The alloying elements can be classified as α (e.g. Al, Sn, O, N) and β (e.g. Nb, Cu, Fe, Cr, Ni, Mo, H) stabilizers – the former favour a peritectoid transformation, while the latter give rise to an eutectoid system (Williams [1970], McDonald [1971], Banerjee and Krishnan [1973], Carpenter [1985], Banerjee *et al.* [1997]).
- *Formation of martensite and Ω phase*: bcc β can be quenched to form hcp martensite in dilute Zr alloys – both lath and acicular martensite have been reported (Banerjee *et al.* [1997]). The Ω transformation, a displacive transformation, occurs when Zr with high concentration of β stabilizers is retained in a $\alpha+\beta$ mixture in a metastable equilibrium (Banerjee *et al.* [1997]). The Ω formation is detrimental to mechanical properties.
- *Formation of intermetallic phases*: Hard intermetallic phases (e.g. Zr_2(Fe, Ni) and Zr(Cr, $Fe)_2$ phases in Zircaloy-2) can form in 'single-phase' Zircaloys and can strongly influence the corrosion and mechanical properties during/subsequent TMP stages (Williams [1970], McDonald [1971], Banerjee [1973], Carpenter [1985], Banerjee *et al.* [1997]).
- *Formation of hydrides*: The life of different fuel channel components is often 'limited' by the formation of hydrides. In typical PHWR operation, the fuel bundles last about 1 year – a life restricted by burn-up, while the life of the pressure tube may range from 10 to 25 years. Ten-year lifetime is expected for older Zircaloy-2 tubes, while at least 2.5 times life is expected in newer Zr–2.5 wt% Nb pressure tubes – an improvement related to lower deuterium absorption of the two-phase alloy. Depending on the amount of hydrogen and on cooling rate, the type of hydrides may range from meta-stable γ (fct – face-centred tetragonal) phase to equilibrium δ (fcc – face-centred cubic) and ε (fct) phases (Weatherly [1981], Perovic *et al.* [1982], Northwood and Kosasih [1983], Dey and Banerjee [1984]).

Formation of γ hydride involves both invariant plane strain criterion and diffusion of hydrogen – the so-called 'bainitic or mixed mode of transformation'. δ Hydride formation, on the other hand, is favoured by slower cooling rates and by higher hydrogen content. Pure diffusional transformation as well as mixed mode (bainitic) transformation have been reported as the possible mechanism(s) of δ hydride formation.

16.1.3 Structure–property correlation in Zr alloys

Between 1943 and 1953, the neutron economy of Zr has improved nearly 580 times. This has been possible through elimination of Hf. Any new alloy design for Zr naturally needs to consider the effects of alloying on neutron economy (no straight relationship between neutron economy and composition/structure exists) and also on the mechanical and corrosion properties. Especially of relevance are mechanical/corrosion properties under reactor working conditions. This section attempts to bring out these properties and also tries to correlate structure–property.

16.1.3.1 Mechanical properties. The performance of Zr tubes and clads are naturally dictated by a range of relevant mechanical properties – ranging from simple tensile properties to the more complex fracture mechanics and creep behaviour (Lemaignan and Motta [1993]). The deformation modes are dominated by slip and/or twinning (Tenckhoff [1988]). Typical slip systems include prismatic $(10\bar{1}0)<\bar{1}2\bar{1}0>$, pyramidal $(10\bar{1}1)<11\bar{2}\bar{3}>$ and basal slip $(0001)<11\bar{2}2>$, and their relative activation can differ significantly with working temperature (Akhtar and Teghtsoonian [1971]). Twinning also can play a major role at low-temperature deformation. The role of twinning is, however, strongly dependent on alloy chemistry and grain size.

The extent of strengthening in Zr alloys is usually controlled through alloying, especially interstitial concentration (e.g. O and N), by controlling the grain size and by using cold-worked (but stress-relieved) structures (Lemaignan and Motta [1993]). Also significant are the nature and severity of deformation heterogeneities (Kiran Kumar [2006]) and strong orientation or texture dependence of the deformation behaviour.

In two-phase alloys, the deformation behaviour depends on the presence and continuity of the 2nd phase bcc β (Kiran Kumar *et al.* [2003, 2004]). An apparent continuity of the softer 2nd phase is reported and can lead to insignificant changes in deformation texture – an effect caused by the absence of effective macroscopic strain in the hcp α phase.

Creep, especially anisotropic creep behaviour, is also an important/relevant mechanical property for Zr-based alloys (Murty and Adams [1985], Matsuo [1987]).

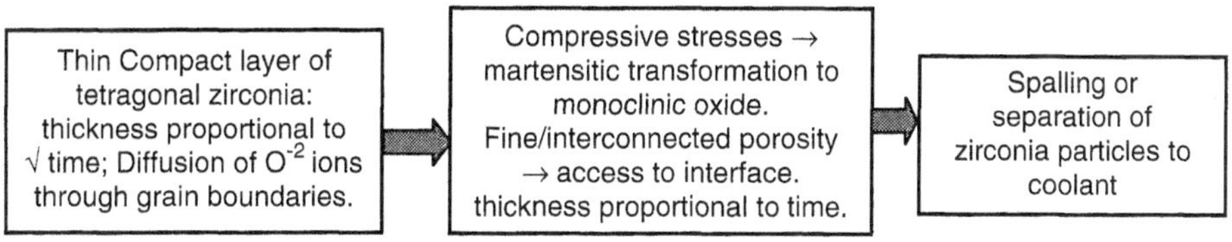

Figure 16.3. The general oxidation behaviour of Zr alloys.

Typical thermal creep rates for Zr alloys are high and even room-temperature grain boundary sliding has been reported at low strain rates. The creep mechanisms of Zr alloys are complex. Anisotropic creep response has been partially explained from crystallographic texture, but unresolved issues remain on the relative activation of slip systems, the grain shape effects and on the effects of dislocation substructure.

16.1.3.2 Corrosion properties. The effects of corrosion on Zr tubes/clads can be generalized as oxidation, hydride embrittlement, stress corrosion cracking and irradiation effects.

16.1.3.2.1 Oxidation. The general oxidation behaviour of Zr alloys (Lemaignan and Motta [1993]) is shown in Figure 16.3. As shown in the flowchart, the transition between a stable/protective oxide layer and a non-protective oxide, determines the oxidation behaviour and kinetics. The 2nd phase particles, intermetallics – see Section 16.1.2.3 – are normally expected to increase the corrosion resistance by corroding slowly. Other specific oxidation problems include nodular oxidation, in BWRs, where oxide nodules nucleate and grow. Though possible effects of microstructure on the formation of the oxide layer is widely acknowledged, the exact link between the two remains largely unknown. The oxidation problem is normally addressed through chemistry of the alloy.

16.1.3.2.2 Hydride embrittlement. The operational performance of Zr clads is largely restricted by problems associated with hydrogen absorption (Weatherly [1981], Perovic *et al.* [1982], Northwood and Kosasih [1983], Dey and Banerjee [1984]) – delayed hydride cracking (DHC), blistering, etc. Detrimental effects of hydrides are often indexed in terms of their morphological orientations. For example, in tubes, hydrides oriented radially would have the worst effects on mechanical properties under hoop loading (Hardie [1972]). Hence the F_n number, representing the fraction of radial hydrides, has usually been taken as a critical quality parameter. Normally, formation of hydrides is expected to be influenced by stresses, temperature gradients and by crystallographic texture – the latter is often argued to be related to the F_n number. A recent study (Mani [2006]), however, did show the predominantly

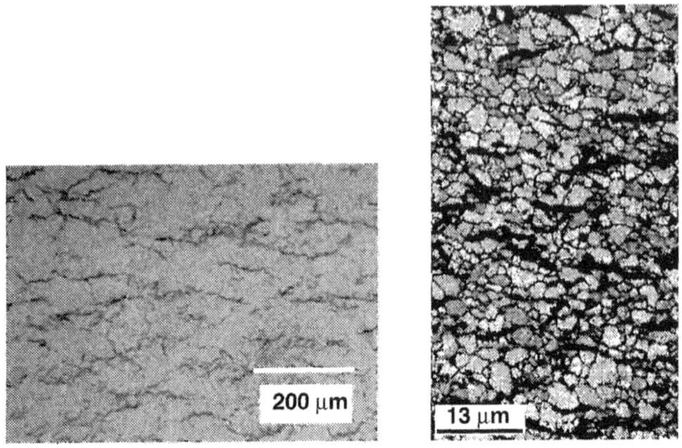

(a) (b)

Figure 16.4. (a) Optical image of δ hydrides in Zircaloy-2. (b) EBSD image of (a) – confidence index (or relative accuracy of indexing) has been used for mapping. The hydrides are clearly identifiable as 'dark features' along 'some' grain boundaries (Mani [2006]).

grain boundary nature of hydrides (see Figure 16.4). A clear preference did emerge between boundaries with or without visible presence of hydrides. Coincident site lattice (CSL) boundaries were, in general, resistant to hydriding. Elastically harder grains or orientations also arrested hydriding.

16.1.3.2.3 Stress corrosion cracking. Pellet-cladding interaction or interaction between UO_2 fuel pellets and Zr clad can lead to failures, especially at significant fuel burnup (Cox [1990]). The pellet-cladding interaction failures are of stress corrosion cracking type – failures through combined factors of stresses and corrosive environment. The stresses are developed through fast variations in linear heat generation rates (thermal power released per length of fuel rod – 15–25 kW/m and the corrosive environment, typically iodine (cesium and cadmium are also suspected), is created in the fuel rod as fission product). Pellet-cladding interaction, strongly affected by crystallographic texture – as cracks are reported to follow the basal planes (Lemaignan and Motta [1993]), is addressed by reducing hoof stresses through reduced reactor power change rates. Claddings have also been adopted to eliminate/reduce pellet-cladding interaction – graphite coating in CANDU fuel clads and co-extruded (pure recrystallized Zr as the internal surface) fuel rods in BWRs (Lemaignan and Motta [1993]).

16.1.3.3 Irradiation effects. Irradiation effects (Lemaignan and Motta [1993]), caused by neutron flux, can be detrimental to both mechanical and corrosion

Table 16.2. A summary of irradiation effects and damages typical of Zr alloys.

Irradiation effect or damage	Description
Zr matrix: Displacement of atoms, changes in dislocation structure and limited void creation	Typically 20 dpa (displacement per atom) in 3 years. Strong changes in nature and type of dislocations. Limited void formation, as compared to stainless steel. The increased defect concentration can lead to increase in strength (YS & UTS) and drop in ductility – changes which are rapidly saturated
2nd phase: Dissolution and reprecipitation; amorphization of precipitatiates	2nd phases (e.g. Zr2(Fe,Ni) and Zr(Cr,Fe)2 in Zircaloys) were reported to undergo such effects. These, in turn, may strongly affect corrosion properties
Irradiation growth: Growth under constant volume – unstressed material under irradiation	Partitioning of interstitial and vacancies to different sinks (e.g. dislocations, grain boundaries). Such sinks are anisotropically distributed – leading to anisotropic irradiation growth. In single crystal: expansion along *a*-axis and contraction along *c*-axis. A similar growth anisotropy has also been reported in polycrystalline commercial Zr alloys. The typical crystallographic texture of Zr tubes/clads usually causes irradiation growth along axial direction. Other than crystallographic texture, alloying additions can also control irradiation growth – low irradiation growth in Zr–1 wt% Nb
Irradiation creep: Slow deformation under irradiation and under external stresses	Affected by dislocation structure, grain shape and, in case of two-phase alloys, distribution of 2nd phase

properties. Table 16.2 tries to summarize the typical in-reactor irradiation effects and also outlines the irradiation damages, in-reactor damages typical in Zr alloys.

16.1.4 TMP of zirconium components

The earlier two sections have tried to bring out the 'metallurgy' of Zr alloys – especially the aspects of 'metallurgy' which are relevant to the TMP of Zr components. TMP should normally be based on two considerations – (I) engineering and (II) structure–property of the finished product. The former mainly deals with dimensional tolerances and absence of macroscopic defects, while the latter takes appropriate developments of microstructure in account. The Zr TMP is mainly designed to meet the engineering considerations, but is also calibrated for appropriate microstructural developments. This point is further highlighted in Table 16.3. As shown in the table, for both fuel clad and pressure tubes, the dimensional specifications are rigid and extremely demanding. Even the metallurgical specifications like yield strength (YS), ultimate tensile strength (UTS) and oxidation rates are rather well defined. On the other hand, specifications on microstructural

Table 16.3. Typical specifications for PHWR Zircaloy-4 fuel clad and Zr–2.5 wt% Nb pressure tube. Both dimensional and metallurgical specifications are given. After Ganguly [2002].

Zr component	Dimensional specifications	Metallugical specifications
Zircaloy-4 fuel clad	Outside diameter: 15.215 ± 0.063 mm; wall: 0.381 ± 0.076 mm; length: 485.8 ± 0.051 mm; straightness: 1 in 1200; surface finish: 0.8 μm (minimum); end squareness: 0.05 mm (maximum) low residual stress	YS: 393 MPa; UTS: 483 MPa; % elongation: 20%; hoof stress: 621 MPa; low residual stress; strong basal fibre in the radial direction; low dislocation density;
Zr–2.5 wt% Nb pressure tube	Outside diameter: 133.15 ± 0.3 mm; wall: 4.7 ± 0.15 mm; length: 6425 ± 0.5 mm; bow/bend: 0.25 mm in 300 mm	YS: above 324 MPa (300°C) and below 586 MPa (room temperature); UTS: above 462 MPa (300°C); % elongation: 14% (300°C); hydrostatic pressure test at 280 kg/cm^2 – no leakage; oxidation test at 400°C for 72 h: less than 35 mg/m^2 weight gain; low residual stress; strong basal fibre in the radial direction; *100% quality check*

indices like crystallographic texture, anisotropic nature of the residual stress are often loosely defined and qualitative. For example, traditionally crystallographic texture in Zr alloys are defined by Kern's factors,[3] which itself is semi-quantitative at best. To summarize, *TMP of Zr components is an area where 'microstructural engineering' has enormous potential. But in a typical TMP sequences, this potential is only appreciated partially.* The reasons for this partial appreciation are:

- Incomplete understanding of structure–property correlation, including understanding/appreciation of structural developments during TMP.
- For functional reactor designs, higher priority is given to dimensional tolerances.
- Often Zr components are over-designed for safety. Marginal property improvements through optimized TMP is shadowed by such over-designs.
- Any major changes in alloy design or in TMP sequence should be validated by actual in-reactors tests, which are expensive and enormously time-consuming.

These, on the other hand, make Zr TMP an evolving subject – subject with enormous potential for continuing research and development. The rest of this section

[3] This was developed (Kearns [1965]) for materials with hexagonal symmetry. It gives effective contribution of the [0001] basal poles to the overall texture. Kearns factor can be described as $\sum V_i \cos\phi_i$, where V_i is the basal pole strength at an angle ϕ_i to a specified reference direction.

deals with two specific TMP sequences – TMP sequences for Zircaloy-4 fuel clad and for Zr–2.5 wt% Nb pressure tube.

16.1.4.1 TMP sequence for Zircaloy-4 fuel clad. Figure 16.5 shows the typical TMP sequence of Zircaloy-4 fuel clad. The alloy is first β-quenched to establish a randomized texture and to get a uniform distribution of the intermetallic 2nd phases and segregated impurities. Copper jacketed[4] rounds are then hot extruded in the two-phase region. Extrusion parameters are selected to promote dynamic

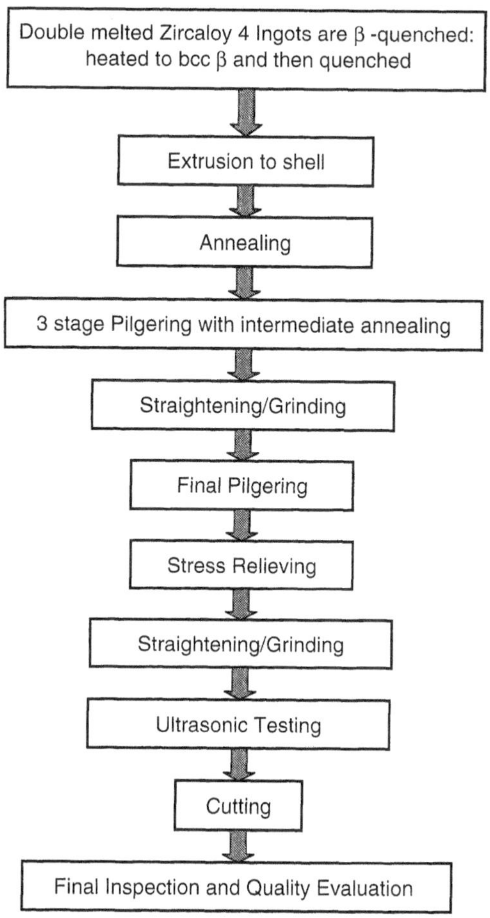

Figure 16.5. TMP sequence of Zircaloy-4 fuel clad. After Ganguly [2002].

[4] Copper jacketed to prevent oxidation. Copper jacket also provides lubrication during hot extrusion.

recrystallization and to prevent grain growth. Subsequent pilgering and intermediate annealing steps are designed to meet exact dimensional tolerances and also to maximize the concentration of basal poles in the radial direction. A final stress relieving is expected to reduce residual stress, but not to eliminate it completely. The final structure consists of elongated (partially recrystallized) grains of a narrow size range. As discussed earlier, the TMP meets the strict dimensional tolerances and the specified property requirements. Freedom from macroscopic TMP defects is often considered more vital than precise control of microstructures.

16.1.4.2 TMP sequence for Zr–2.5 wt% Nb pressure tube. There is nothing sacrosanct about a TMP sequence. Based on the availability of fabrication tools, engineering considerations and considerations of structure property, the TMP sequences can sometimes be modified. An 'example' is included to make this clearer (see Figure 16.6). As shown in the figure, the original CANDU route of processing

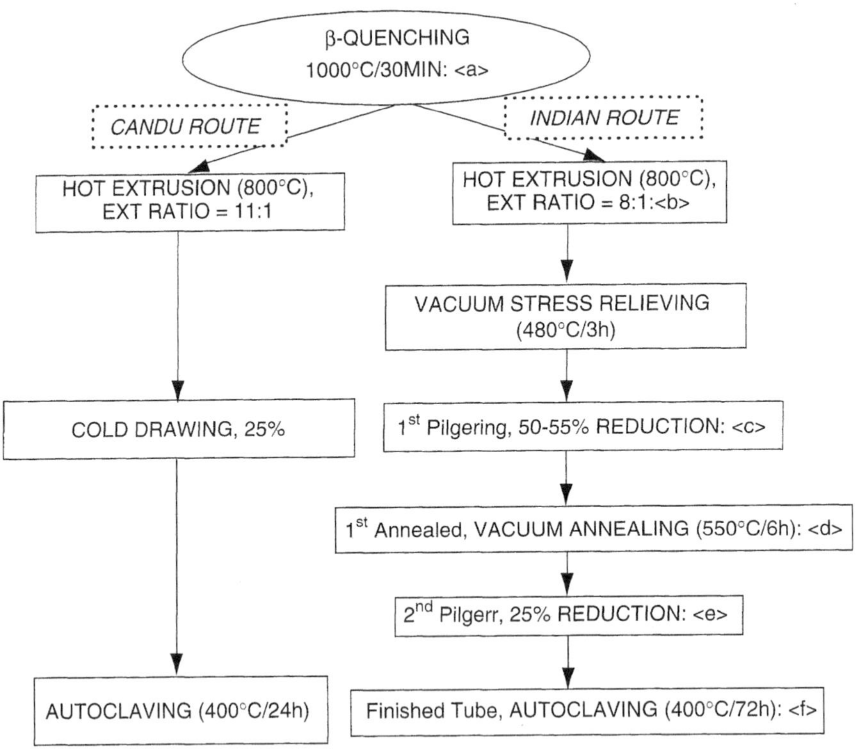

Figure 16.6. Flow sheet for original CANDU and the modified Indian TMP routes of Zr–2.5 wt% Nb pressure tube fabrication (Srivastava *et al.* [1995], Kiran Kumar *et al.* [2004]).

Zr–2.5 wt% Nb pressure tube was modified (Srivastava *et al.* [1995], Kiran Kumar *et al.* [2004]) – modifications adopted from both engineering and structure–property considerations. Some of these 'considerations are highlighted below:

- The two-stage cold pilgering process ensures a better control of the dimensional tolerances of the finished tubes.
- The structural variations along the tube, resulting from hot extrusion and an associated temperature gradient between the ends, are levelled off by additional cold working and annealing treatments.
- The variation in mechanical properties from batch to batch, as a result of variation in exact composition and hot working temperature, can be controlled through a proper selection of annealing temperature in subsequent steps.
- A lower extrusion ratio results in a lower aspect ratio of the hcp α grains thereby increasing the resistance to irradiation growth and reducing circumferential basal pole texture.

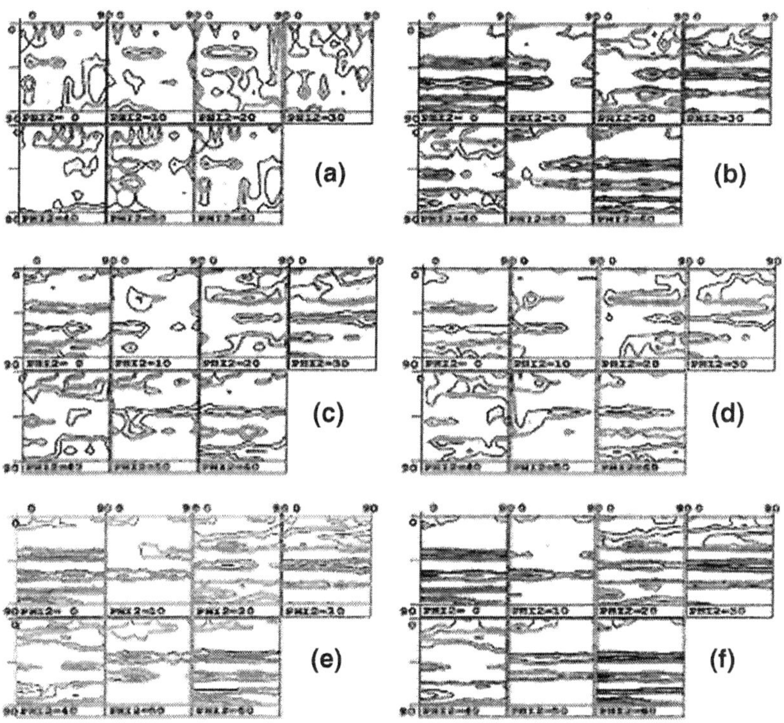

Figure 16.7. ODFs representing the crystallographic texture of the hcp α phase at the different TMP stages, stages as marked in Figure 16.6. Following contour levels (times random) are used for ODF plotting – 1.5, 2.5, 4.5, 5.5 and 8.7.

It is important to point out that in the original CANDU route, the hot extruded texture (hot extrusion parameters are selected from similar considerations as in Zircaloy-4) is literally carried forward, while the microstructure is subjected to minor refinement through cold drawing and autoclaving – the latter provides stress relief and also acts as a quality check for oxidation rate. In the modified route, however, texture changes are associated with each stages (<a> to <f> – as given in Figure 16.6). It is interesting to note that the TMP stages of cold pilgering and annealing are designed in such a way that the texture of the finished tube is identical to that of hot extruded material (see Figure 16.7). Such modifications, as in Figure 16.6, naturally takes its own toll of development time and efforts. But even then it cannot be considered as real 'optimum', at least in terms of an optimum microstructure. Extended knowledge base on hcp Zr may drive the present Zr TMP sequences to such limits – limits where 'microstructural engineering' can be tailored to exploit uncharted potentials.

LITERATURE

Ganguly C., Proceedings of the Symposium Zirconium-2002, P.K. Dey (ed.), BARC, Mumbai, India (2002), p. 1.
Mckay A., "The Making of Atomic Age", Oxford University Press UK (1984).
Northwood D.O. and Kosasih U., Intl. Metal Rev., 28 (1983), 92.

16.2. TITANIUM FORGINGS IN THE AEROSPACE INDUSTRY

16.2.1 Introduction

Titanium is a relatively new, light, metal whose commercial alloys have been developed since about 1950, mostly for high-technology aerospace applications. The relative density of titanium is 4.5, close to half that of steel, and many alloys possess mechanical properties equivalent to those of high-strength steels so that their specific strength to weight ratios are much higher than for steels. Also the material usually exhibits very good corrosion resistance so that it can be used in some very testing environments. Moreover, and in this case very much like steel, many titanium alloys undergo allotropic transformations, usually between the high-temperature bcc (β) and low-temperature hexagonal (α) phases so that a wide variety of microstructures and properties can be developed by heat treatments and TMP.

Titanium alloys are used in the aerospace industry for two major roles: firstly in modern jet engines, particularly in the low to intermediate temperature stages of the compressors – large by-pass engines have front fans of diameter up to 2 m requiring light, high-strength Ti solutions. Most of the other fans and discs in the subsequent moderate temperature sections of the engines are also made of Ti alloys for the same reasons. Secondly, Ti alloys are used extensively as structural members, particularly in the form of forged components for critically stressed parts such as undercarriage components and engine mountings. They have also found large applications in military aircraft for wing boxes and the basic aircraft structure. Electron beam-welded Ti wing sections were probably first developed for the famous F14 (Tomcat) fighter aircraft – and used for about 40 years. The latest fighter planes include about 30–35% by weight of titanium alloys compared with about 10–15% for civil aircraft. Finally, it is worth noting that high-strength Ti alloy fasteners are widely used for both types of aircraft; there are over a million fasteners in a large, modern plane.

These applications impose very demanding in-service conditions on the material which therefore has to possess very high mechanical properties. High-yield strengths are important but so are ductility, fracture toughness and fatigue resistance for the structural components. The same applies to jet engine parts with the additional requirement of creep resistance. These properties are closely linked to the microstructural transformations that occur during and after hot shaping. Ti alloys are often difficult to shape by hot deformation due to the high flow stresses, and reduced 'processing window' imposed by the phase transformations. As often, plastic deformation of the hexagonal phase is difficult due to texture and stability effects. This, together with the cost of the basic metal, leads to relatively high component costs which are being progressively reduced by the development of innovative shaping, forming and joining fabrication techniques.

16.2.2 Some physical metallurgy of Ti alloys

The physical metallurgy of Ti alloys is described in some detail by Lütjering and Williams [2003], Collins [1984] and Polmear [1995]. This section summarizes the essential features necessary to understand the transformations occurring during processing.

Pure titanium possess a hexagonal crystal structure (α-Ti) at room temperature which transforms on heating to the bcc β phase at 882°C; the latter phase is then stable up to the melting point of 1670°C. Ti can be alloyed with a large number of elements to stabilize either the α or β phases or, in many important cases, to generate two-phase α/β alloys. It also often undergoes a martensitic transformation on cooling from the β phase and other more exotic transformations are possible.

Important elements that dissolve preferentially in the α phase and therefore stabilize it are Al and O. The opposite effect is obtained by Mo and V additions which strongly stabilize the β phase; sometimes when in sufficiently high concentrations, 100% β can be retained down to room temperature. Other common alloying elements such as Zr, Sn and Si are neutral with respect to phase stability but are often added for solution and sometimes precipitation strengthening.

Two schematic phase diagrams are given for the α and β type alloys in Figure 16.8. In the case of the α-stabilized alloys rich in Al, an ordering reaction can occur to form a Ti_3X compound (α_2) which has a deleterious effect on ductility; this limits the amount of Al added to industrial alloys. The α/β transus temperature is clearly controlled by the solute additions, particularly for the β stabilizers. According to the temperature and the composition, a wide variety of α/β combinations can be developed. Finally, and as indicated by the Ms line, the β phase can transform to martensite (α') on rapid cooling. Note that the M_s and M_f temperatures are very close to each other.

Table 16.4 gives some common alloy compositions together with their typical room temperature tensile properties (in part from Polmear [1995]).

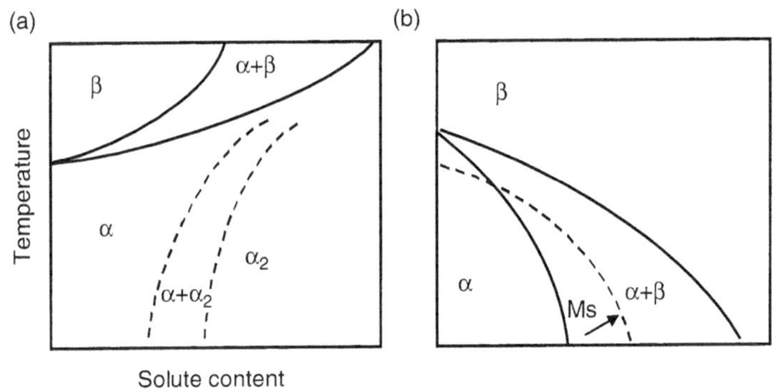

Figure 16.8. Schematic phase diagrams for Ti alloys: (a) α stabilizing solutes (the dotted lines are for the case of Ti–Al) and (b) for β stabilizers showing a typical Ms line.

Table 16.4. Composition and properties of some selected Ti alloys.

Alloy designation	Al	Sn	Zr	Mo	V	Other	Yield stress (MPa)	Tensile stress (MPa)	El (%)	Type
Ti-685	6		5	0.5		0.2Si	900	1050	10	Near α
Ti-6-2-4-2	6	2	4	2		0.2Si	960	1030	15	Near α
Ti-6-2-4-6	6	2	4	6			1170	1270	10	α/β
Ti-6-4 (anneal)	6				4		925	990	14	α/β
Ti-6-4 (aged)	6				4		1100	1170	10	α/β
Ti-17	5	2	2	4		4Cr	1060	1150	6	Near β
10-2-3	3				10	2Fe	1250	1320	8	Near β

16.2.3 Hot working conditions

The alloys are generally cast as cylindrical ingots after consumable arc-melting. Since, in the as-cast state they are susceptible to cracking, hot working is usually carried out in two stages. The β and α/β alloys are first pre-forged, typically by press forging in open dies, at relatively low strain rates and at a temperature above the β transus (1000–1150°C). This first deformation in the β phase takes advantage of the relatively low flow stresses and the good formability of the bcc structure to break up the coarse, brittle as-cast structure. However, significant grain growth can occur in the single-phase field and cooling from the β phase creates a coarse, acicular structure which has to be refined by subsequent processing.

It is then standard practice to hot forge the billets in the two-phase region to get rid of the acicular structure. For the α/β alloys, this is a critical step but requires much higher forging pressures because of the difficulty of deforming the two-phase material. On reheating to the $\alpha+\beta$ domain, the coarse α develops β grains which, by solute partitioning, contain more β stabilizing solutes and thereby deplete the α phase. This mixture is hot deformed to large strains, of order 1.5, at a temperature about 50°C below the β transus so breaking up the α needles into a fine, more equiaxed, morphology. This material can then be recrystallization annealed in the $\alpha+\beta$ field to develop, after cooling, a structure composed of fine equiaxed (or spheroidized) α grains and transformed β. This state is used for the final shaping operations which are often carried out by closed-die forging.

Hot transformation of the α and near α alloys is more difficult since the hexagonal crystal structure is intrinsically less ductile (more texture-dependent and susceptible to plastic instabilities and shear banding). Also when forged in the more workable $\alpha+\beta$ field the temperature range is relatively narrow, so temperature control is critical. Note that Ti possesses a relatively low coefficient of thermal conductivity so that large strain plastic deformation can generate significant temperature increases which are not dissipated rapidly.

As shown by Figure 16.10 the flow stresses are very sensitive to structure, temperature and strain rate. Thus, even for small temperature differences below the forging regime, the flow stresses can rise very steeply with increasing strain rate and decreasing temperature. Note also the lower flow stresses and wider forging temperature window of the near beta Ti-10-2-3 compared with the near alpha Ti-834 (see Figure 16.9).

It can be seen in Table 16.5 that in practice, the forging temperature often covers both the β and $\alpha+\beta$ fields. In many cases, hot working starts in the β phase and is carried out through the transus with most of the deformation being accomplished in the nominally two-phase region. (Given the time to complete the $\beta \rightarrow \alpha+\beta$ transformation in practical closed-die forging most of the metal is in fact forged in the β state.)

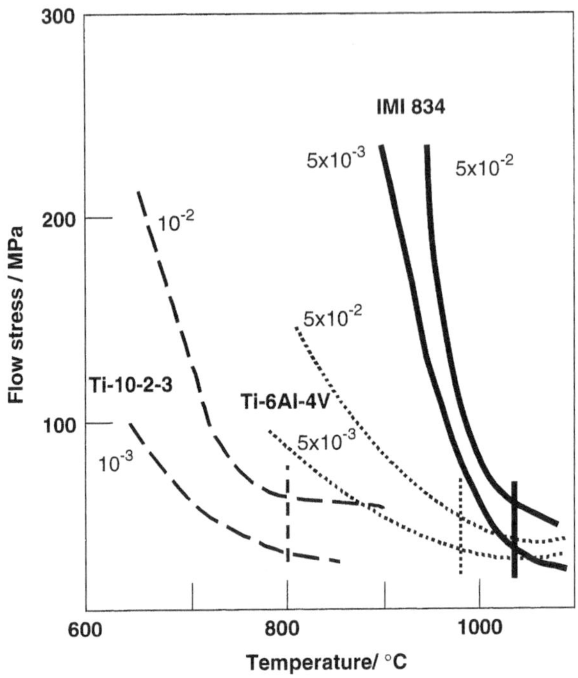

Figure 16.9. Flow stresses of three typical alloys as a function of temperature for two strain rates (from Flower [1990]) Ti-10-2-3 is near-β, Ti–6Al–4V is equiaxed $\alpha+\beta$ and IMI 834 is a near-α alloy (base Ti–5.5Al–4Sn–3.5Zr). The vertical dashed lines indicate the β transus temperature for each alloy. Permission obtained from Maney.

In all cases, the forging requires good lubrication to avoid galling (surface damage during working). The usual lubricants are either graphite suspensions or molybdenum disulphide, but glass lubricants can be necessary for severe high-temperature processing. It should also be noted that Ti and its alloys easily pick up oxygen and hydrogen at high temperatures. Oxygen leads to scaling and eventually subsurface cracks so that descaling or machining is necessary. Hydrogen pick-up is avoided by preferring electrical pre-heating furnaces to oil or gas fired furnaces.

16.2.4 General α/β alloys

The most widely used alloy is Ti-6-4, which accounts for about 50% of all Ti alloy production in the West. It is a general-purpose alloy which can be forged, rolled and extruded to a wide variety of components, with closed-die forgings to make fan blades being a typical example.

The alloy can be shape forged and finally annealed in the two-phase ($\alpha+\beta$) field or the single β-phase region; the resulting microstructures and properties

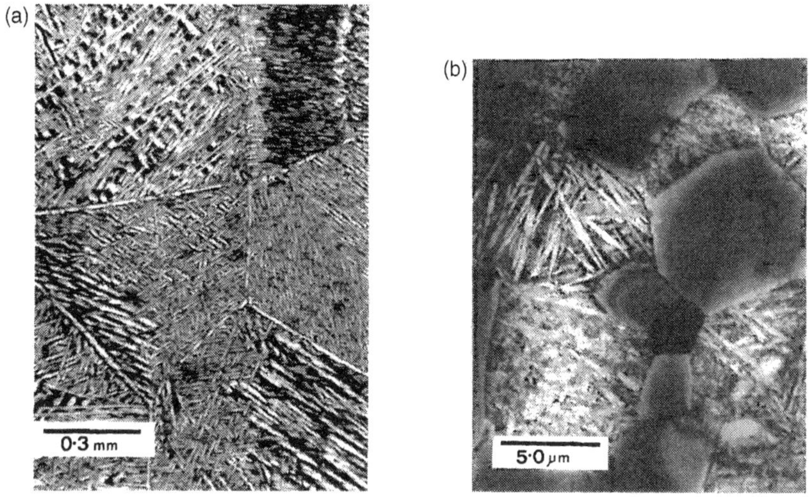

Figure 16.10. Typical microstructures of α-rich Ti alloys (a) the near-α Ti-685 heat treated in the β field (1 h at 1080°C), air cooled and aged to form coarse Widmansttaten α platelets and (b) Ti-6-4 treated in the (α+β) field (950°C 5 h and water quenched) to form primary equiaxed α and fine transformed β. Taken from Hammond and Nutting [1977]. Permission obtained from Maney.

Table 16.5. Typical forging conditions of some Ti alloys. Courtesy of J.-F. Uginet, Aubert & Duval, Pamiers.

Alloy designation	Forgeability	Transus temperature (°C)	Forging domain	Forging temperatures (°C)
Ti-685	xxx	1025	β	1050–950
Ti-6-2-4-2	xx	995	α+β, β	970–850, 1050–1000
Ti-6-2-4-6	xx	945	β	970–850
Ti-6-4	xxx	995	α+β, β	960–850, 1030–1000
Ti-17	xx	890	β	920–800
10-2-3	xxx	800	β or α+β	850–720

are significantly different. Slow to intermediate cooling from the β phase (β annealed) develops a basket weave Widmanstätten α structure, e.g. Figure 16.10a. Hot working in the two phase domain is often, though not necessarily, carried out at a temperature such that the two phases are in roughly equal proportions. After a sufficiently large strain, the acicular α is fragmented and can then be annealed to form equiaxed α. The amount of strain required to break up the initially acicular α (>1 or 1.5) depends upon the thickness of the α lamellae (Weiss [1986]). Also, since there are two phases of comparable volume fractions very little grain growth occurs during post-deformation recrystallization (Zener drag of one phase inhibits growth of the other). After a recrystallization annealing (mill annealed) and subsequent cooling, the structure of Figure 16.10b is produced; it is composed of fine equiaxed α and transformed β, the latter as fine needle-like Widmanstätten α with some retained β. This structure is much finer than that of the β processed alloy. In some cases, the grain size is sufficiently fine (about 10 μm) to allow superplastic deformation and forming (Chapter 6). Both types of structure respond to aging at about 500°C to moderately increase the yield stress.

Fig 16-10a shows the type of coarse basket-weave structure developed after high-temperature annealing in the β field followed by slow cooling to α. The typical microstructure obtained in Ti-6-4 by annealing after hot forging in the $(\alpha+\beta)$ field to develop equiaxed α and β is illustrated in Figure 16.10b; most of the β transforms on cooling to fine Widmanstätten α but about 10–20% can be retained. The exact amount of retained β depends on the solute partitioning that occurs during the $(\alpha+\beta)$ anneal. At lower annealing temperatures, more solute β stabilizers go into the β phase and therefore more β is retained at room temperature.

The primary α phase can therefore be closely controlled in terms of both its morphology (by the quantity of hot deformation) and its relative amount and composition (by the annealing temperature inside the two-phase region). It is generally accepted that the Widmansttäten α structure gives significantly better resistance to crack *propagation* than the more equiaxed α/β structures. Thus the β forged and annealed (2 h at 705°C) alloy possesses a tenacity of 79 MPa$\sqrt{m}$ and a 10^7 cycles to fatigue failure limit of 744 MPa compared with values of 52 MPa$\sqrt{m}$ (tenacity) and 494 MPa (fatigue failure limit) for the $(\alpha+\beta)$ forged and annealed material. Many studies, e.g. Ruppen and McEvily, quoted by Polmear [1995], have shown that this is a consequence of the localized crack branching that occurs at a crack tip in the α-Widmansttäten structure, particularly at low stress intensities; cracks dissipate more mechanical energy by multiple branching in the α lamellae than in equiaxed grains. On the other hand, the $(\alpha+\beta)$ structure is in general more ductile than the Widmansttäten structure and more resistant to crack *initiation*. For example, the low-cycle fatigue behaviour of most alloys is controlled by the number of cycles to plastically induced crack initiation. Consequently, low-cycle fatigue lives of the $(\alpha+\beta)$ mill annealed state are 10–15% above those of the β-annealed Ti-6-4. To optimize for medium temperature, rotating components requiring both types of fatigue and some creep resisitance, it is common practice to aim for a duplex structure containing about 30% primary, equi-axed α and the rest as Widmansttäten α. Some practical annealing conditions are given in Table 16.6.

The Ti-6-2-4-6 is a higher strength α/β alloy used for more specific applications such as helicopter turbines. It is forged in the β field, taking advantage of the reduced flow stresses, to develop a Widmansttäten structure and the best compromise for yield strength, fatigue and creep resistance.

Table 16.6. Annealing conditions of Ti-6-4. Courtesy of J.-F. Uginet, Aubert & Duval, Pamiers.

Annealing conditions	Temperature range	Microstructure
$(\alpha+\beta)$ Anneal	700–730°C	Stress relief, and structure stabilization
Solutionizing and ageing	About 950°C, quench + ageing 700°C relief	High temperature controls the amount of α and ageing for stress
Recrystallization anneal	About 970°C + slow cool	Growth of alpha grains

16.2.5 Near-α alloys

The near-α alloys such as Ti 6-2-4-2 and 685 are usually employed for their higher creep resistance. Despite a high melting temperature, most Ti alloys have a rather disappointing resistance to the prolonged application of stresses at elevated temperatures. The above α/β alloys such as Ti-6-4 are limited to in-service temperatures of about 350°C, but the standard near-α alloys are creep resistant up to about 500 or 550°C – hence their principal role in the components of early stages of jet engines. The creep resistance is due in part to the compact hexagonal crystal structure (and the quasi-absence of non-compact β) but also to the additions of alloying elements such as Si which precipitate out during ageing treatments (and the alloy designation is then rigorously given as Ti 6-2-4-2 S, for example).

These alloys are hot forged in the $\alpha+\beta$ field or the β phase (see Table 16.5) but transform on cooling to almost fully α or α + martensite α' (with a maximum of 2% β). Martensite is obtained after rapid quenching but, since there is no supersaturation of the solute atoms, it only slightly improves the room-temperature mechanical properties. Martensite often leads to inferior creep resistance so that low cooling rates to alpha are usually preferred for these alloys.

As seen in Figure 16.9, the flow stresses of the near-α alloys are significantly higher than those of the other alloy types and forging has to be carried out at much higher temperatures. Also the acicular structures developed in cooled and transformed β are very sensitive to localized plastic deformation by shear banding as analysed in Ti-6-2-4-2 by Semiatin and Lahoti [1981]. This shear banding gives rise to large degrees of flow softening as a result of both adiabatic heating and microstructural changes in the highly sheared acicular matrix. The resulting structure is very heterogeneous. These plastic instability effects are reduced by forging in the $\alpha+\beta$ field which is microstructurally more stable.

The final heat treatment depends on the alloy and its applications. Thus the Ti-685 is usually solution treated in the β field followed by a stabilization anneal in the range 500–590°C to give an acicular α structure. As noted above, it is the latter which possesses very good toughness and creep resistance. On the other hand, the 6-2-4-2 S and the IMI 834 alloys are solutionized in the upper end of the $\alpha+\beta$ field to optimize for both tensile and creep properties.

16.2.6 β and near-β alloys

Recent advances in Ti alloys include the developments of β and near-β alloys which combine the advantages of improved formability with high room temperature yield strength. The alloys develop their strength through a combination of solid solution hardening, cold working and 2nd phase precipitation in the β matrix leading to yield strengths up to 1500 MPa. This class of alloy includes the highest strength to weight ratios of all Ti alloys but initially they are more expensive and therefore their use is restricted to some critical components.

These alloys contain sufficient β stabilizing elements to almost completely stabilize the β phase but insufficient for complete stabilization (termed metastable β alloys). Note that they contain enough β stabilizers to avoid forming martensite on quenching; they therefore lie to the right of the Ms line in Figure 16.8 but to the left of the pure β field at ambient temperatures. They can be cooled down from say, the β forging conditions to β at room temperature, cold formed in a relatively soft bcc condition and aged to precipitate α for final hardening. This class of alloys also includes the β rich α/β alloys which, as opposed to Ti-6-4, contain a majority of the β phase and are forged in the $(\alpha+\beta)$ field.

The β metastable Ti-17 alloy is used for fan discs in the compressor stages of jet engines (service $T<425°C$). The processing route (Uginet [1994]) includes

- upset (or closed die) forging at 840°C ($\alpha+\beta$) and air cooling,
- closed-die forging at about 910°C (β transus + 30°C) then water quench,
- machining,
- solution treatment 800°C + water quench,
- ageing for 8 h at about 600°C and air cool.

The near-β Ti-10-2-3 alloy combines high strength (UTS 1200–1300 MPa) with very good toughness (40–80 MPa$\sqrt{m}$) and fatigue strength (700 MPa fatigue limit). It is mostly used for structural components and has replaced steel for the critical landing gear components of the Boeing 777 with a weight saving of 270 kg. The processing route is particularly well controlled so that from an initial $\alpha+\beta$ billet one has to

- β forge (β transus + 45°C) + water quench,
- closed-die forge (about 20% finishing strain) at the β transus − 50°C + air cool,
- solution treat at 760°C + water quench,

- machine and assemble,
- age according to yield strength and toughness goals, e.g. 510–515°C.

ACKNOWLEDGEMENTS

The author acknowledges Jean-François Uginet of the Aubert and Duval Research Centre, Pamiers for data and pertinent comments.

LITERATURE

The series of Conference Proceedings "Proceedings of the World Conference on Titanium" starting in 1970, then denoted Titanium '92, etc.

Chapter 17

New Technologies

Chapter 17

New Technologies

17.1. SUBMICRON MATERIALS BY SEVERE PLASTIC DEFORMATION

17.1.1 Introduction

Since the pioneering work of Hall [1951] and Petch [1953], material scientists and engineers have been attracted by materials with small grain sizes. A finer grain size increases the strength and the fracture toughness of the material and provides the potential for superplastic deformation at moderate temperatures and high strain rates (cf. Chapter 6). In principle, fine grains can be obtained by classical thermo-mechanical treatments with careful control of the (static) recrystallization parameters or by controlled phase transformation. This leads in general to a grain size above 10 μm or, exceptionally, a few microns in diameter. However, several techniques to obtain submicron or nanosize grains are now available, e.g. vapour deposition, high-energy ball milling, fast solidification and severe plastic deformation (SPD). In the present chapter, we will explore the achievements and possibilities of SPD. This technique uses the natural development of a substructure during warm or cold deformation. After small to moderate deformation, cells or subgrains usually have sizes in the submicron range, but are separated by low-angle boundaries (in the range 1–10°). After large deformation ($\varepsilon > 3$), part of the former cells or subgrains have developed a high-angle boundary and can be considered as 'grains'.

17.1.2 Geometrical dynamic recrystallization

When a material is deformed at elevated temperature and when no 'classical' dynamic recrystallization occurs, subgrains are formed within the original grains (cf. Chapter 5). The fraction of high-angle grain boundaries will increase during straining, mainly because of two effects. The first is a pure geometrical effect due to the increase of the surface/volume ratio. The second reason is that the original high-angle boundaries become serrated due to the surface tension of the intersecting low-angle boundaries (Figure 17.1). When the serrated boundary spacing is reduced to about twice the size of a subgrain, the serrations touch each other and 'grain impingement' occurs. This process is called 'geometrical dynamic recrystallization' (McQueen *et al.* [1985]) and is one of the possible mechanisms for 'continuous recrystallization'. The size of the recrystallized fine grains is in principle equal to the size of the subgrains. The latter can be reduced by lower deformation temperatures and higher strain rates (higher Z). In practice there are, however, limits to the

Z values. For the same distribution of grain sizes (with dimensions in the normal direction between t_{min} and t_{max} as shown in Figure 17.2), a decrease of the subgrain size d, due to higher strain rate, will increase the strain needed to start grain impingement. When the strain rate is too high, this critical strain will never be reached. When the temperature is too low, the substructure will in general not be

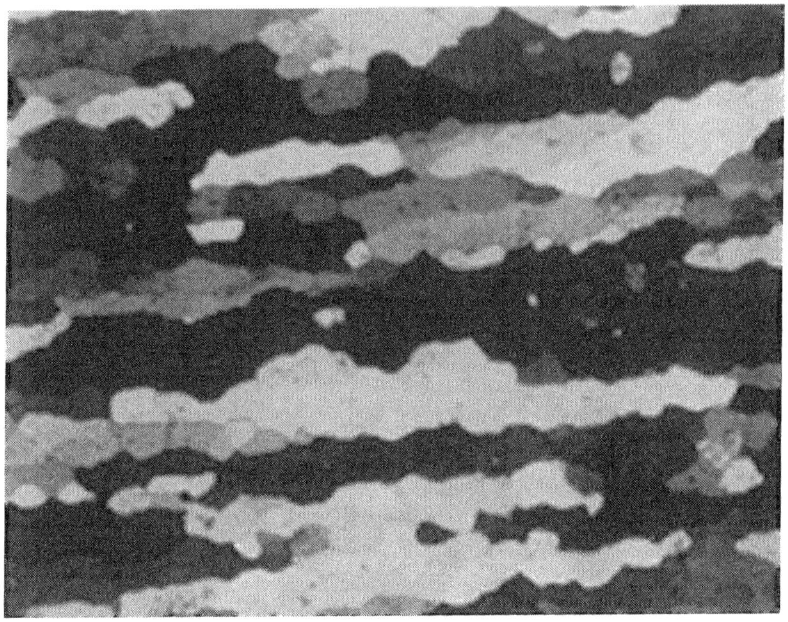

Figure 17.1. Microstructure of an Al alloy AA1050 deformed in plane strain compression at 500°C and a strain rate of 1/s; the grain boundaries are serrated.

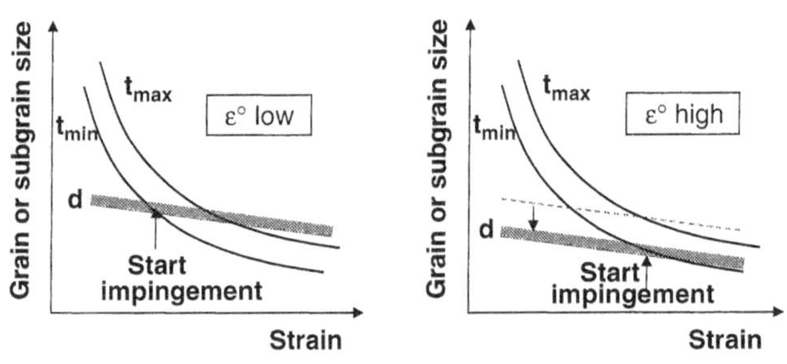

Figure 17.2. Schematic illustration of the influence of strain rate on the subgrain size and the consequences for the start of grain impingement. t_{min} and t_{max} are the minimum and maximum thicknesses of the 'original' grains measured perpendicular to the rolling plane, and d is the subgrain size, both as a function of strain.

equiaxed, and no serrations (no impingement) will occur because of the reduced mobility of the grain boundaries (Humphreys *et al.* [2001a]).

It has been shown for a Al–2Mg (Cr, Fe) alloy that with a conventional rolling mill, grain sizes of 1–2 μm could be obtained, in a 'processing window' between 200 and 400°C (Gholinia *et al.* [2000a]). For higher Mg contents, the processing window is smaller because of the reduced mobility of the grain boundaries. Grains in the submicron range seems to be out of reach by this technique.

17.1.3 Severe plastic deformation

17.1.3.1 Main methods for SPD. Probably the oldest, and certainly the most popular SPD technique at the moment, is *Equal Channel Angular Pressing (ECAP)*, also known as "Equal Channel Angular Extrusion" (ECAE). This technique was originally developed by Segal *et al.* [1981] and Segal [1995]. A billet of the test material is pressed through a die consisting of two channels with identical cross sections, intersecting at an angle ϕ, usually $60° < \phi < 135°$ and often $\phi = 90°$ (Figure 17.3a). Some dies have a rounded corner with angle ψ, others have $\psi = 0$.

The deformation occurs by simple shear parallel to the intersecting plane of the channels. In theory, the shear is concentrated in a narrow band around this plane. However, FE simulations (Kim *et al.* [2000, 2002a]) and experiments with inscribed grids (Gholinia *et al.* [2002a]) have shown that in practice the shear is spread over a deformation zone. The deformation is plane strain, since the strain

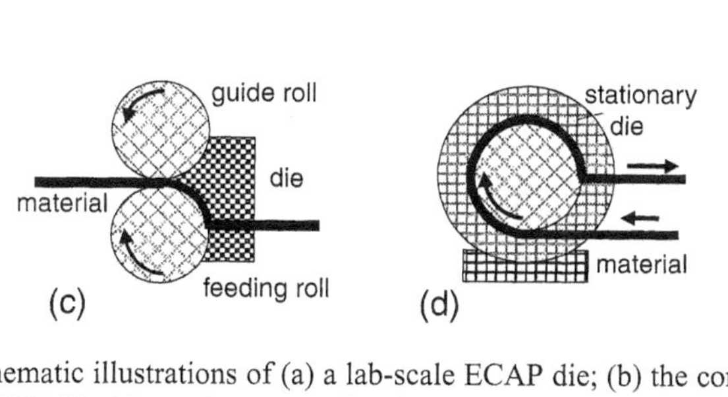

Figure 17.3. Schematic illustrations of (a) a lab-scale ECAP die; (b) the conshearing process (after Saito *et al.* [2000]); (c) continuous confined strip shearing (after Lee *et al.* [2002a]) and (d) the ECAP-conform set-up (after Raab *et al.* [2004]).

in the TD direction is zero. The equivalent strain per pass depends on the angles ϕ and ψ (Iwahashi *et al.* [1996]):

$$\varepsilon_{eq} = 1/\sqrt{3}[2\cot(\psi/2+\phi/2)+\psi\,\mathrm{cosec}(\psi/2+\phi/2)] \qquad (17.1)$$

For $\psi = 0$ and $\phi = 90°$, $\varepsilon_{eq} = 1.155$. Because the channels have an identical cross-section, the dimensions of the billet remain unchanged, and the process can, in principle, be repeated any number of times. If the cross-section has a 4-fold symmetry axis, the sample can be rotated by 90° or 180° about the extrusion direction (ED) before reinsertion in the die. Based on the regular repetition of such a rotation φ around ED, four commonly applied routes have been defined. For route A, φ is 0° (no rotation), for route C it is 180°, for route B_C it is 90°, and for route B_A φ is alternatively $+90°$ and $-90°$. Note that via route C each volume element is restored to its initial shape after every $2n$ passes and for route B_C after every $4n$ passes, where n is an integer. From the point of view of dislocation activity, there is always a change of strain path between 2 ECAP passes. The nature of this change depends on the die angle ϕ and the rotation angle φ. The effect on microstructure and mechanical properties of the four routes, imposing different grain shape changes and different changes of strain path, will be discussed in next section. In general, one can say that ECAP yields grain sizes of 300–500 nm.

In spite of the actual popularity of the technique, some drawbacks of ECAP should be recognized. ECAP is a discontinuous process with limitations in up-scaling potential. Moreover, the volume fraction of useful material (with uniform microstructure and without cracks) can be rather low because only the portion of the billet that has passed through the shear zone, will receive the desired deformation and grain refinement. Barber *et al.* figured out that for a sample with square cross-section and an aspect ratio of 6, after 8 ECAP passes route A, only ~30% of the material is fully worked as intended. For route B, the efficiency is ~45% and for route C it is ~83% (Barber *et al.* [2004]).

To increase the efficiency of ECAP, a number of continuous processes have recently been proposed. They combine the concept of ECAP with classical rolling. In the *conshearing* process (Saito *et al.* [2000]), a large feeding roll and a number of smaller satellite rolls push a sheet into an ECAP die (Figure 17.3b). Guide shoes in between the satellite rolls prevent buckling of the sheet. In *continuous confined strip shearing* (C2S2) (Lee *et al.* [2002a]), a specially designed feeding roll and a guide roll are used to feed a metal strip into the ECAP channel (Figure 17.3c). The initial sheet thickness is slightly reduced by the two rolls, but regains its initial thickness in the outlet of the ECAP die. In the *ECAP-conform* set-up, a groove in a central rotating shaft contains the workpiece (Figure 17.3d). Owing to frictional forces the workpiece is driven into a second channel, as in ECAP. All these continuous techniques are still in an experimental stage, but they open perspectives for a more efficient production of fine-grained materials.

Rolling mills can also be used without an ECAP attachment. In *accumulated roll bonding* (ARB) (Saito *et al.* [1998]), two sheets are stacked together and rolled simultaneously with 50% reduction. When the original contact surface has adequately been degreased and wire-brushed, the two sheets are bonded together (roll bonding). The new sheet is divided in two parts and the whole sequence is repeated. The alignment of the sheets relative to each other can be changed after each cycle. For various materials including steel, aluminium alloys, copper and nickel, grain refinement down to the submicron level has been reported (Tsuji *et al.* [2004]). Nevertheless, considerable technological difficulties in the production of ARB samples have also been noticed. Edge cracks are frequently observed and the thickness reduction of the strip can be inhomogeneous (Kestens *et al.* [2004]).

In *asymmetric rolling* the shear component on the sheet surface is enhanced as much as possible. This can be done using rolls with different diameters or rolls with equal diameter but different rotation speeds. Cui and Ohori [2000] obtained an average grain size of ~2 μm in high-purity aluminium after 90% reduction. The evolution of the structure during asymmetric rolling was attributed to the simultaneous action of compression and shear.

Other SPD techniques that have been proposed rely upon the repetitive bending and straightening of a sheet. Several variants of this method have been proposed, e.g. *repetitive corrugation and straightening* (Huang *et al.* [2001]), *constrained groove pressing* (Shin *et al.* [2002]) and *constrained groove rolling* (Lee and Park [2002b]). Finally, one important lab-scale technique should certainly be mentioned: *high-pressure torsion* (HPT). Small discs (typically 10–20 mm diameter and 0.2–1 mm thick) are strained in torsion under an applied pressure of several GPa. Although in a classical torsion test it is generally assumed that the centre of the sample is not deformed ($\gamma = 2\pi RN/l$ with R the distance from the axis, N the number of turns and l the sample length), numerous investigations show that after several rotations the HPT samples are uniform over the diameter (Valiev *et al.* [2000]), which shows that the real deformation in HPT is more complex than described by the simple analytical expressions used in a classical torsion test. The main advantage of HPT is that extreme grain refinement (up to 100 nm) can be obtained.

17.1.3.2 Microstructural development. Grain refinement by SPD implies the creation of new high-angle grain boundaries (HAB). This can be accomplished by three mechanisms (Gil Sevillano *et al.* [1980]). The first is the elongation of existing grains during plastic deformation, causing an increase in high-angle boundary area. The second is the creation of high-angle boundaries by grain subdivision mechanisms. Finally, an elongated grain can be split up by a localization phenomenon such as a shear band. The second mechanism is probably the most important one and merits some further explanations. The development of new boundaries inside an original grain by deformation has been known for several decades, probably since 1940 (Barrett and Levenson [1940]). It has been described qualitatively in earlier review papers e.g. Langford and Cohen [1969] and Gil Sevillano *et al.* [1980]. The possible mechanisms of formation have more recently been reviewed by Hughes and Hansen [1997]. Grain subdivision starts at low to medium strains when grains break up in cells and cell blocks (Bay *et al.* [1992]). With increasing strain, this substructure evolves towards a lamellar structure. During this process, new high-angle boundaries are generated. This happens by the simultaneous action of a microstructural and a texture mechanism (Hughes and Hansen [1997]). The former starts at low deformations and consist in the accumulation of dislocations in the cell- and cell block boundaries in which the misorientions gradually increase with increasing strain. Some boundaries remain low-angle boundaries but a significant fraction evolves into mediumhigh-angle boundaries mostly in the range 15–30°. The texture mechanism involves the rotation of different parts of a subdivided grain towards different end orientations. This can generate very high misorientations in the range 20–60°.

When the deformation is applied following a continuous strain path, it will result in a fibrous structure with a relatively high fraction of low-angle boundaries as is often found in cold-rolled sheets or in drawn wires. However, several factors can generate a break-up of this fibrous structure. A regular change in strain path, for example, seems to be very effective in generating a granular structure, with a mixture of HAB and LAB. Large non-deformable particles cause inhomogeneous deformation of the adjacent matrix and destroy the planarity of the boundaries. Finally, the formation of shear bands may cause significant displacements of the grain structure (Humphreys *et al.* [1999b]). In fact, it seems that in materials with low-stacking fault energy, the refinement of the microstructure is realized via the formation of shear bands (Valiev *et al.* [2000], Higashida and Morikawa [2004]). Most of the high-angle boundaries retained after SPD have a non-equilibrium character (Wang *et al.* [1996], Morris and Munoz-Morris [2002]). A post-SPD thermal treatment may be necessary for certain applications.

At elevated temperature grain, subdivision is less extensive and the inhomogeneous deformation around particles is smaller (Humphreys and Hatherly [1996]). New high-angle boundaries are formed by a process called 'geometrical dynamic recrystallization', discussed above.

17.1.3.3 Influence of strain path. In most SPD processes (except HPT), the deformation is applied with repetitive changes in strain path. Especially in ECAP, several processing routes are available. The implications on the sample

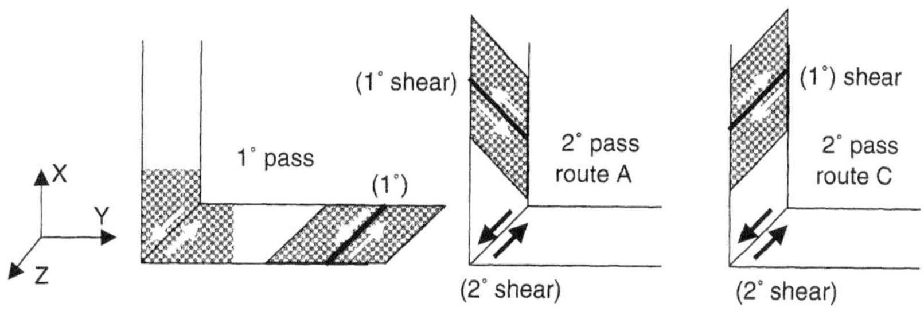

Figure 17.4. Interactions of subsequent shear deformations in the first and second ECAP pass.

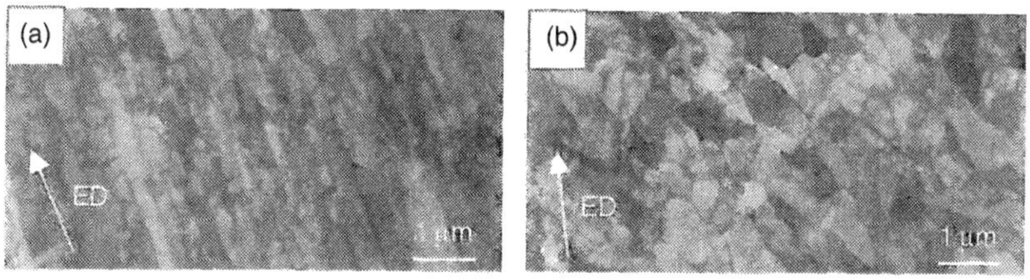

Figure 17.5. TEM micrographs in plane *XY* of Figure 17.4 of IF steel after 8 passes via (a) route A (b) route C. ED is the extrusion direction (De Messemaeker *et al.* [2004]).

distortion have been described in detail by Furukawa *et al.* [2001] and are illustrated in Figure 17.4. A cubic element in the initial billet, is elongated into a rhombohedral shape during the first ECAP pass. The elongation is visible in the (*XY*) plane but not in the (*YZ*) plane. The first shear plane is active as indicated in Fig 17.4a. When the second passage through the die is carried out without rotation (route A), a further elongation in plane (*XY*) occurs. A second shear plane, perpendicular to the first (when $\phi = 90°$) is now active (Figure 17.4b). In further passes, the 1st and 2nd shear planes are active alternatively and further elongation occurs in each pass. When route C is applied, (Figure 17.4c), deformation always occurs along the same shear plane, but alternating in shear direction. The shape of an initial cubic element is restored after each $2N$ passes, with N an integer. In route B, the cubic volume element is elongated in each orthogonal plane after the 2nd pass. For B_A this distortion will increase after each pass, but for route B_C the restoration of the cubic element is observed after each $4N$ passes. In routes B_A and B_C, shear occurs on shear planes intersecting at 120°. In spite of the dynamic character of the (sub)structure development, the choice of processing route has a definite influence on the microstructure. Figure 17.5 illustrates the elongated appearance of the microstructure after route A and the equiaxed nature after route C.

A further explanation for this has been provided by Dupuy and Rauch [2002]. They used the α parameter introduced by Schmitt *et al.* [1985] to quantitatively express the change of strain path from one pass to another:

$$\alpha = \frac{\dot{\varepsilon}_p \dot{\varepsilon}}{\| \dot{\varepsilon}_p \| \, \| \dot{\varepsilon} \|} \tag{17.2}$$

with $\dot{\varepsilon}$ and $\dot{\varepsilon}_p$ the strain rate tensors of the consecutive passes. For the most drastic 'orthogonal' change in strain path, $\alpha=0$. For a monotonic strain $\alpha=1$ and for strain reversal $\alpha=-1$. With the billet rotation in between two passes as a parameter, Dupuys found for different die angles of 90°, 120°(or 60°) and 135° (or 45°), the α-values shown in Figure 17.6. Route C corresponds in all cases to strain reversal. For route B, $-0.5<\alpha<-0.25$ or in all cases 'close to' orthogonal. For route A however, α is strongly dependant on the ECAP die angle ϕ. Dupuy assumed that the development of HAB during ECAP would be less efficient with more negative α-values. Several experimental results published in literature seem to be in agreement with Figure 17.6. For a die angle of 90°, most papers (e.g. Oh-Ishi *et al.* [1998], Langdon *et al.* [2000]) indicate that route B is more efficient than A and C, but it has also been reported that for a die with $\phi = 120°$ route A and B are nearly as effective but C is clearly less effective (Gholinia *et al.* [2000b]).

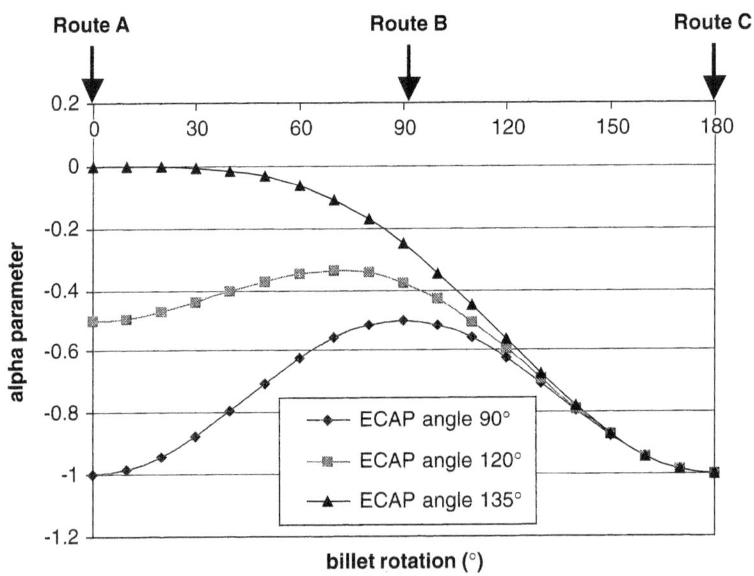

Figure 17.6. Influence of ECAP-billet rotation on the α-parameter defined in formula (17.2).

Not only the microstructure, but also the crystallographic texture of the samples is affected by SPD. Owing to space limitations this topic will not be discussed further. For more information, the reader is referred to some recent studies (Toth *et al.* [2004], De Messemaeker *et al.* [2005], Li *et al.* [2005], Gazder *et al.* [2006]).

17.1.4 Properties of submicron materials obtained by SPD

17.1.4.1 Strength and ductility. One of the major driving forces behind the development of submicron materials is their good strength/ductility balance. As long as the total elongation in an uniaxial tensile test is considered, ECAP samples have a better strength/ductility balance than the same material after different degrees of cold rolling (Zhu and Langdon [2004]). However, as reported by many investigators, ECAP material looses most of its strain hardening capacity and in a tensile test, the major part of the deformation is due to extensive post-uniform deformation. It is also worth mentioning that the high strength of ECAP samples is not 'abnormal'. Results from Horita *et al.* [2000a] show that for an equivalent strain, the ECAP samples exhibit the same strength as cold-rolled material.

17.1.4.2 Superplasticity. As shown in Figure 6.5, a reduction in grain size shifts the domain where grain boundary sliding (superplasticity) is active, to higher strain rates. The production of submicron materials by SPD creates an opportunity to reduce the production time for superplastic forming (typically 20–30 min/part) by a factor of 10 or more, which opens new markets for this forming technique. An alternative strategy is to reduce the deformation temperature. Several experimental results have been published, demonstrating the occurrence of high-strain rate superplasticity and low-temperature superplasticity.

High-strain rate superplasticity (HSR-SP) is usually defined as superplastic deformation at a strain rate higher than 10^{-2}/s. HSR-SP has been observed in the past, but in a restricted range of materials such as some metal–matrix composites or materials produced by powder metallurgy (Higashi [1996]). Valiev *et al.* [1997] have shown that HSR-SP can be obtained in two as-cast Al alloys after ECAP. In a classical Supral 100 alloy (Al–6 Cu–0.4 Zr) a grain size of 0.5 μm was obtained after 8 ECAP passes at 400°C and 4 passes at 200°C. Elongations of 970% at 10^{-2}/s and 300°C or 740% at 10^{-1}/s and 350°C have been reported. With a Russian alloy 1420 (Al−5.5 Mg−2.2 Li−0.12 Zr), subjected to the same ECAP scheme, a grain size of 1.2 μm and elongations of 1180% at 10^{-2}/s and 350°C, or 910% at 10^{-1}/s were obtained. Even at a strain rate of 1/s (350°C), 350% elongation could still be reached. Further studies on this alloy showed that the optimum temperature for HSR-SP was 400°C with an elongation of 950% at a strain rate of 1/s (Lee *et al.* [1999]).

Another well-investigated family of alloys are the Al–Mg–Sc alloys. In Al−3Mg–0.2Sc, a maximum elongation of 2280% was obtained at 400°C and 3.3 10^{-2}/s with an initial grain size of 1.1 μm. Even at a strain rate of 1/s, 210% elongation could be attained (Horita *et al.* [2000b]). In variants with higher Mg content, HSR-SP has also been recorded, e.g. in Al−6Mg–0.3Sc–0.3Mn, where elongations-to-failure up to 2000% have been observed at 450°C and an initial strain rate of 5.6 10^{-2}/s (Musin *et al.* [2004]). HSR-SP has been reported in several other alloys, e.g. in a spray cast Al-7034 (Xu *et al.* [2003]), a number of Mg alloys (Horita *et al.* [2004]) and in a commercial AA2024 alloy (Xu *et al.* [2004]).

These studies also showed that very fine grains are not a sufficient condition to achieve HSR-SP. The fine grains must be stable enough to withstand coarsening at high deformation temperatures. For example, the grain size in classical Al–Mg alloys (AA5xxx), can be refined very efficiently by SPD, but at superplastic deformation temperatures those grains are not stable enough to withstand (dynamic) grain growth. For this reason, the alloys used for superplastic applications contain fine dispersoids that slowdown or inhibit grain growth. In the Supral 100 alloy, Al_3Zr particles are present and in the Russian alloy 1420, β'-Al_3Zr and δ'-Al_3Li; in the Al–3Mg–0.2Sc alloy, pinning of grain boundaries is provided by Al_3Sc particles. The very fine grain size provided by SPD, can also be used to lower the superplastic deformation temperature, at the expense of the higher strain rate. When superplastic deformation occurs below about $0.5T_m$, it is denoted 'low-temperature superplasticity' (LT-SP).

17.1.4.3 Fatigue. The fatigue properties of fine-grained materials have been less intensely studied than their static strength properties. In general, it is observed that the high cycle fatigue (HCF) life of most SPD materials is enhanced compared to their coarse-grained counterparts (Vinogradov and Hashimoto [2004]). In HCF, the elastic component of the strain amplitude is dominant and the fatigue life is dictated by the fracture strength of the material, so that in general the fatigue limit of the material increases with increasing strength (Vinogradov and Hashimoto [2004]), which is in favour of SPD materials. In low cycle fatigue on the other hand, the fatigue resistance of conventional coarse-grained materials, is in general superior to that of SPD material. This is schematically shown in Figure 17.7. The total strain amplitude ($\Delta\varepsilon_t/2$) can be divided into elastic and plastic parts. The elastic part, related to HCF, can in general be expressed by the Basquin law as a function of the fatigue strength σ'_f, Young's modulus E, the fatigue strength exponent b and the number of cycles N_f. The plastic part, related to LCF, is expressed by the Coffin–Manson relation, with ε'_f the fatigue ductility and c the fatigue ductility coefficient (with $c > b$):

$$\Delta\varepsilon_t/2 = \Delta\varepsilon_{el}/2 + \Delta\varepsilon_{pl}/2 = (2N_f)^b(\sigma'_f/\mathrm{E}) + \varepsilon'_f(2N_f)^c \qquad (17.3)$$

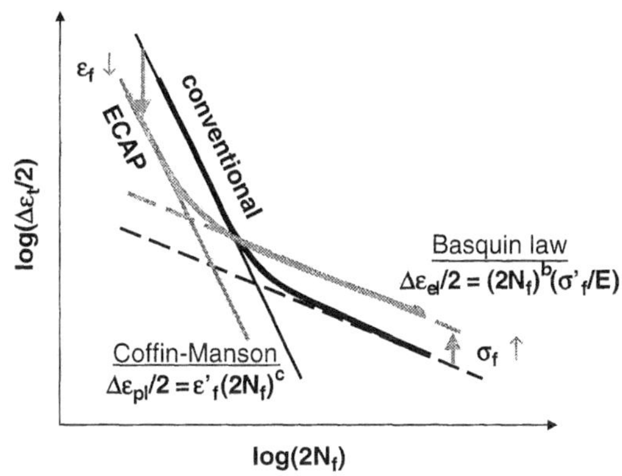

Figure 17.7. Schematic illustration of the influence of ECAP deformation on the fatigue life of a material. Adapted from Mughrabi *et al.* [2004]).

On a $\log(\Delta\varepsilon_t/2)$ vs. $\log(2N_f)$ plot, both relations are represented by straight lines (Figure 17.7). At low cycles to failure, the plastic part of the strain amplitude is much larger than the elastic part, but for a large number of cycles the elastic part is dominant. When a material with a conventional grain size is now deformed by SPD, the strength in general (also σ'_f) will increase and the ductility (ε'_f) will decrease, shifting the fatigue curve downwards for LCF and upwards for HCF.

It has been suggested (Patlan *et al.* [2001]) that the reduced fatigue resistance in LCF is mainly due to the susceptibility to strain localization of ECAP material. Hence, it is reasonable to expect an enhancement of LCF properties after a short annealing treatment. This has indeed been observed in a number of alloys such as AA5056 (Patlan *et al.* [2001]) or Cu (Vinogradov [1998]). Unfortunately, an improvement in LCF goes hand in hand with a decrease in HCF resistance. Hence, for each application, a suitable balance between both must be found.

17.1.4.4 Corrosion. The corrosion behaviour of ultrafine grained materials has received only limited attention in the published literature and it is difficult to formulate general conclusions at this time. Thorpe *et al.* [1988] have shown that for nanostructured materials (not made by SPD) enhanced corrosion properties in comparison with coarse-grained variants may be possible. On the other hand, Rofagha *et al.* [1991] observed no difference in corrosion resistance between nanocrystalline pure Ni and its coarse grained counterparts and a degradation in corrosion resistance was even reported for fine grained Ni–P (Rofagha *et al.* [1993]).

For a 99.6% Cu with a grain size of 200 nm produced by ECAP, it was observed that the corrosion behaviour had not qualitatively changed (Vinogradov *et al.* [1999]), but later (Yamasaki *et al.* [2001]) a better stress corrosion resistance was reported. For a Ti alloy, an increase in corrosion resistance in HCl and H_2SO_4 solutions after 8 ECAP passes, route B_C, was observed. This improvement was related to the rapid formation of passive films at surface crystalline defects like grain boundaries and dislocations (Balyanov *et al.* [2004]). In a commercial purity Al (AA1050), the corrosion resistance was investigated as a function of increasing number of ECAP passes and it was clearly shown that the resistance to pitting corrosion improved with increasing number of passes (Chung *et al.* [2004]). This was related to the decreasing size of Si-containing impurities. They formed local galvanic cells, which decreased in area with increasing ECAP passes.

LITERATURE

Y.T. Zhu, T.G. Langdon, R.Z. Valiev, S.L. Semiatin, D.H. Shin and T.C. Lowe (eds), "Ultrafine Grained Materials III", TMS, Warrendale, USA (2004).
Y.T. Zhu, T.G. Langdon, Z. Horita, M.J. Zehetbauer, S.L. Semiatin and T.C. Lowe (eds), "Ultrafine Grained Materials IV", TMS, Warrendale, USA (2006).
Z. Horita (ed.), "Nanomaterials by severe plastic deformation", Mater. Sci. Forum, 503–504 (2006).

17.2. GRAIN BOUNDARY ENGINEERING FOR LOCAL CORROSION RESISTANCE IN AUSTENITIC STAINLESS STEEL

17.2.1 Introduction

Austenitic stainless steels have excellent resistance to general corrosion. They are, however, prone to localized corrosion – crevice and pitting corrosion, intergranular corrosion (IGC) and stress corrosion cracking (IGSCC) (Cihal [1984]). The last two forms of localized corrosion, IGC and IGSCC, are often[1] caused by 'sensitization' (Uhlig [1971], Cihal [1984], Fontanna [1986], Dayal *et al.* [2005]). 'Sensitization' typically develops when an austenitic stainless steel is welded or heat treated at a temperature range, 450–850°C (Dayal *et al.* [2005]). Some time–temperature combinations lead to formation of Cr-rich carbides or intermetallics at grain boundaries. Growth of Cr-rich phases can create Cr-depleted regions. If the Cr-depletion goes below 12–13 wt%, then the passive film over the depleted regions becomes weak and easily breaks in aggressive solutions. This makes the sensitized austenitic stainless steel prone to IGC and IGSCC. A list of Cr-rich phases, typically linked to 'sensitization', is given in Table 17.1.

17.2.2 Sensitization control

'Sensitization' control is important to a designer – important for both the fabrication schedule and for lifetime predictions (Cihal [1984], Dayal *et al.* [2005]).

[1] Can also be caused by segregation of active elements – such as phosphorous (Cihal [1984]).

Table 17.1. Examples of grain boundary Cr-rich carbides and intermetallic phases (Stickler and Vinckier [1961], Weiss and Stickler [1972], Minami *et al.* [1986]). Though $M_{23}C_6$ is often related to 'sensitization', the same may also be caused by other Cr-rich intermetallic phases.

Phase	Crystal structure	Chemistry (*typical examples of the chemistry are included – though different concentrations of alloying elements are possible*)
$M_{23}C_6$	Complex fcc	$(Cr_{16}Fe_5Mo_2)C_6$
Sigma	bct: $D8_b$	Fe(CrMo)
Chi	bcc A12 (α-Mn type)	$Fe_{36}Cr_{12}Mo_{10}$
Laves	Hexagonal C14 or C36	$(Fe_{0.85}Cr_{0.15})(Mo_{0.3}Nb_{0.4}Si_{0.25}Ti_{0.05})$

Naturally different engineering approaches to sensitization control, i.e. for restricting nucleation and growth of Cr-rich grain boundary phases, have been developed and adopted over the years. These include both the conventional approach of controlling chemistry and heat treatment and the relatively more recent approaches of controlling grain boundary nature (Uhlig [1971], Cihal [1984], Fontanna [1986], Dayal *et al.* [2005]).

17.2.2.1 Sensitization control through chemistry. Chemistry has strong effects on sensitization behaviour (Cihal [1984], Dayal *et al.* [2005]) and several approaches to sensitization control through chemistry modifications have been adopted:

- *Lower carbon content*: Lower carbon content is often used as an effective means of reducing the formation of $M_{23}C_6$. Different austenitic stainless steel grades, L and XL – indicating low and extralow-carbon concentration, have been developed for this specific purpose. The advent of vacuum metallurgy in liquid metal processing has made manufacturing of such L and XL grades easy, though in terms of both cost-economics and strengths such grades are less attractive than their high-carbon counterparts. For example, typical yield and ultimate tensile strength of AISI type 304 and 304L grades are 170 and 480 MPa and 205 and 515, respectively (ASM [1994]) and their costs usually differ by 10–20%.
- *Stabilizing carbon*: Stabilizing carbon by precipitating it with titanium or niobium is also an option often adopted. This is not a 'full-proof' method, as complete elimination of Cr-rich carbides may still be difficult.
- *Alloying additions to prevent grain boundary precipitation*: Alloying elements like cerium added beyond a minimum concentration are known to resist sensitization. This is especially effective in improving resistance to IGSCC (Watanabe *et al.* [2000]). The large-size alloying elements segregate to grain boundaries and resulting strain field makes grain boundary precipitation of Cr-rich phases

difficult. Controlled additions of such elements are now being adopted. The primary problem of this approach is a significant drop in workability, making a wide acceptability difficult.

17.2.2.2 Sensitization control through heat treatment. Solution annealing, or annealing at time–temperature combinations where Cr-rich precipitates are dissolved, is a typical step in any sensitization control. The last step of TMP of austenitic stainless steel almost always involves 'mill annealing', annealing typically above 1050°C and for ½ h or more, to eliminate any Cr-rich phases in the finished product – a 'desensitization' process. Though effective as a TMP stage, this approach often has the risk of coarsening the grain size and cannot prevent or control sensitization during service. It needs to be noted that laser surface processing is fast emerging as a technological step of sensitization control. The exact science behind this is not fully charted – though one may consider a combination of desensitization and changes in grain size and grain boundary nature as the possible cause (Dayal *et al.* [2005]).

17.2.2.3 Sensitization control by 'optimized' grain boundary nature. Sensitization involves two steps: nucleation of grain boundary Cr-rich phases and diffusion-controlled growth of such phases leading to Cr-depleted regions. Correspondingly, the role of grain boundaries is extremely important in sensitization control and two different approaches (Palumbo and Aust [1988], Lin *et al.* [1995], Kumar *et al.* [2002], Shimada *et al.* [2002], Wasnik *et al.* [2002]) have emerged. The approaches depend on two issues – ability to alter grain boundary nature and relationship between sensitization and grain boundary nature. These are discussed separately.

17.2.3 Ability to alter grain boundary nature

Thermomechanical processing can be used to alter the grain boundary nature – the relative concentration of special (e.g. low CSL – see Chapter 2) and random boundaries, the proximity/perfection of the special boundaries to exact CSL nature and the relative connectivity of special/random boundaries. A simple cold rolling treatment followed by recrystallization annealing can generate such changes effectively. Typically, both the concentration and the perfection of CSL increases slightly and then drops significantly with increasing percentages of prior cold work. This behaviour, reported in several independent studies (Palumbo and Aust [1988], Lin *et al.* [1995], Kumar *et al.* [2002], Shimada *et al.* [2002], Wasnik *et al.* [2002]), can be rationalized in terms of annealing twins. After small degrees of cold deformation, recovery and/or, recrystallization is twin dominated and hence an increase in $\Sigma 3$ twin boundaries is observed, while large prior cold working and

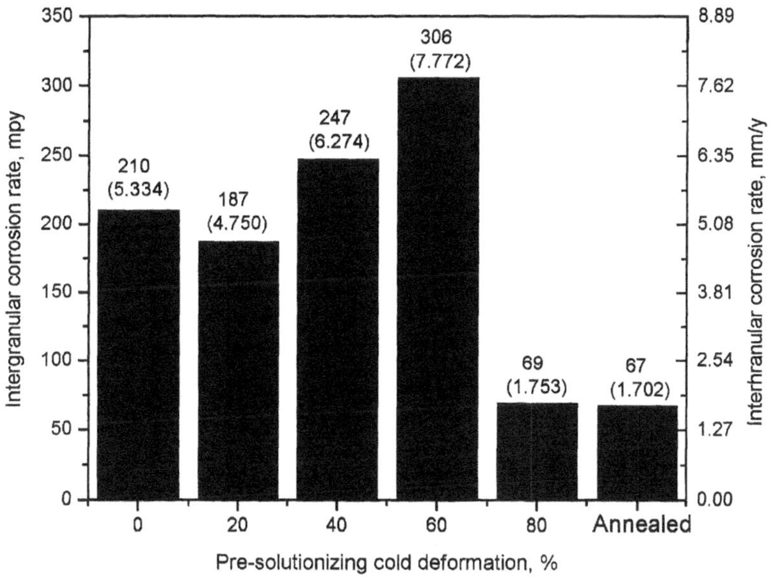

Figure 17.8. Estimated corrosion rates, estimated as per practice B – ASTM A262 [2000] tests, in AISI-type 304 stainless steel. The stainless steel samples were subjected to different degree of prior cold rolling followed by solutionizing and sensitization treatments – treatments leading to significant differences in grain boundary nature and in degree of sensitization. A mill-annealed sample without any sensitization treatment has also been included as a reference (Wasnik *et al.* [2002]).

corresponding domination of strain localizations in recrystallization, effectively randomizes grain boundary nature. The other possibility of altering grain boundary nature is through 'transformation annealing' between austenite (γ) and delta (δ) ferrite – though this has been mostly an uncharted possibility so far.

17.2.4 Sensitization and grain boundary nature

The earlier section has established clear possibilities of altering grain boundary nature through a combination of cold work and recrystallization annealing. This, on the other hand, can strongly affect the degree of sensitization as shown in Figure 17.8. As described in more details elsewhere (Wasnik *et al.* [2002]), prior cold work and the corresponding difference in grain boundary nature can reflect strongly on the degree of sensitization.[2] The data in Figure 17.8 are, however, biased by differences in grain size. To take this, as well as the role of other low-CSL boundaries into account, the concept of effective grain boundary energy (EGBE) has been proposed (Wasnik *et al.* [2002]). EGBE$=((\Sigma\gamma_i f_i)4/d))/\gamma_{max})$,

[2] Estimated by practice B: ASTM A262 (ASTM Book of Standards [2000]) tests.

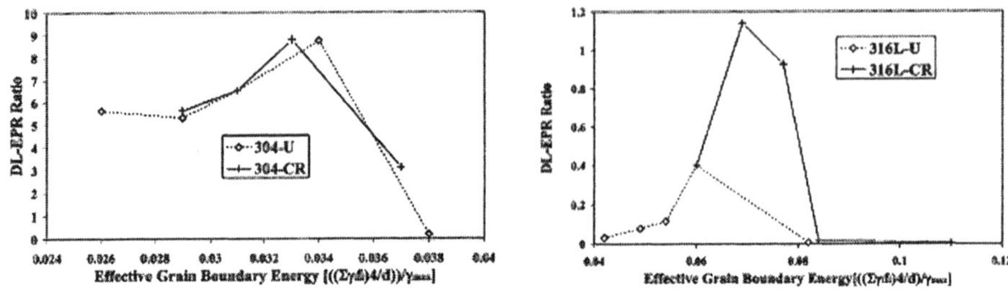

Figure 17.9. Relating effective grain boundary energy (EGBE) with DL-EPR ratio in (a) 304 and (b) 316 L austenitic stainless steel. U and CR represent the respective rolling modes of unidirectional rolling and cross rolling. The DL-EPR ratio is an electrochemical representation of the degree of sensitization, while EGBE considers (as described in the text) the role of all low-CSL boundaries and grain size (Wasnik *et al.* [2002]).

where f_i is the fraction of each class of boundary (such as $\Sigma 3$, $\Sigma 5, \ldots, \Sigma 29$, random) and γ_i is the corresponding energy. The γ_i for respective CSL was calculated using the general formula – $\gamma_{CSL}=(1-1/\sqrt{\Sigma})\gamma_{max}$ (Aleshin *et al.* [1978]), where γ_{max} is the energy of random boundaries and Σ the corresponding CSL notation. As shown in Figure 17.9, the trend in degree of sensitization (or the corresponding DL-EPR ratio) with EGBE is quite clear and consistent. The degree of sensitization increased with randomization of grain boundaries up to a critical value and then dropped significantly. This brings out clear possibilities of 'tailoring' grain boundary nature for improved resistance to sensitization.

Though the coinage of 'grain boundary design' has been relatively recent (Watanabe [1984]), references to different property/performance of grain boundaries dates back to the first classical experiment of Aust and Rutter [1959] and to a range of interesting results of both past and present (Ibe and Lücke [1966], Viswanathan and Bauer [1973], Watanabe *et al.* [1980, 1989], Palumbo and Aust [1988], Lin *et al.* [1995], Kumar *et al.* [2002], Shimada *et al.* [2002], Wasnik *et al.* [2002]). There are serious lacunas in extrapolating exact grain boundary nature from local orientation measurements – i.e. to identify the tilt or twist nature of boundaries (also see Chapter 2). In addition, existing CSL theory (Randle [1996]) is a 2D approximation of the actual 3D grain boundary nature. Even then the approach of grain boundary engineering is emerging as an attractive means of sensitization control – a viable technological approach which can be incorporated in a TMP sequence. Interesting, two exactly opposite directions of grain boundary engineering can be appreciated:

- The approach to increase special boundary concentration (Palumbo and Aust [1988], Lin *et al.* [1995], Kumar *et al.* [2002], Shimada *et al.* [2002]).

A combination of low reduction and high-temperature annealing is effective in increasing special boundary concentrations and connectivity and in turn has been brought out (Palumbo [1997], Palumbo and Aust [1998]) as a technology for producing austenitic stainless steels with improved resistance to sensitization.

- The approach to randomized grain boundaries (Wasnik *et al.* [2002]). This relatively new approach has also shown remarkable improvements in the resistance to sensitization. The problem seems to be in incorporating the large plastic deformation necessary for effective randomization of grain boundaries. Special forming techniques like pilgering and possible incorporation of transformation annealing, between γ austenite and δ ferrite, may offer a practical TMP solution to this approach.

REFERENCES

Aaron H.D., Fainstein D. and Kotler G.R., J. Appl. Phys., 41 (1970), 4401.

Abbruzzese G., Ciancaglioni I. and Campopiano A., Tex. Micro., 8–9 (1988), 401.

Abbruzzese G. and Lücke K., Acta Metall., 34 (1986), 905.

Abe H., Scand. J. Metall., 13 (1984), 226.

Abe M., 'Formable Steel' in "Constitution and Properties of Steel, Materials Science and Technology", R.W. Cahn, P. Haasen and E.J. Kramer (eds), vol. 7, Chapter 7, VCH, New York (1992), pp. 288–333.

Abedrabbo N., Zampaloni M.A. and Pourboghrat F., Int. J. Mech. Sci., 47 (2005), 333.

Ahmetoglu M. and Altan T., J. Mater. Process. Tech. 98 (2000), 25–33.

Akbari G.H., Sellars C.M. and Whitman J.A., Acta Mater., 45 (1997), 5047.

Akef A. and Driver J.H., Mater. Sci. Eng. A, 132 (1991), 245–255.

Akhtar A. and Teghtsoonian A., Acta Metall., 19 (1971), 655.

Alberdi J.M. Ph.D. thesis, Universidad de Navarra, San Sebastien, Spain (1984)

Aldareguia J.M., Garcia-Rosales C. and Gil Sevillano J., Scripta Mater., 41 (1999), 729.

Aleshin A.N., Aristov V.Yu., Bokshtein B.S. and Shvindlerman L.S., Phys. Stat. Sol. (A), 45 (1978), 359.

Almen J.O. and Black P.H., "Residual Stress and Fatigue in Metals", McGraw-Hill, New York (1963).

Altan T., Boulger F.W., Becker J.R., Akgerman N. and Henning H.J., "Forging Equipment, Materials and Practices", Metals and Ceramics Information Centre – A Department of Defence Information Analysis Centre, USA (1973).

Altherthum H., Z. Metall., 14 (1922), 417.

Aluminum – Properties and Physical Metallurgy, 5th Edition, ed J.E. Hatch, (1993) ASM, Ohio, USA.

Amontons G., Mem. Acad. Royal. Sci. (Paris), (1699), 206.

Anand V.B., "Computer Graphics and Geometric Modeling for Engineers", Wiley, New York (1996).

Anderson M.P., Grest G.S. and Srolovitz D.J., Phil. Mag., B59 (1989), 293.

Anderson M.P., Srolovitz D.J., Grest G.S. and Sahani P.S., Acta Metall., 32 (1984), 783–789.

Argon A.S., 'Mechanical properties of single phase crystalline media', in "Physical Metallurgy", R.W. Cahn and P. Haasen (eds), North-Holland, Amsterdam (1996), pp. 1875–1955.

Armstrong R.W., Metall. Trans., 1 (1970), 1169–1176.

Arzt E., Ashby M.F. and Easterling K.E., Metall. Trans., 14A (1983), 211–221.

Asgari S., El-Danaf E., Kalidindi S.R. and Doherty R.D., Metall. Mater. Trans., 28A (1997), 1781–1795.

Ashby M.F., Phil. Mag., 21 (1970), 399–424.

Ashby M.F., Acta Mater., 20 (1972), 887–897.

Ashby M.F., Acta Metall., 37 (1989), 1273.

Ashby M.F., Gandhi C. and Taplin D.M.R., Acta Metall., 27 (1979), 699–729.

ASM Handbook, "Forming and Forging", S.L. Semiatin (ed.), vol. 14, ASM, Ohio (1988a).
ASM Metals Handbook, "Mechanical Testing and Evaluation", 9th edition, vol. 8, ASM, Ohio (1985).
ASM Metals Handbook, "Forming and Forging", 9th edition, vol. 14, ASM, Ohio (1988b).
ASM Speciality Handbook, in "Stainless Steels", J.R. Davies (ed.), ASM, Ohio (1994).
ASTM Book of Standards, vol. 10, ASTM, Philadelphia, USA (2000).
Atkinson H.V., Acta Metall., 36 (1988), 469.
Atkinson H.V. and Rickinson B.A., "Hot Isostatic Processing", Adam Hilger, Bristol (1991).
Aust K.T. and Rutter J.W., Trans. AIME, 215 (1959), 119.
Avitzur B., J. Eng. Ind., 90 (1968), 79
Avitzur B., "Handbook of Metal Forming Processes", Wiley, New York (1983).
Avrami M., J. Chem. Phys., 7 (1939), 1103–1112.
Avrami M., J. Chem. Phys., 8 (1940), 212–224.
Bae C.M. and Nam W.J., Mater. Sci. Tech., (1994), 971.
Balaji P.A., Sundararajan T. and Lal G.K., ASME J. Appl. Mech., 1 (1991), 644.
Balyanov A., Kutnyakova J., Amirkhanova N.A., Stolyarov V.V., Valiev R.Z., Liao X.Z., Zhao Y.H., Jiang Y.B., Xu H.F., Lowe T.C. and Zhu Y.T., Scripta Mater., 51 (2004), 225–229.
Banerjee S., Dey G.K., Srivastava D. and Ranganathan S., Metall. Mater. Trans., 28A (1997), 2201.
Banerjee S. and Krishnan R., Metall. Trans., 4A (1973), 1811.
Barber R.E., Dudo T., Yasskin P.B. and Hartwig T., 'Production yield in ECEA processed material', in "Ultrafine Grained Materials III", Y.T. Zhu, T.G. Langdon, R.Z. Valiev, S.L. Semiatin, D.H. Shin and T.C. Lowe (eds), TMS, Warrandale, USA (2004), pp. 667–672.
Barlat F., Lege D. and Brem J., Int. J. Plasticity, 5 (1991), 51–66.
Barlow K.N., Lancaster P.N. and Maddison R.T., Metals Tech., 11, (1984), 14.
Barrett C.S. and Massalski T.B., "Structure of Metals: Crystallographic Methods, Principles and Data", Pergamon Press, Oxford (1980).
Barret W.F., Brown W. and Hadfield R.A., Sci. Trans. Roy. Dublin Soc., 1 (1900), 67.
Barrett C.S. and Levenson L.H., Trans. AIME, 137 (1940), 112.
Basson F. and Driver J.H., Acta Mater., 48 (2000), 2101–2115.
Basuki A. and Aernoudt E., J. Mater. Process. Tech., 89–90 (1999), 37–43.
Bate P.S., Huang Y. and Humphreys F.J., Acta Mater., 52 (2004), 4281–4289.
Bauri R., Pancholi V., Samajdar I. and Surappa M.K., Sci. Tech. Adv. Mater., 6 (2005), 933.
Bay B., Hansen N., Hughes D.A., and Kuhlmann-Wilsdorf D., Acta Metall. Mater., 40 (1992), 205–219.
Beck P.A., in "Sorby Centennial Symposium on the History of Metallurgy", C.A. Smith (ed.), Met. Soc. Conf., vol. 27, Gordon and Breach, New York (1963), p. 313.

Belyakov A., Sakai T., Miura H., Kaibyshev R. and Tsuzaki K., Acta Mater., 50 (2002), 1547.
Benford J.G., IEEE Trans. Mag., MAG-20 (5) (1984), 1545.
Berger A., Wilbrandt P.J., Ernst F., Klement U. and Haasen P., Prog. Mater. Sci., 32 (1988), 1.
Bhadeshia H.K.D.H., "Bainite in Steels", 2nd edition, IOM Communications Ltd., UK (2001).
Bhardwaj S.A., Proc. Symp. Zirconium-2002, P.K. Dey (ed.), BARC, Mumbai, India (2002), p. 66.
Biba N., Stebounov S. and Lishiny A., J. Mater. Process Tech., 113 (2001), 34.
Bishop J.F.W. and Hill R., Phil. Mag., 42 (1951), 414–427.
Bishop J.F.W., Metall. Rev., 2 (1957), 361.
Bland D.R. and Ford H., Proc. Inst. Mech. Eng. (London), 159 (1948), 144.
Blazynski T.Z. (ed.), "Plasticity and Modern Metal Forming Technology", Elsevier, New York, USA (1989).
Blazynski T.Z., "Metal Forming: Tool Profiles and Flow", Macmillan Press Ltd., London, UK (1976).
Bleck W., J. Metals, 48 (1996), 26–30.
Blum W., 'Creep of Aluminium and Aluminium Alloys' in "Hot Deformation of Aluminum Alloys", T.G. Langdon, H.D. Merchant, J.G. Morris and M.A. Zaidi (eds), The Minerals, Metals and Materials Society, Warrendale, PA (1991), pp. 181–209.
Boehlert C.J., Mater. Sci. Eng., A279 (2000), 118.
Boettinger W.J., Coriell S.R., Greer A.L., Karma A., Kurz W., Rappaz M. and Trivedi, Acta Mater., 48 (2000), 43.
Bogers A.J. and Burgers W.G., Acta Metall., 12 (1964), 255.
Boisse P., Altan T. and Luttervelt K.V. (eds), "Friction and Flow Stress in Forming and Cutting", Kogan Page Ltd., London, UK (2003).
Bollman W., "Crystal Lattices, Interfaces, Matrices – An Extension of Crystallography", ISBN 2-88105-000-X (1982).
Borbely A. and Driver, J.H., Arch. Metall. Mater., 50 (2005), 65.
Borbely A., Driver J.H. and Ungar T., Acta Mater., 48 (2000), 2005–2016.
Borisov A.G., Fedorov O.P. and Moslov V V., J. Crystal Growth, 112 (1991), 463.
Boutin F.R., "Procédé de fabrication de bandes ou feuilles en aluminium présentant une texture en dé prédominante", Patent No. FR 70.40921 (1970).
Bowen A.W., Mater. Sci. Tech., 6 (1990), 1058–1071.
Boyer C.B., 'Historical Review of HIP Equipment', in "Hot Isostatic Pressing – Theory and Applications", M. Koizumi (ed.), Elsevier (1992), p. 465.
Bradford S.A., Trans. AIME, 20 (1964), 1400.
Bramley A.N. and Mynors D.J., Mater. Design, 21 (2000), 279.
Brandon D.G., Acta Metall., 14 (1966), 1479.
Briant C.L., Mater Sci. Tech., 7 (1991), 739.
Bunge H.J., Z. Metall., 56 (1965), 872–874.
Bunge H.J., in "Textures in Research and Practice", J. Grewen and G. Wassermann (eds), Springer, Berlin (1969), pp. 24–35.

Bunge H.J., "Texture Analysis in Materials Science – Mathematical Methods", Butterworths, London (1982).
Bunge H.J., Z. Metallkd., 76 (1985), 92–101.
Burke J.E. and Turnbull D., Prog. Met. Phys., 3 (1952), 220.
Cahn J.W., Acta Metall., 4 (1956), 449.
Cahn J.W., Trans. AIME, 242 (1968), 166.
Cahn R.W., J. Inst. Metals, 76 (1949), 121.
Calladine C.R., "Plasticity for Engineers – Theory and Applications", Horwood, Chichester, England (2000).
Cao J., Yao H., Karafillis A. and Boyce M.C., Int. J. Plas., 16 (2000), 1105.
Carleer B., van der Kevie G., de Winter L. and van Veldhuizen B., J. Mater. Process. Tech., 104 (2000), 158.
Carlson P., PhD Thesis, ed.: Dean of faculty of Sci. & Tech., Upsala University, ISBN: 91-554-6136-0, (2005).
Carpenter G.J.C., Can. Metall. Quart., 24 (1985), 251.
Carpenter H.C.H. and Elam C.F., J. Inst. Metall., 24 (2) (1920), 83.
Cast Iron Soil Pipe and Fittings Handbook (ed.), Cast Iron Pipe Institute, Tennessee, USA (1994).
Chaudhury A., M.Tech. Thesis, IIT, Bombay, India (2005).
Cheadle B.A., "Zirconium in Nuclear Industry", ASTM STP 633, ASTM, Philadelphia, USA (1977).
Cheong B.H., Lin J. and Ball A.A., J. Mater. Process. Tech., 134 (2003), 10.
Chihab K., Estrin Y., Kubin L.P. and Vergnol J., Scripta Metall., 21 (1987), 203.
Childs T.H.C., Met. Technol., 1 (1974), 306.
Chin G.Y., Hosford W.F. and Mendorf D.R., Proc. R. Soc., A309 (1969), 433–456.
Chin G.Y., Mammel W.L. and Dolan M.T., Trans. Met. Soc. AIME, 239 (1967), 1854–1855.
Chmelik, Ziegenbein A., Neuhäuser and Luckáč, Mater. Sci. Engg., A324 (2002), 200.
Cho H., Jo H-H., Lee S-G., Kim B-M. and Kim Y-J., J. Mater. Process. Tech., 130–131, (2002), 416.
Christian J.W., "The Theory of Transformations in Metals and Alloys", Part I, Pergamon Press, UK (1975).
Christian J.W., Metall. Trans., 13A (1982), 509.
Christian J.W. and Mahajan S., Prog. Mater. Sci., 39 (1995), 1–157.
Chung M.K., Choi Y.S., Kim J.G., Kim Y.M. and Lee J.C., Mater. Sci. Eng., A366 (2004), 282–291.
Cicalè S., Samajdar I., Verlinden B., Abbruzzese G. and Van Houtte P., ISIJ Intl., 42 (2002), 770.
Cihal V., "Intergranular Corrosion of Steel and Alloys", Materials Science Monographs, vol. 18, Elsevier, Amsterdam (1984).
Cocks A.C.F and Ashby M.F., Met. Sci., 16 (1982), 465.
Collins E.W., "The Physical Metallurgy of Titanium Alloys", ASM-International, Metals Park, Ohio (1984).
Cornfield G.C. and Johnson R.H., Int. J. Mech. Sci., 12 (1970), 479.

Cottrell A.H., "Dislocations and Plastic Flow in Crystals", Clarendon Press, Oxford (1953).
Coulomb C.A., Mem. Math. Phys., Acd. Roy. Sci. (Paris), 10 (1785), 161.
Courtney T.H., "Mechanical Behaviour of Materials", 2nd edition, McGraw-Hill (2000).
Couturier C., Doherty R., Maurice C. and Fortunier R., Acta Mater., 53 (2005), 977–989.
Couturier G., Maurice C. and Fortunier R., Phil. Mag., 83 (2003), 3387–3405.
Cox B., J. Nucl. Mater., 179 (1990), 249.
Crumbach M., Pomana G., Wagner P. and Gottstein G., 'A Taylor type deformation texture model considering grain interaction and material properties. Part 1-Fundamentals' in "Recrystallisation and Grain Growth", G. Gottstein and D.A. Molodov (eds), Proceedings of the First Joint Conference, Springer, Berlin (2001), pp. 1053–1060.
Cui Q. and Ohori K., Mater. Sci. Tech., 16 (2000), 1095–1101.
Daaland O. and Nes E., Acta Mater., 44 (1996), 1389–1411.
Davies R.W., Khaleel M.A., Kinsel W.C. and Hbib, H.M., J. Eng. Mater. Tech., 124 (2002), 125.
Dayal R.K., Parvathavarthini N. and Raj B., Int. Mater. Rev., 50 (2005), 129.
De Messemaeker J., Verlinden B. and Van Humbeeck J., 'Mechanical behaviour of IF steel processed by equal channel angular pressing' in "Ultrafine Grained Materials III", Y.T. Zhu, T.G. Langdon, R.Z. Valiev, S.L. Semiatin, D.H. Shin and T.C. Lowe (eds), TMS (2004), 595–600.
De Messemaeker J., Verlinden B. and Van Humbeeck J., Acta Mater., 53 (2005), 4245–4257.
Derby B., Acta Metall., 39 (1991), 955.
Derby B. and Ashby M.F., Scripta Metall., 21 (1987), 879.
Deshpande N.U., Gokhale A.M., Denzer D.K. and Liu J., Metall. Mater. Trans., 29A (1998), 1191–1201.
Dey G.K. and Banerjee S., J. Nucl. Mater., 125 (1984), 219.
Dieter G., 'Mechanical Metallurgy', 2nd edition, McGraw-Hill, Tokyo (1976).
Dieter G.E., "Mechanical Metallurgy", SI Metric edition, McGraw-Hill (1988).
Dillamore I.L. and Katoh H., Metall. Sci., 8 (1974), 73–83.
Dillamore I.L., Roberts J.G. and Bush A.C., Met. Sci., 13 (1979), 73.
Dillamore I.L., Smith C.J.E. and Watson T.W., Met. Sci. J., 1 (1967), 49.
Dimitrov O., Fromageau R. and Dimitrov C., 'Effects of Trace Impurities on Recrystallization Phenomena' in "Recrystallization of Metallic Materials", F. Haessner (ed.), Riederer, Stutgart (1978), p. 137.
Dodd B. and Bai Y., "Ductile Fracture and Ductility – With Applications to Metalworking", Academic Press (1987).
Dodd B. and Boddington P., J. Mech. Work. Tech., 3 (1980), 239.
Doege E. and Bohnsack R., J. Mater. Process. Tech., 98 (2000), 165.
Doherty R.D., Scripta Metall., 19 (1985), 927.
Doherty R.D., 'Nucleation Processes in Recrystallization: History and Current Status' in "Solid to Solid Phase Transformations in Inorganic Materials",

J. Howe, D. Laughlin, J. Lee, U. Dahmen and W. Soffa (eds), TMS, Warrendale (2005).
Doherty R.D. and Baumann S.F., 'Structural origin of solute enhanced strain hardening of Aluminium alloys' in "Aluminum Alloys for Packaging", J.G. Morris, H.D. Merchant, E.J. Westerman and P.L. Morris (eds), TMS-AIME, Warrendale, PA (1993a), pp. 369–391.
Doherty R.D, Chen L.-C. and Samajdar. I., Mater. Sci. Eng., 257A (1998), 18–36.
Doherty R.D., Hughes D.A., Humphreys F.J., Jonas J.J., Juul Jensen D., Kassner M.E., King W.E., McNelley T.R., McQueen H.J. and Rollett A.D., Mater. Sci. Eng., A238 (1997), 219.
Doherty R.D., Liu J. and Sanders Jr., R.E., "Aluminum Alloy Sheet for Food and Beverage Container", US Patent # 5,192,378 (1994).
Doherty R.D. and McBride J.K., 'Solute enhanced strain hardening of Aluminium alloys' in "Aluminum Alloys for Packaging", J.G. Morris, H.D. Merchant, E.J. Westerman and P.L. Morris (eds), TMS-AIME, Warrendale, PA (1993b), pp. 347–368.
Doherty R.D., Samajdar I., Necker C., Vatne H.E. and Nes E., 'Nucleation of recrystallization in cold and hot deformed polycrystals' in "Microstructural and Crystallographic Aspects of Recrystallization", N. Hansen, D. Juul Jensen, Y.L. Liu and B. Ralph (eds), 16th Riso International Symposium on Materials Science, Riso National Laboratory, Roskilde, Denmark (1995), pp. 1–23.
Dorward R.C. and Beerntsen D.J., Metall. Mater. Trans., 26A (1995), 2481–2484.
Draney C., Froes F.H. and Hebeison J., Mat. Tech. and Adv. Perf. Mat., 19.2 (2004), p. 140.
Driver J.H., Perocheau F. and Maurice C., Mater. Sci. Forum, 331–337 (2000), 43–56.
Duckham A., Knutsen R.D. and Engler O., Acta Mater., 49 (2001), 2739–2749.
Duggan B.J., Hutchinson, W.B. and Hatherley M., Scripta Metall., 12 (1978), 1293.
Duggan B.J., Lücke K., Köhlhoff G. and Lee C.S., Acta Metall., 41 (1993), 1921.
Dupouy G. and Perrier, F., J. de Microscopie, 1 (1967), 167.
Dupuy L. and Rauch E.F., Mater. Sci. Eng., A337 (2002), 241–247.
Dutta B., Samajdar I. and Surappa M.K., Mater. Sci. Tech., 14 (1998), 36.
Dzubinsky M. and Kovac F., Scripta Mater., 45 (2001), 1205.
Edelson B.I. and Baldwin W.M., Trans. ASM, 55 (1962), 230–250.
Ekelund S., "The Analysis of Factors Influencing Rolling Pressure and Power Consumption in the Hot Rolling of Steel", Steel, 93 (8–14) (1933) (translated from Jernkontorets Ann, February 1927).
Ekstrom H.E., Mishin O.V. and Ostensson L., Proc. ICAA, 9 (2004), 252–257.
El-Danaf E., Kalidindi S.R. and Doherty R.D., Metall. Mater. Trans., 30A (1999), 1223–1233.
El-Danaf E., Kalidindi S.R., Doherty R.D. and Necker C.T., Acta Mater., 48 (2000), 2665–2673.
Embury J.D. and Fisher R.M., Acta Metall., 14 (1966), 147–159.
Engler O. and Lücke K., Scripta Metall. Mater., 27 (1992), 1527–1532.
English A.T. and Chin G.Y., Acta Metall., 13 (1965), 1013–1016.
Eridon J.M., "ASM Handbook", vol. 15 – Casting, ASM, Ohio (1988).

Esche S.K., Kinzel G.L. and Altan T., Proc. Numisheet '96, J.K. Lee *et al* (ed.), The Ohio State University, Deaborn, Michigan, USA (1996), p. 270.
Estrin Y., Kubin L.P., 'Continuum models for materials with microstructures', ed. Mühlhaus H.B., Wiley, New York, (1995), 395.
Estrin Y. and Mecking H., Acta Metall., 32 (1984), 57–70.
Etter A.L., Baudin T. and Penelle R., Scripta Mater., 47 (2002), 725.
Evenschor P.D. and Hauk V., Z. Metall., 66 (1975), 164.
Ewing J.A. and Rosenhain W., Phil. Trans. Roy. Soc., 193A (1900), 353.
Feltham P., Acta Metall., 5 (1957), 97.
Fields D.S. and Backofen W.A., Proc. ASTM, 57 (1957), 1259–1272.
Fields R.J., Weerasooriya T. and Ashby M.F., Metall. Trans., 11A (1980), 333–347.
Fleischer R.L., Acta Metall., 11 (1963), 203–209.
Flemings M.C., "Solidification Processing", McGraw-Hill, New York (1974).
Flower H.M., Mater. Sci. Tech., 6 (1990), 1082.
Fontanna M.G., "Corrosion Engineering", 3rd edition, McGraw-Hill Book Company, New Delhi (1986).
Fukagawa T., Okada H. and Maehara Y., ISIJ. Int., 11 (1994), 906.
Fukuda M., J. Iron Steel Inst. Jpn., 53 (1967), 559–561.
Fullman R.F., "Metal Interfaces", ASM, Cleveland, OH (1952), p. 179.
Furu T., Doctoral Thesis, NTH-Trondheim, Norway (1992).
Furu T., Marthinsen K. and Nes E., Mater. Sci. Tech., 6 (1990), 1093–1102.
Furu T., Orsund R. and Nes E., Acta Metall. Mater., 43 (1995), 2209–2232.
Furugen M. and Hayashi C., J. Mech. Working Tech., 10 (1984), 273.
Furukawa M., Horita Z., Nemoto M. and Langdon T.G., J. Mater. Sci., 36 (2001), 2835–2843.
Gandhi C. and Ashby M.F., Acta Metall., 27 (1979), 1565–1602.
Ganguly C., Proc. Symp. Zirconium-2002, P.K. Dey (ed.), BARC, Mumbai, India (2002), p. 1.
Gardner K.J. and Grimes R., Met. Sci., 13 (1979), 216.
Gawne D.T. and Higgins G.T., in "Texture in Research and Practice", G. Wassermann and J. Grewen (eds), Springer, Berlin (1969), p. 319.
Gazder A.A., Dalla Torre F., Gu C.F., Davies C.H.J. and Pereloma E.V., Mater. Sci. Eng., 415 (2006), 126–139.
Gholinia A., Bate P. and Prangnell P.B., Acta Mater., 50 (2002a), 2121–2136.
Gholinia A., Humphreys F.J. and Prangnell P.B., Mater. Sci. Tech., 16 (2000a), 1251–1255.
Gholinia A., Humphreys F.J. and Prangnell P.B., Acta Mater., 50 (2002b), 4461.
Gholinia A., Pragnell P.B. and Markushev M.V., Acta Mater., 48 (2000b), 1115–1130.
Ghosh A. and Malik A.K., "Manufacturing Science", East West Press Pvt. Ltd., India (1985).
Gil Sevillano J., PhD Thesis, Katholieke Universiteit Leuven (1974).
Gil Sevillano J., 'Flow stress and work hardening' in "Materials Science and Technology; A comprehensive Review", R.W. Cahn, P. Haasen and E.J. Kramer (eds), vol. 6, VCH, Weinheim (1993), pp. 19–88.

Gil Sevillano J., Van Houtte P. and Aernoudt E., Prog. Mater. Sci., 25 (1980), 69–412.
Girault E., "Bainitic Transformation in TRIP-Assisted Steels and its Influence on Mechanical Properties", PhD Thesis Department, MTM-KULeuven (ISBN 90-5682-172-5) (1999).
Girault E., Mertens A., Jacques P., Houbaert Y., Verlinden B. and Van Humbeeck J., Scripta Mater., 44 (2001), 885–892.
Gittus J., "Creep, Viscoelasticity and Creep Fracture in Solids", Applied Science Publishers, London (1975).
Gladman T., "The Physical Metallurgy of Microalloyed Steels", The Institute of Materials, London (1997).
Glez J.C. and Driver J.H., Acta Mater., 51 (2003), 2989–3003.
Godet S., Harlet P., Delannay F. and Jacques P.J., Steel Res., 73 (2002), 280–286.
Gokhale A.M. and Dehoff R.T., Metall. Trans., 16A (1989), 559.
Golding E.W., "Electrical Measurements and Measuring Instruments", 5th edition, Pitman, London (1963).
Goodwin G.M., Metall. Ital., 60 (1968), 767.
Gordon P. and Vandermeer R., "Recrystallization, Grain Growth and Textures", ASM, Metals Park, Ohio (1966), p. 205.
Goss N.P., Patent No. US1965559 (1934).
Gottstein G. and Mecking H., in "Preferred Orientations in Deformed Metals and Rocks: An Introduction to Modern Texture Analysis", Academic Press, INC Orlando (1985), p. 183.
Groenbaek J. and Birker T., J. Mater. Process. Tech., 98 (2000), 155.
Groma I., Phys. Rev., B57 (1998), 7535.
Grong O. and Shercliff H.R., Prog. Mater. Sci., 47 (2002), 163–282.
Gudmundsson H., Brooks D. and Wert J.A., Acta Metall., 39 (1991), 19.
Guiglionda G, PhD Thesis, Ecole des Mines de St Etienne (2003).
Guiglionda G., Borbely A. and Driver J.H., Acta Mater., 52 (2004), 3413.
Gurland J. and Plateau J., Trans. ASM, 56 (1963), 442–454.
Gurson A.L., J. Eng. Mater. Tech., 99 (1977), 2–15.
Hähner P., Mater. Sci. Engg., A164 (1993), 23.
Hähner P. and Egidio R., Acta Mater., 51 (2003), 3385.
Hall E.O., Proc. Phys. Soc. Lond., 643 (1951), 747–753.
Hamilton C.H., "Superplasticity", AGARD lecture series No. 168, Agard (ISBN 92-835-0525-5) (1989), pp. 2.1–2.23.
Hammond C. and Nutting J., Met. Sci., 11 (1977), 474.
Hansen N., Metall. Mater. Trans., 32A (2001), 2917–2935.
Harase J., Shimizu R. and Dingley D.J., Acta Metall. Mater., 39 (1991), 763.
Hardie D., J. Nucl. Mater., 42 (1972), 317.
Hardwick D. and McG.Tegart W.J., J. Inst. Metals, 90 (1961–1962), 17.
Hardy W.B., Collected Scientific Papers, Cambridge University Press, Cambridge, UK (1936).
Hart E.W., Acta Metall., 15 (1967), 351–355.
Hatherly M. and Hutchinson B., "Introduction to Textures in Metals", vol. 5, Institution of Metallurgists, Chameleon Press (1979).

Hauk V. and Oudelhoven R., Z. Metallk., 79 (1988), 41.
Hauk V. and Vaessen G., Z. Metallk., 76 (1985), 102.
Hayakawa Y. and Szpunar J.A., Acta Mater., 45 (1997), 4713.
Hayashi Y. and Takatani M., Proc. 13th Biennial Congress – International Deep Drawing Research Group, Melbourne, (1984), 62.
Hayes J.S., Prangnell P.B. and Bate P.S., in "Ultrafine Grained Materials", Zhu *et al.* (eds), TMS, Warrendale (2002), p. 495.
Hecker S.S. and Stout M.G., 'Strain hardening of heavily cold worked metals' in "Deformation, Processing and Structure", G. Krauss (ed.), ASM, St. Louis (1982), pp. 1–47.
Heindlhofer K., "Evaluation of Residual Stresses", McGraw-Hill, New York (1948).
Held J.F., "Mechanical Working and Steel Processing IV", The Metallurgical Society of AIME, New York (1965), Paper 3.
Henry S., Minghetti T. and Rappaz M., Acta Mater., 18 (1998), 6431.
Hesselbarth H.W. and Göbel I.R., Acta Metall. Mater., 39 (1991), 2135–2143.
Higashi K., Mabuchi M. and Langdon T.G., ISIJ Intl., 36 (1996), 1423–1438.
Higashida K. and Morikawa T., in "Nanomaterials by Severe Plastic Deformation", M.J. Zehetbauer and R.Z. Valiev (eds), Wiley-VCH (2004), pp. 517–522.
Hill R., Proc. Roy. Soc. A, 193 (1948), 281–297.
Hillert M., Acta Metall., 13 (1965), 227–238.
Hirsch J. and Lucke K., Acta Metall., 36 (1988), 2863–2882.
Hirth J.P. and Lothe J., "Theory of Dislocations", McGraw-Hill, New York (1968).
Hitchcock J.H., "Elastic Deformation of Rolls During Cold Rolling" ASME Research Publication Roll Neck Bearings (1935).
Hiwatashi S., Takahashi M., Sakuma Y. and Usuda M., Proc. 26th Int. Symp. on Automotive Technology and Automation, Aachen, Germany (1993), pp. 263–268.
Hiwatashi S., Van Bael, B., Van Houtte P. and Teodosiu C., Int. J. Plasticity, 14 (1998), 647.
Hjelen J., Ørsund R. and Nes E., Acta Metall., 39 (1991), 1377.
Hockett J.E. and Sherby O.D., J. Mech. Phys. Solids, 23 (1975), 87–98.
Hoffman O. and Sachs G., "Introduction to the Theory of Plasticity for Engineers", McGraw-Hill, New York (1953).
Holm. E.A., Miodownik M.A. and Rollett A., Acta Mater., 51 (2003), 2701–2716.
Honda K. and Kaya S., Sci. Repts. Tohoku Univ., 15 (1926), 15.
Hondros E.D., in "Interfaces Conference", R.C. Gifkins (ed.), Butterwrts, UK (1969).
Honeff H. and Mecking H., Proc. ICOTOM 5, G. Gottstein and K. Lücke (eds), Springer, Berlin (1978), pp. 265–275.
Honeycombe R.W.K. and Bhadeshia H.K.D.H., "Steels", 2nd edition, Edward Arnold, UK (1995).
Horita Z., Fujinami T., Nemoto M. and Langdon T.G., Metall. Trans., 31A (2000a), 691–701.
Horita Z., Furukawa M., Nemoto M., Barnes A.J. and Langdon T.G., Acta Mater., 48 (2000b), 3633–3640.
Horita Z., Matsubara K., Miyahara Y. and Langdon T.G., 'Production of superplastic Mg alloys by severe plastic deformation' in "Nanomaterials by Severe

Plastic Deformation", M.J. Zehetbauer and R.Z. Valiev (eds), Wiley (2004), pp. 711–716.
Hornbogen E., Acta Metall., 32 (1984), 615.
Hosford W.F., Trans. AIME, 230 (1964), 12.
Hosford W.F., "The Mechanics of Crystals and Textured Polycrystals", Oxford University Press (1993).
Hosford W.F. and Caddell R.M., "Metal Forming, Mechanics and Metallurgy", Prentice-Hall, Englewood Cliffs, NJ, USA (1983).
Hosford W.F. and Duncan J.L., JOM, 51 (1999), 34.
Hu H., Rath B.B. and Vandermeer R.A., in "Recrystallization '90", T. Chandra (ed.), TMS (1990).
Huang J.Y., Zhu Y.T., Jiang H. and Lowe T.C., Acta Mater., 49 (2001), 1497–1505.
Huang Y. and Humphreys F.J., Acta Mater., 48 (2000), 2017.
Hughes D.A. and Hansen N., Metall. Trans., 24A (1993), 2021–2037.
Hughes D.A. and Hansen N., Acta Mater., 45 (1997), 3871–3886.
Hull D., 'Growth of twins and associated dislocation phenomena' in "Deformation Twinning" R.E. Reed-Hill, J.P. Hirth and H.C. Rogers (eds), Gordon and Breach, New York (1963), pp. 121–155.
Hultgren F.A., "Blast furnace and steel plant", 56 (1968), 149.
Huml P. and Fogelholm R., J. Mater. Process. Tech., 42 (1994), 167.
Humphreys F.J., Acta Mater., 45 (1997), 4231–4240.
Humphreys F.J., Mater. Sci. Tech., 15 (1999a), 37–44.
Humphreys F.J. and Bate P.S., Mater. Sci. Forum, 357–359 (2001), 477–488.
Humphreys F.J. and Hatherly M., "Recrystallization and Related Annealing Phenomena", Pergamon Press (1996).
Humphreys F.J. and Hatherly M., "Recrystallization and Related Annealing Phenomena", 2nd edition, Elsevier (2004).
Humphreys F.J., Prangnell P.B., Bowen J.R., Gholinia A. and Harris C., Phil. Trans. Roy. Soc. Lond., A357 (1999b), 1663–1681.
Humphreys F.J., Prangnell P.B. and Priestner R., Cur. Opi. Sol. Stat. Mater. Sci., 5 (2001), 15.
Hurley P.J. and Humphreys F.J., Acta Mater., 51 (2003), 3779–3793.
Hutchinson B., Int. Metal Rev., 29 (1984), 25–42.
Hutchinson B., Mater. Sci. Forum, 157–162 (1994), 1917–1928.
Hutchinson W.B., Acta Metall., 37 (1989), 1047.
Hutchinson W.B., Int. Mater. Rev., 29 (1984), 25.
Ibe G. and Lücke K., in "Recrystallisation, Grain Growth and Textures", H. Margolin (ed.), ASM, USA (1966), p. 434.
Inakazu N., Kaneko Y. and Inoue H., Mater. Sci. Forum, 157–162 (1994), 715–720.
Inokuti Y. and Maeda C., ISIJ Intl., 24 (1984), 655.
Ishinaga N., J. Mater. Process. Tech., 71 (1997), 100.
Itami A., Takahashi M. and Ushioda K., ISIJ Intl., 35 (1995), 1121–1127.
Iwahashi Y., Wang J., Horita Z., Nemoto M. and Langdon T.G., Scripta Metall., 35 (1996), 143–146.

Jacques P., Girault E., Mertens A., Verlinden B., Van Humbeeck J. and Delannay F., ISIJ Int., 41 (2001), 1068–1074.
Jeffrey C.B. and Linga Murty, K., “Symposium on Zirconium Alloys for Reactor Component (ZARC-91)”, BARC, Library and Information Services, BARC, Mumbai, India (1991), p. 1.
Jenko M., Korousic B., Mandrino D. and Presern V., Vacuum, 57 (2000), 295.
Johari O. and Thomas G., “The Stereographic Projection and its Applications”, Techniques of Metal Research, vol. IIA, Interscience Publishers (Wiley), New York (1969).
Johnson W.A. and Mehl R.F., AIME, 135 (1939), 416–458.
Jones H., J. Met. Sci., 5 (1971), 15.
Juul Jensen D., Acta Metall. Mater., 43 (1995), 4117.
Kalidindi S.R., Int. J. Plasticity, 14 (1998a), 1265–1277.
Kalidindi S.R., J. Mech. Phys. Solids, 46 (1998b), 267–290.
Kalischer S., Ber. 14, (1881), 2747.
Kallend J.S. and Huang Y.C., Met. Sci., 18 (1984), 381.
Kalpakjian S., “Manufacturing Engineering and Technology”, Addison-Wesley, Reading (Mass.) USA (1995).
Kalwa C., Hillenbrand H.-G. and Gräf M., “High Strength Steel Pipes: New Developments and Applications”, www.europipe.com/download/EP_TP45_02en.pdf (accessed February 2007).
Kalwa G. and Kaup K., ‘Steels for Line Pipe’ in “Steel: A Handbook for Materials Research and Engineering, Vol 2: Applications”, Verein Deutscher Eisenhüttenleute (ed.), Springer, Berlin (1993), pp. 573–583.
Kamma C. and Hornbogen E., J. Mater. Sci. Lett., 11 (1976), 2340.
Kang C.G., Kim N.H. and Kim B.M., J. Mater. Process. Tech., 100 (2000), 53.
Kawasaki K., Nagai T. and Nakashima K., Phil. Mag. B, 60 (1989), 399–421.
Kearns J.J., “Thermal Expansion and Preferred orientation in Zircaloy (LWBR Development Program)”, Bettis Atomic Power Laboratory, Pittsburgh, PA (1965).
Keeler S.P., Sheet Met. Ind. 42 (1965), 683.
Kestens L., Reis A.C. and Houbaert Y., ‘The role of surface shear during accumulative roll bonding of an interstitial steel’, in “Ultrafine Grained Materials III”, Zhu Y.T., T.G. Langdon, R.Z. Valiev, S.L. Semiatin, D.H. Shin and T.C. Lowe (eds), TMS (2004), pp. 285–290.
Kim H.S., Seo M.H. and Hong S.I., Mater. Sci. Eng., A291 (2000), 86–90.
Kim H.S., Seo M.H., Hong S.I., Lee H.R., Chun B.S. and Lee K.H., Mater. Sci. Forum, 386–388 (2002), 421–426.
Kim N.H., Kang C.G. and Kim B.M., Intl. J. Mech. Sci., 43 (2001), 1507.
Kim Y.H. and Park J.J., J. Mater. Process. Tech., 130–131 (2002a), 42.
Kiran Kumar M., PhD Thesis, IIT, Bombay, India (2006).
Kiran Kumar M., Samajdar I., Venkatramani N., Dey G.K., Tewari R., Srivastava D. and Banerjee S., Acta Mater., 51(2003), 625.
Kiran Kumar M., Vanitha C., Samajdar I., Dey G.K., Tewari R., Srivastava D. and Banerjee S., J. Nucl. Mater., 335 (2004), 48.

Kiran Kumar M., Vanitha C., Samajdar I., Dey G.K., Tewari R., Srivastava D. and Banerjee S., Mater. Sci. Tech., 22 (2006), 331.
Klocker H. and Tvergaard V., Int. J. Fracture, 106 (2000), 259–276.
Ko D-C., Kim B-M., J. Mater. Process. Tech., 102 (2000), 19.
Koç M. and Altan T., J. Mater. Process. Tech., 108 (2001), 384.
Kocks U.F., J. Eng. Mater. Tech. (ASME.H), 98 (1976), 76.
Kocks U.F., Argon A.S. and Ashby M.F., "Thermodynamics and Kinetics of Slip", Pergamon Press, Oxford (1975).
Kocks U.F. and Mecking H., Prog. Mater. Sci., 48 (2003), 171–273.
Kocks U.F., Tomé C.N. and Wenk H.-R., "Texture and Anisotropy: Preferred Orientations in Polycrystals and their Effect on Materials Properties", Cambridge University Press, Cambridge, UK (1998).
Kolmogorov A.N., Izv. Akad. Nauk. USSR-Ser-Matemat., 1 (3) (1937), 355–359.
Köster U. and Hornbogen E., Z. Metall., 59 (1968), 792.
Kovács Z., Lendvai J. and Vörös, Mater. Sci., Engg., A279 (2000), 179.
Kowalski B., Lacey A.J. and Sellars C.M., Mater. Sci. Tech., 19 (2003), 1564–1570.
Kozasu I., 'Processing – Thermomechanical Controlled Processing' in "Materials Science and Technology", R.W. Cahn, P. Haasen and E.J. Kramer (eds), vol. 7, Chapter 5, VCH, New York (1992), pp. 183–217.
Kozmanov Y.D. and Mukhaeva E.V., Fiz. Met. i. Metalloved, 25 (1968), 689.
Kraft F.F., Chakkingal U., Baker G. and Wright R.N., J. Mater. Process. Tech., 60 (1996), 171–178.
Krutenat R.C., Barnard R.S., Dismukes J.P., Kear B.H. and Witzke H., Patent No. US5200145 (1991).
Kuhlmann D., Masing G. and Raffelsieper J., Z. Metall., 40 (1949), 241.
Kuhlmann-Wilsdorf D., Mater. Sci. Eng., A113 (1989), 1.
Kuhlmann-Wilsdorf D., Acta Mater., 47 (1999a), 1697–1712.
Kuhlmann-Wilsdorf D., Phil. Mag., 79 (1999b), 955–1008.
Kumar M., Schwartz A.J., King W.E., Acta Mater., 50 (2002), 2599.
Kurz W. and Fisher D.J., "Fundamentals of Solidification", 3rd edition, Trans. Tech., Switzerland (1989).
Laasraoui A. and Jonas J.J., Metall. Trans. A, 22 (1991), 1545–1558.
Lagneborg R., 'Creep: Mechanisms and Theories' in "Creep and Fatigue in High Temparature Alloys", ed. J. Bressers, Applied Science Publishers, London (1981).
Lagneborg R., Zajac S. and Hutchinson B., Scripta Metall., 29 (1993), 159–164.
Laird C. and Aaronson H.I., Acta Metall., 17 (1969), 505.
Lamarsh J., "Introduction to Nuclear Engineering", Addison-Wesley, UK (1983).
Langdon T.G., Furukawa M., Horita Z. and Nemoto M., 'Microstructural Evolution during Processing by Severe Plastic Deformation' in "Investigations and Applications of Severe Plastic Deformation", T.C. Lowe and R.Z. Valiev (eds.), Kluwer Academic Publishers, Dordrecht, Netherlands (2000), pp. 149–154.
Langford G., Metall. Trans., 8A (1977), 861–875.
Langford G. and Cohen M., Trans. ASM, 62 (1969), 623–638.
Languillaume J., Kapelski G. and Baudelet B., Acta Mater., 45 (1997), 1201–1212.

Lapovok R., J. Mater. Process. Tech., 80–81 (1998), 608.
Lebensohm R.A. and Tomé C.N., Acta Metall. Mater., 41 (1993), 2611–2624.
Lee C.S. and Duggan B.J., Acta Metall. Mater., 41 (1993), 2691–2699.
Lee E.W. and McNelly T.R., Mater. Sci. Eng., 93 (1987), 45.
Lee J.C., Seok H.K. and Suh J.Y, Acta Mater., 50 (2002a), 4005–4019.
Lee J.W. and Park J.J., J. Mater. Process. Tech., 130–131 (2002b), 208–213.
Lee P.S., Piehler H.R., Adams B.L., Jarvis G., Hampel H. and Rollett A.D., J. Mater. Process. Tech., 80–81 (1998), 315.
Lee S., Berbon P.B., Furukawa M., Horita Z., Nemoto M., Tsenev N.K., Valiev R.Z. and Langdon T.G., Mater. Sci. Eng., A272 (1999), 63–72.
Lemaignan C. and Motta A.T., in "Materials Science & Technology", R.W. Cahn, P. Hassen and E.J. Krammer (eds), vol. 10B, Part II, VCH, Weinheim (1993), p. 1.
Leslie W.C., Ricket R.L., Dotson C.L. and Watson C.S., Trans. ASM, 46 (1954), 1470–1479.
Lesuer D.R., Syn C.K., Whittenberger J.D. and Sherby O.D., Metall. Mater. Trans. A, 30 (1999), 1559–1568.
Li S., Beyerlein I.J., Alexander D.J. and Vogel S.V., Acta Mater., 53 (2005), 2111–2125.
Li Y., Yang K. and Ni Z., J. Mater. Process. Tech., 71 (1997), 414.
Li Y.H. and Sellars C.M., J. Mater. Process. Tech., 80–81 (1998), 282.
Li Y.H. and Sellars C.M., Mater. Sci. Tech., 17 (2001), 1615.
Lin P., Palumbo G., Erb U. and Aust K.T., Scripta Metall. Mater., 33 (1995), 1327.
Liu J. and Kulak M., Mater. Sci. Forum, 331–337 (2000), 127–140.
Liu Q. and Hansen N., Proc. Roy. Soc., A454 (1998a), 2555.
Liu Q., Maurice Cl., Driver J. and Hansen N., Metall. Trans., 29A (1998b), 2333–2344.
Livatyali H. and Larris S.J., J. Mater. Process. Tech., 153–154 (2004), 913.
Llewellyn D.T., "Steels: Metallurgy and Applications", Butterworth-Heinemann, Oxford (ISBN 0 7506 20862) (1994).
Lloyd D.J., Court S.A, Gatenby K.M., Mater. Sci. Tech., 13 (1997), 660.
Lloyd D.J. and Kenny D., Metall. Trans., 13A (1982), 1445–1452.
Loganathan C. and Narayanasamy R., Mater. Sci. Engg., A406 (2005), 229.
Lovato M.L. and Stout M.G., Metall. Trans. A, 23A (1992), 935–951.
Lücke K. and Stüwe H.P., 'On the theory of grain boundary motion' in "Recovery and Recrystallization in Metals", L. Himmel (ed.), Interscience Publications, New York (1963), p. 171.
Lütjering G. and Williams J.C. "Titanium", Springer (2005).
Lyudkovsky G., Rastogi P.K. and Bala M., J. Metals, 22 (1986), 18.
Macherauch E. and Hauk V., "Residual Stresses in Science and Technology", DGM, Germany (1986).
Mahajan S., 'Deformation twinning', in "Encyclopedia of Materials: Science and Technology", K.H.J. Buschow, R.W. Cahn, M. Flemmings, B. ILschner, E. Kramer, S. Mahajan and P. Veyssière (eds.), Elsevier, London (2002).
Male A.T. and Cockroft M.G., J. Inst. Metals, 93 (1964), 489–494.
Manabe K. and Amino M., J. Mater. Process. Tech., 123 (2002), 285.

Mandrino D. and Jenko M., Vacuum, 61 (2001), 157.
Mani K.V., Sain A., Samajdar I., Dey G.K., Srivastava D., Neogi S., Tiwari R. and Banerjee S., Acta Mater., 54 (2006), 2665.
Marciniak Z., Duncan J.L. and Hu S.J., "Mechanics of Sheet Metal Forming", Butterworth-Heinemann, Oxford, UK (2002).
Marshall G.J., Mater. Sci. Forum, 217–222 (1996), 19–30.
Marshall G.J., 'Simulation of commercial hot rolling by laboratory plane strain compression and its application to Aluminium industry challenges' in "Hot Deformation of Aluminum Alloys II", T. Bieler, L.A. Lalli and S.R. MacEwen (eds), TMS (1998), pp. 367–382.
Martin J.W., "Micromechanisms in Particle-Hardened Alloys", Cambridge University Press, Cambridge (1980).
Martin J.W., Doherty R.D. and Cantor B., "Stability of Microstructures in Metallic Systems", 2nd edition, Cambridge University Press, UK (1997).
Marx V., Reher F.R. and Gottstein G., Acta Mater., 47 (1999), 1219–1230.
Masahiro M., Shinchi U., Kazumichi K. and Munekatsu F., 3rd International Conference on Steel Tolling Technology of Pipe and Tube and their Applications, September 2–6, ISIJ, Tokyo, Japan (1985), p. 351.
Masuda S., Fujita F., Kamata M., Yoshino M., Ariizumi T., Okami Y., Takada Y., and Inagaki J., Patent No. US4986341 (1988).
Matsuo M., ISIJ Intl., 29 (10) (1989), 809.
Matsuo Y., J. Nucl. Sci. Tech., 24 (1987), 111.
Maurice C. and Driver J.H., Acta Mater., 45 (1997), 4627–4649.
Maurice C., Piot D., Klocker H. and Driver J.H., Metall. Mater. Trans., 36A (2005), 1039–1047.
Maurice Cl., 'Numerical modelling of grain growth: current status' in "Recrystallisation and Grain Growth", Proceedings of the First Joint Conference, G. Gottstein and D.A. Molodov (eds), Springer, Berlin (2001), pp. 123–134.
Maurice Cl. and Driver J.H., Acta Metall. Mater., 41 (1993), 1653–1664.
Maurice Cl. and Humphreys F.J., '2- and 3-D curvature driven vertex simulations of grain growth' in "Proc. 3rd Int. Conf. on Grain Growth", H. Weiland, B.L. Adams and A.D. Rollett (eds), TMS, Warrendale (1998), 81–90.
McCormick P.G., Acta Metall., 36 (1988), 3061.
McDonald N.R. and Aust J., Inst. Metals., 16 (1971), 179.
Mckay A., "The Making of Atomic Age", Oxford University press, UK (1984).
McQueen H.J., Metall. Mater. Trans., 33A, (2002), 345–362.
McQueen H.J., Blum W., Zhu Q. and Demuth V., 'Hot working and creep of single phase Al alloys' in "Advances in Hot Deformation Textures and Microstructures", T.R. Bieler and J.J. Jonas (eds), TMS, Warrendale (1994), pp. 235–250.
McQueen H.J. and Jonas J.J., 'Hot workability testing techniques' in "Metal Forming: Interrelation between Theory and Practice", A.L. Hoffmanner (ed.), Plenum Press, New York (1971), pp. 393–428.
McQueen H.J. and Kassner M.E., Scripta Metall., 51 (2004), 461–465.
McQueen H.J., Knustad O., Ryum N. and Solberg J.K., Scripta Metall., 19 (1985), 73–78.

McQueen H.J. and Mecking H., Z. Metall., 78 (1987), 387.
McQueen M.J., Solberg J.K., Ryum N. and Nes E., Phil. Mag., A60 (1989), 473.
McQueen H.J., Wong W.A. and Jonas J.J., Can. J. Phys., 45 (1967), 1225–1234.
Mecking H. and Kocks U.F., Acta Metall., 29 (1981), 1865–1875.
Mehl R.F. and Hagel W.C., in "The Austenite-Pearlite Reaction", B. Chalmers and R. King (eds), Progress in Metal Physics, vol. 6, Pergamon Press, UK (1956).
"Metalforming: Modern Machines, Methods and Tooling for Engineers and Operating Personnel", American Machinist (ed.), McGraw-Hill, New York, USA (1982).
Metals Handbook, 9th edition, "Mechanical Testing", vol. 8, ASM, Metals Park, Ohio (1985).
Metals Handbook, "Forming and Forging", 9th edition, vol. 14, ASM, Ohio, USA (1988a).
Metals Handbook, "Surface Cleaning, Finishing & Coating", 9th edition, vol. 5, ASM, Ohio, USA (1988b).
Mfnshchikov A.M., Fridman D.S., Khaidukov I.F. and Zelenyi N.I., Steel in the USSR, 17(1984), 424.
Mielnik E.M., "Metalworking Science and Engineering", McGraw-Hill (1991).
Miller R.W., "Lubricants and their Applications", McGraw-Hill (1993).
Minami Y., Kimura H. and Hara Y.I., Mater. Sci. Tech., 2 (1986), 795.
Miodownik M.A., Holm E.A. and Hassold G.N., Scripta Mater., 42 (2000), 1173–1177.
Möller U.J. and Boor U., "Lubricants in Operation" (Translated and Published from German), Ipswich Book Company, Suffolk, UK (1996).
Montmitonnet P., Farrugia D., Aubin J.L. and Delamare F., Wear, 152 (1992), 327.
Montmitonnet P., Loge R., Hamery M., Chastel Y., Doudoux J.L. and Aubin J.L., J. Mater. Process. Tech., 125–126 (2002), 814.
Morere B., Maurice C, Shahani R. and Driver J., Metall. Mater. Trans., 32A (2001), 625–632.
Morgan E., 'Tinplate and Modern CanmakingTechnology', Pergamon Press (ISBN 0-08-028681-X), Oxford, UK (1985).
Morris D.G. and Munoz-Morris M.A., Acta Mater., 50 (2002), 4047–4060.
Moses A.J., IEEE Proc., A137 (1990), 233.
Mughrabi H., Mater. Sci. Eng., 85 (1987), 15–31.
Mukherjee A.K., 'Plastic Deformation and Fracture of Materials' in "Materials Science & Technology", vol. 6: R.W. Cahn, P. Haasen, E.J. Kramer and H. Mughrabi (eds), Wiley-VCH (1992), p. 407.
Mukherjee A.K., Bird J.E. and Dorn J.E., Trans. Am. Soc. Metals, 62 (1969), 155.
Mullins W.W. and Vinals J., Acta Metall., 37 (1989a), 991.
Mullins W.W., Acta Metall., 37 (1989b), 2979.
Mulot S., Hacquin A., Montmitonnet P. and Aubin J.L., J. Mater. Process. Tech., 60 (1996), 505.
Mura T., "Micromechanics of Defects in Solids", Matrinus Nijhoff Publishers, Dordrecht (1987).
Murr L.E., "Interfacial Phenomena in metals and alloys", Addison-Wesley, Reading, MA (1975).

Murr L.E., "Interfacial Phenomena in Metal and Alloys", Addison-Wesley, UK (1991).
Murty K.L. and Adams B.L., Mater. Sci. Eng., 70 (1985), 179.
Musin F., Kaibyshev R., Motohashi Y. and Itoh G., Scripta Mater., 50 (2004), 511–516.
Nabarro F.R.N., "Theory of Crystal Dislocations", Oxford University Press, Oxford (1967).
Nakaoka K., Takada Y. and Tanaka Y., Patent No. US4832762 (1986).
Nakazima K., Kikuma T. and Hasuka K., "Study on the Formability of Steel Sheets", Yawata Technical Report No. 264 (1968), p. 8517.
Narita K. and Yashimiro Y., IEEE Trans. Mag., MAG-16 (1980), 728.
Naujoks W. and Fabel D.C., "Forging Handbook", ASM, USA (1953).
Necker C.T., Doherty R.D. and Rollett A.D., Proc. ReX'96, Monterey, T.R. McNelly (ed.), Monterey Institute of Advanced Studies (1997), Monterey (Calif.), USA pp. A1–A8.
Needleman A. and Rice J.R., Acta Metall., 28 (1980), 1315.
Nelson D.V., Makino A. and Fuch's E.A., Opt. Laser Eng., 27 (1997), 3.
Nes E., Acta Metall. Mater., 43 (1995), 2189–2207.
Nes E., Prog. Mater. Sci., 41 (1998), 129–193.
Nicolaou P.D. and Semiatin S.L., Acta Mater., 47 (1999), 3679.
Nicolaou P.D. and Semiatin S.L., Acta Mater., 48 (2000), 3441.
Northwood D.O., Mater. Design, 6 (1985), 58.
Northwood D.O. and Kosasih U., Int. Met. Rev., 28 (1983), 92.
Novotny S. and Hein P., J. Mater. Process. Tech., 115 (2001), 65.
Oda Y., Tanaka Y., Yamagami N., Chino A. and Yamada K., NKK Technical Rev., 87 (2002), 12.
Oh-Ishi K., Horita Z., Furukawi M., Nemoto M. and Langdon T.G., Metall. Mater. Trans., 29A (1998), 2011.
Ørsund R. and Nes E., Scripta Metall. Mater., 22 (1988), 671.
Osgood W.R. (ed.), "Residual Stress in Metal and Metal Construction", Reinhold Publishing Company, New York (1954).
Palumbo G., Patent No. US5702543 (1997).
Palumbo G., Patent No. US5817193 (1998).
Palumbo G. and Aust K.T., Scripta Metall., 22 (1988), 847.
Patel J.R. and Cohen M., Acta Metall., 1 (1953), 531.
Patlan V., Vinogradov A., Higashi K. and Kitagawa K., Mater. Sci. Eng., A300 (2001), 171–182.
Pavlovskii B.G., Ilnitskii A.A., Dovgal A.I. and Menshchikov A.M., Steel in the USSR, 14 (1984), 285.
Peck J.F. and Thomas D.A., Trans. Met. Soc. AIME, 221 (1961), 1240.
Percy J., "Metallurgy – Iron & Steel", London (1864), p. 9.
Perocheau F., PhD Thesis, Ecole des Mines de St Etienne (1999) (data provided by L. Poizat, Pechiney Research Centre).
Perocheau F. and Driver J.H., Int. J. Plasticity, 16 (2000), 73–89.

Perocheau F. and Driver J., Int. J. Plasticity, 18 (2002), 185–202.
Perovic V., Weatherly G.C. and Simpson C.J., Scripta Metall., 16 (1982), 409.
Petch N.J., J. Iron Steel Inst. Lond., 173 (1953), 25.
Philippe M.-J., Mater. Sci. Forum, 157–162 (1994), 1337–1350.
Phillips V.A., Trans. Met. Soc. AIME, 236 (1966), 1302.
Pickering F.B., 'High Strength Low Alloy Steels' in "Materials Science and Technology", R.W. Cahn, P. Haasen and E.J. Kramer (eds), vol. 7, Chapter 8, VCH, New York (1992), pp. 335–399.
Pilling J., J. Mater. Sci. Tech., 1 (1985), 461.
Piot D., Ecole des Mines St Etienne, Private Communication (2004).
Plyatskovskii O.A., Chernyavskii A.A., Kozlovskii A.I. and Ksenz A., Steel in the USSR, 16 (1986), 492.
Polmear I.J., "Light Alloys – Metallurgy of the light metals", 3rd edition, Butterworth-Heinemann, Oxford (1995).
Porter D.A. and Easterling K.E., "Phase Transformations in Metals and Alloys", Van Nostrand Reinhold, UK (1986).
Poulsen H.F., Margulies L., Schmidt S. and Winther G., Acta Mater., 51 (2003), 3821–3830.
PremKumar R., Samajdar I., Viswanathan N.N., Singal V. and Seshadri V., J. Mag. Mag. Mater., 264 (2003), 75.
Pugh H.L.D., ASTM Special Technical Publication (1964), p. 374.
Raab G.J., Valiev R.Z., Lowe T.C. and Zhu Y.T., Mater. Sci. Eng., A382 (2004), 30–34.
Raabe D., Phil. Mag. A, 79 (1999), 2339–2358.
Raj S.J. and Pharr G.M., Mater. Sci. Eng., 81 (1986), 217–237.
Rajmohan N., Hayakawa Y. and Szpunar J.A., Acta Mater., 45 (1997), 2485.
Randall S.N. and Prieur H., Iron Steel Engineers, (August) (1967), 109.
Randle V., "The Role of Coincident Site Lattice in Grain Boundary Engineering", The Institute of Materials, UK (1996).
Randle V. and Engler O., "Texture Analysis. Macrotexture, Microtexture and Orientation Mapping", Gordon and Breach, Australia (2000).
Ray R. and Ashby M.F., Metall. Trans., vol 2 (1971), 1113.
Ray R.K. and Jonas J.J., Int. Mater. Rev., 35 (1990), 1–36.
Ray R.K., Jonas J.J., Butrón-Guillén M.P. and Savoie J., ISIJ Intl., 34 (1994), 927–942.
Ray R.K., Jonas J.J. and Hook R.E., Int. Mater. Rev., 39 (1994), 129.
Read H.G., Reynolds Jr., W.T., Hono K. and Tarui T., Scripta Mater., 37 (1997), 1221–1230.
Reddy N.V., Dixit P.M. and Lal G.K., J. Mater. Process. Tech., 55 (1995), 331.
Reed-Hill R.E. and Abbaschian, "Physical Metallurgy Principles", 3rd edition, PWS-Kent Publishing Co., USA (1992).
Rendigs K.-H., Mater. Sci. Forum, 242 (1997), 11–24.
Renouard M. and Wintenberger M., C.R. Acad. Sci. Paris, B283 (1976), 237–240.
Rhines F.N. and Craig K.R., Metall. Trans., 5 (1974), 413.
Rice J.R. and Tracy D.M., J. Mech. Phys. Solids, 17 (1969) 201–217.
Ricks R.A., Phil. Trans. Roy. Soc. Lond., 357 (1999), 1513–1529.

Ridha A.A. and Hutchinsen W.B., Acta Metall., 30 (1982), 1929–1939.

Ridley N., in “Superplasticity”, AGARD lecture series No. 168, Agard (ISBN 92-835-0525-5) (1989), pp. 4.1–4.14.

Rittel D., Aharonov R., Feigin G. and Roman I., Acta Metall. Mater., 39 (1991), 719.

Roberts W., in “Deformation, Processing and Structure”, G. Krauss (ed.), ASM, St. Louis (1982), pp. 109–180.

Roberts W.L., “Hot Rolling of Steel”, Marcel Dekker, New York (1983), p. 419.

Robinson J.M., Int. Mater. Rev., 39 (1994), 217–227.

Roe R.-J., J. Appl. Phys., 36 (1965), 2024–2031.

Rofagha R., Erb U., Ostander D., Palumbo G. and Aust K.T., Nanostruct. Mater., 2 (1993), 1.

Rofagha R., Langer R., El-Sherik A.M., Erb U., Palumbo G. and Aust K.T., Scripta Metall., 25 (1991), 2867–2872.

Rollett A.D. and Kocks U.F., Solid State Phenom., 35–36 (1994), 1–18.

Rollett A.D., Storch M.L., Hilinski E.J. and Goodman S.R., Metall. Mater. Trans., 32A (2001), 2595.

Rose R.M., Shepard L.A. and Wulff J., “Structure and Properties of Materials”, Wiley, UK (1966).

Rossard C. and Blain P., Mém. Sci. Rev. Métall., 57 (1960), 190.

Ryan N.D. and McQueen H.J., “Stainless Steels 87”, The Institute of Metals, London (1987), 498–507.

Saintfort P., Sigli C., Raynaud G.M. and Gomiero Ph., Mater. Sci. Forum, 242 (1997), 25–32.

Saito Y., Tsuji N., Utsunomiya H., Sakai T. and Hong R.G., Scripta Mater., 39 (1998), 1221–1227.

Saito Y., Utsunomiya H., Suzuki H. and Sakai T., Scripta Mater., 42 (2000), 1139–1144.

Samajdar I., Cicalè S., Verlinden B., Van Houtte P. and Abbruzzesse G., Scripta Mater., 39 (8) (1998a), 1083.

Samajdar I. and Doherty R.D., Scripta Metall. Mater., 32 (1995), 845.

Samajdar I. and Doherty R.D., Acta Mater., 46 (1998b), 3145.

Samajdar I., Ratchev P., Verlinden B. and Aernoudt E., Acta Mater., 49 (2001), 1759–1769.

Samajdar I., Rabet L., Verlinden B. and Van Houtte P., ISIJ Intl., 3 (1998c), 539.

Samajdar I., Ratchev P., Verlinden B., Van Houtte P. and De Smet P., Mater. Sci. Eng., A247 (1998d), 58.

Samajdar I., Verlinden B., Kestens L. and Van Houtte P., Acta Mater., 47 (1999), 55.

Samajdar I., Verlinden B. and Van Houtte P., Acta Mater., 46 (1998e), 2751–2763.

Samajdar I., Verlinden B., Van Houtte P. and Vanderschueren D., Mater. Sci. Eng., A238 (1997a), 343–350.

Samajdar I., Verlinden B., Van Houtte P. and Vanderschueren D., Scripta Mater., 37 (1997b), 869.

Santner J.S., Metall. Trans., 9A (1978), 769–779.

Savart F., Chim. Phys., 41 (1829), 61.

Schey J.A., in "Techniques of Metals Research", R.F. Bunshah (ed.), vol. 1, Interscience Publishers, New York (1968), p. 1409.
Schey J.A., "Tribology in Metalworking – Friction, Lubrication and Wear", ASM, Metals Park, OH, USA (1983).
Schey J.A., "Metal Deformation Processes: Friction and Lubrication", Marcel Dekker, New York (1970).
Schmitt J.H., Aernoudt E. and Baudelet B., Mater. Sci. Eng., 75 (1985), 13–20.
Scholz F., Driver J.H. and Woldt E., Scripta Mater., 40 (1999), 949.
Schoppa A., Louis H., Pude F. and Rad Ch. von, J. Mag. Mag. Mater., 254–255 (2003), 370.
Schoppa A., Schneider J. and Wuppermann C.D., J. Mag. Mag. Mater., 215–216 (2000), 74.
Segal V.M., Mater. Sci. Eng., A197 (1995), 157–164.
Segal V.M., Reznikov V.I., Drobyshevkij A.E. and Kopylov V.I., Metally, 1 (1981), 115–123.
Séjournet J., "Friction and Lubrication in Metal Processing", ASME, New York (1966), p. 163.
Séjournet J. and Delcroix J., "Glass lubrification in extrusion of steel", Lubr. Eng., 11 (1955), 389.
Sellars C.M. and McG.Tegart W.J., Rév. Mét., 63 (7) (1966), 31–746.
Sellars C.M. and McG.Tegart W.J., Int. Metall. Rev., 17 (158) (1972), 1–24.
Semiatin S.L. and Jonas J.J., "Formability and Workability of Metals; Plastic Instability and Flow Localization", ASM, Ohio (1984).
Semiatin S.L. and Lahoti G.D., Metall. Trans., 12A (1981), 1705.
Seo Y.H. and Kang C.G., Comp. Sci. Tech., 59 (1999), 643.
Servi I.S. and Turnbull D., Acta Metall., 14 (1966), 161.
Shackelford J.F. and Alexander W., "Materials Science and Engineering Handbook", 3rd edition, CRC Press, USA (2001).
Sheasby J.S., Boggs W.E. and Turkdogan E.T., Met. Sci., 18 (1984), 127.
Shi H., McLaren A.J., Sellars C.M., Shahani R. and Bolingbrooke R., Mater. Sci. Tech., 13 (1997a), 210–216.
Shi H., McLaren A.J., Sellars C.M., Shahani R., and Bolingbroke R., J. Test. Evaluat., 25 (1997b), 61–73.
Shimada M., Kokawa H., Wang Z.J., Sato Y.S. and Karibe I., Acta Mater., 50 (2002), 2331.
Shimanaka H., Ito Y., Matsumara K. and Fukuda B., J. Mag. Mag. Mater., 26 (1982), 57.
Shin D.H., Park J.J., Kim Y.S. and Park K.T., J. Mater. Sci. Eng., 328 (2002), 98–103.
Shin J., Lee Z., Lee T. and Lavernia E.J., Scripta Mater., 45 (2001), 725.
Shoen J.W., Patent No. US5061326 (1990).
Siebel E., Stahl und Eisen, 66–67 (11–12) (1947), 171.
Siebel E. and Neumann F.W., Stahl und Eisen, 74 (1954), 139 (in German).
Silcock J., Heal T. and Hardy H.J., Inst. Metals, 82 (1953), 239.
Singer C., Holmyard E.J. and Hall A.R. (eds), "A History of Technology", vol. 1, Clarendon Press, Oxford (1954).

Siwecki T. and Engberg G., 'Recrystallization controlled rolling of steels' in "Thermomechanical Processing in Theory, Modeling & Practice", B. Hutchinson (ed.), ASM, Ohio (1997), pp. 121–143.

Smallman R.E. and Bishop R.J., "Modern Physical Metallurgy and Materials Engineering", 6th edition, Butterworth-Heinemann, UK (1999).

Smith D.A. and Pond R.C., Int. Metal Rev., 21 (1976), 61.

Sorby H.C., J. Iron Steel Inst., 31.1 (1887), 253.

Spencer A.J.M., "Continuum Mechanics", Longman (1980).

Srivastava D., Dey G.K. and Banerjee S., Metall. Mater. Trans., 26A (1995), 2707.

Srolovitz D.J., Grest G.S. and Anderson M.P., Acta Metall., 34 (1986), 1833–1845.

Stachowicz F., J. Mech. Work. Tech., 13 (1986), 229.

Staley J.T., 'History of Wrought-Aluminum-Alloy Development' in "Aluminum Alloys – Contemporary Research and Applications", A.K. Vasudevan and R.D. Doherty (eds), Academic Press (1989).

Staley J.T., Doherty R.D. and Jaworski A., Metall. Trans., 24A (1993), 2417–2427.

Stephenson E.T., J. Appl. Phys., 57 (1) (1985), 4226.

Stickler R. and Vinckier A., Trans. ASM, 54 (1961), 362.

Stone M.D., AISE Yearly Proceedings, AISE, Pittsburgh (1953), p. 115.

Stout M.G., Kallend J.S., Kocks U.F., Przystupa M.A. and Rollet A.D., Proc. 8° Int. Conf. on Textures of Materials, J.S. Kallend and G. Gottstein (eds), TMS, Warrendale (1988), pp. 479–484.

Stowell M.J., Livesey D.W. and Ridley N., Acta Mater., 32 (1984), 35.

Stowell M.J., Met. Sci., 17 (1983), 1.

Stüwe H.P., Acta Metall., 13 (1965), 1337–1342.

Suéry M., Proc Int. Conf. on Superplasticity, Grenoble 1985, B. Baudelet and M. Suéry (eds), CNRS, Paris (1985), pp. 9.1–9.19.

Sun W., Tieu A.K., Jiang Z., Lu C. and Zhu H., J. Mater. Process. Tech., 140 (2003), 76.

Sun Z.N., Li J.B., Wang Z.G. and Li W.J., Acta Metall. Mater., 40 (1992), 2961.

Sundaram C.V., Trans. IIM, 39 (1986), 12.

Szacinski A.M. and Thomson P.F., Mater. Sci. Tech., 7 (1991), 37.

Takada Y., Abe M., Masuda S. and Inagaki J., J. Appl. Phys., 64 (1988), 5357.

Takahashi M. and Okamoto A., Sumitomo Metals, 27 (1974), 40–49.

Takashi N. and Harasse J., Mater. Sci. Forum, 204–206 (1996), 143.

Takatani H., Gandin Ch.-A. and Rappaz M., Acta Mater., 48 (2000), 675.

Talat "Training in Aluminium Application Technologies", CD-ROM, European Aluminium Association (Brussels), version 2.0 (1999).

Taylor A., "X-ray Metallography", Wiley, New York (1961).

Taylor B., "Metals Handbook", 9th edition, vol. 8, ASM (1985), pp. 547–570.

Taylor G.I., Proc. Roy. Soc. Lond., 145A (1934), 362.

Taylor G.I., J. Inst. Metals, 62 (1938), 307–324.

Tenckhoff E., "Deformation Mechanisms, Texture and Anisotropy in Zirconium and Zircalloys", ASTM STP 966, ASTM, Philadelphia, USA (1988).

Thomason P.F., "Ductile Fracture of Metals", Pergamon Press, Oxford (1990).

Thompson A.W., Metall. Trans., 8A (1977), 833.

Thomson P.F. and Burman N.M., Mater. Sci. Engg., 45 (1980), 95.
Thorpe S.J., Ramaswami B. and Aust A.T., J. Electrochem. Soc., 135 (1988), 2162–2170.
Tomé C.N., Lebensohn R.A. and Kocks U.F., Acta Metall. Mater., 39 (1991), 2667–2680.
Toth L.S., Arruffat Massion R., Germain L., Baik S.C. and Suwas S., Acta Mater., 52 (2004), 1885–1898.
Tricot R., "Thermomechanical Processing of Titanium Alloys for the Aerospace Industry", Matériaux et Techniques (1988), p. 47.
Trink W., Mawhinney M.H., Shannon R.A., Reed R.J. and Garvev J.R., "Industrial Furnaces", Willey, Hoboken, NJ (2004).
Tsuji N., Kamikawa N., Kim H.W. and Minamino Y., 'Fabrication of bulk nanostructured materials by ARB (accumulative roll bonding)' in "Ultrafine Grained Materials III", Y.T. Zhu, T.G. Langdon, R.Z. Valiev, S.L. Semiatin, D.H. Shin and T.C. Lowe (eds), TMS (2004), pp. 219–226.
Turczyn S., J. Mater. Process. Tech., 60 (1996), 275.
Tvergaard V., Int. J. Fracture, 18 (1982), 237–252.
Uginet J.F., 'Processing of Near-Beta Ti Alloys for High Strength Applications' in "Beta Titanium Alloys", A. Vassel, D. Eylon and Y. Combres (eds), SF2M, Paris (1994), pp. 33–40.
Uhlig H.H., "Corrosion and Corrosion Control: An Introduction to Corrosion Science and Engineering", Wiley, New York (1971).
Ungar T. and Borbely A., Appl. Phys. Lett., 69 (1996), 3173.
Ushioda K and Hutchinson W.B., ISIJ., 29 (1989) 10, 862.
Ushioda K., Hutchinson W., Agren J. and von Schlippenbach U., Mater. Sci. Tech., 2 (1986), 807–815.
Valiev R.Z., Islamgaliev R.K. and Alexandrov I.V., Prog. Mater. Sci., 45 (2000), 103–189.
Valiev R.Z., Salimonenko D.A., Tsenev N.K., Berbon P.B. and Langdon T.C., Scripta Mater., 37 (1997), 1945–1950.
Van der Winden M.R. and Sellars C.M., Mater. Sci. Forum, 217–222 (1996), 385–390.
Van Houtte P., Acta Mater., 26 (1978), 591–604.
Van Houtte P., Textures Microstr., 8–9 (1988), 313–350.
Van Houtte P., Acta Metall. Mater., 43 (1995), 2859.
Van Houtte P. and De Buyser L., Acta Metall. Mater., 41 (2) (1993), 323.
Van Houtte P., Li S., Seefeldt M. and Delannay L., Int. J. Plasticity, 21 (2005), 589–624.
van Linden van den Heuvel J.H.L., in "Aluminium: Technologie voor de 21° eeuw", L. Katgerman (ed.), T.U. Delft (ISBN 90-802879-2-X) (1996).
Vandermeer R.A. and Rath B.B., Met. Trans., 20A (1989), 391.
Vanitha C., PhD Thesis, IIT, Bombay (2007).
Van Vlack L.H., Int. Met. Rev., Sept. (1977), 207.
Vasudevan A.K. and Doherty R.D., Acta Metall., 35 (1987), 1193–1219.
Vatne H.E., Furu T., Orsund R. and Nes E., Acta Mater., 44 (1996a), 4463–4473.

Vatne H.E., Shahani R. and Nes E., Acta Mater., 44 (1996b), 4447–4462.
Vazquez V. and Altan T., J. Mater. Process. Tech., 98 (2000), 212.
Venables J.A., 'Deformation twinning in fcc metals' in "Deformation Twinning", R.E. Reed-Hill, J.P. Hirth and H.C. Rogers (eds), Gordon and Breach, New York (1963), pp. 77–116.
Vergne C., Boher C., Gras R. and Levaillant C., Wear, 260 (2006), 957.
Verlinden B. and Voith K., Proc. 15th RISO Conf. on Numerical Predictions of Deformation Processes and the Behavior of Real Materials, Anderson *et al.* (eds), Riso Laboratories, Denmark (1994), pp. 601–606.
Verlinden B., Bocher Ph., Girault E. and Aernoudt E., Scripta Mater., 45 (2001), 909–916.
Versnyder F.L. and Shank M.E., Mater. Sci. Eng., 6 (1970), 213.
Vinogradov A., Scripta Mater., 39 (1998), 797–805.
Vinogradov A. and Hashimoto S., 'Fatigue of Severely Deformed Metals' in "Nanomaterials by Severe Plastic Deformation", M.J. Zehetbauer and R.Z. Valiev (eds), Wiley-VCH, Weinheim, Germany (2004), pp. 663–676.
Vinogradov A., Mimaki T., Hashimoto S. and Valiev R., Scripta Mater., 41 (1999), 319–326.
Virnich K.H. and Lücke K., Proc. 5° Int. Conf. on Textures of Materials, G. Gottstein and K. Lücke (eds), Springer, Berlin (1978), pp. 397–404.
Viswanathan R. and Bauer C.L., Acta Metall., 21 (1973), 1099.
Voce E.J., Inst. Metall., 74 (1948), 537–562.
Voith M., 3rd International Conference on Steel Toolling Technology of Pipe and Tube and their Applications, September 2–6, ISIJ, Tokyo, Japan, (1985), p. 398.
von Kármán T.Z., Angew. Math. Mech., 5 (1925), 139.
Wagner P., Engler O. and Lücke K., Acta Metall. Mater., 43 (1995), 3799.
Wagoner R.H., Chan K.S. and Keeler S.P. (ed.), "Forming Limit Diagrams: Concepts, Methods and Applications", The Minerals, Metals and Materials Society, Warrendale, USA (1989).
Wagoner R.H. and Chenot, J.-L. "Fundamentals of Metal Forming", Wiley (1997).
Wahl H.P. and Wassermann G., Z. Metall., 61 (1970), 326.
Wang J., Iwahashi Y., Horita Z., Furukawa M., Nemoto M., Valiev R.Z. and Langdon T.G., Acta Mater., 44 (1996), 2973–2982.
Wang Y.D., Lin Peng R. and McGreevy R.L., Phil. Mag. Lett., 81 (3) (2001), 153.
Wang Y.N. and Huang J.C., Mater. Chem. Phys., 81 (2003), 11–26.
Warrington D.H. and Boon M., Acta Metall., 23 (1975), 599.
Wasnik D.N., Dey G.K., Kain V. and Samajdar I., Scripta Mater., 49 (2003), 135.
Wasnik D.N., Kain V., Samajdar I., Verlinden B. and De P.K., Acta Mater., 50 (2002), 4587.
Watanabe T., Res. Mech., 11 (1984), 47.
Watanabe T., Fujii H., Oikawa H. and Arai K.I., Acta Metall., 37 (1989), 941.
Watanabe T., Kitamura S. and Karashima S., Acta Metall., 28 (1980), 455.
Watanabe Y., Kain V., Tonozuka T., Shoiji T., Kondo T. and Masuyama F., Scripta Mater., 42 (2000), 312.
Watkins M.T., Int. Metall. Rev., 18 (1973), 1.

Weatherly G.C., Acta Metall., 29 (1981), 501.
Weiss B. and Stickler R., Metall. Trans., 4 (1972), 851.
Weiss I., Froes F.H., Eylon D. and Welch G.E., Metall. Trans., 17A (1986), 1935.
Welzel U., Ligot J., Lamparter P., Vermeulen A.C. and Mittemeijer E.J., J. Appl. Crystall., 38 (2005), 1.
Whiteley R., Trans. ASM, 52 (1960), 154.
Williams C.D., Reactor Technology, 13 (1970), 147.
Wilkens M., "Fundamental Aspects of Dislocation Theory", vol II, Nat. Bur. Stand. (US) Spec. Publ. N° 317, Washington, DC (1970).
Wistreich J.G., Metall. Rev., 3 (1958), 97.
Withers P.J. and Bhadeshia H.K.D.H., Mater. Sci. Tech., 17 (2001a), 355.
Withers P.J. and Bhadeshia H.K.D.H., Mater. Sci. Tech., 17 (2001b), 366.
Wong C.C., Dean T.A. and Lin J., Int. J. Mach. Tool. Manu., 43 (2003), 1419.
Wright R.N., Wire Tech., 4 (1976), 57.
Wright R.N., Wire J. Int., 32 (1999), 118.
Xu C., Furukawa M., Horita Z. and Langdon T.G., Acta Mater., 51 (2003), 6139–6149.
Xu C., Furukawa M., Horita Z. and Langdon T.G., 'Achieving a superplastic forming capability through severe plastic deformation' in "Nanomaterials by Severe Plastic Deformation", M.J. Zehetbauer and R.Z. Valiev (eds), Wiley-VCH, Weinheim, Germany (2004), pp. 701–710.
Yamasaki T., Miyamoto H., Mimaki T., Vinogradov A. and Hashimoto S., Mater. Sci. Eng., A318 (2001), 122–128.
Yoshida H., Matsui T., Otani T. and Mandai K., Annals CIRP, 24 (1975), 191.
Yu C.Y., Sun P.L., Kao P.W. and Chang C.P., Mater. Sci. Eng., 366 (2004), 310.
Zhu Y.T. and Langdon T.G., JOM, 56 (10) (2004), 58–63.
Zelin M., Acta Mater., 50 (2002), 4431.
Zienkiewicz O.C., Taylor R.L. and Zhu J.Z., "The Finite Element Method, its Basis and Fundamentals", 6th edition, Elsevier (2005).

Index

Suffix ‘f’ after the page number indicates a reference to a figure or table caption.

Kolibri 360 GmbH
Ferdinand-Jühlke-Straße 7
99095 Erfurt, Deutschland
produktsicherheit@kolibri360.de